Study and Solutions Guide for

PRECALCULUS

THIRD EDITION

Larson/Hostetler

Dianna L. Zook

Indiana University
Purdue University at Fort Wayne, Indiana

D. C. Heath and Company

Lexington, Massachusetts Toronto

Published simultaneously in Canada.

Printed in the United States of America.

International Standard Book Number: 0-669-28312-6

10 9 8 7 6 5

TO THE STUDENT

The *Study and Solutions Guide for Precalculus* is a supplement to the text by Roland E. Larson and Robert P. Hostetler.

As a mathematics instructor, I often have students come to me with questions about the assigned homework. When I ask to see their work, the reply often is "I didn't know where to start." The purpose of the *Study Guide* is to provide brief summaries of the topics covered in the textbook and enough detailed solutions to problems so that you will be able to work the remaining exercises.

This *Study Guide* is the result of the efforts of Richard Bambauer, Lisa Bickel, Linda Bollinger, Laurie Brooks, Patti Jo Campbell, Darin Johnson, Linda Kifer, Deanna Larson, Timothy Larson, Amy Marshall, John Musser, Scott O'Neil, Louis Rieger, Paula Sibeto Steinhart, and Evelyn Wedzikowski. I would like to thank my husband Edward L. Schlindwein for his support during the several months I worked on this project.

If you have any corrections or suggestions for improving this *Study Guide*, I would appreciate hearing from you.

Good luck with your study of precalculus.

Dianna L. Zook
Indiana University,
Purdue University
at Fort Wayne, Indiana 46805

STUDY STRATEGIES

- Attend all classes and come prepared. Have your homework completed. Bring the text, paper, pen or pencil, and a calculator (scientific or graphing) to each class.
- Read the section in the text that is to be covered before class. Make notes about any questions that you have and, if they are not answered during the lecture, ask them at the appropriate time.
- Participate in class. As mentioned above, ask questions. Also, do not be afraid to answer questions.
- Take notes on all definitions, concepts, rules, formulas and examples. After class, read your notes and fill in any gaps, or make notations of any questions that you have.
- DO THE HOMEWORK!!! You learn mathematics by doing it yourself. Allow **at least** two hours outside of each class for homework. Do not fall behind.
- Seek help when needed. Visit your instructor during office hours and come prepared with specific questions; check with your school's tutoring service; find a study partner in class; check additional books in the library for more examples—just do something before the problem becomes insurmountable.
- Do not cram for exams. Each chapter in the text contains a chapter review and this study guide contains a practice test at the end of each chapter. (The answers are at the back of the study guide.) Work these problems a few days before the exam and review any areas of weakness.

CONTENTS

Chapter 1 Review of Basic Algebra 1

Chapter 2 Functions and Graphs 53

Chapter 3 Polynomial and Rational Functions 115

Chapter 4 Exponential and Logarithmic Functions 178

Chapter 5 Trigonometry 211

Chapter 6 Analytic Trigonometry 263

Chapter 7 Additional Applications of Trigonometry 297

Chapter 8 Systems of Equations and Inequalities 327

Chapter 9 Matrices and Determinants 367

Chapter 10 Sequences, Counting Principles, and Probability 408

Chapter 11 Some Topics in Analytic Geometry 446

Solutions to Chapter Practice Tests 500

CHAPTER 1

Review of Basic Algebra

Section 1.1 The Real Number System **2**

Section 1.2 Exponents and Radicals **9**

Section 1.3 Polynomials: Special Products and Factoring **16**

Section 1.4 Fractional Expressions **21**

Section 1.5 Solving Equations **26**

Section 1.6 Solving Inequalities **35**

Section 1.7 Algebraic Errors and Some Algebra of Calculus **42**

Review Exercises **46**

Practice Test **52**

SECTION 1.1

The Real Number System

You should be able to identify and use the following properties of real numbers.

For all real numbers a, b, c, and d:

- Closure Property
  (a) Addition: $a + b$ is a real number.
  (b) Multiplication: $a \cdot b$ is a real number.

- Commutative Property
  (a) Addition: $a + b = b + a$
  (b) Multiplication: $a \cdot b = b \cdot a$

- Associative Property
  (a) Addition: $(a + b) + c = a + (b + c)$
  (b) Multiplication: $(ab)c = a(bc)$

- Identity Property
  (a) Addition: 0 is the identity; $a + 0 = 0 + a = a$.
  (b) Multiplication: 1 is the identity; $a \cdot 1 = 1 \cdot a = a$.

- Inverse Property
  (a) Addition: $-a$ is the inverse of a; $a + (-a) = -a + a = 0$.
  (b) Multiplication: $1/a$ is the inverse of a, $a \neq 0$; $a(1/a) = (1/a)a = 1$.

- Distributive Property
  (a) Left: $a(b + c) = ab + ac$
  (b) Right: $(a + b)c = ac + bc$

- Properties of Negatives
  (a) $(-1)a = -a$
  (b) $-(-a) = a$
  (c) $(-a)b = a(-b) = -ab$
  (d) $(-a)(-b) = ab$
  (e) $-(a + b) = (-a) + (-b) = -a - b$

- Properties of Equality
  (a) Reflexive: $a = a$
  (b) Symmetric: If $a = b$, then $b = a$.
  (c) Transitive: If $a = b$ and $b = c$, then $a = c$.
  (d) Substitution: If $a = b$, a can be replaced by b in any statement involving a or b.
  (e) If $a = b$, then $a + c = b + c$, and $a - c = b - c$.
  (f) If $a = b$, then $ac = bc$, and $a/c = b/c$, $c \neq 0$.

- Cancellation Laws
  (a) If $a + c = b + c$, then $a = b$.
  (b) If $ac = bc$, then $a = b$, $c \neq 0$.

- Subtraction: $a - b = a + (-b)$

- Division: $a \div b = a(1/b) = a/b$, $b \neq 0$

- Properties of Fractions ($b \neq 0$, $d \neq 0$)
  (a) Equivalent Fractions: $a/b = c/d$ if and only if $ad = bc$.
  (b) Rule of Signs: $-a/b = a/-b = -(a/b)$ and $-a/-b = a/b$
  (c) Equivalent Fractions: $a/b = ac/bc$, $c \neq 0$
  (d) Addition and Subtraction
    1. Like Denominators: $(a/b) \pm (c/b) = (a \pm c)/b$
    2. Unlike Denominators: $(a/b) \pm (c/d) = (ad \pm bc)/bd$
  (e) Multiplication: $(a/b) \cdot (c/d) = ac/bd$
  (f) Division: $(a/b) \div (c/d) = (a/b) \cdot (d/c) = ad/bc$ if $c \neq 0$.

- Properties of Zero
  (a) $a \pm 0 = a$
  (b) $a \cdot 0 = 0$
  (c) $0 \div a = 0/a = 0$, $a \neq 0$
  (d) If $ab = 0$, then $a = 0$ or $b = 0$.
  (e) $a/0$ is undefined.

- Definition of Absolute Value:

$$|a| = \begin{cases} a, & \text{if } a \geq 0 \\ -a, & \text{if } a < 0 \end{cases}$$

- Properties of Absolute Value
  (a) $|a| \geq 0$
  (b) $|-a| = |a|$
  (c) $|ab| = |a||b|$
  (d) $|a/b| = |a|/|b|$, $b \neq 0$
  (e) $|a + b| \leq |a| + |b|$, Triangle Inequality

- Distance Between Two Points on the Real Line
  $d(a, b) = |b - a| = |a - b|$

Solutions to Selected Exercises

3. Determine which numbers are (a) natural numbers, (b) integers, (c) rational numbers, and (d) irrational numbers.

$\{-\pi, -\frac{1}{3}, \frac{6}{3}, \frac{1}{2}\sqrt{2}, -7.5\}$

Solution:

Since $\frac{6}{3} = 2$, we have:

(a) natural number: $\{\frac{6}{3}\}$

(b) integer: $\{\frac{6}{3}\}$

(c) rational numbers: $\{-\frac{1}{3}, \frac{6}{3}, -7.5\}$

(d) irrational numbers: $\{-\pi, \frac{1}{2}\sqrt{2}\}$

9. Identify the property illustrated in the equation $2(x+3) = 2x+6$.

Solution:

By the Distributive Property, we have $2(x+3) = 2 \cdot x + 2 \cdot 3 = 2x + 6$.

13. Identify the properties illustrated in the equation $x(3y) = (x \cdot 3)y = (3x)y$.

Solution:

$$\begin{aligned} x(3y) &= (x \cdot 3)y && \text{by the Associative Property of Multiplication} \\ &= (3x)y && \text{by the Commutative Property of Multiplication} \end{aligned}$$

17. Use the properties of zero to evaluate, if possible, the following expression. If the expression is undefined, state why.

$$\frac{8}{-9+(6+3)}$$

Solution:

$\dfrac{8}{-9+(6+3)} = \dfrac{8}{-9+9} = \dfrac{8}{0}$ which is undefined since the denominator is zero.

19. Perform the indicated operations:

$10 - 6 - 2$

Solution:

$$\begin{aligned} 10 - 6 - 2 &= (10-6) - 2 \\ &= 4 - 2 = 2 \end{aligned}$$

21. Perform the indicated operation:

$$2\left(\frac{77}{-11}\right)$$

Solution:

$$2\left(\frac{77}{-11}\right) = 2(-7) = -14$$

27. Perform the indicated operations:

$\frac{4}{5} \times \frac{1}{2} \times \frac{3}{4}$

Solution:

$\frac{4}{5} \times \frac{1}{2} \times \frac{3}{4} = \frac{1}{5} \times \frac{1}{2} \times \frac{3}{1} = \frac{3}{10}$

29. Perform the indicated operation:

$12 \div \frac{1}{4}$

Solution:

$12 \div \frac{1}{4} = 12 \times \frac{4}{1} = 12 \times 4 = 48$

31. Identify the property used in each step of the solution process of the equation $3x + 15 = 0$.

Solution:

$3x + 15 = 0$	Given equation
$3x + 15 - 15 = 0 - 15$	Subtraction Property of Equality
$3x = -15$	Additive Inverse Property
$\dfrac{3x}{3} = \dfrac{-15}{3}$	Division Property of Equality
$x = -5$	Multiplicative Inverse Property

37. Plot the two real numbers $\frac{5}{6}$ and $\frac{2}{3}$ on the real number line and place the appropriate inequality sign between them.

Solution:

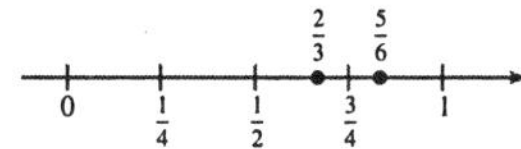

$\frac{5}{6} > \frac{2}{3}$

43. Describe the subset of real numbers represented by the inequality $-1 \leq x < 0$, and sketch the subset on the real number line.

Solution:

$-1 \leq x < 0$ represents all negative real numbers greater than or equal to -1 and less than 0.

45. Use inequality notation to denote the expression "x is negative".

Solution:

"x is negative" can be written as $x < 0$.

49. Use inequality notation to denote the expression "the annual rate of inflation, r, is expected to be at least 3.5% but no more than 6%."

Solution:

The expression "the annual rate of inflation, r, is expected to be at least 3.5% but no more than 6%" can be written

$$3.5\% \le r \le 6\%$$

or

$$0.035 \le r \le 0.06$$

53. Evaluate:

(a) $\dfrac{-5}{|-5|}$ (b) $-3 - |-3|$

Solution:

(a) $\dfrac{-5}{|-5|} = \dfrac{-5}{-(-5)} = \dfrac{-5}{5} = -1$ (b) $-3 - |-3| = -3 - (-(-3)) = -3 - 3 = -6$

55. Place the correct symbol ($<$, $>$, or $=$) between the pair of real numbers.

(a) $|-3| \;\square\; -|-3|$ (b) $|-4| \;\square\; |4|$

Solution:

(a) Since $|-3| = -(-3) = 3$ and $-|-3| = -3$, we have $|-3| > -|-3|$.
(b) Since $|-4| = -(-4) = 4$ and $|4| = 4$, we have $|-4| = |4|$.

59. Find the distance between the points $-\frac{5}{2}$ and 0 on the real line.

Solution:

$$d\left(-\tfrac{5}{2},\ 0\right) = \left|0 - \left(-\tfrac{5}{2}\right)\right| = \left|\tfrac{5}{2}\right| = \tfrac{5}{2}$$

63. Find the distance between the points 9.34 and -5.65 on the real line.

Solution:

$$\begin{aligned} d(9.34,\ -5.65) &= |9.34 - (-5.65)| \\ &= |9.34 + 5.65| \\ &= |14.99| \\ &= 14.99 \end{aligned}$$

69. Use absolute value notation to describe the situation "y is at least six units from 0."

Solution:

Since $d(y,\ 0) = |y - 0| = |y|$ and $d(y,\ 0) \ge 6$, we have $|y| \ge 6$.

73. The accounting department of a company is checking to see whether the actual expenses of a department differ from the budgeted expenses by more than \$500 or by more than 5%. Complete the table and determine whether the actual expense passes the "budget variance test."

	Budgeted Expense, b	Actual Expense, a	$\|a-b\|$	$0.05b$
Taxes	\$37,640.00	\$37,335.80	____	____

Solution:

$$|a-b| = |37{,}335.80 - 37{,}640.00| = |-304.20| = \$304.20$$

$$0.05b = 0.05(37{,}640.00) = \$1882.00$$

Since the difference is less than \$500 and 5% of the budgeted expenses, it passes the "budget variance test."

79. One worker can assemble a component in 7 days and a second worker can do the same task in 5 days. If they work together, what fraction of a component can they assemble in 2 days?

Solution:

One worker assembles $\frac{1}{7}$ of a component each day. The other worker assembles $\frac{1}{5}$ of a component each day. Together, in 2 days, they assemble

$$2\left(\frac{1}{7}+\frac{1}{5}\right) = 2\left(\frac{5+7}{35}\right) = 2\left(\frac{12}{35}\right) = \frac{24}{35}$$

of a component.

81. Use a calculator to order the following real numbers, from smallest to largest.

$\frac{7071}{5000}, \frac{584}{413}, \sqrt{2}, \frac{47}{33}, \frac{127}{90}$

Solution:

$\frac{7071}{5000} = 1.4142$

$\frac{584}{413} = 1.414043584$

$\sqrt{2} = 1.414213562$

$\frac{47}{33} = 1.4\overline{242}$

$\frac{127}{90} = 1.41\overline{1}$

$\frac{127}{90} < \frac{584}{413} < \frac{7071}{5000} < \sqrt{2} < \frac{47}{33}$

85. Use a calculator to find the decimal form of $\frac{41}{333}$. If it is a nonterminating decimal, write the repeating pattern.

Solution:

$\frac{41}{333} = 0.123123123\ldots$

87. Determine whether the statement "The reciprocal of a nonzero integer is an integer" is true or false.

Solution:

"The reciprocal of a nonzero integer is an integer" is **false**. In fact, it is only true for the integers 1 and -1. To see that it is not true in general, pick any other integer and take its reciprocal.

SECTION 1.2

Exponents and Radicals

- You should know the following properties of exponents.

  (a) $a^m a^n = a^{m+n}$ (b) $\dfrac{a^m}{a^n} = a^{m-n}$

  (c) $\dfrac{1}{a^n} = a^{-n}$ (d) $a^0 = 1,\ a \neq 0$

  (e) $(ab)^m = a^m b^m$ (f) $(a^m)^n = a^{mn}$

  (g) $\left(\dfrac{a}{b}\right)^m = \dfrac{a^m}{b^m}$ (h) $|a^2| = |a|^2 = a^2$

- You should know the following properties of radicals.

  (a) $\sqrt[n]{b} = a$ if and only if $a^n = b$ (b) $\sqrt[n]{a^m} = (\sqrt[n]{a})^m$

  (c) $\sqrt[n]{a} \cdot \sqrt[n]{b} = \sqrt[n]{ab}$ (d) $\dfrac{\sqrt[n]{a}}{\sqrt[n]{b}} = \sqrt[n]{\dfrac{a}{b}},\ b \neq 0$

  (e) $\sqrt[m]{\sqrt[n]{a}} = \sqrt[mn]{a}$ (f) $(\sqrt[n]{a})^n = a$

  (g) If n is even, $\sqrt[n]{a^n} = |a|$. (h) If n is odd, $\sqrt[n]{a^n} = a$.

- You should know the definitions of rational exponents.

  (a) $a^{1/n} = \sqrt[n]{a}$

  (b) $a^{m/n} = (\sqrt[n]{a})^m = \sqrt[n]{a^m}$ where m and n are positive integers and m/n is reduced.

- You should be able to simplify radicals.

  (a) Remove all possible factors from the radical sign.
  (b) Rationalize the denominator.
  (c) Reduce the index as far as possible.

- You should be able to write numbers in scientific notation.

Solutions to Selected Exercises

5. Evaluate $6x^0 - (6x)^0$ when $x = 10$.

Solution:

$$\begin{aligned} 6(10)^0 - (6 \cdot 10)^0 &= 6 \cdot 1 - (60)^0 \\ &= 6 - 1 \\ &= 5 \end{aligned}$$

9. Simplify $6y^2(2y^4)^2$.

Solution:

$$\begin{aligned} 6y^2(2y^4)^2 &= 6y^2(2)^2(y^4)^2 \\ &= 6y^2(4)(y^8) \\ &= 24y^{10} \end{aligned}$$

13. Simplify:

$$\frac{12(x+y)^3}{9(x+y)}$$

Solution:

$$\frac{12(x+y)^3}{9(x+y)} = \frac{3 \cdot 4(x+y)^{3-1}}{3 \cdot 3}$$

$$= \frac{4(x+y)^2}{3}$$

15. Simplify $(-2x^2)^3(4x^3)^{-1}$.

Solution:

$$(-2x^2)^3(4x^3)^{-1} = \frac{(-2x^2)^3}{4x^3}$$

$$= \frac{-8x^6}{4x^3}$$

$$= -2x^3$$

19. Simplify $(4a^{-2}b^3)^{-3}$.

Solution:

$$(4a^{-2}b^3)^{-3} = (4)^{-3}(a^{-2})^{-3}(b^3)^{-3} = 4^{-3}a^6b^{-9} = \frac{a^6}{4^3b^9} = \frac{a^6}{64b^9}$$

25. Evaluate (a) $\left(\sqrt[3]{-125}\right)^3$ and (b) $\sqrt[4]{562^4}$. (Do not use a calculator.)

Solution:

(a) $$\left(\sqrt[3]{-125}\right)^3 = \left[(-125)^{1/3}\right]^3$$

$$= (-125)^{3/3}$$

$$= (-125)^1$$

$$= -125$$

(b) $$\sqrt[4]{562^4} = (562^4)^{1/4}$$

$$= (562)^{4/4}$$

$$= (562)^1$$

$$= 562$$

29. Use the properties of radicals to simplify each expression.

(a) $\sqrt{75x^2y^{-4}}$

(b) $\sqrt{5(x-y)^3}$

Solution:

(a) $$\sqrt{75x^2y^{-4}} = \sqrt{25x^2y^{-4}(3)}$$

$$= \sqrt{25}\sqrt{x^2}\sqrt{(y^{-2})^2}\sqrt{3}$$

$$= 5|x|y^{-2}\sqrt{3}$$

$$= \frac{5|x|\sqrt{3}}{y^2}$$

(b) $$\sqrt{5(x-y)^3} = \sqrt{(x-y)^2 5(x-y)}$$

$$= \sqrt{(x-y)^2}\sqrt{5(x-y)}$$

$$= (x-y)\sqrt{5(x-y)}$$

Note: We do not have $|x-y|$ here since $(x-y)$ must be positive for $(x-y)^3$ to be under the radical.

31. Use the properties of radicals to simplify each expression.

(a) $\sqrt{5x^2y}\sqrt{3y}$

(b) $\dfrac{\sqrt{54a^2}}{\sqrt{2a^4}}$

Solution:

(a) $\sqrt{5x^2y}\sqrt{3y} = |x|\sqrt{5y}\sqrt{3y}$
$= |x|\sqrt{15y^2} = |x|y\sqrt{15}$

(b) $\dfrac{\sqrt{54a^2}}{\sqrt{2a^4}} = \sqrt{\dfrac{54a^2}{2a^4}}$
$= \sqrt{\dfrac{27}{a^2}} = \dfrac{\sqrt{9 \cdot 3}}{\sqrt{a^2}} = \dfrac{3\sqrt{3}}{|a|}$

35. Simplify and/or combine the given radicals.

(a) $2\sqrt{4y} - 2\sqrt{9y} + 10\sqrt{y}$

(b) $6\sqrt[3]{32a} + 5\sqrt[3]{500a}$

Solution:

(a) $2\sqrt{4y} - 2\sqrt{9y} + 10\sqrt{y} = 2(2\sqrt{y}) - 2(3\sqrt{y}) + 10\sqrt{y}$
$= 4\sqrt{y} - 6\sqrt{y} + 10\sqrt{y}$
$= (4 - 6 + 10)\sqrt{y}$
$= 8\sqrt{y}$

(b) $6\sqrt[3]{32a} + 5\sqrt[3]{500a} = 6\sqrt[3]{8 \cdot 4a} + 5\sqrt[3]{125 \cdot 4a}$
$= 6(2\sqrt[3]{4a}) + 5(5\sqrt[3]{4a})$
$= 12\sqrt[3]{4a} + 25\sqrt[3]{4a}$
$= (12 + 25)\sqrt[3]{4a}$
$= 37\sqrt[3]{4a}$

39. Rewrite each expression by rationalizing the denominator. Simplify your answer.

(a) $\dfrac{5}{\sqrt[3]{(5x)^2}}$

(b) $\dfrac{3}{\sqrt[4]{(3x)^3}}$

Solution:

(a) $\dfrac{5}{\sqrt[3]{(5x)^2}} = \dfrac{5}{\sqrt[3]{(5x)^2}} \cdot \dfrac{\sqrt[3]{5x}}{\sqrt[3]{5x}} = \dfrac{5\sqrt[3]{5x}}{\sqrt[3]{(5x)^3}} = \dfrac{5\sqrt[3]{5x}}{5x} = \dfrac{\sqrt[3]{5x}}{x}$

(b) $\dfrac{3}{\sqrt[4]{(3x)^3}} = \dfrac{3}{\sqrt[4]{(3x)^3}} \cdot \dfrac{\sqrt[4]{3x}}{\sqrt[4]{3x}} = \dfrac{3\sqrt[4]{3x}}{\sqrt[4]{(3x)^4}} = \dfrac{3\sqrt[4]{3x}}{3x} = \dfrac{\sqrt[4]{3x}}{x}$

41. Rewrite each expression by rationalizing the denominator. Simplify your answer.

(a) $\dfrac{3}{\sqrt{5}+\sqrt{6}}$ (b) $\dfrac{8}{\sqrt{2}-2\sqrt{3}}$

Solution:

(a)
$$\begin{aligned}\frac{3}{\sqrt{5}+\sqrt{6}} &= \frac{3}{\sqrt{5}+\sqrt{6}} \cdot \frac{\sqrt{5}-\sqrt{6}}{\sqrt{5}-\sqrt{6}} \\ &= \frac{3(\sqrt{5}-\sqrt{6})}{(\sqrt{5})^2-(\sqrt{6})^2} \\ &= \frac{3(\sqrt{5}-\sqrt{6})}{5-6} \\ &= \frac{3(\sqrt{5}-\sqrt{6})}{-1} \\ &= -3(\sqrt{5}-\sqrt{6}) \\ &= 3(\sqrt{6}-\sqrt{5})\end{aligned}$$

(b)
$$\begin{aligned}\frac{8}{\sqrt{2}-2\sqrt{3}} &= \frac{8}{\sqrt{2}-2\sqrt{3}} \cdot \frac{\sqrt{2}+2\sqrt{3}}{\sqrt{2}+2\sqrt{3}} \\ &= \frac{8(\sqrt{2}+2\sqrt{3})}{(\sqrt{2})^2-(2\sqrt{3})^2} \\ &= \frac{8(\sqrt{2}+2\sqrt{3})}{2-12} \\ &= \frac{8(\sqrt{2}+2\sqrt{3})}{-10} \\ &= -\frac{4(\sqrt{2}+2\sqrt{3})}{5}\end{aligned}$$

45. Rewrite each expression by rationalizing the numerator. Simplify your answer.

(a) $\dfrac{\sqrt{3}-\sqrt{2}}{x}$ (b) $\dfrac{\sqrt{15}+\sqrt{3}}{12}$

Solution:

(a)
$$\begin{aligned}\frac{\sqrt{3}-\sqrt{2}}{x} &= \frac{\sqrt{3}-\sqrt{2}}{x} \cdot \frac{\sqrt{3}+\sqrt{2}}{\sqrt{3}+\sqrt{2}} \\ &= \frac{(\sqrt{3})^2-(\sqrt{2})^2}{x(\sqrt{3}+\sqrt{2})} \\ &= \frac{3-2}{x(\sqrt{3}+\sqrt{2})} \\ &= \frac{1}{x(\sqrt{3}+\sqrt{2})}\end{aligned}$$

(b)
$$\begin{aligned}\frac{\sqrt{15}+\sqrt{3}}{12} &= \frac{\sqrt{15}+\sqrt{3}}{12} \cdot \frac{\sqrt{15}-\sqrt{3}}{\sqrt{15}-\sqrt{3}} \\ &= \frac{(\sqrt{15})^2-(\sqrt{3})^2}{12(\sqrt{15}-\sqrt{3})} \\ &= \frac{15-3}{12(\sqrt{15}-\sqrt{3})} \\ &= \frac{12}{12(\sqrt{15}-\sqrt{3})} \\ &= \frac{1}{\sqrt{15}-\sqrt{3}}\end{aligned}$$

51. Find the rational exponent form for $\sqrt[3]{-216} = -6$.

Solution:

Rational Exponent Form: $(-216)^{1/3} = -6$

57. Evaluate (a) $36^{1/2}$ and (b) $16^{3/2}$ without using a calculator.

Solution:

(a) $36^{1/2} = \sqrt{36} = 6$

(b) $16^{3/2} = (16^{1/2})^3 = (\sqrt{16})^3 = (4)^3 = 64$

59. Evaluate (a) $64^{-2/3}$ and (b) $\left(\frac{9}{4}\right)^{-1/2}$ without using a calculator.

Solution:

(a) $64^{-2/3} = \dfrac{1}{64^{2/3}} = \dfrac{1}{(\sqrt[3]{64})^2} = \dfrac{1}{4^2} = \dfrac{1}{16}$

(b) $\left(\dfrac{9}{4}\right)^{-1/2} = \left(\dfrac{4}{9}\right)^{1/2} = \sqrt{\dfrac{4}{9}} = \dfrac{\sqrt{4}}{\sqrt{9}} = \dfrac{2}{3}$

Note: $\left(\dfrac{a}{b}\right)^{-n} = \dfrac{1}{(a/b)^n} = \dfrac{1}{a^n/b^n} = \dfrac{b^n}{a^n} = \left(\dfrac{b}{a}\right)^n$

63. Use fractional exponents to verify $\sqrt[6]{(x+1)^4} = \sqrt[3]{(x+1)^2}$.

Solution:

$$\begin{aligned}\sqrt[6]{(x+1)^4} &= (x+1)^{4/6}\\ &= (x+1)^{2/3}\\ &= \sqrt[3]{(x+1)^2}\end{aligned}$$

65. Write $\sqrt{50}\sqrt[3]{2}$ as a single radical.

Solution:

$$\begin{aligned}\sqrt{50}\sqrt[3]{2} &= \sqrt{25 \cdot 2}\sqrt[3]{2}\\ &= 5\sqrt{2}\sqrt[3]{2}\\ &= 5(2)^{1/2}(2)^{1/3}\\ &= 5(2)^{1/2+1/3}\\ &= 5(2)^{3/6+2/6}\\ &= 5(2)^{5/6}\\ &= 5\sqrt[6]{2^5}\end{aligned}$$

67. Write $\sqrt{x}/\sqrt[3]{x}$ as a single radical.

Solution:

$$\begin{aligned}\frac{\sqrt{x}}{\sqrt[3]{x}} &= \frac{x^{1/2}}{x^{1/3}}\\ &= x^{(1/2)-(1/3)}\\ &= x^{(3/6)-(2/6)}\\ &= x^{1/6}\\ &= \sqrt[6]{x}\end{aligned}$$

71. Write the number in scientific notation.

Land Area of Earth:
57,500,000 square miles

Solution:

$57{,}500{,}000.00 = 5.75 \times 10^7$
(decimal moves 7 places to the left)

75. Write the number in decimal form.

U.S. Daily Coca-Cola Consumption: 5.24×10^8 servings

Solution:

$5.24 \times 10^8 = 524{,}000{,}000.00$ (decimal moves 8 places to the right)

79. Use a calculator to evaluate the given expressions. (Round your answers to three decimal places.)

(a) $2400(1+0.06)^{20}$

(b) $750\left(1+\dfrac{0.11}{365}\right)^{800}$

(c) $\dfrac{(2.414 \times 10^4)^6}{(1.68 \times 10^5)^5}$

(d) $(9.3 \times 10^6)^3(6.1 \times 10^{-4})^4$

Solution:

(a) $2400(1+0.06)^{20} = 2400(1.06)^{20} \approx 7697.125$

2400 [×] 1.06 [y^x] 20 [=]

(b) $750\left(1+\dfrac{0.11}{365}\right)^{800} \approx 954.448$

[(] 1 [+] 0.11 [÷] 365 [)] [y^x] 800 [×] 750 [=]

(c) $\dfrac{(2.414 \times 10^4)^6}{(1.68 \times 10^5)^5} = \dfrac{(2.414)^6 \times 10^{24}}{(1.68)^5 \times 10^{25}} = \dfrac{(2.414)^6}{(1.68)^5 \times 10} \approx 1.479$

2.414 [y^x] 6 [÷] [(] 1.68 [y^x] 5 [×] 10 [)] [=]

(d) $(9.3 \times 10^6)^3(6.1 \times 10^{-4})^4 = (9.3)^3 \times 10^{18} \times (6.1)^4 \times 10^{-16}$

$= (9.3)^3(6.1)^4(10)^2 \approx 111{,}369{,}991.292$

9.3 [y^x] 3 [×] 6.1 [y^x] 4 [×] 100 [=]

81. The speed of light is 11,160,000 miles per minute. The distance from the sun to the earth is 93,000,000 miles. Find the time it takes for light to travel from the sun to the earth.

Solution:

$$\text{Time} = \frac{\text{Distance}}{\text{Rate}} = \frac{93{,}000{,}000 \text{ miles}}{11{,}160{,}000 \text{ miles/minute}} = 8\tfrac{1}{3} \text{ minutes}$$

83. The balance A after t years in an account earning an annual interest rate of r compounded n times per year is $A = P(1 + \frac{r}{n})^{nt}$ where P is the original deposit. Complete the following table for \$500 deposited in an account earning 12% compounded daily. [Note that $r = 0.12$ implies an interest rate of 12%.]

Solution:

$A = 500$ [×] [(] 1 [+] .12 [÷] 365 [)] [y^x] [(] 365 [×] t [)] [=]

Fill in the value of t for each calculation.

t	5	10	20
A	$500\left(1+\dfrac{0.12}{365}\right)^{365\times 5}$ $\approx \$910.97$	$500\left(1+\dfrac{0.12}{365}\right)^{365\times 10}$ $\approx \$1659.73$	$500\left(1+\dfrac{0.12}{365}\right)^{365\times 20}$ $\approx \$5509.41$

t	30	40	50
A	$500\left(1+\dfrac{0.12}{365}\right)^{365\times 30}$ $\approx \$18,288.29$	$500\left(1+\dfrac{0.12}{365}\right)^{365\times 40}$ $\approx \$60,707.30$	$500\left(1+\dfrac{0.12}{365}\right)^{365\times 50}$ $\approx \$201,515.59$

87. A funnel is filled with water to a height of h. The time, t, it takes for the funnel to empty is

$$t = 0.03[12^{5/2} - (12 - h)^{5/2}], \quad 0 \le h \le 12.$$

Find t for $h = 7$ centimeters.

Solution:

$$t = 0.03[12^{5/2} - (12 - 7)^{5/2}]$$
$$= 0.03[12^{2.5} - 5^{2.5}]$$
$$\approx 13.29 \text{ units of time}$$

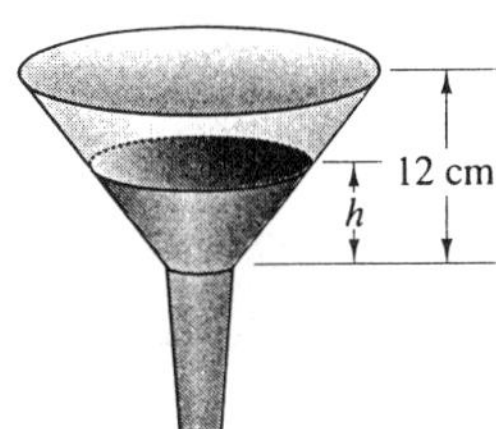

12 [y^x] 2.5 [−] 5 [y^x] 2.5 [=] [×] .03 [=]

SECTION 1.3

Polynomials: Special Products and Factoring

- You should be able to add, subtract, and multiply polynomials.
- You should know the following special products.
  (a) $(u+v)(u-v) = u^2 - v^2$
  (b) $(u+v)^2 = u^2 + 2uv + v^2$
  (c) $(u-v)^2 = u^2 - 2uv + v^2$
  (d) $(u+v)^3 = u^3 + 3u^2v + 3uv^2 + v^3$
  (e) $(u-v)^3 = u^3 - 3u^2v + 3uv^2 - v^3$
- You should be able to factor polynomials.
  (a) Remove any common factors.
  (b) Know and be able to use the following factoring formulas.
    1. $u^2 - v^2 = (u+v)(u-v)$
    2. $u^2 + 2uv + v^2 = (u+v)^2$
    3. $u^2 - 2uv + v^2 = (u-v)^2$
    4. $u^3 + v^3 = (u+v)(u^2 - uv + v^2)$
    5. $u^3 - v^3 = (u-v)(u^2 + uv + v^2)$
- Be able to factor, if possible, trinomials of the form $ax^2 + bx + c$.
- Be able to factor by grouping.

Solutions to Selected Exercises

5. Perform the indicated operations and write the result in standard form.

$(6x+5) - (8x+15)$

Solution:

$$(6x+5) - (8x+15) = 6x + 5 - 8x - 15 = 6x - 8x + 5 - 15 = -2x - 10$$

9. Perform the indicated operations and write the result in standard form.

$(15x^2 - 6) - (-8x^3 - 14x^2 - 17)$

Solution:

$$(15x^2 - 6) - (-8x^3 - 14x^2 - 17) = 15x^2 - 6 + 8x^3 + 14x^2 + 17 = 8x^3 + 29x^2 + 11$$

15. Perform the indicated operations and write the result in standard form.

$(-2x)(-3x)(5x+2)$

Solution:

$$(-2x)(-3x)(5x+2) = 6x^2(5x+2) = 6x^2(5x) + 6x^2(2) = 30x^3 + 12x^2$$

19. Perform the indicated operations and write the result in standard form.

$(x+3)(x^2-3x+9)$

Solution:

$$\begin{aligned}(x+3)(x^2-3x+9) &= (x+3)(x^2) + (x+3)(-3x) + (x+3)(9)\\ &= x^3 + 3x^2 - 3x^2 - 9x + 9x + 27\\ &= x^3 + 27\end{aligned}$$

25. Perform the indicated operations and write the result in standard form.

$(x+\sqrt{5})(x-\sqrt{5})(x+4)$

Solution:

$$\begin{aligned}(x+\sqrt{5})(x-\sqrt{5})(x+4) &= (x^2-5)(x+4) && \text{Special Product}\\ &= x^3 + 4x^2 - 5x - 20 && \text{FOIL Method}\end{aligned}$$

27. Find the product: $(2x-5y)^2$

Solution:

$$(2x-5y)^2 = (2x)^2 - 2(2x)(5y) + (5y)^2 = 4x^2 - 20xy + 25y^2$$

29. Find the product: $[(x-3)+y]^2$

Solution:

$$\begin{aligned}[(x-3)+y]^2 &= (x-3)^2 + 2(x-3)y + y^2\\ &= x^2 - 6x + 9 + 2xy - 6y + y^2\\ &= x^2 + 2xy + y^2 - 6x - 6y + 9\end{aligned}$$

33. Find the product: $[(m-3)+n][(m-3)-n]$

Solution:

$$\begin{aligned}[(m-3)+n][(m-3)-n] &= (m-3)^2 - n^2\\ &= m^2 - 6m + 9 - n^2\\ &= m^2 - n^2 - 6m + 9\end{aligned}$$

39. Find the product: $(2x-y)^3$

Solution:

$$(2x-y)^3 = (2x)^3 - 3(2x)^2y + 3(2x)y^2 - y^3 = 8x^3 - 12x^2y + 6xy^2 - y^3$$

43. Remove the common factor: $(x-1)^2 + 6(x-1)$.

Solution:

$$(x-1)^2 + 6(x-1) = (x-1)[(x-1)+6] = (x-1)(x+5)$$

45. Factor $16y^2 - 9$.

Solution:

$$16y^2 - 9 = (4y)^2 - (3)^2 = (4y+3)(4y-3)$$

47. Factor $(x-1)^2 - 4$.

Solution:

$$(x-1)^2 - 4 = [(x-1)+2][(x-1)-2] = (x+1)(x-3)$$

49. Factor $x^2 - 4x + 4$.

Solution:

$$x^2 - 4x + 4 = x^2 - 2(2)(x) + (2)^2 = (x-2)^2$$

53. Factor $s^2 - 5s + 6$.

Solution:

$s^2 - 5s + 6 = (s-2)(s-3)$ since $(-2)(-3) = 6$ and $(-2) + (-3) = -5$.

55. Factor $x^2 - 30x + 200$.

Solution:

$x^2 - 30x + 200 = (x-10)(x-20)$ since $(-10)(-20) = 200$ and $(-10) + (-20) = -30$.

57. Factor $9z^2 - 3z - 2$.

Solution:

$$9z^2 - 3z - 2 = (3z+1)(3z-2)$$

Check by using the FOIL Method.

61. Factor $x^3 - 8$.

Solution:

$$\begin{aligned} x^3 - 8 &= x^3 - 2^3 \\ &= (x-2)(x^2 + 2x + (2)^2) \\ &= (x-2)(x^2 + 2x + 4) \end{aligned}$$

65. Factor $x^3 - x^2 + 2x - 2$ by grouping.

Solution:

$$x^3 - x^2 + 2x - 2 = x^2(x-1) + 2(x-1)$$
$$= (x-1)(x^2+2)$$

69. Factor $6 + 2x - 3x^3 - x^4$ by grouping.

Solution:

$$6 + 2x - 3x^3 - x^4 = 2(3+x) - x^3(3+x)$$
$$= (3+x)(2-x^3)$$

71. Completely factor $x^3 - 4x^2$.

Solution:

$$x^3 - 4x^2 = x^2(x-4)$$

75. Completely factor $9x^2 + 10x + 1$.

Solution:

$$9x^2 + 10x + 1 = (9x+1)(x+1)$$

Check by using the FOIL Method.

79. Completely factor $2(x+1)(x-3)^2 - 3(x+1)^2(x-3)$.

Solution:

$$2(x+1)(x-3)^2 - 3(x+1)^2(x-3) = (x+1)(x-3)[2(x-3) - 3(x+1)]$$
$$= (x+1)(x-3)[2x - 6 - 3x - 3]$$
$$= (x+1)(x-3)(-x-9)$$
$$= -(x+1)(x-3)(x+9)$$

85. Completely factor $2t^3 - 16$.

Solution:

$$2t^3 - 16 = 2(t^3 - 8) = 2(t^3 - 2^3) = 2(t-2)(t^2 + 2t + 4)$$

91. The cylindrical shell shown in the figure has a volume of $\pi R^2 h - \pi r^2 h$. (a) Factor the expression for the volume. (b) From the result of part (a) show that the volume can be expressed as

2π(average radius)(thickness of the shell)h.

Solution:

(a) $V = \pi R^2 h - \pi r^2 h = \pi h(R^2 - r^2) = \pi h(R+r)(R-r)$

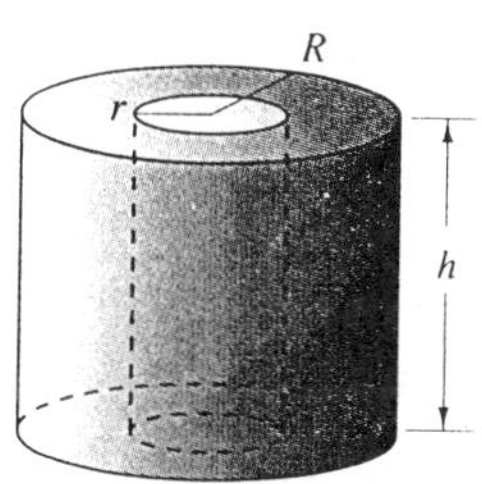

(b) $V = \pi h(R+r)(R-r)$

$= \pi(R+r)(R-r)h$

$= 2\pi\left(\dfrac{R+r}{2}\right)(R-r)h$

$= 2\pi$(average radius)(thickness of the shell)h

95. Construct a "geometric factoring model" to represent the factorization $2x^2 + 7x + 3 = (2x + 1)(x + 3)$.

Solution:

$$2x^2 + 7x + 3 = (2x + 1)(x + 3)$$

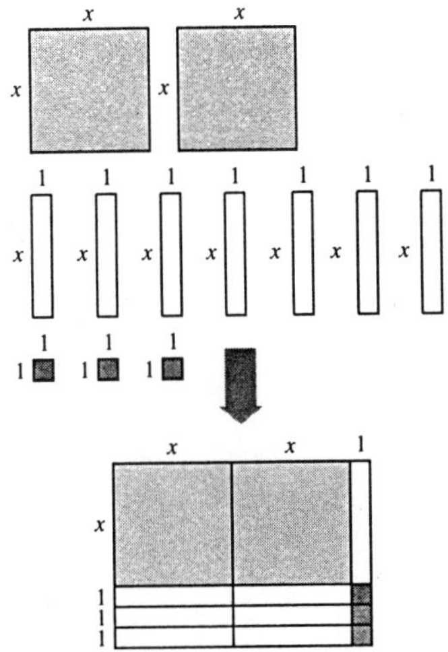

SECTION 1.4

Fractional Expressions

- You should know that a rational expression is the quotient of two polynomials.
- You should be able to simplify rational expressions by reducing them to lowest terms. This may involve factoring both the numerator and the denominator.
- You should be able to add, subtract, multiply, and divide rational expressions.
- You should be able to simplify compound fractions.

Solutions to Selected Exercises

1. Find the domain of $3x^2 - 4x + 7$.

Solution:

The domain of the **polynomial** $3x^2 - 4x + 7$ is the set of all real numbers.

5. Find the domain of $\dfrac{1}{x-2}$.

Solution:

The domain of $\dfrac{1}{x-2}$ is the set of all real numbers except $x = 2$, which would produce an undefined division by zero.

9. Find the domain of $\sqrt{x+1}$.

Solution:

The domain of $\sqrt{x+1}$ is the set of real numbers greater than or equal to -1 since $x + 1 \geq 0$ when $x \geq -1$.

13. Find the factor that makes the two fractions equivalent.

$$\frac{x+1}{x} = \frac{(x+1)(\qquad\qquad)}{x(x-2)}$$

Solution:

$$\frac{x+1}{x} = \frac{x+1}{x} \cdot \frac{x-2}{x-2} = \frac{(x+1)(x-2)}{x(x-2)},\ x \neq 2$$

19. Reduce to lowest terms.

$$\frac{3xy}{xy+x}$$

Solution:

$$\frac{3xy}{xy+x} = \frac{3xy}{x(y+1)} = \frac{3y}{y+1}, \quad x \neq 0$$

23. Reduce to lowest terms.

$$\frac{x^3+5x^2+6x}{x^2-4}$$

Solution:

$$\frac{x^3+5x^2+6x}{x^2-4} = \frac{x(x+2)(x+3)}{(x+2)(x-2)} = \frac{x(x+3)}{x-2}, \quad x \neq -2$$

27. Reduce to lowest terms.

$$\frac{2-x+2x^2-x^3}{x-2}$$

Solution:

$$\frac{2-x+2x^2-x^3}{x-2} = \frac{(2-x)+x^2(2-x)}{x-2}$$

$$= \frac{(2-x)(1+x^2)}{x-2} = \frac{-(x-2)(x^2+1)}{x-2} = -(x^2+1), \quad x \neq 2$$

29. Reduce to lowest terms.

$$\frac{z^3-8}{z^2+2z+4}$$

Solution:

$$\frac{z^3-8}{z^2+2z+4} = \frac{(z-2)(z^2+2z+4)}{z^2+2z+4} = z-2$$

33. Simplify

$$\frac{(x-9)(x+7)}{x+1} \cdot \frac{x}{9-x}.$$

Solution:

$$\frac{(x-9)(x+7)}{x+1} \cdot \frac{x}{9-x} = -\frac{(9-x)(x+7)x}{(x+1)(9-x)} = -\frac{x(x+7)}{x+1}, \quad x \neq 9$$

37. Simplify

$$\frac{t^2-t-6}{t^2+6t+9} \cdot \frac{t+3}{t^2-4}.$$

Solution:

$$\frac{t^2-t-6}{t^2+6t+9} \cdot \frac{t+3}{t^2-4} = \frac{(t+2)(t-3)}{(t+3)(t+3)} \cdot \frac{t+3}{(t+2)(t-2)} = \frac{t-3}{(t+3)(t-2)}, \quad t \neq -2$$

39. Simplify

$$\frac{x^2 + xy - 2y^2}{x^3 + x^2y} \cdot \frac{x}{x^2 + 3xy + 2y^2}.$$

Solution:

$$\begin{aligned}\frac{x^2 + xy - 2y^2}{x^3 + x^2y} \cdot \frac{x}{x^2 + 3xy + 2y^2} &= \frac{(x+2y)(x-y)}{x^2(x+y)} \cdot \frac{x}{(x+2y)(x+y)} \\ &= \frac{x-y}{x(x+y)^2},\ x \neq -2y\end{aligned}$$

43. Simplify

$$\frac{\left(\dfrac{x^2}{(x+1)^2}\right)}{\left(\dfrac{x}{(x+1)^3}\right)}.$$

Solution:

$$\frac{\left(\dfrac{x^2}{(x+1)^2}\right)}{\left(\dfrac{x}{(x+1)^3}\right)} = \frac{x^2}{(x+1)^2} \div \frac{x}{(x+1)^3} = \frac{x^2}{(x+1)^2} \cdot \frac{(x+1)^3}{x} = x(x+1),\quad x \neq 0,\ -1$$

47. Simplify

$$6 - \frac{5}{x+3}.$$

Solution:

$$6 - \frac{5}{x+3} = \frac{6(x+3)}{x+3} - \frac{5}{x+3} = \frac{6(x+3)-5}{x+3} = \frac{6x+18-5}{x+3} = \frac{6x+13}{x+3}$$

53. Simplify

$$\frac{1}{x^2 - x - 2} - \frac{x}{x^2 - 5x + 6}.$$

Solution:

$$\begin{aligned}\frac{1}{x^2 - x - 2} - \frac{x}{x^2 - 5x + 6} &= \frac{1}{(x-2)(x+1)} - \frac{x}{(x-2)(x-3)} \\ &= \frac{(x-3) - x(x+1)}{(x+1)(x-2)(x-3)} \\ &= \frac{-x^2 - 3}{(x+1)(x-2)(x-3)} = -\frac{x^2+3}{(x+1)(x-2)(x-3)}\end{aligned}$$

57. Simplify

$$\frac{\left(\frac{x}{2}-1\right)}{(x-2)}.$$

Solution:

$$\frac{\left(\frac{x}{2}-1\right)}{(x-2)} = \frac{\left(\frac{x}{2}-1\right)}{(x-2)} \cdot \frac{2}{2} = \frac{(x-2)}{2(x-2)} = \frac{1}{2}, \quad x \neq 2$$

61. Simplify

$$\frac{\left(\frac{x+3}{x-3}\right)^2}{\frac{1}{x+3}+\frac{1}{x-3}}.$$

Solution:

(a) Combining method:

$$\frac{\left(\frac{x+3}{x-3}\right)^2}{\frac{1}{x+3}+\frac{1}{x-3}} = \frac{\frac{(x+3)^2}{(x-3)^2}}{\frac{(x-3)+(x+3)}{(x+3)(x-3)}} = \frac{(x+3)^2}{(x-3)^2} \cdot \frac{(x+3)(x-3)}{2x} = \frac{(x+3)^3}{2x(x-3)}$$

(b) LCD method:

$$\frac{\frac{(x+3)^2}{(x-3)^2}}{\frac{1}{x+3}+\frac{1}{x-3}} \cdot \frac{(x+3)(x-3)^2}{(x+3)(x-3)^2} = \frac{(x+3)^3}{(x-3)^2+(x+3)(x-3)}$$

$$= \frac{(x+3)^3}{(x^2-6x+9)+(x^2-9)} = \frac{(x+3)^3}{2x^2-6x} = \frac{(x+3)^3}{2x(x-3)}$$

65. Simplify

$$\frac{\left(\sqrt{x}-\frac{1}{2\sqrt{x}}\right)}{\sqrt{x}}.$$

Solution:

$$\frac{\left(\sqrt{x}-\frac{1}{2\sqrt{x}}\right)}{\sqrt{x}} = \frac{\left(\sqrt{x}-\frac{1}{2\sqrt{x}}\right)}{\sqrt{x}} \cdot \frac{2\sqrt{x}}{2\sqrt{x}} = \frac{2x-1}{2x}, \quad x > 0$$

67. Simplify

$$\frac{\dfrac{t^2}{\sqrt{t^2+1}} - \sqrt{t^2+1}}{t^2}.$$

Solution:

$$\frac{\dfrac{t^2}{\sqrt{t^2+1}} - \sqrt{t^2+1}}{t^2} = \frac{\dfrac{t^2}{\sqrt{t^2+1}} - \sqrt{t^2+1}}{t^2} \cdot \frac{\sqrt{t^2+1}}{\sqrt{t^2+1}}$$

$$= \frac{t^2 - (t^2+1)}{t^2\sqrt{t^2+1}}$$

$$= -\frac{1}{t^2\sqrt{t^2+1}}$$

71. Rationalize the numerator of

$$\frac{\sqrt{x+2} - \sqrt{x}}{2}.$$

Solution:

$$\frac{\sqrt{x+2} - \sqrt{x}}{2} = \frac{\sqrt{x+2} - \sqrt{x}}{2} \cdot \frac{\sqrt{x+2} + \sqrt{x}}{\sqrt{x+2} + \sqrt{x}}$$

$$= \frac{(x+2) - x}{2(\sqrt{x+2} + \sqrt{x})} = \frac{2}{2(\sqrt{x+2} + \sqrt{x})} = \frac{1}{\sqrt{x+2} + \sqrt{x}}$$

75. Find the average of $\dfrac{x}{3}$ and $\dfrac{2x}{5}$.

Solution:

$$\text{Average} = \frac{\left(\dfrac{x}{3} + \dfrac{2x}{5}\right)}{2} = \frac{\left(\dfrac{x}{3} + \dfrac{2x}{5}\right)}{2} \cdot \frac{15}{15} = \frac{5x+6x}{30} = \frac{11x}{30}$$

79. When two resistors are connected in parallel, the total resistance is given by

$$\frac{1}{\dfrac{1}{R_1} + \dfrac{1}{R_2}}.$$

Simplify this compound fraction.

Solution:

$$\frac{1}{\dfrac{1}{R_1} + \dfrac{1}{R_2}} = \frac{1}{\dfrac{1}{R_1} + \dfrac{1}{R_2}} \cdot \frac{R_1R_2}{R_1R_2} = \frac{R_1R_2}{R_2 + R_1}$$

SECTION 1.5

Solving Equations

- To generate equivalent equations you may:
  (a) Remove symbols of grouping.
  (b) Combine like terms.
  (c) Add or subtract the same quantity to both sides.
  (d) Multiply or divide both sides by the same nonzero quantity.
  (e) Interchange the two sides of the equation.

- If you have multiplied or divided both sides by a variable, check for extraneous solutions.

- To solve quadratic equations:
  (a) Factor, if possible.
  (b) Take the square root of both sides, if possible.
  (c) Complete the square.
  (d) Use the Quadratic Formula

$$x = \frac{-b \pm \sqrt{b^2 - 4ac}}{2a}.$$

- The discriminant of a quadratic equation can be used to determine the type of solutions.
  (a) If $b^2 - 4ac > 0$, then there are two distinct real solutions.
  (b) If $b^2 - 4ac = 0$, then there is one repeated real solution.
  (c) If $b^2 - 4ac < 0$, then there are no real solutions.

- You should be able to solve some polynomials of higher degree by:
  (a) factoring and solving equations of quadratic type.
  (b) factoring by grouping.

- You should be able to solve equations involving radicals—always check for extraneous solutions.

- You should be able to solve equations involving absolute value—again, check for extraneous solutions.

Solutions to Selected Exercises

3. Determine whether the equation $-6(x-3)+5=-2x+10$ is an identity or a conditional equation.

Solution:

$$\begin{aligned}
-6(x-3)+5 &= -2x+10\\
-6x+18+5 &= -2x+10\\
-6x+23 &= -2x+10\\
-4x+23 &= 10\\
-4x &= -13\\
x &= \tfrac{13}{4} \qquad \text{Conditional}
\end{aligned}$$

9. Determine whether the value of x is a solution of the equation $3x^2+2x-5=2x^2-2$.
(a) $x=-3$ (b) $x=1$ (c) $x=4$ (d) $x=-5$

Solution:

(a) $3(-3)^2+2(-3)-5 \stackrel{?}{=} 2(-3)^2-2$

$$16=16$$

$x=-3$ **is** a solution.

(b) $3(1)^2+2(1)-5 \stackrel{?}{=} 2(1)^2-2$

$$0=0$$

$x=1$ **is** a solution.

(c) $3(4)^2+2(4)-5 \stackrel{?}{=} 2(4)^2-2$

$$51 \neq 30$$

$x=4$ is **not** a solution.

(d) $3(-5)^2+2(-5)-5 \stackrel{?}{=} 2(-5)^2-2$

$$60 \neq 48$$

$x=-5$ is **not** a solution.

13. Solve the equation $2(x+5)-7=3(x-2)$.

Solution:

$$\begin{aligned}
2(x+5)-7 &= 3(x-2)\\
2x+10-7 &= 3x-6\\
2x+3 &= 3x-6\\
-x+3 &= -6\\
-x &= -9\\
x &= 9
\end{aligned}$$

15. Solve the equation:

$$\frac{5x}{4}+\frac{1}{2}=x-\frac{1}{2}$$

Solution:

$$\begin{aligned}
\frac{5x}{4}+\frac{1}{2} &= x-\frac{1}{2}\\
4\left(\frac{5x}{4}+\frac{1}{2}\right) &= 4\left(x-\frac{1}{2}\right)\\
5x+2 &= 4x-2\\
x+2 &= -2\\
x &= -4
\end{aligned}$$

17. Solve the equation $0.25x + 0.75(10 - x) = 3$.

Solution:

$$\begin{aligned} 0.25x + 0.75(10 - x) &= 3 \\ 100[0.25x + 0.75(10 - x)] &= 100(3) \\ 25x + 75(10 - x) &= 300 \\ 25x + 750 - 75x &= 300 \\ -50x + 750 &= 300 \\ -50x &= -450 \\ x &= 9 \end{aligned}$$

19. Solve the equation $x + 8 = 2(x - 2) - x$, if possible.

Solution:

$$\begin{aligned} x + 8 &= 2(x - 2) - x \\ x + 8 &= 2x - 4 - x \\ x + 8 &= x - 4 \\ 8 &\neq -4 \qquad \text{Not possible} \end{aligned}$$

Thus, the equation has no solution.

23. Solve the equation:

$$\frac{5x - 4}{5x + 4} = \frac{2}{3}$$

Solution:

$$\begin{aligned} \frac{5x - 4}{5x + 4} &= \frac{2}{3} \\ 3(5x - 4) &= 2(5x + 4) \qquad \text{Cross multiply} \\ 15x - 12 &= 10x + 8 \\ 5x &= 20 \\ x &= 4 \end{aligned}$$

27. Solve the equation:

$$\frac{1}{x - 3} + \frac{1}{x + 3} = \frac{10}{x^2 - 9}$$

Solution:

$$\begin{aligned} \frac{1}{x - 3} + \frac{1}{x + 3} &= \frac{10}{x^2 - 9} \\ \frac{(x + 3) + (x - 3)}{(x - 3)(x + 3)} &= \frac{10}{x^2 - 9} \\ (x^2 - 9)\left(\frac{2x}{x^2 - 9}\right) &= \left(\frac{10}{x^2 - 9}\right)(x^2 - 9) \\ 2x &= 10 \\ x &= 5 \end{aligned}$$

29. Solve the equation:

$$\frac{7}{2x+1} - \frac{8x}{2x-1} = -4$$

Solution:

$$\frac{7}{2x+1} - \frac{8x}{2x-1} = -4$$

$$(2x+1)(2x-1)\left[\frac{7}{2x+1} - \frac{8x}{2x-1}\right] = -4(2x+1)(2x-1)$$

$$7(2x-1) - 8x(2x+1) = -4(4x^2-1)$$

$$14x - 7 - 16x^2 - 8x = -16x^2 + 4$$

$$-16x^2 + 6x - 7 = -16x^2 + 4$$

$$6x - 7 = 4$$

$$6x = 11$$

$$x = \frac{11}{6}$$

31. Solve the equation $(x+2)^2 + 5 = (x+3)^2$.

Solution:

$$(x+2)^2 + 5 = (x+3)^2$$

$$x^2 + 4x + 4 + 5 = x^2 + 6x + 9$$

$$4x + 9 = 6x + 9$$

$$4x = 6x$$

$$-2x = 0$$

$$x = 0$$

37. Solve $x^2 - 2x - 8 = 0$ by factoring.

Solution:

$$x^2 - 2x - 8 = 0$$

$$(x+2)(x-4) = 0$$

$$x = -2 \quad \text{or} \quad x = 4$$

41. Solve $2x^2 = 19x + 33$ by factoring.

Solution:

$$2x^2 = 19x + 33$$

$$2x^2 - 19x - 33 = 0$$

$$(2x+3)(x-11) = 0$$

$$2x + 3 = 0 \text{ or } x - 11 = 0$$

$$2x = -3 \text{ or } x = 11$$

$$x = -\tfrac{3}{2}$$

43. Solve $3x^2 = 36$ by extracting square roots.

Solution:

$$3x^2 = 36$$

$$x^2 = 12$$

$$x = \pm\sqrt{12}$$

$$x = \pm 2\sqrt{3}$$

47. Solve $x^2 + 4x - 32 = 0$ by completing the square.

Solution:

$$x^2 + 4x - 32 = 0$$

$$x^2 + 4x = 32$$

$$x^2 + 4x + 4 = 32 + 4$$

$$(x + 2)^2 = 36$$

$$x + 2 = \pm 6$$

$$x = -2 \pm 6$$

$$x = 4 \quad \text{or} \quad x = -8$$

49. Solve $9x^2 - 18x + 3 = 0$ by completing the square.

Solution:

$$9x^2 - 18x + 3 = 0 \qquad \text{Divide both sides by 9.}$$

$$x^2 - 2x + \frac{1}{3} = 0$$

$$x^2 - 2x = -\frac{1}{3}$$

$$x^2 - 2x + 1 = -\frac{1}{3} + 1$$

$$(x - 1)^2 = \frac{2}{3}$$

$$x - 1 = \pm\sqrt{\frac{2}{3}}$$

$$x = 1 \pm \sqrt{\frac{2}{3}} = 1 \pm \frac{\sqrt{2}}{\sqrt{3}} \cdot \frac{\sqrt{3}}{\sqrt{3}} = 1 \pm \frac{\sqrt{6}}{3}$$

53. Use the Quadratic Formula to solve $16x^2 + 8x - 3 = 0$.

Solution:

$$16x^2 + 8x - 3 = 0; \; a = 16, \; b = 8, \; c = -3$$

$$x = \frac{-8 \pm \sqrt{8^2 - 4(16)(-3)}}{2(16)} = \frac{-8 \pm \sqrt{64 + 192}}{32} = \frac{-8 \pm \sqrt{256}}{32} = \frac{-8 \pm 16}{32}$$

$$x = \frac{-8 + 16}{32} = \frac{1}{4}$$

$$x = \frac{-8 - 16}{32} = -\frac{3}{4}$$

57. Use the Quadratic Formula to solve $12x - 9x^2 = -3$.

Solution:

$$12x - 9x^2 = -3$$

$$0 = 9x^2 - 12x - 3$$

$$0 = 3(3x^2 - 4x - 1) \qquad \text{Divide both sides by 3.}$$

$$0 = 3x^2 - 4x - 1$$

$$a = 3, \ b = -4, \ c = -1$$

$$x = \frac{-(-4) \pm \sqrt{(-4)^2 - 4(3)(-1)}}{2(3)}$$

$$= \frac{4 \pm \sqrt{16 + 12}}{6} = \frac{4 \pm \sqrt{28}}{6} = \frac{4 \pm 2\sqrt{7}}{6} = \frac{2(2 \pm \sqrt{7})}{6} = \frac{2 \pm \sqrt{7}}{3}$$

61. Use the Quadratic Formula to solve $(y - 5)^2 = 2y$.

Solution:

$$(y - 5)^2 = 2y$$

$$y^2 - 10y + 25 = 2y$$

$$y^2 - 12y + 25 = 0$$

$$a = 1, \ b = -12, \ c = 25$$

$$y = \frac{-(-12) \pm \sqrt{(-12)^2 - 4(1)(25)}}{2(1)}$$

$$= \frac{12 \pm \sqrt{144 - 100}}{2} = \frac{12 \pm \sqrt{44}}{2} = \frac{12 \pm 2\sqrt{11}}{2} = \frac{2(6 \pm \sqrt{11})}{2} = 6 \pm \sqrt{11}$$

65. Find all the solutions of $x^3 - 2x^2 - 3x = 0$.

Solution:

$$x^3 - 2x^2 - 3x = 0$$

$$x(x^2 - 2x - 3) = 0$$

$$x(x + 1)(x - 3) = 0$$

$$x = 0, \ x = -1, \ \text{or } x = 3$$

69. Find all the solutions of $x^4 + 5x^2 - 36 = 0$.

Solution:

$$\begin{aligned} x^4 + 5x^2 - 36 &= 0 \\ (x^2)^2 + 5x^2 - 36 &= 0 \qquad \text{Quadratic type} \\ (x^2 + 9)(x^2 - 4) &= 0 \end{aligned}$$

$$x^2 + 9 = 0 \Rightarrow x^2 = -9 \Rightarrow \text{No real solutions}$$
$$x^2 - 4 = 0 \Rightarrow x^2 = 4 \quad \Rightarrow x = \pm 2$$

73. Find all the solutions of $\sqrt{x} + \sqrt{x - 20} = 10$.

Solution:

$$\begin{aligned} \sqrt{x} &= 10 - \sqrt{x - 20} \\ (\sqrt{x})^2 &= (10 - \sqrt{x - 20})^2 \\ x &= 100 - 20\sqrt{x - 20} + x - 20 \\ -80 &= -20\sqrt{x - 20} \\ 4 &= \sqrt{x - 20} \\ 4^2 &= (\sqrt{x - 20})^2 \\ 16 &= x - 20 \\ 36 &= x \end{aligned}$$

77. Find all solutions of $|2x - 1| = 5$.

Solution:

$$|2x - 1| = 5$$
$$\begin{aligned} 2x - 1 &= -5 \quad \text{or} \quad & 2x - 1 &= 5 \\ 2x &= -4 & 2x &= 6 \\ x &= -2 & x &= 3 \end{aligned}$$

81. Complete the square on the quadratic portion of

$$\frac{1}{x^2 - 4x - 12}.$$

Solution:

$$\begin{aligned} \frac{1}{x^2 - 4x - 12} &= \frac{1}{x^2 - 4x + 4 - 4 - 12} \\ &= \frac{1}{(x - 2)^2 - 16} \end{aligned}$$

85. Use the following information about a possible negative income tax for a family of two adults and two children. The plan would guarantee the poor a minimum income while encouraging families to increase their private income. (See the figure in the textbook.)

Family's earned income: $I = x$

Government payment: $G = 8000 - \frac{1}{2}x, \quad 0 \le x \le 16{,}000$

Spendable income: $S = I + G$

The spendable income is \$11,800, find the earned income, x.

Solution:

$$\begin{aligned} S &= I + G \\ &= x + \left(8000 - \tfrac{1}{2}x\right) \text{ for } 0 \le x \le 16{,}000 \\ &= \tfrac{1}{2}x + 8000 \end{aligned}$$

Since $S = 11{,}800$, we have

$$\begin{aligned} 11{,}800 &= \tfrac{1}{2}x + 8000 \\ 3800 &= \tfrac{1}{2}x \\ 7600 &= x. \end{aligned}$$

The earned income is \$7,600.

89. An open box is to be made from a square piece of material by cutting 2-inch squares from each corner and turning up the sides (see figure). The volume of the finished box is to be 200 cubic inches. Find the size of the original piece of material.

Solution:

$$\begin{aligned} \text{Volume} &= (\text{length})(\text{width})(\text{height}) \\ 200 &= (x)(x)(2) \\ 200 &= 2x^2 \\ 100 &= x^2 \\ x^2 &= 100 \\ x &= 10 \end{aligned}$$

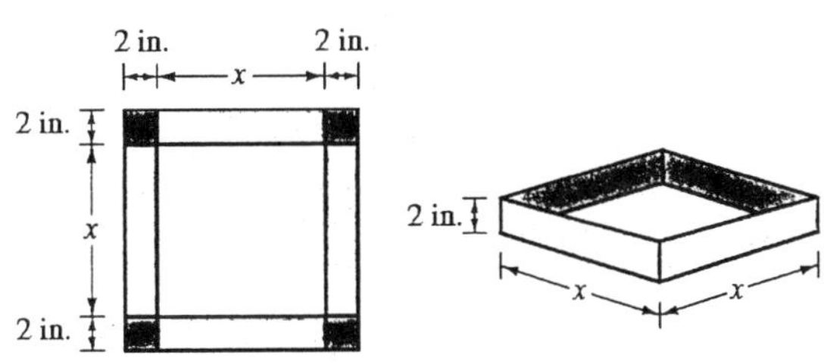

Since x must be positive, the size of the original piece is $(x + 4)$ by $(x + 4)$ or 14 inches by 14 inches.

93. The cost for producing x units of a product is given by $C = 0.125x^2 + 20x + 5000$. Determine the number of units produced if the cost is \$14,000.

Solution:

$$C = 0.125x^2 + 20x + 5000$$

$$14{,}000 = 0.125x^2 + 20x + 5000 \qquad \text{Replace } C \text{ with } 14{,}000.$$

$$0 = 0.125x^2 + 20x - 9000$$

$$0 = x^2 + 160x - 72{,}000 \qquad \text{Multiply both sides by 8.}$$

$$0 = (x - 200)(x + 360)$$

Choosing the positive value for x, we have $x = 200$ units.

SECTION 1.6

Solving Inequalities

- The procedures for solving a linear inequality are similar to those used for solving linear equations. However, when you multiply or divide both sides of an inequality by a negative number, *reverse* the inequality.
- You should be able to solve inequalities involving absolute value.
  (a) If $|x| < a$, then $-a < x < a$.
  (b) If $|x| > a$, then $x < -a$ OR $x > a$.
- You should be able to solve polynomial inequalities, or inequalities involving fractions.
  (a) Find the critical numbers.
  (b) Check the test intervals.

Solutions to Selected Exercises

7. Determine whether the values of x satisfy the inequality $0 < \dfrac{x-2}{4} < 2$.

(a) $x = 4$ (b) $x = 10$ (c) $x = 0$ (d) $x = \dfrac{7}{2}$

Solution:

(a) $x = 4$

$$0 \overset{?}{<} \frac{4-2}{4} \overset{?}{<} 2$$

$$0 < \frac{1}{2} < 2, \quad x = 4$$

$x = 4$ **is** a solution.

(b) $x = 10$

$$0 \overset{?}{<} \frac{10-2}{4} \overset{?}{<} 2$$

$$0 < 2 \not< 2$$

$x = 10$ is **not** a solution.

(c) $x = 0$

$$0 \overset{?}{<} \frac{0-2}{4} \overset{?}{<} 2$$

$$0 \not< -\frac{1}{2} < 2$$

$x = 0$ is **not** a solution.

(d) $x = \dfrac{7}{2}$

$$0 \overset{?}{<} \frac{(7/2)-2}{4} \overset{?}{<} 2$$

$$0 < \frac{3}{8} < 2$$

$x = \frac{7}{2}$ **is** a solution.

9. Solve $4x < 12$ and sketch the solution on the real number line.

Solution:

$$4x < 12$$

$$x < 3$$

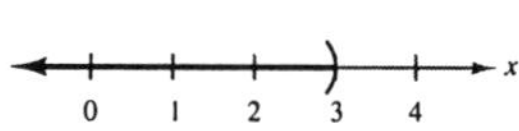

15. Solve $4(x+1) < 2x+3$ and sketch the solution on the real number line.

Solution:

$$4(x+1) < 2x+3$$

$$4x+4 < 2x+3$$

$$2x < -1$$

$$x < -\tfrac{1}{2}$$

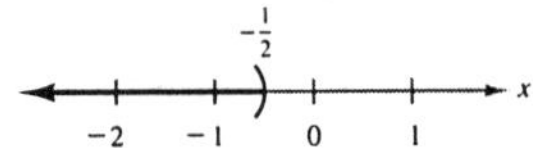

21. Solve

$$-4 < \frac{2x-3}{3} < 4$$

and sketch the solution on the real number line.

Solution:

$$-4 < \frac{2x-3}{3} < 4$$

$$-12 < 2x-3 < 12$$

$$-9 < 2x < 15$$

$$-\frac{9}{2} < x < \frac{15}{2}$$

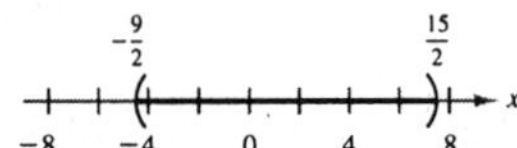

27. Solve

$$\left|\frac{x}{2}\right| > 3$$

and sketch the solution on the real number line.

Solution:

$$\left|\frac{x}{2}\right| > 3$$

$$\frac{x}{2} < -3 \quad \text{or} \quad \frac{x}{2} > 3$$

$$x < -6 \quad \text{or} \quad x > 6$$

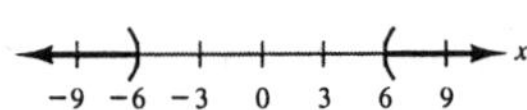

33. Solve

$$\left|\frac{x-3}{2}\right| \geq 5$$

and sketch the solution on the real number line.

Solution:

$$\left|\frac{x-3}{2}\right| \geq 5$$

$$\frac{x-3}{2} \leq -5 \quad \text{or} \quad \frac{x-3}{2} \geq 5$$

$$x-3 \leq -10 \quad \text{or} \quad x-3 \geq 10$$

$$x \leq -7 \quad \text{or} \quad x \geq 13$$

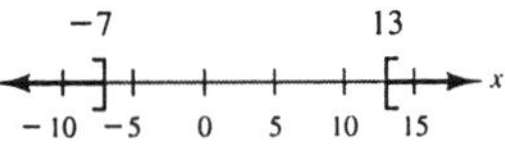

37. Solve $2|x+10| \geq 9$ and sketch the solution on the real number line.

Solution:

$$2|x+10| \geq 9$$

$$|x+10| \geq \tfrac{9}{2}$$

$$x+10 \leq -\tfrac{9}{2} \quad \text{or} \quad x+10 \geq \tfrac{9}{2}$$

$$x \leq -\tfrac{29}{2} \quad \text{or} \quad x \geq -\tfrac{11}{2}$$

$-\frac{29}{2}$ $\quad$ $-\frac{11}{2}$

− 16 − 12 −8 −4 0 x

43. Use absolute value notation to define the pair of intervals on the real line.

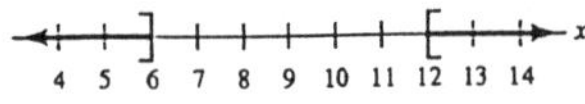

Solution:

The intervals consist of all points that are at least three units away from 9.

$$|x-9| \geq 3$$

47. Solve the inequality $x^2 > 4$ and give the answer in interval notation.

Solution:

$$x^2 > 4$$

$$x^2 - 4 > 0$$

$$(x+2)(x-2) > 0$$

Critical numbers: $x = \pm 2$
Test intervals: $(-\infty, -2)$, $(-2, 2)$, $(2, \infty)$
Solution intervals: $(-\infty, -2) \cup (2, \infty)$

51. Solve the inequality $x^2 + 4x + 4 \geq 9$ and give the answer in interval notation.

Solution:

$$x^2 + 4x + 4 \geq 9$$

$$x^2 + 4x - 5 \geq 0$$

$$(x + 5)(x - 1) \geq 0$$

Critical numbers: $x = -5, \ x = 1$
Test intervals: $(-\infty, \ -5), \ (-5, \ 1), \ (1, \ \infty)$
Solution intervals: $(-\infty, \ -5] \cup [1, \ \infty)$

53. Solve the inequality $3(x - 1)(x + 1) > 0$ and give the answer in interval notation.

Solution:

$$3(x - 1)(x + 1) > 0$$

Critical numbers: $x = -1, \ x = 1$
Test intervals: $(-\infty, \ -1), \ (-1, \ 1), \ (1, \ \infty)$
Solution intervals: $(-\infty, \ -1) \cup (1, \ \infty)$

57. Solve the inequality $4x^3 - 6x^2 < 0$ and give the answer in interval notation.

Solution:

$$4x^3 - 6x^2 < 0$$

$$2x^2(2x - 3) < 0$$

Critical numbers: $x = 0, \ x = \frac{3}{2}$
Test intervals: $(-\infty, \ 0), \ (0, \ \frac{3}{2}), \ (\frac{3}{2}, \ \infty)$
Solution intervals: $(-\infty, \ 0) \cup (0, \ \frac{3}{2})$
Note: $x = 0$ is *not* a solution.

61. Solve the inequality $1/x > x$ and give the answer in interval notation.

Solution:

$$\frac{1}{x} > x$$

$$\frac{1}{x} - x > 0$$

$$\frac{1 - x^2}{x} > 0$$

$$\frac{(1+x)(1-x)}{x} > 0$$

Critical numbers: $x = -1,\ x = 0,\ x = 1$
Test intervals: $(-\infty,\ -1),\ (-1,\ 0),\ (0,\ 1),\ (1,\ \infty)$
Solution intervals: $(-\infty,\ -1) \cup (0,\ 1)$

67. Find the interval on the real number line for which the radicand is nonnegative for $\sqrt{x-5}$.

Solution:

The radicand of $\sqrt{x-5}$ is $x - 5$.

$$x - 5 \geq 0$$

$$x \geq 5$$

Therefore, the interval is $[5,\ \infty)$.

73. Find the intervals on the real number line for which the radicand is nonnegative for $\sqrt{x^2 - 7x + 12}$.

Solution:

The radicand is $x^2 - 7x + 12$.

$$x^2 - 7x + 12 \geq 0$$

$$(x-3)(x-4) \geq 0$$

Critical numbers: $x = 3,\ x = 4$
Test intervals: $(-\infty,\ 3),\ (3,\ 4),\ (4,\ \infty)$
Solution intervals: $(-\infty,\ 3] \cup [4,\ \infty)$

77. Use a calculator to solve the inequality and round each number in your answer to two decimal places.

$$\frac{1}{2.3x - 5.2} > 3.4$$

Solution:

$$\frac{1}{2.3x - 5.2} > 3.4$$

$$\frac{1}{2.3x - 5.2} - 3.4 > 0$$

$$\frac{1 - 3.4(2.3x - 5.2)}{2.3x - 5.2} > 0$$

$$\frac{18.68 - 7.82x}{2.3x - 5.2} > 0$$

Critical numbers: $x = \frac{18.68}{7.82} \approx 2.39$, $x = \frac{5.2}{2.3} \approx 2.26$
Test intervals: $(-\infty, 2.26)$, $(2.26, 2.39)$, $(2.39, \infty)$
Solution interval: $(2.26, 2.39)$

81. The revenue for selling x units of a product is $R = 115.95x$. The cost of producing x units is $C = 95x + 750$. In order to obtain a profit, the revenue must be *greater than* the cost. For what values of x will this product return a profit?

Solution:

$$R > C$$

$$115.95x > 95x + 750$$

$$20.95x > 750$$

$$x > 35.7995$$

$$x \geq 36 \text{ units}$$

This product will return a profit if the number of units, x, is greater than or equal to 36.

85. A rectangular playing field with a perimeter of 100 meters is to have an area of at least 500 square meters. Within what bounds must the length of the rectangle lie?

Solution:

$$2L + 2W = 100$$

$$W = \frac{100 - 2L}{2} = 50 - L$$

$$LW \geq 500$$

$$L(50 - L) \geq 500$$

$$50L - L^2 \geq 500$$

$$0 \geq L^2 - 50L + 500$$

By the Quadratic Formula, the critical numbers are:

$$x = \frac{-(-50) \pm \sqrt{(-50)^2 - 4(1)(500)}}{2(1)} = \frac{50 \pm \sqrt{500}}{2} = \frac{50 \pm 10\sqrt{5}}{2} = 25 \pm 5\sqrt{5}$$

Solution interval: $[25 - 5\sqrt{5},\ 25 + 5\sqrt{5}]$ or 13.8 meters $\leq L \leq$ 36.2 meters

SECTION 1.7

Algebraic Errors and Some Algebra of Calculus

- You should be able to recognize and avoid the common algebraic errors listed in this section.
- You should be able to "unsimplify" algebraic expressions by the following methods:
  (a) Unusual Factoring
  (b) Inserting Factors or Terms
  (c) Rewriting with Negative Exponents
  (d) Writing a Fraction as a Sum of Terms.

Solutions to Selected Exercises

3. Find and correct any errors in $5z + 3(x - 2) = 5z + 3x - 2$.

Solution:

$$\begin{aligned} 5z + 3(x-2) &= 5z + 3x - 6 \quad \text{By the Distributive Law} \\ &\neq 5z + 3x - 2 \end{aligned}$$

5. Find and correct any errors in $-\dfrac{x-3}{x-1} = \dfrac{3-x}{1-x}$.

Solution:

$$\begin{aligned} -\frac{x-3}{x-1} &= \frac{-(x-3)}{x-1} \\ &= \frac{3-x}{x-1} \quad \text{Only the numerator is multiplied by } (-1). \\ &\neq \frac{3-x}{1-x} \end{aligned}$$

11. Find and correct any errors in $\sqrt{x+9} = \sqrt{x} + 3$.

Solution:

$$\sqrt{x+9} \neq \sqrt{x} + 3$$

The root of a sum does not equal the sum of the roots.

$$\sqrt{x} + 3 = \sqrt{(\sqrt{x}+3)^2} = \sqrt{x + 6\sqrt{x} + 9} \neq \sqrt{x+9}$$

$\sqrt{x+9}$ cannot be simplified.

15. Find and correct any errors in

$$\frac{1}{x+y^{-1}} = \frac{y}{x+1}.$$

Solution:

$$\frac{1}{x+y^{-1}} = \frac{1}{x+(1/y)} \bullet \frac{y}{y} = \frac{y}{xy+1} \neq \frac{y}{x+1}$$

19. Find and correct any errors in $\sqrt[3]{x^3+7x^2} = x^2\sqrt[3]{x+7}$.

Solution:

$$\sqrt[3]{x^3+7x^2} = \sqrt[3]{x^2(x+7)} = \sqrt[3]{x^2}\sqrt[3]{x+7} \neq x^2\sqrt[3]{x+7}$$

Radicals apply to every factor of the radicand.

23. Find and correct any errors in

$$\frac{1}{2y} = \left(\frac{1}{2}\right)y.$$

Solution:

$$\frac{1}{2y} = \frac{1}{2} \bullet \frac{1}{y} = \left(\frac{1}{2}\right)\frac{1}{y} \neq \left(\frac{1}{2}\right)y$$

Use the definition for multiplying fractions.

27. Insert the required factor in the parentheses.

$$\tfrac{1}{3}x^3 + 5 = (\qquad)(x^3+15)$$

Solution:

$$\begin{aligned}\tfrac{1}{3}x^3 + 5 &= \tfrac{1}{3}x^3 + \tfrac{1}{3}(15)\\ &= \left(\tfrac{1}{3}\right)(x^3+15)\end{aligned}$$

31. Insert the required factor in the parentheses.

$$\frac{1}{\sqrt{x}(1+\sqrt{x})^2} = (\quad)\frac{1}{(1+\sqrt{x})^2}\left(\frac{1}{2\sqrt{x}}\right)$$

Solution:

$$\frac{1}{\sqrt{x}(1+\sqrt{x})^2} = \frac{1}{\sqrt{x}} \bullet \frac{1}{(1+\sqrt{x})^2} = (2)\left(\frac{1}{2\sqrt{x}}\right)\frac{1}{(1+\sqrt{x})^2} = (2)\frac{1}{(1+\sqrt{x})^2}\left(\frac{1}{2\sqrt{x}}\right)$$

35. Insert the required factor in the parentheses.

$$\frac{3}{x}+\frac{5}{2x^2}-\frac{3}{2}x=(\quad)(6x+5-3x^3)$$

Solution:

$$\frac{3}{x}+\frac{5}{2x^2}-\frac{3}{2}x=\frac{6x}{2x^2}+\frac{5}{2x^2}-\frac{3x^3}{2x^2}=\left(\frac{1}{2x^2}\right)(6x+5-3x^3)$$

37. Insert the required factor in the parentheses.

$$\frac{x^2}{1/12}-\frac{y^2}{2/3}=\frac{12x^2}{(\quad)}-\frac{3y^2}{(\quad)}$$

Solution:

$$\frac{x^2}{1/12}-\frac{y^2}{2/3}=x^2\left(\frac{12}{1}\right)-y^2\left(\frac{3}{2}\right)=\frac{12x^2}{1}-\frac{3y^2}{2}$$

41. Insert the required factor in the parentheses.

$$\frac{x^2}{\sqrt{x^2+1}}-\sqrt{x^2+1}=\frac{1}{\sqrt{x^2+1}}(\quad)$$

Solution:

$$\begin{aligned}\frac{x^2}{\sqrt{x^2+1}}-\sqrt{x^2+1}&=\frac{x^2}{\sqrt{x^2+1}}-\frac{\sqrt{x^2+1}}{1}\cdot\frac{\sqrt{x^2+1}}{\sqrt{x^2+1}}\\&=\frac{x^2-(x^2+1)}{\sqrt{x^2+1}}=\frac{-1}{\sqrt{x^2+1}}=\frac{1}{\sqrt{x^2+1}}(-1)\end{aligned}$$

43. Insert the required factor in the parentheses.

$$\frac{1}{10}(2x+1)^{5/2}-\frac{1}{6}(2x+1)^{3/2}=\frac{(2x+1)^{3/2}}{15}(\quad)$$

Solution:

$$\begin{aligned}\frac{1}{10}(2x+1)^{5/2}-\frac{1}{6}(2x+1)^{3/2}&=\frac{3}{30}(2x+1)^{3/2}(2x+1)^1-\frac{5}{30}(2x+1)^{3/2}\\&=\frac{1}{30}(2x+1)^{3/2}[3(2x+1)-5]\\&=\frac{1}{30}(2x+1)^{3/2}(6x-2)\\&=\frac{1}{30}(2x+1)^{3/2}2(3x-1)\\&=\frac{1}{15}(2x+1)^{3/2}(3x-1)\end{aligned}$$

47. Write the following as a sum of two or more terms.

$$\frac{4x^3 - 7x^2 + 1}{x^{1/3}}$$

Solution:

$$\frac{4x^3 - 7x^2 + 1}{x^{1/3}} = \frac{4x^3}{x^{1/3}} - \frac{7x^2}{x^{1/3}} + \frac{1}{x^{1/3}}$$

$$= 4x^{3-1/3} - 7x^{2-1/3} + \frac{1}{x^{1/3}} = 4x^{8/3} - 7x^{5/3} + \frac{1}{x^{1/3}}$$

51. Simplify the expression.

$$\frac{-2(x^2-3)^{-3}(2x)(6x+1)^3 - 3(6x+1)^2(6)(x^2-3)^{-2}}{[(6x+1)^3]^2}$$

Solution:

$$\frac{-2(x^2-3)^{-3}(2x)(6x+1)^3 - 3(6x+1)^2(6)(x^2-3)^{-2}}{[(6x+1)^3]^2}$$

$$= \frac{2(x^2-3)^{-3}(6x+1)^2[-2x(6x+1) - 9(x^2-3)]}{(6x+1)^6}$$

$$= \frac{2[-12x^2 - 2x - 9x^2 + 27]}{(x^2-3)^3(6x+1)^4}$$

$$= \frac{2(-21x^2 - 2x + 27)}{(x^2-3)^3(6x+1)^4}$$

$$= \frac{-2(21x^2 + 2x - 27)}{(x^2-3)^3(6x+1)^4}$$

55. Simplify the expression.

$$\frac{(3x+2)^{3/4}(2x+3)^{-2/3}(2) - (2x+3)^{1/3}(3x+2)^{-1/4}(3)}{[(3x+2)^{3/4}]^2}$$

Solution:

$$\frac{(3x+2)^{3/4}(2x+3)^{-2/3}(2) - (2x+3)^{1/3}(3x+2)^{-1/4}(3)}{[(3x+2)^{3/4}]^2}$$

$$= \frac{(3x+2)^{-1/4}(2x+3)^{-2/3}[2(3x+2) - 3(2x+3)]}{(3x+2)^{6/4}}$$

$$= \frac{6x + 4 - 6x - 9}{(3x+2)^{1/4}(2x+3)^{2/3}(3x+2)^{6/4}}$$

$$= \frac{-5}{(2x+3)^{2/3}(3x+2)^{7/4}}$$

REVIEW EXERCISES FOR CHAPTER 1

Solutions to Selected Exercises

3. Use absolute value notation to describe the expression "the distance between x and 7 is at least 4."

Solution:

Since the distance between x and 7 is

$$d(x,\ 7) = |x - 7|$$

and since the distance is at least 4, we have

$$d(x,\ 7) \geq 4$$

$$|x - 7| \geq 4.$$

7. Perform the indicated operations on $(3^2/5^2)^{-3}$ without the aid of a calculator.

Solution:

$$\left(\frac{3^2}{5^2}\right)^{-3} = \left(\frac{5^2}{3^2}\right)^{3} = \frac{5^6}{3^6} = \frac{15{,}625}{729}$$

13. Write the number in scientific notation.

Daily U.S. Consumption of Dunkin' Donuts: 2,740,000

Solution:

$2{,}740{,}000.00 = 2.74 \times 10^6$ (decimal moves 6 places to the left.)

17. Use a calculator to evaluate the given expression. Round your answer to three decimal places.

(a) $1800(1 + 0.08)^{24}$ (b) $0.0024(7{,}658{,}400)$

Solution:

(a) $1800(1 + 0.08)^{24} = 1800(1.08)^{24}$

$\approx 11{,}414.125$

1800 [×] 1.08 [y^x] 24 [=]

(b) $0.0024(7{,}658{,}400) = 18{,}380.160$

.0024 [×] 7658400 [=]

23. Describe the error and make the necessary correction.

$$-x^2(-x^2 + 3) = x^4 + 3x^2$$

Solution:

The sign on the second term is incorrect.

$$-x^2(-x^2 + 3) = (-x^2)(-x^2) + (-x^2)(3) = x^4 - 3x^2$$

29. Simplify $\sqrt{50} - \sqrt{18}$.

Solution:

$$\sqrt{50} - \sqrt{18} = \sqrt{25 \cdot 2} - \sqrt{9 \cdot 2} = 5\sqrt{2} - 3\sqrt{2} = (5-3)\sqrt{2} = 2\sqrt{2}$$

33. Perform the required operations and/or simplify $(x^2 - 2x + 1)(x^3 - 1)$.

Solution:

$$\begin{aligned}(x^2 - 2x + 1)(x^3 - 1) &= (x^2 - 2x + 1)x^3 - (x^2 - 2x + 1)(1) \\ &= x^5 - 2x^4 + x^3 - x^2 + 2x - 1\end{aligned}$$

35. Perform the required operations and/or simplify.

$$\frac{x^2 - 4}{x^4 - 2x^2 - 8} \cdot \frac{x^2 + 2}{x^2}$$

Solution:

$$\frac{x^2 - 4}{x^4 - 2x^2 - 8} \cdot \frac{x^2 + 2}{x^2} = \frac{x^2 - 4}{(x^2 - 4)(x^2 + 2)} \cdot \frac{x^2 + 2}{x^2} = \frac{1}{x^2}$$

39. Perform the required operations and/or simplify.

$$\frac{1}{x - 1} - \frac{1}{x + 2}$$

Solution:

$$\begin{aligned}\frac{1}{x - 1} - \frac{1}{x + 2} &= \frac{1}{x - 1} \cdot \frac{x + 2}{x + 2} - \frac{1}{x + 2} \cdot \frac{x - 1}{x - 1} \\ &= \frac{(x + 2) - (x - 1)}{(x - 1)(x + 2)} = \frac{x + 2 - x + 1}{(x - 1)(x + 2)} = \frac{3}{(x - 1)(x + 2)}\end{aligned}$$

45. Perform the required operations and/or simplify.

$$\frac{1}{x - 2} + \frac{1}{(x - 2)^2} + \frac{1}{x + 2}$$

Solution:

$$\begin{aligned}\frac{1}{x - 2} + \frac{1}{(x - 2)^2} + \frac{1}{x + 2} &= \frac{(x - 2)(x + 2) + (x + 2) + (x - 2)^2}{(x + 2)(x - 2)^2} \\ &= \frac{(x^2 - 4) + (x + 2) + (x^2 - 4x + 4)}{(x + 2)(x - 2)^2} = \frac{2x^2 - 3x + 2}{(x + 2)(x - 2)^2}\end{aligned}$$

49. Simplify the compound fraction.

$$\frac{\dfrac{3a}{(a^2/x)-1}}{\dfrac{a}{x}-1}$$

Solution:

$$\frac{\dfrac{3a}{(a^2/x)-1}}{\dfrac{a}{x}-1} = \frac{\left[\dfrac{3a}{(a^2/x)-1}\right]\dfrac{x}{x}}{\dfrac{a-x}{x}} = \frac{3ax}{a^2-x} \cdot \frac{x}{a-x} = \frac{3ax^2}{(a^2-x)(a-x)}$$

51. Insert the missing factor for $x^3 - x^2 + 2x - 2 = (x-1)(\quad)$.

Solution:

$$x^3 - x^2 + 2x - 2 = x^2(x-1) + 2(x-1) = (x-1)(x^2+2)$$

55. Insert the missing factor for

$$\frac{t}{\sqrt{t+1}} - \sqrt{t+1} = \frac{1}{\sqrt{t+1}}(\quad).$$

Solution:

$$\frac{t}{\sqrt{t+1}} - \sqrt{t+1} = \frac{t}{\sqrt{t+1}} - \frac{\sqrt{t+1}}{1} \cdot \frac{\sqrt{t+1}}{\sqrt{t+1}}$$

$$= \frac{t}{\sqrt{t+1}} - \frac{t+1}{\sqrt{t+1}} = \frac{t-(t+1)}{\sqrt{t+1}} = \frac{-1}{\sqrt{t+1}} = \frac{1}{\sqrt{t+1}}(-1)$$

59. Solve the equation

$$3\left(1 - \frac{1}{5t}\right) = 0.$$

Solution:

$$3\left(1 - \frac{1}{5t}\right) = 0$$

$$1 - \frac{1}{5t} = 0$$

$$1 = \frac{1}{5t}$$

$$5t = 1$$

$$t = \frac{1}{5}$$

65. Solve the equation $5x^4 - 12x^3 = 0$.

Solution:

$$5x^4 - 12x^3 = 0$$

$$x^3(5x-12) = 0$$

$$x^3 = 0 \quad \text{or} \quad 5x - 12 = 0$$

$$x = 0 \quad \text{or} \quad x = \tfrac{12}{5}$$

69. Solve the equation $\sqrt{x+4} = 3$.

Solution:

$$\begin{aligned}\sqrt{x+4} &= 3\\(\sqrt{x+4})^2 &= (3)^2\\x+4 &= 9\\x &= 5\end{aligned}$$

71. Solve the equation $\sqrt{2x+3} + \sqrt{x-2} = 2$.

Solution:

$$\begin{aligned}\sqrt{2x+3} + \sqrt{x-2} &= 2\\(\sqrt{2x+3})^2 &= (2-\sqrt{x-2})^2\\2x+3 &= 4-4\sqrt{x-2}+x-2\\x+1 &= -4\sqrt{x-2}\\(x+1)^2 &= (-4\sqrt{x-2})^2\\x^2+2x+1 &= 16(x-2)\\x^2-14x+33 &= 0\\(x-3)(x-11) &= 0\end{aligned}$$

$x = 3$, extraneous OR $x = 11$, extraneous

No solution. Always check your answers in the original equation.

77. Solve the inequality $x^2 - 4 \le 0$.

Solution:

$$\begin{aligned}x^2 - 4 &\le 0\\(x+2)(x-2) &\le 0\end{aligned}$$

Critical numbers: $x = 2,\ x = -2$
Test intervals: $(-\infty, -2), (-2, 2), (2, \infty)$
Solution interval: $[-2, 2]$

81. Solve the inequality.

$$\left|x - \tfrac{3}{2}\right| \ge \tfrac{3}{2}$$

Solution:

$$\left|x - \tfrac{3}{2}\right| \ge \tfrac{3}{2}$$

$$x - \tfrac{3}{2} \le -\tfrac{3}{2} \quad \text{or} \quad x - \tfrac{3}{2} \ge \tfrac{3}{2}$$

$$x \le 0 \quad \text{or} \quad x \ge 3$$

$(-\infty, 0] \cup [3, \infty)$

83. Find the domain of $\sqrt{2x-10}$ by finding the interval on the real number line for which the radicand is nonnegative.

Solution:

The radicand of $\sqrt{2x-10}$ is $2x-10$. Solve $2x-10 \geq 0$ for the domain.

$$2x - 10 \geq 0$$
$$2x \geq 10$$
$$x \geq 5$$

The domain of $\sqrt{2x-10}$ is the interval $[5, \infty)$.

87. A car radiator contains 10 quarts of a 30% antifreeze solution. How many quarts will have to be replaced with pure antifreeze if the resulting solution is to be 50% antifreeze?

Solution:

Let x = the number of quarts of pure antifreeze.

$$30\% \text{ of } (10-x) + 100\% \text{ of } x = 50\% \text{ of } 10 \text{ quarts}$$
$$0.30(10-x) + 1.00x = 0.50(10)$$
$$3 - 0.30x + x = 5$$
$$0.70x = 2$$
$$x = \frac{2}{0.70} \approx 2.86 \text{ quarts}$$

89. The distance from a spacecraft to the horizon is 1000 miles. Find x, the altitude of the craft, as shown in the figure. Assume that the radius of the earth is 4000 miles.

Solution:

$$1000^2 + 4000^2 = (4000+x)^2 \qquad \text{Pythagorean Theorem}$$
$$1000^2 + 4000^2 = 4000^2 + 8000x + x^2$$
$$0 = x^2 + 8000x - 1000^2$$
$$x = \frac{-8000 \pm \sqrt{8000^2 - 4(-1000^2)}}{2}$$

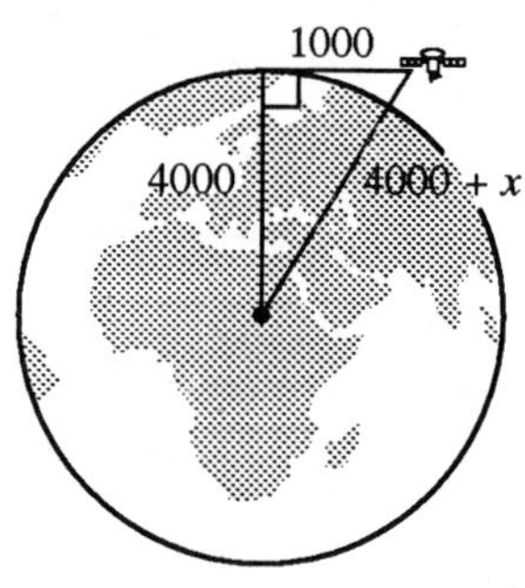

Considering only the positive root,

$$x = \frac{-8000 + 2000\sqrt{17}}{2} = -4000 + 1000\sqrt{17}$$

$x \approx 123$ miles

93. The revenue for selling x units of a product is $R = 125.95x$. The cost of producing x units is $C = 92x + 1200$. In order to obtain a profit, the revenue must be greater than the cost. For what values of x will this product return a profit?

Solution:

$$R > C$$

$$125.95x > 92x + 1200$$

$$33.95x > 1200$$

$$x > 35.346$$

$$x \geq 36 \text{ units}$$

This product will return a profit if the number of units, x, is greater than or equal to 36.

Practice Test for Chapter 1

1. Simplify $4 + 3(18 - 11)$.

2. Simplify $\left(\frac{4}{15} \div 2\right) - \left(5 \times \frac{8}{15}\right)$.

3. Use absolute value notation to describe the expression "the distance between x and -6 is less than 4."

4. Write 0.0000439 in scientific notation.

5. Simplify $(3x^2y^{-1})^2(4x^{-2}y)^{-1}$.

6. Simplify $\sqrt[3]{81x^5y^6}$.

7. Rationalize the denominator of $4/\sqrt[3]{2}$ and simplify.

8. Multiply $(x+3)(x^2 - 4x - 7)$ and simplify.

9. Factor $x^4 - 81$ completely.

10. Factor $x^5 - 4x^3 - x^2 + 4$ completely.

11. Factor $8x^2 + 6x - 9$ completely.

12. Reduce $\dfrac{8x^3 + 8x^2y}{x^2y + xy^2}$ to lowest terms.

13. Simplify $\dfrac{3x}{x^2 - x - 6} - \dfrac{2}{x-3}$.

14. Simplify $\dfrac{\dfrac{1}{x+1} - \dfrac{1}{x}}{\dfrac{1}{x^2+x}}$.

15. Simplify $\dfrac{x^2 - 49}{x^2 + 6x - 7} \cdot \dfrac{x^2 - 1}{x}$.

16. Solve $\dfrac{1}{x+2} - \dfrac{3}{x-4} = \dfrac{5}{x^2 - 2x - 8}$.

17. Solve $(x + 12)^2 = 20$ by taking the square root of both sides.

18. Solve $3x^2 + 6x + 2 = 0$ by completing the square.

19. Solve $2x^2 - 3x - 5 = 0$ by the Quadratic Formula.

20. Solve $-3 \leq \dfrac{4-x}{2} \leq 5$.

21. Solve $|x - 15| \geq 10$.

22. Solve $x^3 - 9x \leq 0$.

23. Solve $\dfrac{4}{x+3} > \dfrac{-1}{x-2}$.

24. Insert the required factor in the parentheses: $x^2(1-2x)^{4/3} - 3x(1-2x)^{1/3} = x(1-2x)^{1/3}(\quad)$.

25. True or False: $\sqrt{36 + x^2} = 6 + x$. Explain.

CHAPTER 2

Functions and Graphs

Section 2.1 The Cartesian Plane **54**

Section 2.2 Graphs of Equations **60**

Section 2.3 Lines in the Plane **67**

Section 2.4 Functions **75**

Section 2.5 Graphs of Functions **80**

Section 2.6 Combinations of Functions **87**

Section 2.7 Inverse Functions **91**

Section 2.8 Variation and Mathematical Models **98**

Review Exercises **103**

Practice Test **113**

SECTION 2.1

The Cartesian Plane

- You should be able to plot points.
- You should know that the distance between (x_1, y_1) and (x_2, y_2) in the plane is

  $$d = \sqrt{(x_2 - x_1)^2 + (y_2 - y_1)^2}.$$

- You should know that the midpoint of the line segment joining (x_1, y_1) and (x_2, y_2) is

  $$\left(\frac{x_1 + x_2}{2}, \frac{y_1 + y_2}{2}\right).$$

Solutions to Selected Exercises

3. Sketch the square with vertices $(2, 4)$, $(5, 1)$, $(2, -2)$, and $(-1, 1)$.

Solution:

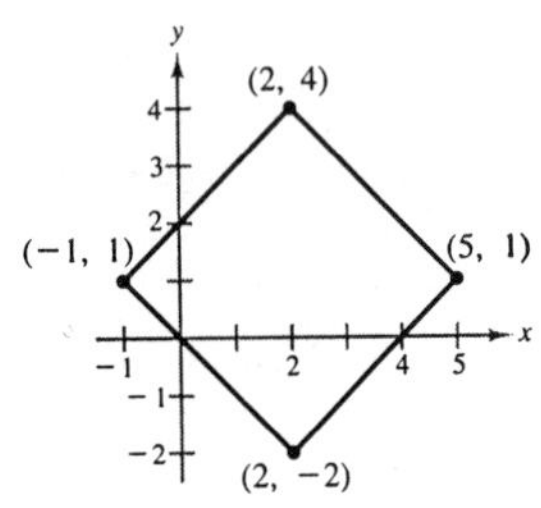

5. Find the coordinates of the vertices of the figure in its new position.

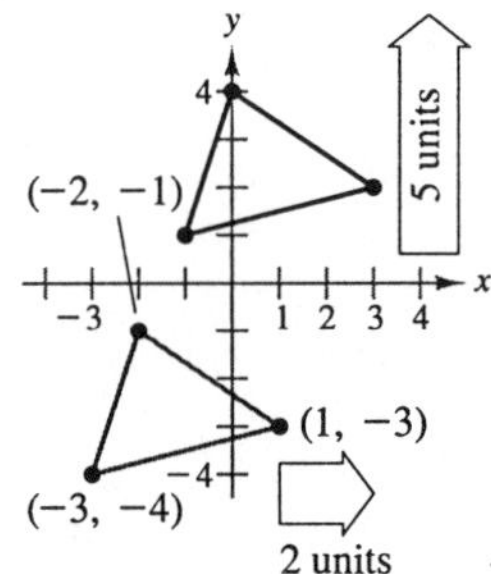

Solution:

$$(-2 + 2, \ -1 + 5) = (0, \ 4)$$

$$(-3 + 2, \ -4 + 5) = (-1, \ 1)$$

$$(1 + 2, \ -3 + 5) = (3, \ 2)$$

9. Find the distance between the points $(-3, -1)$ and $(2, -1)$.

Solution:

Since the points $(-3, -1)$ and $(2, -1)$ lie on the same vertical line, the distance between the points is given by the absolute value of the difference of their x–coordinates.

$$d = |-3 - 2| = 5$$

13. For the indicated triangle (a) find the length of the two sides of the right triangle and use the Pythagorean Theorem to find the length of the hypotenuse, and (b) use the Distance Formula to find the length of the hypotenuse of the triangle.

Solution:

(a) $a = |-3 - 7| = 10$

$b = |4 - 1| = 3$

$c = \sqrt{10^2 + 3^2} = \sqrt{109}$

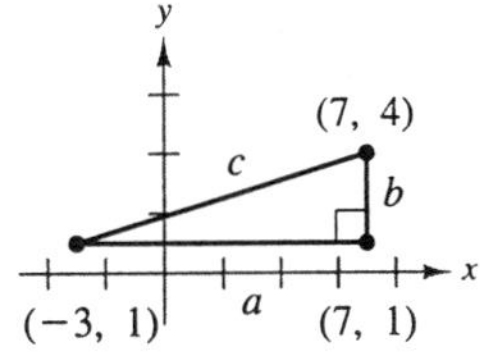

(b) $c = \sqrt{(7 - (-3))^2 + (4 - 1)^2}$

$= \sqrt{10^2 + 3^2}$

$= \sqrt{109}$

17. (a) Plot the points $(-4, 10)$ and $(4, -5)$, (b) find the distance between the points, and (c) find the midpoint of the line segment joining the points.

Solution:

(a)

(b) $d = \sqrt{(-4 - 4)^2 + (10 - (-5))^2}$

$= \sqrt{(-8)^2 + (15)^2}$

$= \sqrt{289} = 17$

(c) $m = \left(\dfrac{-4 + 4}{2}, \dfrac{10 + (-5)}{2}\right)$

$= \left(0, \dfrac{5}{2}\right)$

23. (a) Plot the points (6.2, 5.4), and (−3.7, 1.8), (b) find the distance between the points, and (c) find the midpoint of the line segment joining the points.

Solution:

(a)

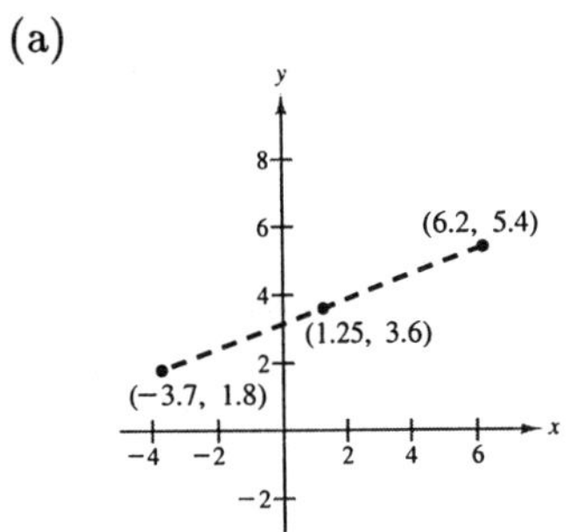

(b) $d = \sqrt{(6.2-(-3.7))^2 + (5.4-1.8)^2}$

$= \sqrt{(9.9)^2 + (3.6)^2}$

$= \sqrt{110.97} \approx 10.5342$

(c) $m = \left(\dfrac{6.2+(-3.7)}{2}, \dfrac{5.4+1.8}{2}\right)$

$= (1.25,\ 3.6)$

27. Use the Midpoint Formula to estimate the sales of a company for 1991. Assume the annual sales followed a linear pattern.

Year	1989	1993
Sales	\$520,000	\$740,000

Solution:

$$\frac{520{,}000 + 740{,}000}{2} = 630{,}000$$

The estimated sales for 1991 is \$630,000.

29. Show that the points (4, 0), (2, 1), and (−1, −5) form the vertices of a right triangle.

Solution:

$d_1 = \sqrt{(-1-2)^2 + (-5-1)^2} = \sqrt{45}$

$d_2 = \sqrt{(2-4)^2 + (1-0)^2} = \sqrt{5}$

$d_3 = \sqrt{(4-(-1))^2 + (0-(-5))^2} = \sqrt{50}$

Since ${d_1}^2 + {d_2}^2 = {d_3}^2$, we can conclude by the Pythagorean Theorem that the triangle is a right triangle.

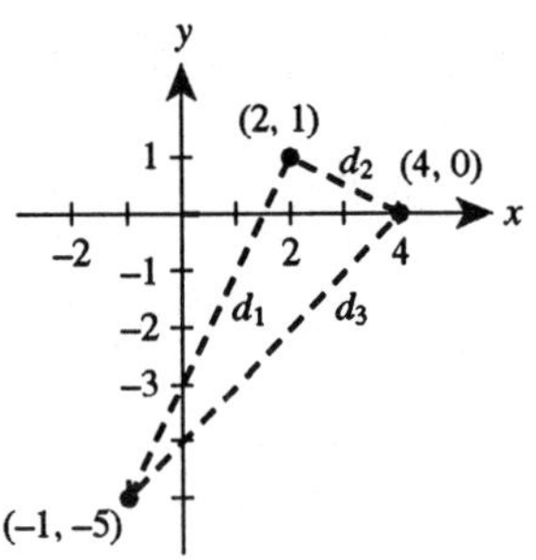

33. Find x so that the distance between (1, 2) and (x, −10) is 13.

Solution:

$$\sqrt{(x-1)^2+(-10-2)^2}=13$$
$$\sqrt{x^2-2x+1+144}=13$$
$$x^2-2x+145=169$$
$$x^2-2x-24=0$$
$$(x+4)(x-6)=0$$
$$x=-4 \quad \text{or} \quad x=6$$

37. Find a relationship between x and y so that (x, y) is equidistant from the points (4, −1) and (−2, 3).

Solution:

The distance between (4, −1) and (x, y) is equal to the distance between (−2, 3) and (x, y).

$$\sqrt{(x-4)^2+(y+1)^2}=\sqrt{(x+2)^2+(y-3)^2}$$
$$(x-4)^2+(y+1)^2=(x+2)^2+(y-3)^2$$
$$x^2-8x+16+y^2+2y+1=x^2+4x+4+y^2-6y+9$$
$$-12x+8y+4=0$$
$$3x-2y-1=0 \qquad \text{Divide both sides by } -4$$

41. Determine the quadrant(s) in which (x, y) is located so that the conditions $x > 0$ and $y > 0$ are satisfied.

Solution:

$x > 0 \rightarrow x$ lies in Quadrant I or in Quadrant IV.

$y > 0 \rightarrow y$ lies in Quadrant I or in Quadrant II.

$x > 0$ and $y > 0 \rightarrow (x, y)$ lies in Quadrant I.

45. Determine the quadrant(s) in which (x, y) is located so that the condition $y < -5$ is satisfied.

Solution:

$y < -5 \rightarrow y$ is negative $\rightarrow y$ lies in either Quadrant III or Quadrant IV.

51. Use the Midpoint Formula twice to find the three points that divide the line segment joining (x_1, y_1) and (x_2, y_2) into four parts.

Solution:

The midpoint of the given line segment is $\left(\dfrac{x_1+x_2}{2}, \dfrac{y_1+y_2}{2}\right)$.

The midpoint between (x_1, y_1) and $\left(\dfrac{x_1+x_2}{2}, \dfrac{y_1+y_2}{2}\right)$ is

$$\left(\frac{x_1+\dfrac{x_1+x_2}{2}}{2}, \frac{y_1+\dfrac{y_1+y_2}{2}}{2}\right) = \left(\frac{3x_1+x_2}{4}, \frac{3y_1+y_2}{4}\right).$$

The midpoint between $\left(\dfrac{x_1+x_2}{2}, \dfrac{y_1+y_2}{2}\right)$ and (x_2, y_2) is

$$\left(\frac{\dfrac{x_1+x_2}{2}+x_2}{2}, \frac{\dfrac{y_1+y_2}{2}+y_2}{2}\right) = \left(\frac{x_1+3x_2}{4}, \frac{y_1+3y_2}{4}\right).$$

Thus, the three points are

$$\left(\frac{3x_1+x_2}{4}, \frac{3y_1+y_2}{4}\right), \left(\frac{x_1+x_2}{2}, \frac{y_1+y_2}{2}\right), \quad \text{and} \quad \left(\frac{x_1+3x_2}{4}, \frac{y_1+3y_2}{4}\right).$$

55. The normal temperature y (Fahrenheit) for Duluth, Minnesota for each month of the year is given in the table. The months are numbered 1 through 12, with 1 corresponding to January. *(Source: NOAA)* Plot the points.

x	1	2	3	4	5	6	7	8	9	10	11	12
y	6	12	23	38	50	59	65	63	54	44	28	14

Solution:

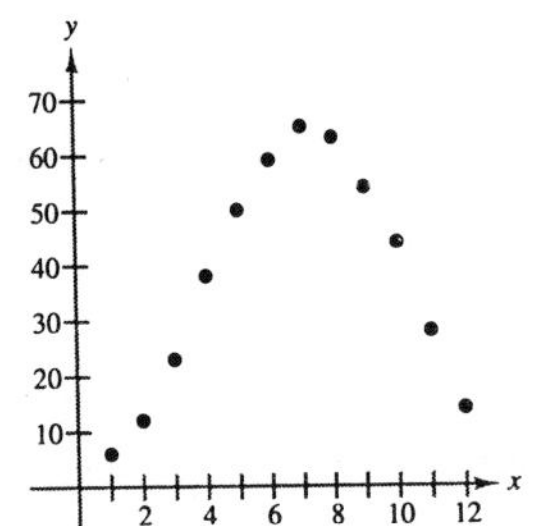

59. Find the percentage increase in the cost of a 30-second spot from Super Bowl I to Super Bowl XXV.

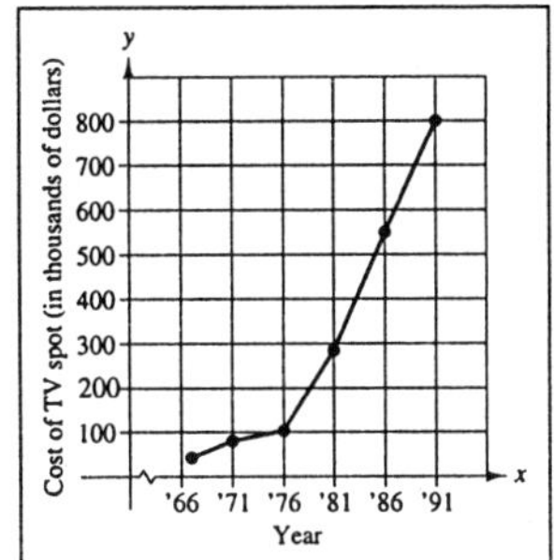

Solution:

$$\frac{800{,}000 - 42{,}500}{42{,}500} \approx 17.8235$$

$$= 1782.35\%$$

63. Prove that the diagonals of the parallelogram in the figure bisect each other.

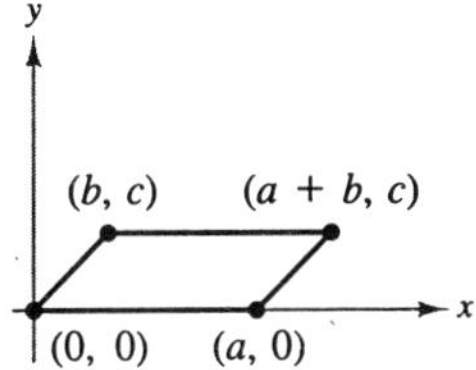

Solution:

The midpoint of the diagonal connecting $(0,\ 0)$ and $(a+b,\ c)$ is

$$\left(\frac{a+b}{2},\ \frac{c}{2}\right).$$

The midpoint of the diagonal connecting $(a,\ 0)$ and $(b,\ c)$ is

$$\left(\frac{a+b}{2},\ \frac{c}{2}\right).$$

Thus, the diagonals bisect each other.

SECTION 2.2

Graphs of Equations

- You should be able to use the point-plotting method of graphing.
- You should be able to find x- and y-intercepts.
  (a) To find the x-intercepts, let $y = 0$ and solve for x.
  (b) To find the y-intercepts, let $x = 0$ and solve for y.
- You should be able to test for symmetry.
  (a) To test for x-axis symmetry, replace y with $-y$.
  (b) To test for y-axis symmetry, replace x with $-x$.
  (c) To test for origin symmetry, replace x with $-x$ and y with $-y$.
- You should know the standard equation of a circle with center (h, k) and radius r:

  $$(x-h)^2 + (y-k)^2 = r^2$$

Solutions to Selected Exercises

5. Determine whether the points (a) $(1, \frac{1}{5})$, and (b) $(2, \frac{1}{2})$ lie on the graph of the equation $x^2y - x^2 + 4y = 0$.

Solution:

(a) $\left(1, \frac{1}{5}\right)$ lies on the graph since $(1)^2\left(\frac{1}{5}\right) - (1)^2 + 4\left(\frac{1}{5}\right) = \frac{1}{5} - 1 + \frac{4}{5} = 0$.

(b) $\left(2, \frac{1}{2}\right)$ lies on the graph since $(2)^2\left(\frac{1}{2}\right) - (2)^2 + 4\left(\frac{1}{2}\right) = 2 - 4 + 2 = 0$.

7. Find the constant C so that the ordered pair $(2, 6)$ is a solution point of the equation $y = x^2 + C$.

Solution:

$$y = x^2 + C$$
$$6 = (2)^2 + C$$
$$6 = 4 + C$$
$$C = 2$$

11. Complete the table (shown in the textbook) and use the solution points to sketch the graph of the equation $2x + y = 3$.

Solution:

$$2x + y = 3$$
$$y = -2x + 3$$

x	-4	-2	0	2	4
y	11	7	3	-1	-5
(x, y)	$(-4, 11)$	$(-2, 7)$	$(0, 3)$	$(2, -1)$	$(4, -5)$

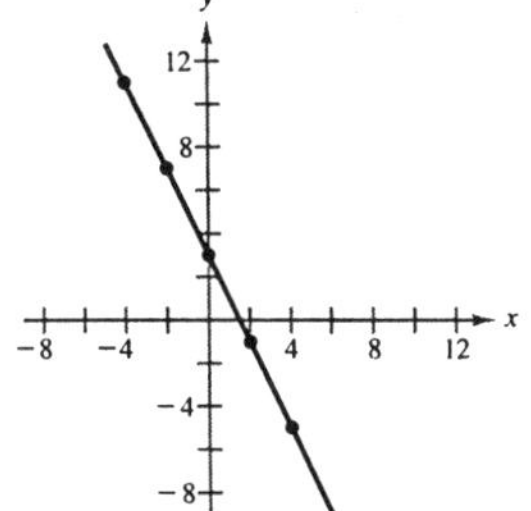

15. Find the x- and y-intercepts of the graph of the equation $y = x^2 + x - 2$.

Solution:

Let $y = 0$. Then $0 = x^2 + x - 2 = (x+2)(x-1)$ and $x = -2$ or $x = 1$.
x-intercepts: $(-2, 0)$ and $(1, 0)$

Let $x = 0$. Then $y = -2$.
y-intercept: $(0, -2)$

19. Find the x- and y-intercepts of the graph of the equation $xy - 2y - x + 1 = 0$.

Solution:

Let $y = 0$. Then $-x + 1 = 0$ and $x = 1$.
x-intercept: $(1, 0)$

Let $x = 0$. Then $-2y + 1 = 0$ and $y = \frac{1}{2}$.
y-intercept: $(0, \frac{1}{2})$

23. Check for symmetry with respect to both axes and the origin for $x - y^2 = 0$.

Solution:

By replacing y with $-y$, we have

$$x - (-y)^2 = 0$$
$$x - y^2 = 0$$

which is the original equation. Replacing x with $-x$ or replacing both x and y with $-x$ and $-y$ does not yield equivalent equations. Thus, $x - y^2 = 0$ is symmetric with respect to the x-axis.

27. Check for symmetry with respect to both axes and the origin for

$$y = \frac{x}{x^2+1}.$$

Solution:

Replacing x with $-x$ or y with $-y$ does not yield equivalent equations. Replacing x with $-x$ and y with $-y$ yields the following.

$$-y = \frac{-x}{(-x)^2+1}$$

$$-y = \frac{-x}{x^2+1} \qquad \text{Multiply both sides by } -1.$$

$$y = \frac{x}{x^2+1}$$

Thus, $y = \dfrac{x}{x^2+1}$ is symmetric with respect to the origin.

Note: An equation is symmetric with respect to the origin if it is symmetric with respect to both the x-axis and the y-axis. Also, if an equation is symmetric with respect to the origin, then one of the following is true:

1. The equation has both x-axis and y-axis symmetry or
2. The equation has neither x-axis nor y-axis symmetry.

31. Use symmetry to sketch the complete graph.

Solution:

$$y = -x^3 + x$$

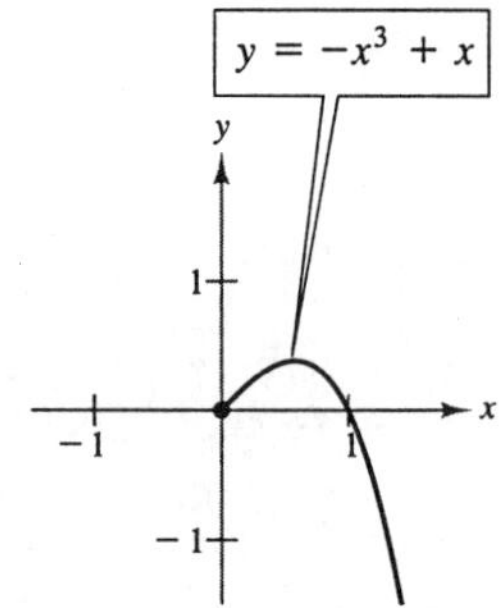

Origin Symmetry

37. Match $y = x^3 - x$ with its graph.

Solution:

$y = x^3 - x$

x-intercepts: $(-1, 0)$, $(0, 0)$, $(1, 0)$
y-intercept: $(0, 0)$
Symmetry: Origin
Matches graph (e)

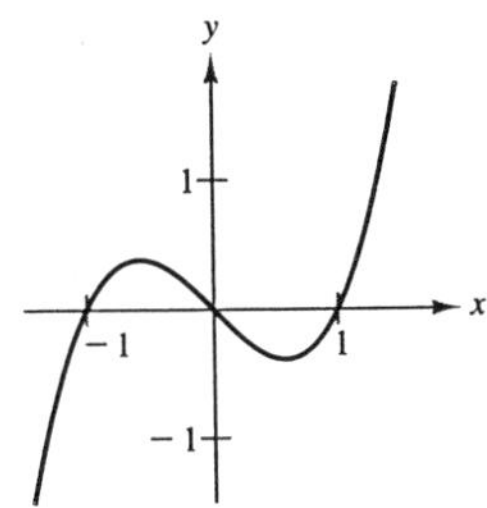

39. Sketch the graph of $y = -3x + 2$. Identify any intercepts and test for symmetry.

Solution:

$y = -3x + 2$

x-intercept: $\left(\frac{2}{3}, 0\right)$

y-intercept: $(0, 2)$

No symmetry

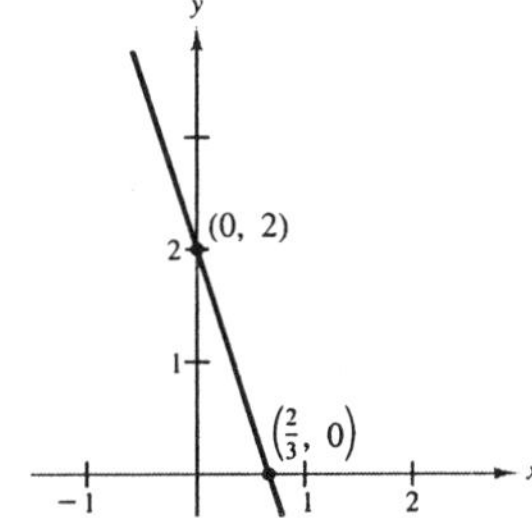

43. Sketch the graph of $y = x^2 - 4x + 3$. Identify any intercepts and test for symmetry.

Solution:

$y = x^2 - 4x + 3 = (x - 1)(x - 3)$

x-intercepts: $(1, 0)$, $(3, 0)$

y-intercept: $(0, 3)$

No symmetry

x	-1	0	1	2	3	4
y	8	3	0	-1	0	3

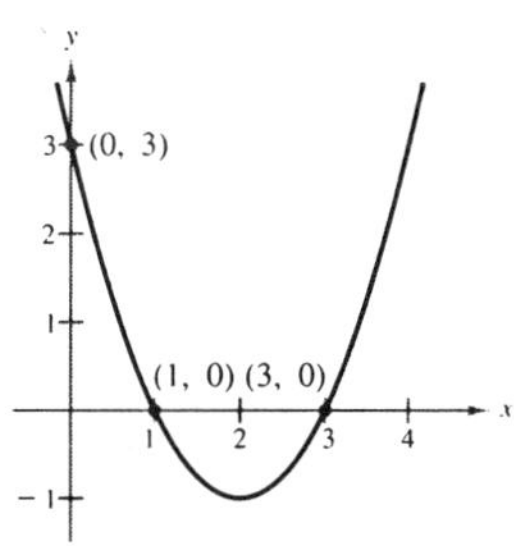

49. Sketch the graph of $y = \sqrt{x-3}$. Identify any intercepts and test for symmetry.

Solution:

$y = \sqrt{x-3}$
x-intercept: $(3, 0)$
No y-intercept
No symmetry

x	3	4	7	12
y	0	1	2	3

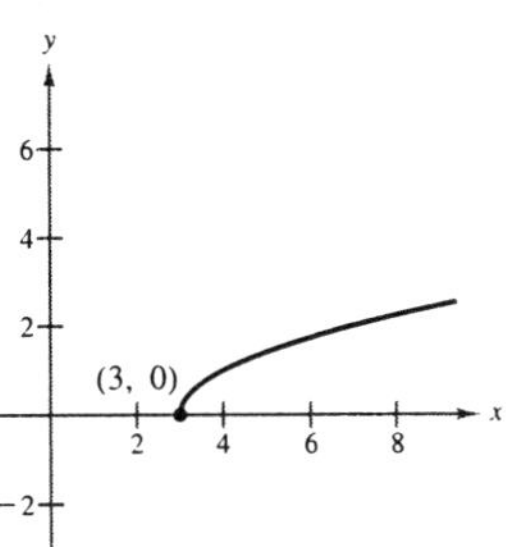

Note: The domain is $[3, \infty)$ and the range is $[0, \infty)$.

55. Sketch the graph of $x = y^2 - 1$. Identify any intercepts and test for symmetry.

Solution:

$x = y^2 - 1$
x-intercept: $(-1, 0)$
y-intercepts: $(0, -1)$, $(0, 1)$
x-axis symmetry

x	-1	0	3
y	0	± 1	± 2

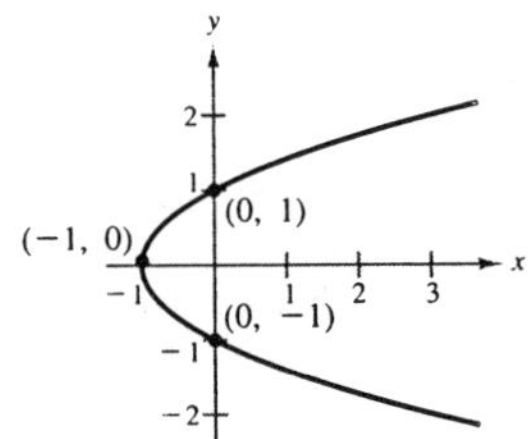

59. Find the standard form of the equation of the circle with center $(0, 0)$ and radius 3.

Solution:

$$(x-0)^2 + (y-0)^2 = 3^2$$
$$x^2 + y^2 = 9$$

61. Find the standard form of the equation of the circle with center $(2, -1)$ and radius 4.

Solution:

$$(x-2)^2 + (y+1)^2 = 4^2$$
$$(x-2)^2 + (y+1)^2 = 16$$

63. Find the standard form of the equation of the circle with center $(-1, 2)$ and passing through $(0, 0)$.

Solution:

$$(x+1)^2 + (y-2)^2 = r^2$$
$$(0+1)^2 + (0-2)^2 = r^2 \Rightarrow r^2 = 5$$
$$(x+1)^2 + (y-2)^2 = 5$$

67. Find the center and radius, and sketch the graph of the equation.

$x^2 + y^2 - 2x + 6y + 6 = 0$

Solution:

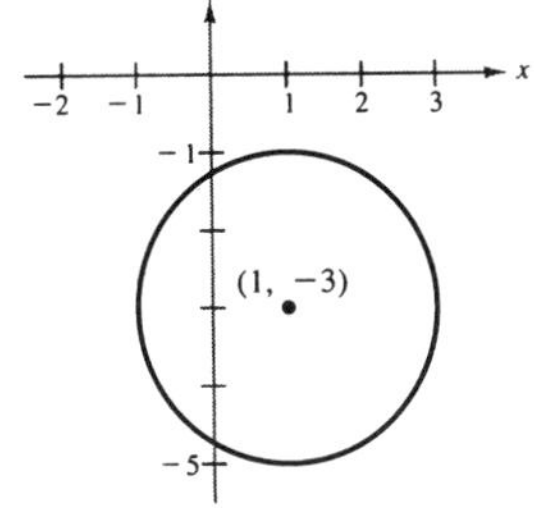

$$x^2 + y^2 - 2x + 6y + 6 = 0$$
$$(x^2 - 2x + 1^2) + (y^2 + 6y + 3^2) = -6 + 1 + 9$$
$$(x-1)^2 + (y+3)^2 = 4$$

Center: $(1, -3)$
Radius: 2

69. Find the center and radius, and sketch the graph of the equation.

$x^2 + y^2 - 2x + 6y + 10 = 0$

Solution:

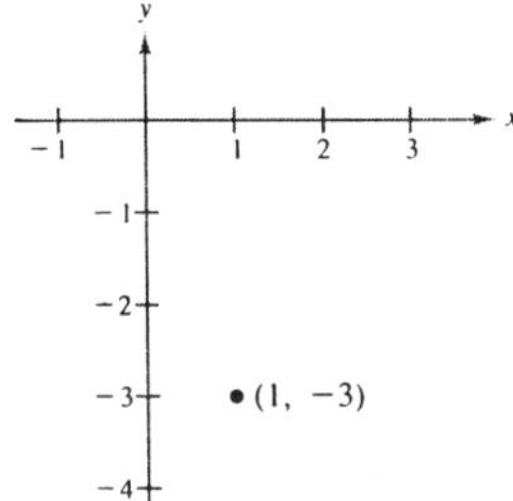

$$x^2 + y^2 - 2x + 6y + 10 = 0$$
$$(x^2 - 2x + 1^2) + (y^2 + 6y + 3^2) = -10 + 1 + 9$$
$$(x-1)^2 + (y+3)^2 = 0$$

Center: $(1, -3)$
Radius: 0
The graph is a single point, the center $(1, -3)$.

73. Find the center and radius, and sketch the graph of the equation.

$16x^2 + 16y^2 + 16x + 40y - 7 = 0$

Solution:

$$16x^2 + 16y^2 + 16x + 40y - 7 = 0$$
$$x^2 + y^2 + x + \tfrac{5}{2}y - \tfrac{7}{16} = 0 \quad \text{Divide both sides by 16.}$$
$$\left(x^2 + x + \left(\tfrac{1}{2}\right)^2\right) + \left(y^2 + \tfrac{5}{2}y + \left(\tfrac{5}{4}\right)^2\right) = \tfrac{7}{16} + \tfrac{1}{4} + \tfrac{25}{16}$$
$$\left(x + \tfrac{1}{2}\right)^2 + \left(y + \tfrac{5}{4}\right)^2 = \tfrac{36}{16}$$
$$\left(x + \tfrac{1}{2}\right)^2 + \left(y + \tfrac{5}{4}\right)^2 = \tfrac{9}{4}$$

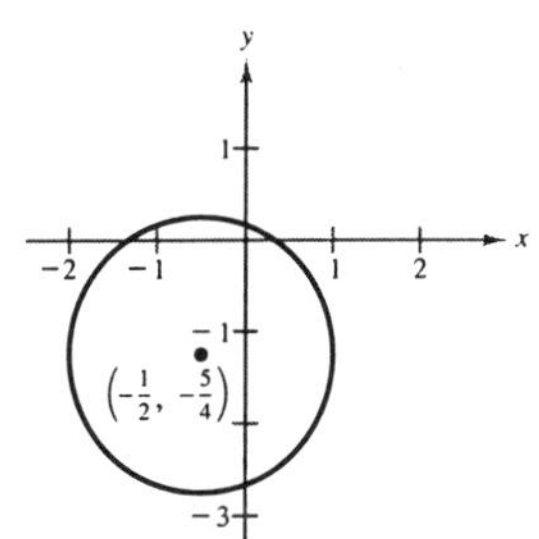

Center: $\left(-\tfrac{1}{2}, -\tfrac{5}{4}\right)$
Radius: $\tfrac{3}{2}$

77. The following table gives the per capita federal debt for the United States for selected years from 1950 to 1990. *(Source: U.S. Treasury Department)*

Year	1950	1960	1970	1980	1985	1990
Per capita debt	\$1688	\$1572	\$1807	\$3981	\$7614	\$12,848

A mathematical model for the per capita debt during this period is $y = 0.40t^3 - 9.42t^2 + 1053.24$ where y represents per capita debt and t is the time in years with $t = 0$ corresponding to 1950. (a) Sketch a graph to compare the given data and the model for that data; (b) use the model to predict y for the year 1994, and (c) for the year 2000.

Solution:

$y = 0.40t^3 - 9.42t^2 + 1053.24$

(a)

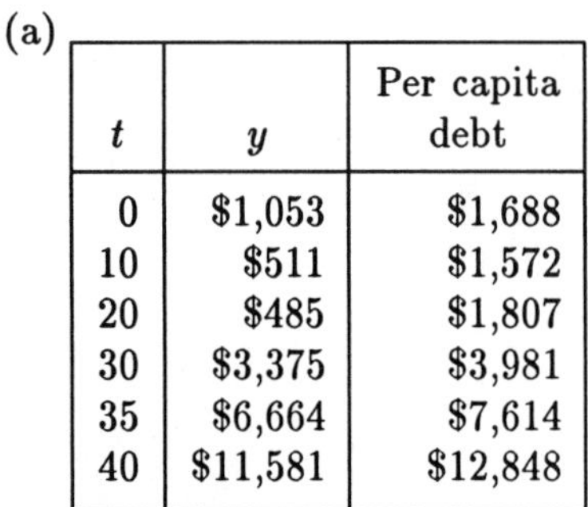

t	y	Per capita debt
0	\$1,053	\$1,688
10	\$511	\$1,572
20	\$485	\$1,807
30	\$3,375	\$3,981
35	\$6,664	\$7,614
40	\$11,581	\$12,848

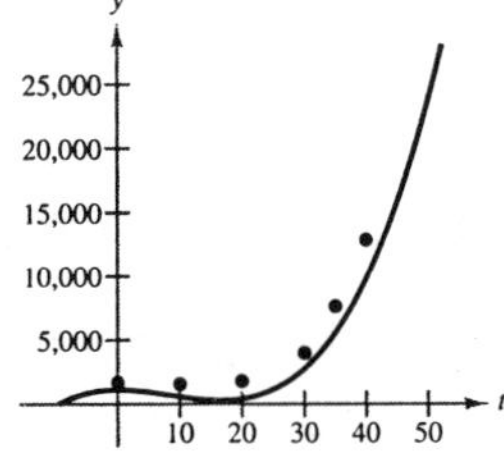

(b) When $t = 44$, $y \approx \$16{,}890$.

(c) When $t = 50$, $y \approx \$27{,}503$.

SECTION 2.3

Lines in the Plane

You should know the following important facts about lines.

- The slope of the line through (x_1, y_1) and (x_2, y_2) is

$$m = \frac{y_2 - y_1}{x_2 - x_1}.$$

- (a) If $m > 0$, the line rises from left to right.
  (b) If $m = 0$, the line is horizontal.
  (c) If $m < 0$, the line falls from left to right.
  (d) If m is undefined, the line is vertical.

- Equations of Lines
  (a) Point-Slope: $y - y_1 = m(x - x_1)$
  (b) Two-Point: $y - y_1 = \frac{y_2 - y_1}{x_2 - x_1}(x - x_1)$
  (c) Slope-Intercept: $y = mx + b$
  (d) General: $Ax + By + C = 0$
  (d) Vertical: $x = a$
  (e) Horizontal: $y = b$

- Given two distinct nonvertical lines

$$L_1 : y = m_1x + b_1 \quad \text{and} \quad L_2 : y = m_2x + b_2$$

  (a) L_1 is parallel to L_2 if and only if $m_1 = m_2$ and $b_1 \neq b_2$.
  (b) L_1 is perpendicular to L_2 if and only if $m_1 = -1/m_2$.

Solutions to Selected Exercises

3. Estimate the slope from its graph.

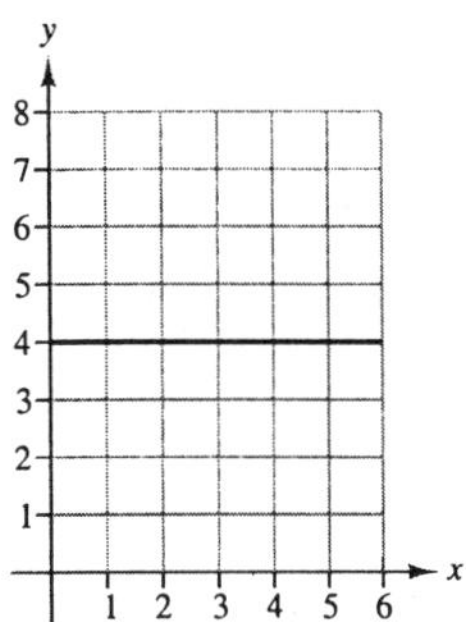

Solution:

Since the line is horizontal, it has a slope of zero.

9. Plot the points $(-3, -2)$ and $(1, 6)$ and find the slope of the line passing through the points.

Solution:

$$m = \frac{6-(-2)}{1-(-3)} = \frac{8}{4} = 2$$

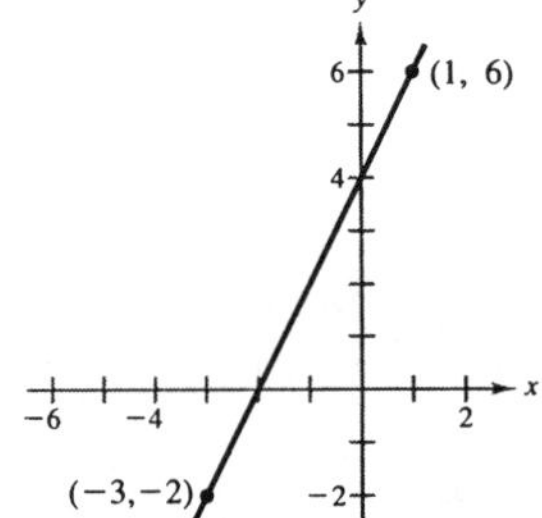

11. Plot the points $(-6, -1)$ and $(-6, 4)$ and find the slope of the line passing through the points.

Solution:

$$m = \frac{4-(-1)}{-6-(-6)} = \frac{5}{0}$$

The slope is undefined.

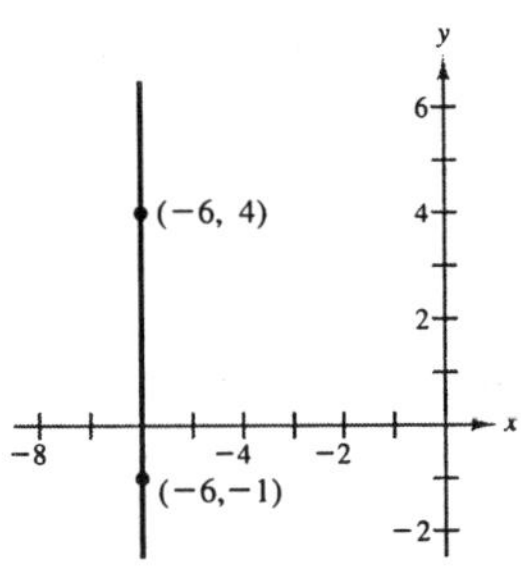

15. Use the point $(2, 1)$ on the line and the slope $m = 0$ of the line to find three additional points that the line passes through. (The solution is not unique.)

Solution:

Since $m = 0$, the line is horizontal, and since the line passes through $(2, 1)$, all other points on the line will be of the form $(x, 1)$. Three additional points are: $(0, 1)$, $(1, 1)$, $(3, 1)$.

19. Use the point $(-8,\ 1)$ on the line and the undefined slope of the line to find three additional points that the line passes through. (The solution is not unique.)

Solution:

Since m is undefined, the line is vertical, and since the line passes through $(-8,\ 1)$, all other points on the line will be of the form $(-8,\ y)$. Three additional points are: $(-8,\ -1)$, $(-8,\ 0)$, $(-8,\ 2)$.

21. Determine if the lines L_1 and L_2 passing through the given pairs of points are parallel, perpendicular, or neither.

$L_1 : (0,\ -1),\ (5,\ 9)$

$L_2 : (0,\ 3),\ (4,\ 1)$

Solution:

The slope of L_1 is $m_1 = \dfrac{9-(-1)}{5-0} = \dfrac{10}{5} = 2.$

The slope of L_2 is $m_2 = \dfrac{1-3}{4-0} = -\dfrac{2}{4} = -\dfrac{1}{2}.$

Since m_1 and m_2 are negative reciprocals of each other, the lines are perpendicular.

25. When driving down a mountain road, you notice signs warning of a "12% grade". This means that the slope of the road is $-12/100$. Determine the amount of horizontal change in your position if you note from elevation markers that you have descended 2000 feet vertically.

Solution:

$$\text{Slope} = \frac{\text{Rise}}{\text{Run}}$$

$$-\frac{12}{100} = -\frac{2000}{y}$$

$$-12y = -200{,}000$$

$$y = 16{,}666\tfrac{2}{3} \text{ feet} \approx 3.16 \text{ miles}$$

27. Find the slope and y-intercept, if possible, of the line specified by $5x - y + 3 = 0$. Sketch a graph of the line.

Solution:

$$5x - y + 3 = 0$$
$$-y = -5x - 3$$
$$y = 5x + 3$$

Slope: $m = 5$
y-intercept: $(0, 3)$

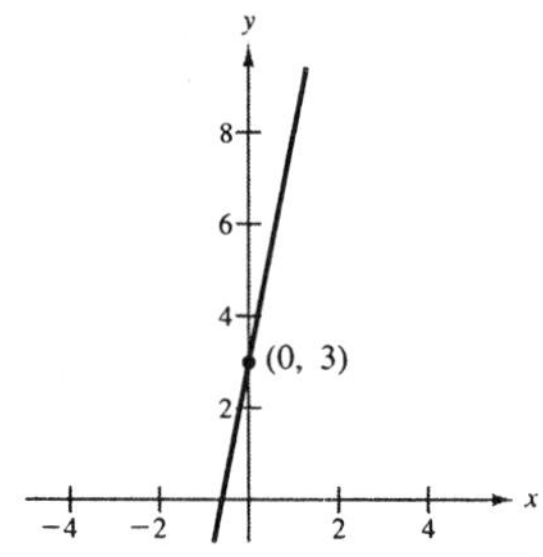

29. Find the slope and y-intercept, if possible, of the line specified by $5x - 2 = 0$. Sketch a graph of the line.

Solution:

$$5x - 2 = 0$$
$$5x = 2$$
$$x = \frac{2}{5} \quad \text{Vertical line}$$

Slope: Undefined
y-intercept: None

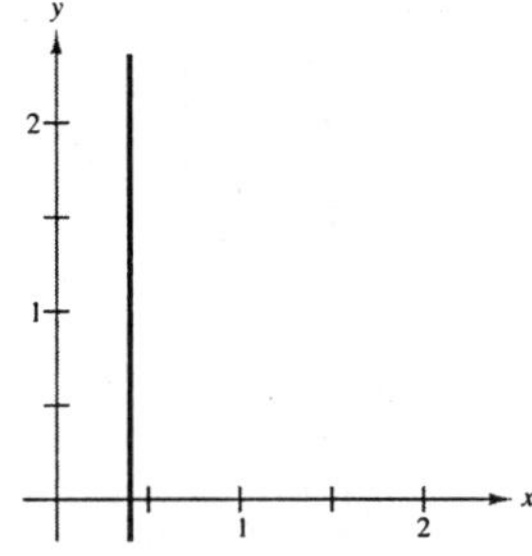

35. Find an equation for the line passing through the points $(2, \frac{1}{2})$, $(\frac{1}{2}, \frac{5}{4})$.

Solution:

$$m = \frac{\frac{5}{4} - \frac{1}{2}}{\frac{1}{2} - 2} = \frac{\frac{3}{4}}{-\frac{3}{2}} = -\frac{1}{2}$$

$$y - \frac{1}{2} = -\frac{1}{2}(x - 2)$$
$$2y - 1 = -(x - 2)$$
$$2y - 1 = -x + 2$$
$$x + 2y - 3 = 0$$

39. Find an equation for the line passing through the points (1, 0.6), and (−2, −0.6).

Solution:

$$m = \frac{-0.6 - 0.6}{-2 - 1} = 0.4$$

$$y - 0.6 = 0.4(x - 1)$$

$$y - 0.6 = 0.4x - 0.4$$

$$y = 0.4x + 0.2 \quad \text{or} \quad 2x - 5y + 1 = 0$$

43. Find an equation of the line that passes through the point (−3, 6) and has a slope of $m = -2$. Sketch a graph of the line.

Solution:

$$y - 6 = -2\,[x - (-3)]$$

$$y - 6 = -2x - 6$$

$$y = -2x$$

$$2x + y = 0$$

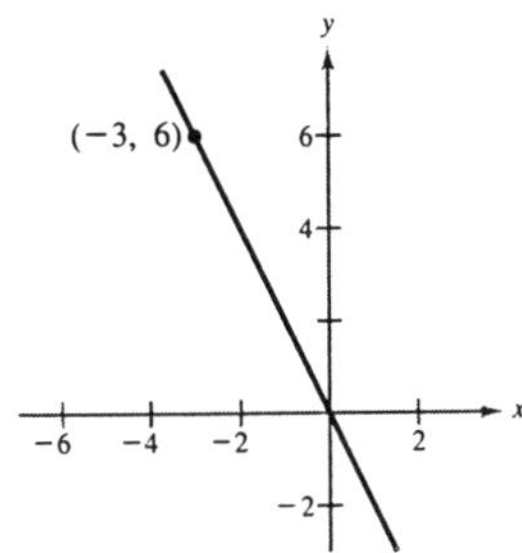

47. Find an equation of the line that passes through the point (6, −1) and has an undefined slope. Sketch a graph of the line.

Solution:

Since the slope is undefined, the line is vertical and since the line passes through (6, −1), its equation is $x = 6$ or $x - 6 = 0$.

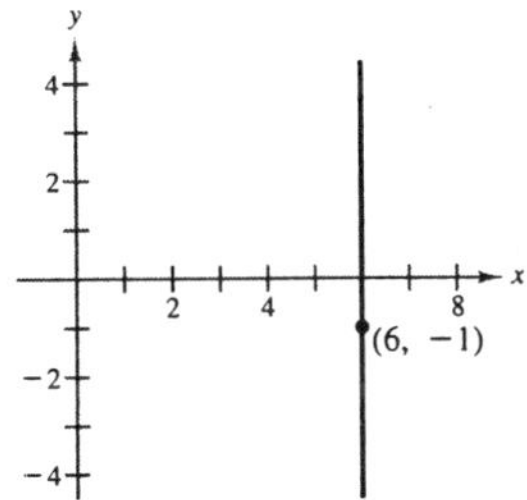

53. Use the intercept form to write an equation of the line with x-intercept $(-\frac{1}{6}, 0)$ and y-intercept $(0, -\frac{2}{3})$.

Solution:

x-intercept: $(-\frac{1}{6}, 0)$

y-intercept: $(0, -\frac{2}{3})$

$$\frac{x}{-1/6} + \frac{y}{-2/3} = 1$$

$$-6x - \frac{3}{2}y = 1$$

$$-12x - 3y = 2$$

$$12x + 3y = -2$$

$$12x + 3y + 2 = 0$$

59. Write the equation of the line through the point $(-6, 4)$, (a) parallel to the line $3x + 4y = 7$, and (b) perpendicular to the line $3x + 4y = 7$.

Solution:

$$3x + 4y = 7$$

$$4y = -3x + 7$$

$$y = -\tfrac{3}{4}x + \tfrac{7}{4}$$

The slope of the given line is $m_1 = -3/4$.

(a) The slope of the parallel line is $m_2 = m_1 = -3/4$.

$$y - 4 = -\tfrac{3}{4}[x - (-6)]$$

$$y - 4 = -\tfrac{3}{4}x - \tfrac{9}{2}$$

$$y = -\tfrac{3}{4}x - \tfrac{1}{2}$$

$$4y = -3x - 2$$

$$3x + 4y + 2 = 0$$

(b) The slope of the perpendicular line is $m_2 = -1/m_1 = 4/3$.

$$y - 4 = \tfrac{4}{3}[x - (-6)]$$

$$y - 4 = \tfrac{4}{3}x + 8$$

$$y = \tfrac{4}{3}x + 12$$

$$3y = 4x + 36$$

$$4x - 3y + 36 = 0$$

63. The 1990 value of a product is \$2540 and the rate at which the value is expected to change during the next five years is a \$125 increase per year. Write a linear equation for the dollar value V of the product in terms of the year t. (Let $t = 0$ represent 1990.)

Solution:

Value = \$2540 + (\$125)(the number of years t after 1990)

$$V = 125t + 2540, \quad 0 \le t \le 5$$

67. Match the description "A person is paying \$10 per week to a friend to repay a \$100 loan" with a graph. Determine the slope and how it is interpreted in the situation.

Solution:

A person is paying \$10 per week to a friend to repay a \$100 loan. Matches graph (b). The slope is $m = -10$. This represents the decrease in the amount of the loan each week.

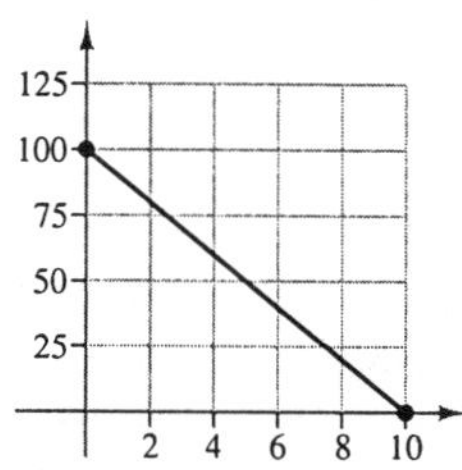

71. Find the equation of the line giving the relationship between the temperature in degrees Celsius, C, and degrees Fahrenheit, F. Use the fact that water freezes at 0° Celsius (32° Fahrenheit) and boils at 100° Celsius (212° Fahrenheit).

Solution:

Using the points (0, 32) and (100, 212), we have

$$m = \frac{212 - 32}{100 - 0} = \frac{180}{100} = \frac{9}{5}$$

$$F - 32 = \frac{9}{5}(C - 0)$$

$$F = \frac{9}{5}C + 32.$$

75. A small business purchases a piece of equipment for \$875. After five years the equipment will be outdated and have no value. Write a linear equation giving the value V of the equipment during the five years it will be used.

Solution:

Using the points (0, 875) and (5, 0), where the first coordinate represents the year t and the second coordinate represents the value V, we have

$$m = \frac{0 - 875}{5 - 0} = -175$$

$$V = -175t + 875, \quad 0 \le t \le 5.$$

77. A store is offering a 15% discount on all items in its inventory. Write a linear equation giving the sale price S for an item with a list price L.

Solution:

Sale Price = List Price − 15% of the List Price

$$S = L - 0.15L$$

$$S = 0.85L$$

81. A contractor purchases a piece of equipment for \$36,500. The equipment requires an average expenditure of \$5.25 per hour for fuel and maintenance, and the operator is paid \$11.50 per hour.

(a) Write a linear equation giving the total cost C of operating this equipment for t hours. (Include the purchase cost for the equipment.)

(b) If customers are charged \$27 per hour of machine use, write an equation for the revenue R derived from t hours of use.

(c) Use the formula for profit ($P = R - C$) to write an equation for the profit derived from t hours of use.

(d) *Break-Even Point* Use the result of part (c) to find the number of hours this equipment must be used to yield a profit of 0 dollars.

Solution:

(a) $$\begin{aligned} C &= 36{,}500 + 5.25t + 11.50t \\ &= 16.75t + 36{,}500 \end{aligned}$$

(b) $R = 27t$

(c) $$\begin{aligned} P &= R - C \\ &= 27t - (16.75t + 36{,}500) \\ &= 10.25t - 36{,}500 \end{aligned}$$

(d) $$\begin{aligned} 0 &= 10.25t - 36{,}500 \\ 36{,}500 &= 10.25t \\ t &\approx 3561 \text{ hours} \end{aligned}$$

SECTION 2.4

Functions

- Given a set or an equation, you should be able to determine if it represents a function.
- Given a function, you should be able to do the following.
  (a) Find the domain.
  (b) Evaluate it at specific values.

Solutions to Selected Exercises

1. Determine which of the sets of ordered pairs represents a function from A to B. Give reasons for your answers.

$A = \{0, 1, 2, 3\}$ and $B = \{-2, -1, 0, 1, 2\}$

(a) $\{(0, 1), (1, -2), (2, 0), (3, 2)\}$
(b) $\{(0, -1), (2, 2), (1, -2), (3, 0), (1, 1)\}$
(c) $\{(0, 0), (1, 0), (2, 0), (3, 0)\}$
(d) $\{(0, 2), (3, 0), (1, 1)\}$

Solution:

(a) Each element of A is matched with exactly one element of B, so it does represent a function.
(b) The element 1 in A is matched with two elements, -2 and 1 of B, so it does not represent a function.
(c) Each element of A is matched with exactly one element of B, so it does represent a function.
(d) The element 2 in A is not matched with an element of B, so it does not represent a function.

3. Determine if y is a function of x for $x^2 + y^2 = 4$.

Solution:

y is not a function of x since some values of x give two values for y. For example, if $x = 0$, then $y = \pm 2$.

7. Determine if y is a function of x for $2x + 3y = 4$.

Solution:

$$2x + 3y = 4$$
$$y = \tfrac{1}{3}(4 - 2x)$$

y is a function of x. No value of x yields more than one value of y.

13. Evaluate the function $f(x) = 2x - 3$ at the specified value of the independent variable and simplify.

(a) $f(1)$ (b) $f(-3)$ (c) $f(x-1)$

Solution:

(a) $f(1) = 2(1) - 3$
$= -1$

(b) $f(-3) = 2(-3) - 3$
$= -9$

(c) $f(x-1) = 2(x-1) - 3$
$= 2x - 5$

17. Evaluate the function $f(y) = 3 - \sqrt{y}$ at the specified value of the independent variable and simplify.

(a) $f(4)$ (b) $f(0.25)$ (c) $f(4x^2)$

Solution:

(a) $f(4) = 3 - \sqrt{4} = 1$

(b) $f(0.25) = 3 - \sqrt{0.25} = 2.5$

(c) $f(4x^2) = 3 - \sqrt{4x^2}$
$= 3 - 2|x|$

21. Evaluate the function $f(x) = |x|/x$ at the specified value of the independent variable and simplify.

(a) $f(2)$ (b) $f(-2)$ (c) $f(x-1)$

Solution:

(a) $f(2) = \dfrac{|2|}{2} = 1$

(b) $f(-2) = \dfrac{|-2|}{-2} = -1$

(c) $f(x-1) = \dfrac{|x-1|}{x-1}$

23. Evaluate the function at the specified value of the independent variable and simplify.

$$f(x) = \begin{cases} 2x+1, & x < 0 \\ 2x+2, & x \geq 0 \end{cases}$$

(a) $f(-1)$ (b) $f(0)$ (c) $f(2)$

Solution:

(a) $f(-1) = 2(-1) + 1 = -1$

(b) $f(0) = 2(0) + 2 = 2$

(c) $f(2) = 2(2) + 2 = 6$

27. Find all real values x such that $f(x) = 0$ for $f(x) = x^2 - 9$.

Solution:

$x^2 - 9 = 0$

$x^2 = 9$

$x = \pm 3$

31. Find the domain of $h(t) = 4/t$.

Solution:

The domain includes all real numbers except 0, i.e. $t \neq 0$.

35. Find the domain of $f(x) = \sqrt[4]{1-x^2}$.

Solution:

Choose x-values for which $1 - x^2 \geq 0$. The domain is $-1 \leq x \leq 1$.

39. Assume that the domain of $f(x) = x^2$ is the set $A = \{-2, -1, 0, 1, 2\}$. Determine the set of ordered pairs representing the function f.

Solution:

$$\{(-2,\ f(-2)),\ (-1,\ f(-1)),\ (0,\ f(0)),\ (1,\ f(1)),\ 2,\ f(2))\}$$
$$= \{(-2,\ 4),\ (-1,\ 1),\ (0,\ 0),\ (1,\ 1),\ (2,\ 4)\}$$

45. Find the value(s) of x for which $f(x) = g(x)$ where $f(x) = \sqrt{3x} + 1$ and $g(x) = x + 1$.

Solution:

$$\begin{aligned} f(x) &= g(x) \\ \sqrt{3x} + 1 &= x + 1 \\ \sqrt{3x} &= x \\ 3x &= x^2 \\ 0 &= x^2 - 3x \\ 0 &= x(x-3) \\ x = 0 \quad &\text{or} \quad x = 3 \end{aligned}$$

47. For $f(x) = x^2 - x + 1$, find $\dfrac{f(2+h) - f(2)}{h}$ and simplify your answer.

Solution:

$$\begin{aligned} f(x) &= x^2 - x + 1 \\ f(2+h) &= (2+h)^2 - (2+h) + 1 \\ &= 4 + 4h + h^2 - 2 - h + 1 \\ &= h^2 + 3h + 3 \\ f(2) &= (2)^2 - 2 + 1 = 3 \\ f(2+h) - f(2) &= h^2 + 3h \\ \frac{f(2+h) - f(2)}{h} &= h + 3 \end{aligned}$$

53. Express the area A of a circle as a function of its circumference C.

Solution:

$$A = \pi r^2, \quad C = 2\pi r$$

$$r = \frac{C}{2\pi}$$

$$A = \pi\left(\frac{C}{2\pi}\right)^2 = \frac{C^2}{4\pi}$$

55. A right triangle is formed in the first quadrant by the x- and y-axes and a line through the point (1, 2), as shown in the figure. Write the area of the triangle as a function of x, and determine the domain of the function.

Solution:

$$A = \frac{1}{2}bh = \frac{1}{2}xy$$

Since $(0, y)$, $(1, 2)$ and $(x, 0)$ all lie on the same line, the slopes between any pair are equal.

$$\frac{2-y}{1-0} = \frac{0-2}{x-1}$$

$$2 - y = -\frac{2}{x-1}$$

$$y = \frac{2}{x-1} + 2$$

$$y = \frac{2x}{x-1}$$

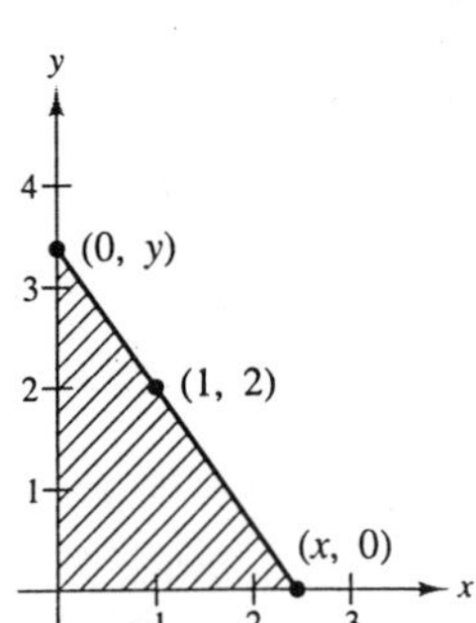

Therefore,

$$A = \frac{1}{2}x\left(\frac{2x}{x-1}\right) = \frac{x^2}{x-1}.$$

The domain of A includes x-values such that $x^2/(x-1) > 0$. Using methods of Section 2.8, we find that the domain is $x > 1$.

59. A balloon carrying a transmitter ascends vertically from a point 2000 feet from the receiving station (see figure). Let d be the distance between the balloon and the receiving station. Express the height of the balloon as a function of d. What is the domain of the function?

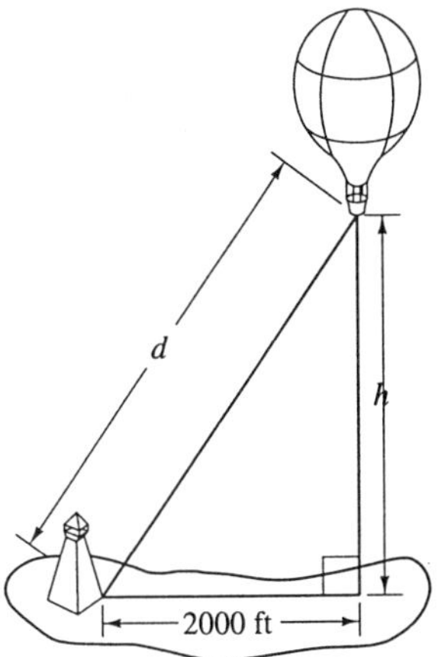

Solution:

By the Pythagorean Theorem we have

$$h^2 + 2000^2 = d^2$$

$$h^2 = d^2 - 2000^2$$

$$h = \sqrt{d^2 - 2000^2}.$$

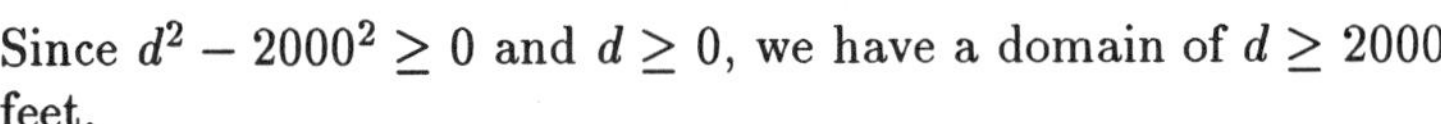

Since $d^2 - 2000^2 \geq 0$ and $d \geq 0$, we have a domain of $d \geq 2000$ feet.

61. A company produces a product for which the variable cost is \$12.30 per unit and the fixed costs are \$98,000. The product sells for \$17.98. Let x be the number of units produced and sold.

(a) Write the total cost C as a function of the number of units produced.

(b) Write the revenue R as a function of the number of units sold.

(c) Write the profit P as a function of the number of units sold.
(**Note:** $P = R - C$.)

Solution:

(a) Cost = variable costs + fixed costs

$$C = 12.30x + 98{,}000$$

(b) Revenue = price per unit × number of units

$$R = 17.98x$$

(c) Profit = Revenue − Cost

$$P = 17.98x - (12.30x + 98{,}000)$$

$$P = 5.68x - 98{,}000$$

SECTION 2.5

Graphs of Functions

- You should be able to determine the domain and range of a function from its graph.
- You should be able to use the vertical line test for functions.
- You should know that the graph of $f(x) = c$ is a horizontal line through $(0, c)$.
- You should be able to determine when a function is constant, increasing, or decreasing.
- You should know that f is
  (a) Odd if $f(-x) = -f(x)$.
  (b) Even if $f(-x) = f(x)$.
- You should know the basic types of transformations.
  (a) Vertical and horizontal shifts
  (b) Reflections in the coordinate axes

Solutions to Selected Exercises

5. Determine the domain and range of the function $f(x) = \sqrt{25 - x^2}$.

Solution:

From the graph we see that the x-values do not extend beyond $x = -5$ (on the left) and $x = 5$ (on the right). The domain is $[-5, 5]$. Similarly, the y-values do not extend beyond $y = 0$ and $y = 5$. The range is $[0, 5]$.

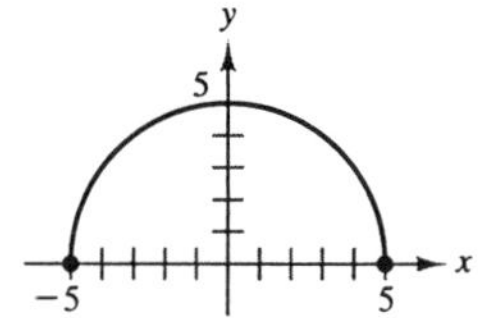

7. Use the vertical line test to determine if y is a function of x where $y = x^2$.

Solution:

Since no vertical line would ever cross the graph more than one time, y *is* a function of x.

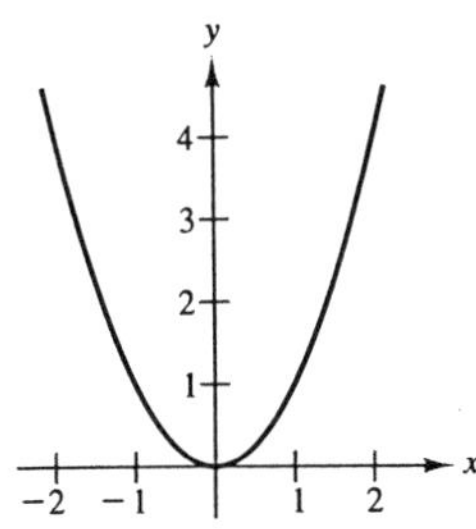

11. Use the vertical line test to determine if y is a function of x where $x^2 = xy - 1$.

Solution:

Since no vertical line would ever cross the graph more than one time, y *is* a function of x.

$$y = \frac{x^2 + 1}{x}$$

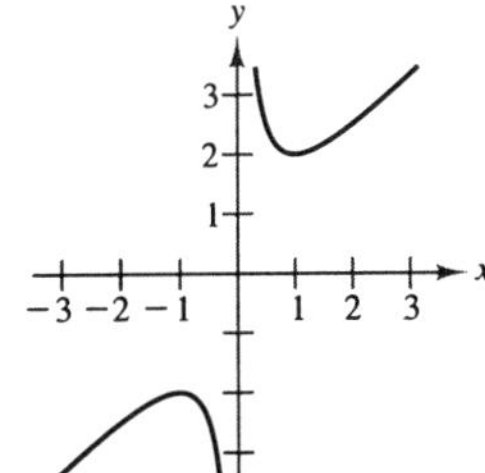

15. (a) Determine the intervals over which the function is increasing, decreasing, or constant, and (b) determine if the function is even, odd, or neither for $f(x) = x^3 - 3x^2$.

Solution:

(a) By its graph we see that f is increasing on $(-\infty, 0)$ and $(2, \infty)$ and is decreasing on $(0, 2)$.

(b) $f(-x) = (-x)^3 - 3(-x)^2$

$= -x^3 - 3x^2$

$f(-x) \neq f(x)$ and $f(-x) \neq -f(x)$, so the function is neither odd nor even.

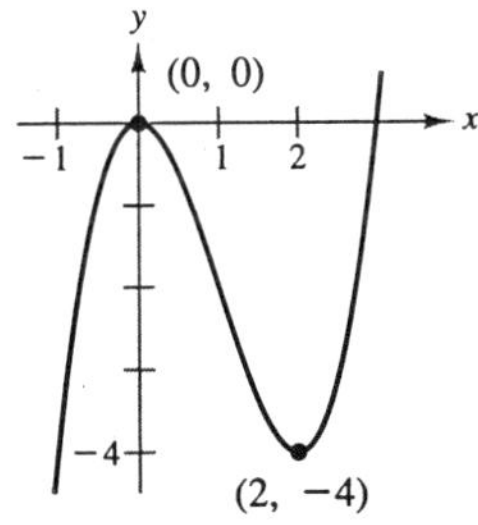

19. (a) Determine the intervals over which the function is increasing, decreasing, or constant, and (b) determine if the function is even, odd, or neither for $f(x) = x\sqrt{x+3}$.

Solution:

(a) By its graph we see that f is increasing on $(-2, \infty)$ and decreasing on $(-3, -2)$.

(b) $f(-x) = -x\sqrt{-x+3}$
$f(-x) \neq f(x)$ and $f(-x) \neq -f(x)$, so the function is neither odd nor even.

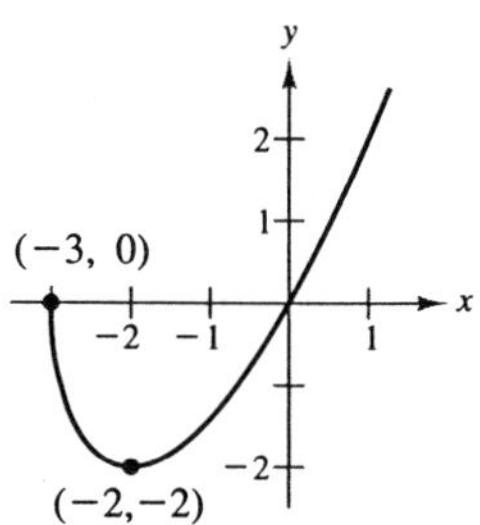

23. Determine whether $g(x) = x^3 - 5x$ is even, odd, or neither.

Solution:

$$\begin{aligned} g(x) &= x^3 - 5x \\ g(-x) &= (-x)^3 - 5(-x) \\ &= -x^3 + 5x \\ &= -(x^3 - 5x) \\ &= -g(x) \end{aligned}$$

Therefore, g is odd.

27. Sketch the graph of $f(x) = 3$ and determine whether the function is odd, even, or neither.

Solution:

$f(x) = 3$
Domain: $(-\infty, \infty)$
Range: $\{3\}$
y-intercept: $(0, 3)$
y-axis symmetry
$f(-x) = 3 = f(x)$
Therefore, f is even.

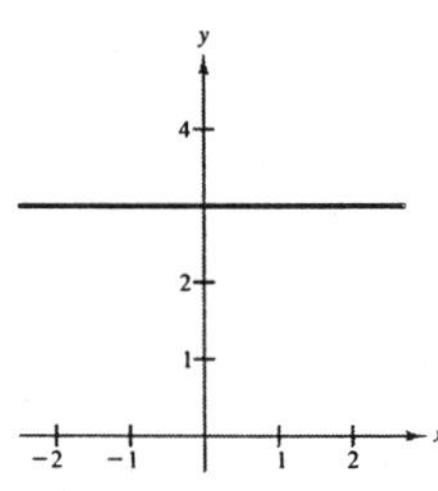

31. Sketch the graph of $g(s) = s^3/4$ and determine whether the function is odd, even, or neither.

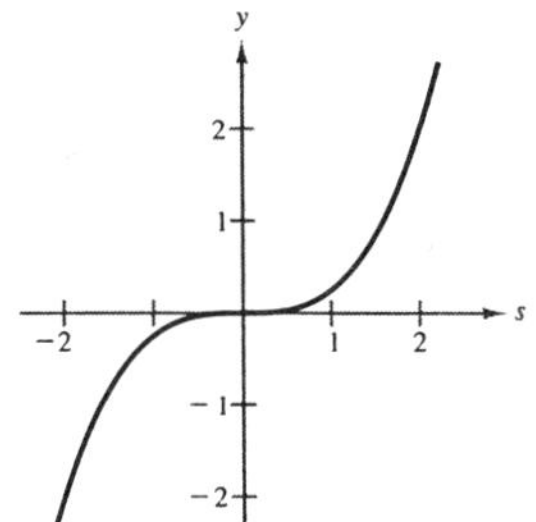

Solution:

$g(s) = s^3/4$
Intercept: (0, 0)
Origin symmetry
Domain: $(-\infty,\ \infty)$
Range: $(-\infty,\ \infty)$
$g(-s) = -g(s)$
Therefore, g is odd.

35. Sketch the graph of $g(t) = (t-1)^2 + 2$ and determine whether the function is odd, even, or neither.

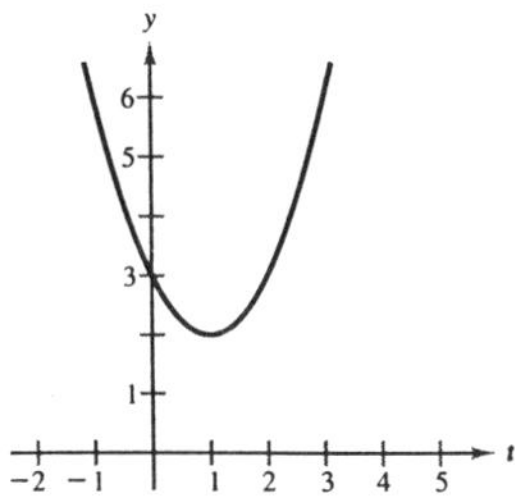

Solution:

$g(t) = (t-1)^2 + 2$
Intercept: (0, 3)
No symmetries
Domain: $(-\infty,\ \infty)$
Range: $[2,\ \infty)$
Neither odd nor even

37. Sketch the graph of

$$f(x) = \begin{cases} x+3, & \text{if } x \le 0 \\ 3, & \text{if } 0 < x \le 2 \\ 2x-1, & \text{if } x > 2 \end{cases}$$

and determine whether the function is odd, even, or neither.

Solution:

For $x \le 0$, $f(x) = x + 3$.
For $0 < x \le 2$, $f(x) = 3$.
For $x > 2$, $f(x) = 2x - 1$.
Thus, the graph of f is as shown.

$$f(-x) = \begin{cases} -x+3, & \text{if } x \le 0 \\ 3, & \text{if } 0 < x \le 2 \\ -2x-1, & \text{if } x > 2 \end{cases}$$

So, f is neither odd nor even.

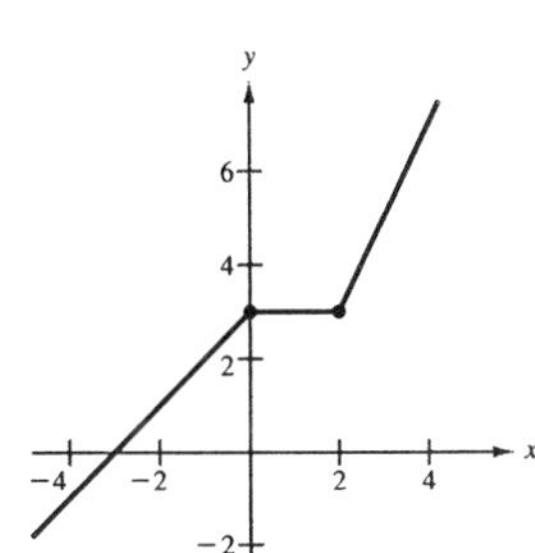

43. Sketch the graph of $f(x) = x^2 - 9$ and determine the interval(s), if any, on the real axis for which $f(x) \geq 0$.

Solution:

$f(x) = x^2 - 9$
x-intercepts: $(-3,\ 0)$, $(0,\ 3)$
y-intercept: $(0,\ -9)$
y-axis symmetry
Domain: $(-\infty,\ \infty)$
Range: $[-9,\ \infty)$
$f(x) \geq 0$ on the intervals $(-\infty,\ -3]$ and $[3,\ \infty)$.

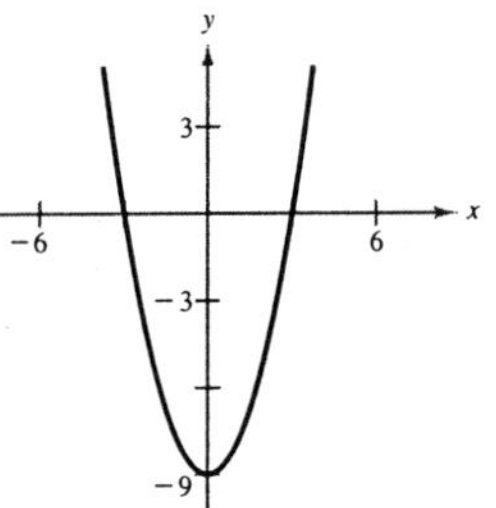

47. Sketch the graph of $f(x) = x^2 + 1$ and determine the interval(s), if any, on the real axis for which $f(x) \geq 0$.

Solution:

$f(x) = x^2 + 1$
x-intercept: None
y-intercept: $(0,\ 1)$
y-axis symmetry
Domain: $(-\infty,\ \infty)$
Range: $[1,\ \infty)$
$f(x) \geq 0$ for all real numbers.

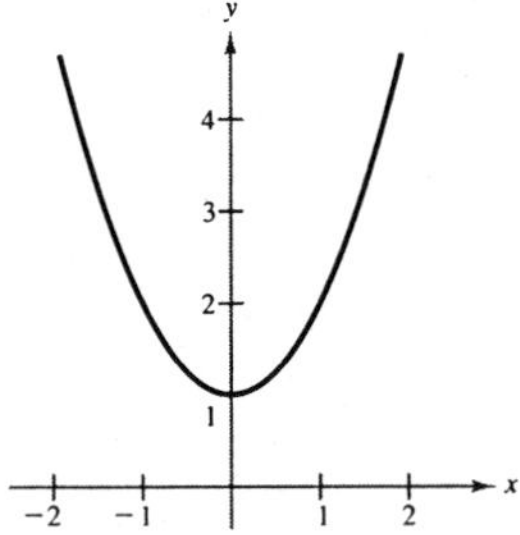

51. Sketch (on the same set of coordinate axes) a graph of f for $c = -2,\ 0$ and 2.

(a) $f(x) = \frac{1}{2}x + c$ (b) $f(x) = \frac{1}{2}(x - c)$ (c) $f(x) = \frac{1}{2}(cx)$

Solution:

(a) $f(x) = \frac{1}{2}x + c$

$c = -2 : f(x) = \frac{1}{2}x - 2$

$c = \ \ 0 : f(x) = \frac{1}{2}x$

$c = \ \ 2 : f(x) = \frac{1}{2}x + 2$

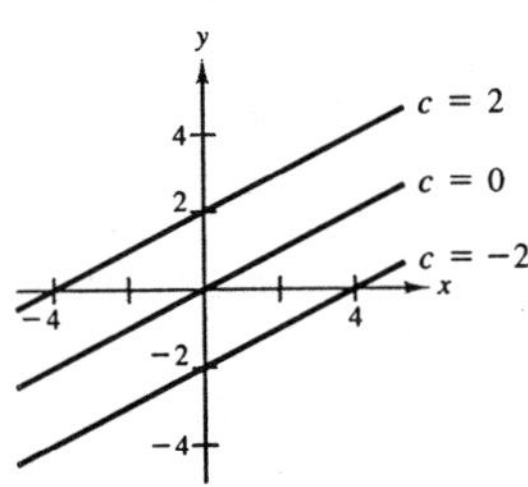

–CONTINUED ON NEXT PAGE–

51. –CONTINUED–

(b) $f(x) = \frac{1}{2}(x - c)$

$c = -2 : f(x) = \frac{1}{2}(x + 2) = \frac{1}{2}x + 1$

$c = \ \ 0 : f(x) = \frac{1}{2}x$

$c = \ \ 2 : f(x) = \frac{1}{2}(x - 2) = \frac{1}{2}x - 1$

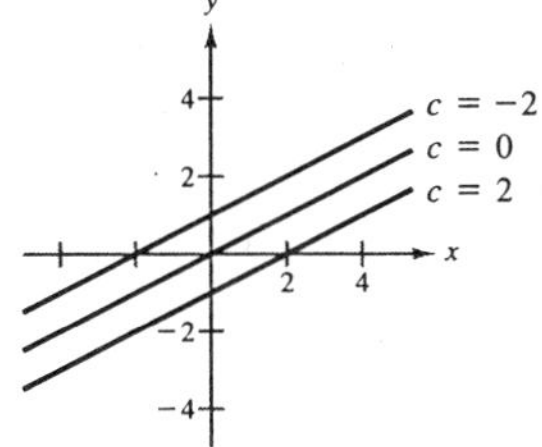

(c) $f(x) = \frac{1}{2}(cx)$

$c = -2 : f(x) = \frac{1}{2}(-2x) = -x$

$c = \ \ 0 : f(x) = \frac{1}{2}(0x) = 0$

$c = \ \ 2 : f(x) = \frac{1}{2}(2x) = x$

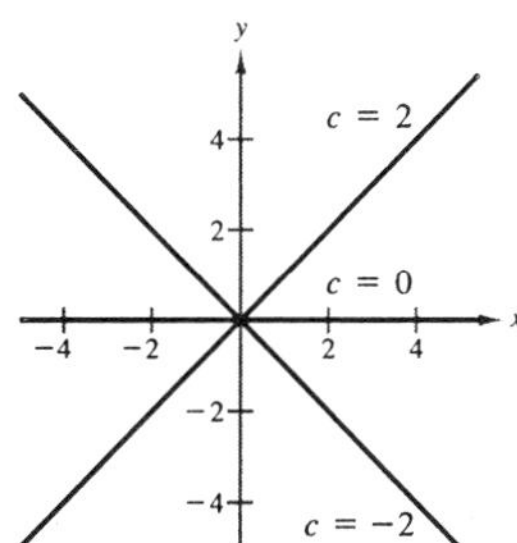

55. Use the graph of $f(x) = x^2$ to write formulas for the functions whose graphs are shown.

(a)

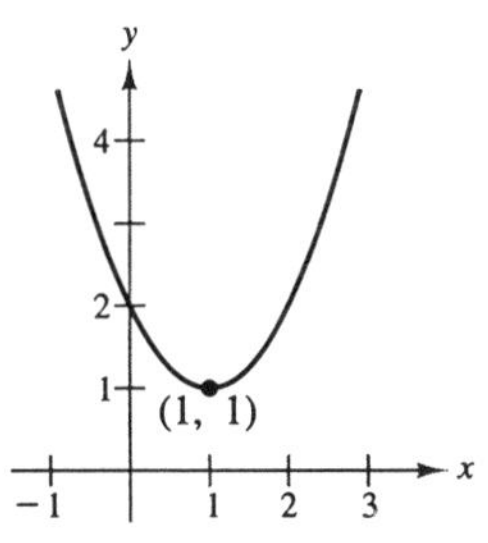

(b)

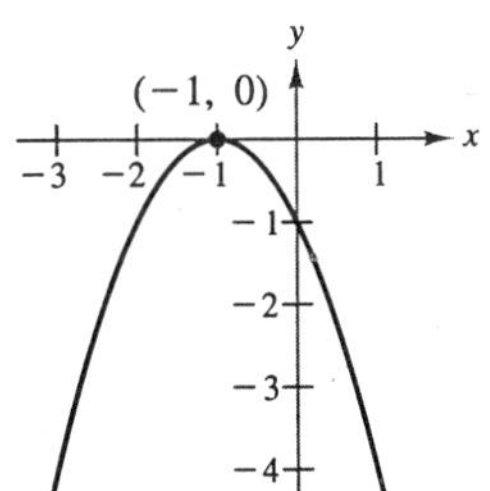

Solution:

(a) Horizontal shift one unit to the right and vertical shift one unit upward:

$h(x) = f(x - 1) + 1 = (x - 1)^2 + 1$

(b) Horizontal shift one unit to the left and reflection in the x-axis:

$h(x) = -f(x + 1) = -(x + 1)^2$

59. The marketing department for a company estimates that the demand for a product is given by $p = 100 - 0.0001x$ where p is the price per unit and x is the number of units. The cost of producing x units is given by $C = 350{,}000 + 30x$, and the profit for producing and selling x units is given by $P = R - C = xp - C$. Sketch the graph of the profit function and estimate the number of units that would produce a maximum profit.

Solution:

$$\begin{aligned} P &= R - C = xp - C \\ &= x(100 - 0.0001x) - (350{,}000 + 30x) \\ &= -0.0001x^2 + 70x - 350{,}000 \end{aligned}$$

$x = 350{,}000$ units would produce a maximum profit.

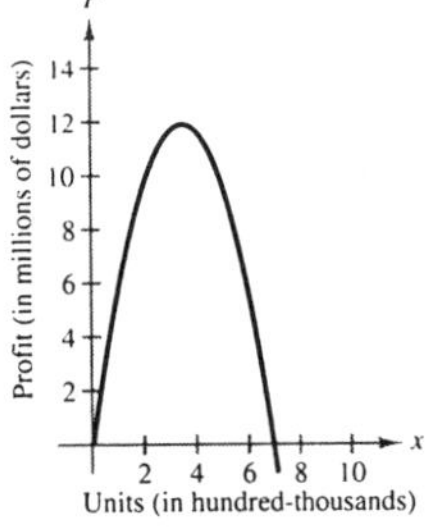

63. Write the height h of the given rectangle as a function of x.

Solution:

$$\begin{aligned} h &= \text{top} - \text{bottom} \\ &= (4x - x^2) - x^2 \\ &= 4x - 2x^2 \end{aligned}$$

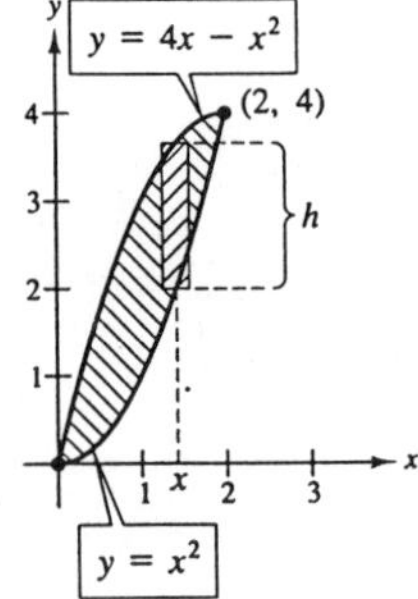

67. Prove that a function of the following form is odd.

$$f(x) = a_{2n+1}x^{2n+1} + a_{2n-1}x^{2n-1} + \ldots + a_3x^3 + a_1x$$

Solution:

$$\begin{aligned} f(x) &= a_{2n+1}x^{2n+1} + a_{2n-1}x^{2n-1} + \ldots + a_3x^3 + a_1x \\ f(-x) &= a_{2n+1}(-x)^{2n+1} + a_{2n-1}(-x)^{2n-1} + \ldots + a_3(-x)^3 + a_1(-x) \\ &= -a_{2n+1}x^{2n+1} - a_{2n-1}x^{2n-1} - \ldots - a_3x^3 - a_1x = -f(x) \end{aligned}$$

Therefore, $f(x)$ is odd.

SECTION 2.6

Combinations of Functions

- Given two functions, f and g, you should be able to form the following functions (if defined):
  1. Sum: $(f+g)(x) = f(x) + g(x)$
  2. Difference: $(f-g)(x) = f(x) - g(x)$
  3. Product: $(fg)(x) = f(x)g(x)$
  4. Quotient: $(f/g)(x) = f(x)/g(x),\ g(x) \neq 0$
  5. Composition of f with g: $(f \circ g)(x) = f(g(x))$
  6. Composition of g with f: $(g \circ f)(x) = g(f(x))$

Solutions to Selected Exercises

5. Find (a) $(f+g)(x)$, (b) $(f-g)(x)$, (c) $(fg)(x)$, and (d) $(f/g)(x)$. What is the domain of f/g?

Solution:

$$f(x) = x^2 + 5, \quad g(x) = \sqrt{1-x}$$

(a) $(f+g)(x) = f(x) + g(x)$

$= x^2 + 5 + \sqrt{1-x}$

(b) $(f-g)(x) = f(x) - g(x)$

$= x^2 + 5 - \sqrt{1-x}$

(c) $(fg)(x) = f(x)g(x) = (x^2+5)\sqrt{1-x}$

(d) $\left(\dfrac{f}{g}\right)(x) = \dfrac{f(x)}{g(x)} = \dfrac{x^2+5}{\sqrt{1-x}},\ x < 1$

The domain of f/g is $(-\infty, 1)$.

9. Evaluate $(f+g)(3)$ for $f(x) = x^2 + 1$ and $g(x) = x - 4$.

Solution:

$$(f+g)(3) = f(3) + g(3) = [(3)^2 + 1] + (3-4) = 10 - 1 = 9$$

13. Evaluate $(f-g)(2t)$ for $f(x) = x^2 + 1$ and $g(x) = x - 4$.

Solution:

$$(f-g)(2t) = f(2t) - g(2t) = [(2t)^2 + 1] - [(2t) - 4] = 4t^2 + 1 - 2t + 4 = 4t^2 - 2t + 5$$

17. Evaluate $(f/g)(5)$ for $f(x) = x^2 + 1$ and $g(x) = x - 4$.

Solution:

$$\left(\frac{f}{g}\right)(5) = \frac{f(5)}{g(5)} = \frac{(5)^2 + 1}{5 - 4} = 26$$

19. Evaluate $(f/g)(-1) - g(3)$ for $f(x) = x^2 + 1$ and $g(x) = x - 4$.

Solution:

$$\left(\frac{f}{g}\right)(-1) - g(3) = \frac{f(-1)}{g(-1)} - g(3) = \frac{(-1)^2 + 1}{-1 - 4} - (3 - 4) = -\frac{2}{5} + 1 = \frac{3}{5}$$

23. Find (a) $f \circ g$, (b) $g \circ f$, and (c) $f \circ f$ for $f(x) = 3x + 5$ and $g(x) = 5 - x$.

Solution:

(a) $f \circ g = f(g(x))$
$= f(5 - x)$
$= 3(5 - x) + 5$
$= 20 - 3x$

(b) $g \circ f = g(f(x))$
$= g(3x + 5)$
$= 5 - (3x + 5)$
$= -3x$

(c) $f \circ f = f(f(x))$
$= f(3x + 5)$
$= 3(3x + 5) + 5$
$= 9x + 20$

25. Find (a) $f \circ g$ and (b) $g \circ f$ for $f(x) = \sqrt{x + 4}$ and $g(x) = x^2$.

Solution:

(a) $f \circ g = f(g(x))$
$= f(x^2)$
$= \sqrt{x^2 + 4}$

(b) $g \circ f = g(f(x))$
$= g(\sqrt{x + 4})$
$= (\sqrt{x + 4})^2$
$= x + 4$

29. Find (a) $f \circ g$ and (b) $g \circ f$ for $f(x) = \sqrt{x}$ and $g(x) = \sqrt{x}$.

Solution:

(a) $f \circ g = f(g(x))$
$= f(\sqrt{x}) = \sqrt{\sqrt{x}} = \sqrt[4]{x}$

(b) Same as (a)

31. Find (a) $f \circ g$ and (b) $g \circ f$ for $f(x) = |x|$ and $g(x) = x + 6$.

Solution:

(a) $f \circ g = f(g(x))$
$= f(x+6) = |x+6|$

(b) $g \circ f = g(f(x))$
$= g(|x|) = |x| + 6$

33. Use the graphs of f and g to evaluate (a) $(f+g)(3)$ and (b) $(f/g)(2)$.

Solution:

(a) $(f+g)(3) = f(3) + g(3)$
$= 2 + 1 = 3$

(b) $\left(\frac{f}{g}\right)(2) = \frac{f(2)}{g(2)} = \frac{0}{2} = 0$

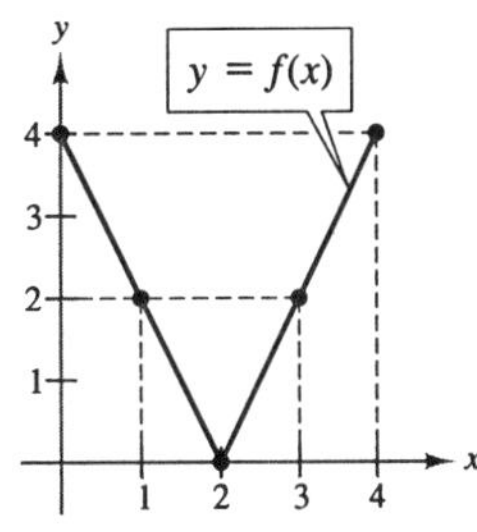

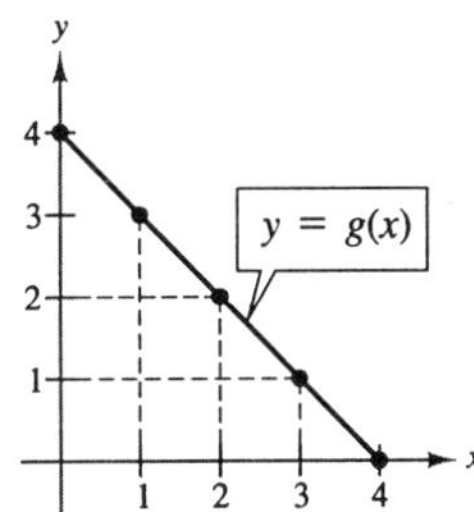

37. Find two functions f and g such that $(f \circ g)(x) = h(x)$ for $h(x) = (2x+1)^2$.

Solution:

Let $f(x) = x^2$ and $g(x) = 2x + 1$, then $(f \circ g)(x) = h(x)$. This is not a unique solution. For example, if $f(x) = (x+1)^2$ and $g(x) = 2x$, then $(f \circ g)(x) = h(x)$ as well.

41. Find two functions f and g such that $(f \circ g)(x) = h(x)$ for $h(x) = 1/(x+2)$.

Solution:

Let $f(x) = 1/x$ and $g(x) = x + 2$, then $(f \circ g)(x) = h(x)$. Again, this is not a unique solution. Other possibilities are:

$f(x) = \frac{1}{x+2}$ and $g(x) = x$ OR $f(x) = \frac{1}{x+1}$ and $g(x) = x + 1$ OR

$f(x) = \frac{1}{x^2+2}$ and $g(x) = \sqrt{x}$

47. Determine the domain of (a) f, (b) g, and (c) $f \circ g$ for $f(x) = 3/(x^2 - 1)$ and $g(x) = x + 1$.

Solution:

(a) The domain of $f(x) = 3/(x^2 - 1)$ includes all real numbers except $x = \pm 1$.

(b) The domain of $g(x) = x + 1$ includes all real numbers.

(c) $f \circ g = f(g(x)) = f(x+1) = \dfrac{3}{(x+1)^2 - 1} = \dfrac{3}{x^2 + 2x} = \dfrac{3}{x(x+2)}$

The domain of $f \circ g$ includes all real numbers except $x = 0$ and $x = -2$.

51. A pebble is dropped into a calm pond, causing ripples in the form of concentric circles (see figure). The radius (in feet) of the outer ripple is given by $r(t) = 0.6t$, where t is time in seconds after the pebble strikes the water. The area of the circle is given by the function $A(r) = \pi r^2$. Find and interpret $(A \circ r)(t)$.

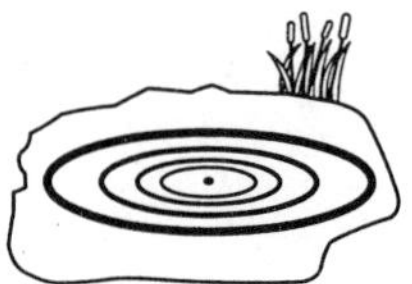

Solution:

$(A \circ r)(t) = A(r(t)) = A(0.6t) = \pi(0.6t)^2 = 0.36\pi t^2$

This equation gives the area as a function of time.

55. Prove that the product of two odd functions is an even function.

Solution:

Let $f(x)$ and $g(x)$ be two odd functions and define $h(x) = f(x)g(x)$. Then

$$\begin{aligned} h(-x) &= f(-x)g(-x) \\ &= [-f(x)][-g(x)] \qquad \text{since } f(x) \text{ and } g(x) \text{ are odd} \\ &= f(x)g(x) \\ &= h(x) \end{aligned}$$

Thus, h is even.

Solutions to Selected Exercises

5. Match the polynomial function $f(x) = -2x^2 - 8x - 9$.

Solution:

$f(x) = -2x^2 - 8x - 9$

$f(x) \to -\infty$ as $x \to \infty$

$f(x) \to -\infty$ as $x \to -\infty$

Parabola opens downward

Matches (b)

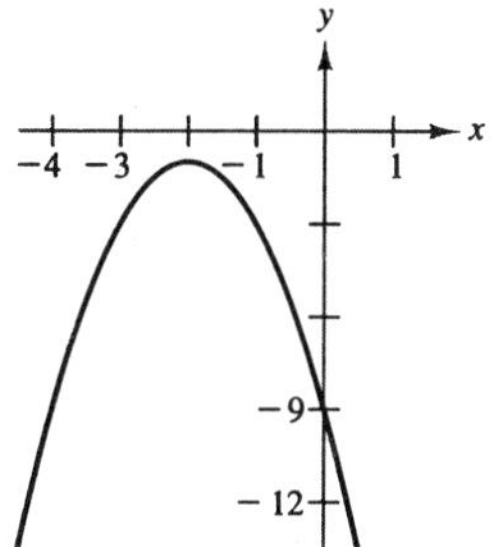

9. Match the polynomial function $f(x) = 3x^4 + 4x^3$.

Solution:

$f(x) = 3x^4 + 4x^3 = x^3(3x + 4)$

Zeros: $0,\ -\frac{4}{3}$

$f(x) \to \infty$ as $x \to \infty$

$f(x) \to \infty$ as $x \to -\infty$

Matches (d)

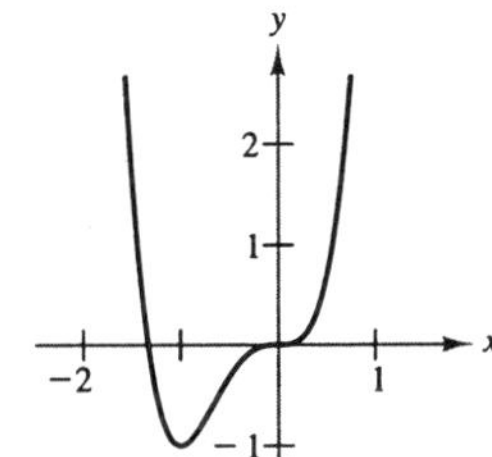

13. Determine the right-hand and left-hand behavior of the graph of $g(x) = 5 - \frac{7}{2}x - 3x^2$.

Solution:

$g(x) = 5 - \frac{7}{2}x - 3x^2 = -3x^2 - \frac{7}{2}x + 5$

Even degree with leading coefficient of -3

Left: $g(x) \to -\infty$ as $x \to -\infty$, so the graph moves down to the left.

Right: $g(x) \to -\infty$ as $x \to +\infty$, so the graph moves down to the right.

17. Determine the right-hand and left-hand behavior of the graph of $f(x) = 6 - 2x + 4x^2 - 5x^3$.

Solution:

$$f(x) = 6 - 2x + 4x^2 - 5x^3 = -5x^3 + 4x^2 - 2x + 6$$

Odd degree with a negative leading coefficient of -5

Left: $f(x) \to \infty$ as $x \to -\infty$, so the graph moves up to the left.

Right: $f(x) \to -\infty$ as $x \to \infty$, so the graph moves down to the right.

23. Find all the real zeros of $h(t) = t^2 - 6t + 9$.

Solution:

$$\begin{aligned} h(t) &= t^2 - 6t + 9 \\ 0 &= t^2 - 6t + 9 \\ 0 &= (t - 3)^2 \\ t &= 3 \end{aligned}$$

27. Find all the real zeros of $f(x) = 3x^2 - 12x + 3$.

Solution:

$$\begin{aligned} f(x) &= 3x^2 - 12x + 3 \\ 0 &= 3(x^2 - 4x + 1) \\ x &= \frac{4 \pm \sqrt{12}}{2} \quad \text{by the Quadratic Formula} \\ &= \frac{4 \pm 2\sqrt{3}}{2} \\ &= 2 \pm \sqrt{3} \end{aligned}$$

31. Find all the real zeros of $g(t) = \frac{1}{2}t^4 - \frac{1}{2}$.

Solution:

$$\begin{aligned} g(t) &= \tfrac{1}{2}t^4 - \tfrac{1}{2} \\ 0 &= \tfrac{1}{2}(t^4 - 1) \\ 0 &= \tfrac{1}{2}(t^2 + 1)(t^2 - 1) \\ 0 &= \tfrac{1}{2}(t^2 + 1)(t + 1)(t - 1) \\ t &= \pm 1 \end{aligned}$$

35. Find all the real zeros of $f(x) = 5x^4 + 15x^2 + 10$.

Solution:

$$f(x) = 5x^4 + 15x^2 + 10$$
$$0 = 5(x^4 + 3x^2 + 2)$$
$$0 = 5(x^2 + 1)(x^2 + 2)$$

No real zeros

37. Find a polynomial function that has the zeros 0 and 10.

Solution:

$$f(x) = (x - 0)(x - 10)$$
$$f(x) = x^2 - 10x$$

Note: $f(x) = a(x - 0)(x - 10) = ax(x - 10)$ has the zeros 0 and 10 for all real numbers a.

43. Find a polynomial function that has the zeros 4, -3, 3 and 0.

Solution:

$$\begin{aligned} f(x) &= (x - 4)(x + 3)(x - 3)(x - 0) \\ &= (x - 4)(x^2 - 9)x \\ &= x^4 - 4x^3 - 9x^2 + 36x \end{aligned}$$

Note: $f(x) = a(x^4 - 4x^3 - 9x^2 + 36x)$ has these zeros for all real numbers a.

47. Sketch the graph of $f(x) = -\frac{3}{2}$.

Solution:

$f(x) = -\frac{3}{2}$ is a horizontal line.

y-intercept: $(0, -\frac{3}{2})$

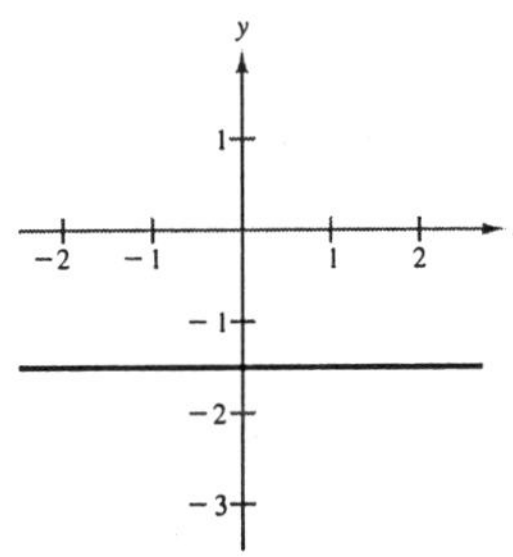

51. Sketch the graph of $f(x) = x^3 - 3x^2$.

Solution:

$f(x) = x^3 - 3x^2 = x^2(x - 3)$

Zeros: 0 and 3

Right: Moves up

Left: Moves down

x	0	1	2	3	−1
$f(x)$	0	−2	−4	0	−4

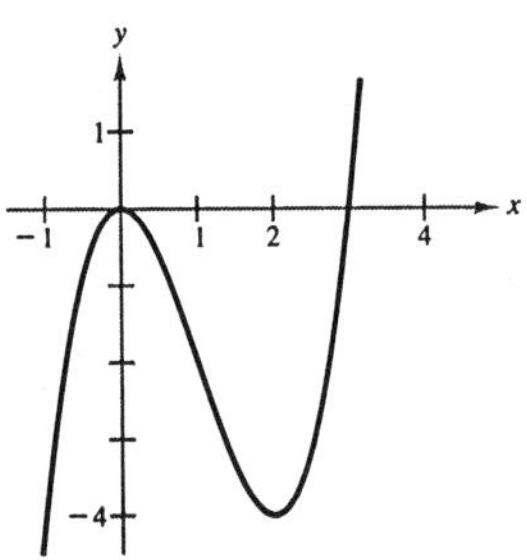

55. Sketch the graph of $g(t) = -\frac{1}{4}(t - 2)^2(t + 2)^2$.

Solution:

$g(t) = -\frac{1}{4}(t - 2)^2(t + 2)^2$

Zeros: 2 and −2

Right: Moves down

Left: Moves down

t	−3	−2	−1	0	1	2	3
$g(t)$	$-\frac{25}{4}$	0	$-\frac{9}{4}$	−4	$-\frac{9}{4}$	0	$-\frac{25}{4}$

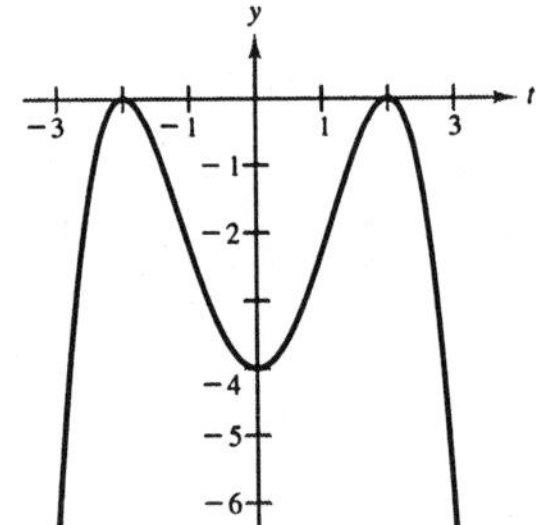

59. Sketch the graph of $h(x) = \frac{1}{3}x^3(x - 4)^2$.

Solution:

$h(x) = \frac{1}{3}x^3(x - 4)^2$

Zeros: 0 and 4

Right: Moves up

Left: Moves down

x	−1	0	1	2	3	4	5
$h(x)$	$-\frac{25}{3}$	0	3	$\frac{32}{3}$	9	0	$\frac{125}{3}$

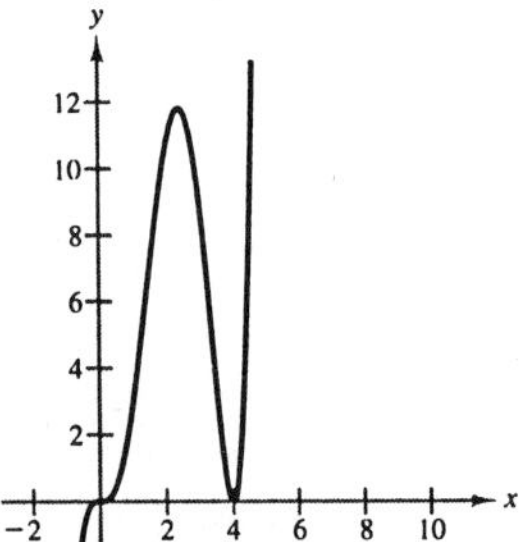

65. Follow the procedure given in Section 4.2 in the textbook to estimate the zero of $f(x) = x^4 - 10x^2 - 11$ in the interval [3, 4]. (Give your approximation to the nearest tenth.)

Solution:

$f(x) = x^4 - 10x^2 - 11, \quad [3, 4]$

x	3	3.1	3.2	3.3	3.4	3.5
$f(x)$	−20	−14.748	−8.542	−1.308	7.034	16.563

x	3.6	3.7	3.8	3.9	4.0
$f(x)$	27.362	39.516	53.114	68.244	85

The zero lies between 3.3 and 3.4. It is closer to 3.3.

69. The total revenue for a soft drink company is related to its advertising expense by the function

$$R = \frac{1}{50{,}000}(-x^3 + 600x^2), \quad 0 \le x \le 400$$

where R is the total revenue in millions of dollars and x is the amount spent on advertising (in 10,000's of dollars). Use the graph of this function shown on the figure to estimate the point on the graph at which the function is increasing most rapidly. This is the *point of diminishing returns* because any expense above this amount will yield less return per dollar invested in advertising.

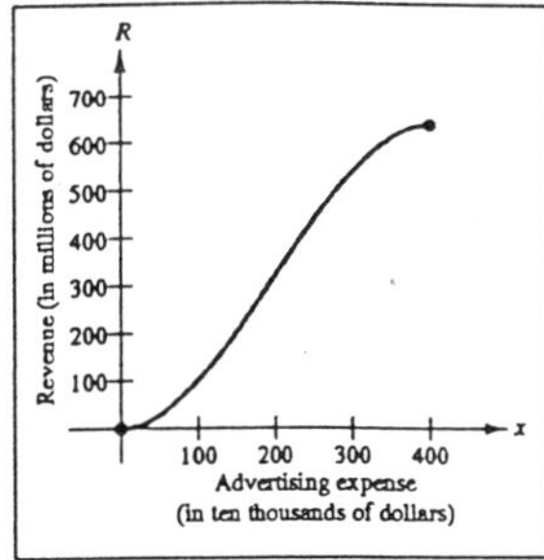

Solution:

The point of diminishing returns (where the graph changes from curving upward to curving downward) occurs when $x = 200$. The point is (200, 320) which corresponds to spending \$2,000,000 on advertising to obtain a revenue of \$320 million.

SECTION 3.3

Polynomial Division and Synthetic Division

You should know the following basic techniques and principles of polynomial division.

- The Division Algorithm (Long Division of Polynomials)
- Synthetic Division
- $f(k)$ is equal to the remainder of $f(x)$ divided by $(x-k)$.
- $f(k)=0$ if and only if $(x-k)$ is a factor of $f(x)$.

Solutions to Selected Exercises

5. Divide $x^4+5x^3+6x^2-x-2$ by $x+2$ using long division.

Solution:

$$\begin{array}{r} x^3+3x^2-1 \\ x+2\,\overline{)\,x^4+5x^3+6x^2-x-2} \\ \underline{x^4+2x^3} \\ 3x^3+6x^2 \\ \underline{3x^3+6x^2} \\ -x-2 \\ \underline{-x-2} \\ 0 \end{array}$$

Thus,

$$\frac{x^4+5x^3+6x^2-x-2}{x+2}=x^3+3x^2-1.$$

7. Divide $7x+3$ by $x+2$ using long division.

Solution:

$$\begin{array}{r} 7 \\ x+2\,\overline{)\,7x+3} \\ \underline{7x+14} \\ -11 \end{array}$$

Thus, $\dfrac{7x+3}{x+2}=7-\dfrac{11}{x+2}$.

11. Divide $x^4 + 3x^2 + 1$ by $x^2 - 2x + 3$ using long division.

Solution:

$$
\begin{array}{r}
x^2 + 2x + 4 \\
x^2 - 2x + 3 \overline{) \; x^4 + 0x^3 + 3x^2 + 0x + 1} \\
\underline{x^4 - 2x^3 + 3x^2 } \\
2x^3 + 0x^2 + 0x \\
\underline{2x^3 - 4x^2 + 6x } \\
4x^2 - 6x + 1 \\
\underline{4x^2 - 8x + 12} \\
2x - 11
\end{array}
$$

Thus, $\dfrac{x^4 + 3x^2 + 1}{x^2 - 2x + 3} = x^2 + 2x + 4 + \dfrac{2x - 11}{x^2 - 2x + 3}$.

15. Divide $3x^3 - 17x^2 + 15x - 25$ by $x - 5$ using synthetic division.

Solution:

$$
\begin{array}{r|rrrr}
5 & 3 & -17 & 15 & -25 \\
 & & 15 & -10 & 25 \\
\hline
 & 3 & -2 & 5 & 0
\end{array}
$$

Thus,

$$\frac{3x^3 - 17x^2 + 15x - 25}{x - 5} = 3x^2 - 2x + 5.$$

19. Divide $-x^3 + 75x - 250$ by $x + 10$ using synthetic division.

Solution:

$$
\begin{array}{r|rrrr}
-10 & -1 & 0 & 75 & -250 \\
 & & 10 & -100 & 250 \\
\hline
 & -1 & 10 & -25 & 0
\end{array}
$$

Thus,

$$\frac{-x^3 + 75x - 250}{x + 10} = -x^2 + 10x - 25.$$

23. Divide $10x^4 - 50x^3 - 800$ by $x - 6$ using synthetic division.

Solution:

$$
\begin{array}{r|rrrrr}
6 & 10 & -50 & 0 & 0 & -800 \\
 & & 60 & 60 & 360 & 2160 \\
\hline
 & 10 & 10 & 60 & 360 & 1360
\end{array}
$$

Thus, $\dfrac{10x^4 - 50x^3 - 800}{x - 6} = 10x^3 + 10x^2 + 60x + 360 + \dfrac{1360}{x - 6}$.

27. Divide $-3x^4$ by $x - 2$ using synthetic division.

Solution:

2	-3	0	0	0	0
		-6	-12	-24	-48
	-3	-6	-12	-24	-48

Thus,

$$\frac{-3x^4}{x-2} = -3x^3 - 6x^2 - 12x - 24 - \frac{48}{x-2}.$$

31. Divide $4x^3 + 16x^2 - 23x - 15$ by $x + \frac{1}{2}$ using synthetic division.

Solution:

$-\frac{1}{2}$	4	16	-23	-15
		-2	-7	15
	4	14	-30	0

Thus,

$$\frac{4x^3 + 16x^2 - 23x - 15}{x + \frac{1}{2}} = 4x^2 + 14x - 30.$$

35. Use synthetic division to show that $x = \frac{1}{2}$ is a solution of $2x^3 - 15x^2 + 27x - 10 = 0$, and use the result to factor the polynomial completely.

Solution:

$\frac{1}{2}$	2	-15	27	-10
		1	-7	10
	2	-14	20	0

$$\begin{aligned} 2x^3 - 15x^2 + 27x - 10 &= \left(x - \tfrac{1}{2}\right)(2x^2 - 14x + 20) \\ &= \left(x - \tfrac{1}{2}\right)2(x^2 - 7x + 10) = (2x-1)(x-2)(x-5) \end{aligned}$$

39. Use synthetic division to show that $x = 1 + \sqrt{3}$ is a solution of $x^3 - 3x^2 + 2 = 0$, and use the result to factor the polynomial completely.

Solution:

$1+\sqrt{3}$	1	-3	0	2
		$1+\sqrt{3}$	$1-\sqrt{3}$	-2
	1	$-2+\sqrt{3}$	$1-\sqrt{3}$	0

$$\begin{aligned} x^3 - 3x^2 + 2 &= \left[x - (1+\sqrt{3})\right]\left[x^2 + (-2+\sqrt{3})x + 1 - \sqrt{3}\right] \\ &= (x - 1 - \sqrt{3})(x^2 - 2x + \sqrt{3}x + 1 - \sqrt{3}) \\ &= (x - 1 - \sqrt{3})\left[(x^2 - 2x + 1) + \sqrt{3}(x-1)\right] \\ &= (x - 1 - \sqrt{3})\left[(x-1)^2 + \sqrt{3}(x-1)\right] \\ &= (x - 1 - \sqrt{3})(x-1)\left[(x-1) + \sqrt{3}\right] \\ &= \left(x - 1 - \sqrt{3}\right)\left(x - 1 + \sqrt{3}\right)(x-1) \end{aligned}$$

41. Express the function $f(x) = x^3 - x^2 - 14x + 11$ in the form $f(x) = (x - k)q(x) + r$ for $k = 4$, and demonstrate that $f(k) = r$.

Solution:

$$\begin{array}{r|rrrr} 4 & 1 & -1 & -14 & 11 \\ & & 4 & 12 & -8 \\ \hline & 1 & 3 & -2 & 3 \end{array}$$

$$f(x) = (x - 4)(x^2 + 3x - 2) + 3$$

$$r = 3$$

$$\begin{aligned} f(4) &= 4^3 - 4^2 - 14(4) + 11 \\ &= 64 - 16 - 56 + 11 \\ &= 3 \end{aligned}$$

45. Use synthetic division to find the required function values of $f(x) = 4x^3 - 13x + 10$.

(a) $f(1)$ (b) $f(-2)$ (c) $f\left(\frac{1}{2}\right)$ (d) $f(8)$

Solution:

(a)
$$\begin{array}{r|rrrr} 1 & 4 & 0 & -13 & 10 \\ & & 4 & 4 & -9 \\ \hline & 4 & 4 & -9 & 1 \end{array}$$

Thus, $f(1) = 1$.

(b)
$$\begin{array}{r|rrrr} -2 & 4 & 0 & -13 & 10 \\ & & -8 & 16 & -6 \\ \hline & 4 & -8 & 3 & 4 \end{array}$$

Thus, $f(-2) = 4$.

(c)
$$\begin{array}{r|rrrr} \frac{1}{2} & 4 & 0 & -13 & 10 \\ & & 2 & 1 & -6 \\ \hline & 4 & 2 & -12 & 4 \end{array}$$

Thus, $f\left(\frac{1}{2}\right) = 4$.

(d)
$$\begin{array}{r|rrrr} 8 & 4 & 0 & -13 & 10 \\ & & 32 & 256 & 1944 \\ \hline & 4 & 32 & 243 & 1954 \end{array}$$

Thus, $f(8) = 1954$.

49. Use synthetic division to find the required function values of $f(x) = x^3 - 2x^2 - 11x + 52$.

(a) $f(5)$ (b) $f(-4)$ (c) $f(1.2)$ (d) $f(2)$

Solution:

(a)
$$\begin{array}{r|rrrr} 5 & 1 & -2 & -11 & 52 \\ & & 5 & 15 & 20 \\ \hline & 1 & 3 & 4 & 72 \end{array}$$

Thus, $f(5) = 72$.

(b)
$$\begin{array}{r|rrrr} -4 & 1 & -2 & -11 & 52 \\ & & -4 & 24 & -52 \\ \hline & 1 & -6 & 13 & 0 \end{array}$$

Thus, $f(-4) = 0$.

(c)
$$\begin{array}{r|rrrr} 1.2 & 1 & -2 & -11 & 52 \\ & & 1.2 & -0.96 & -14.352 \\ \hline & 1 & -0.8 & -11.96 & 37.648 \end{array}$$

Thus, $f(1.2) = 37.648$.

(d)
$$\begin{array}{r|rrrr} 2 & 1 & -2 & -11 & 52 \\ & & 2 & 0 & -22 \\ \hline & 1 & 0 & -11 & 30 \end{array}$$

Thus, $f(2) = 30$.

53. Simplify the rational expression

$$\frac{x^3 + 3x^2 - x - 3}{x + 1}.$$

Solution:

$$\begin{array}{r|rrrr} -1 & 1 & 3 & -1 & -3 \\ & & -1 & -2 & 3 \\ \hline & 1 & 2 & -3 & 0 \end{array}$$

Thus, $\dfrac{x^3 + 3x^2 - x - 3}{x + 1} = x^2 + 2x - 3 = (x + 3)(x - 1)$.

57. The horsepower y developed by a compact car engine is approximated by the model

$$y = -1.42x^3 + 5.04x^2 + 32.45x - 0.75, \quad 1 \le x \le 5$$

where x is the engine speed in thousands of revolutions per minute. Note on the graph that there are two engine speeds that develop 110 horsepower, one of which is 5000 rpm. Approximate the other engine speed.

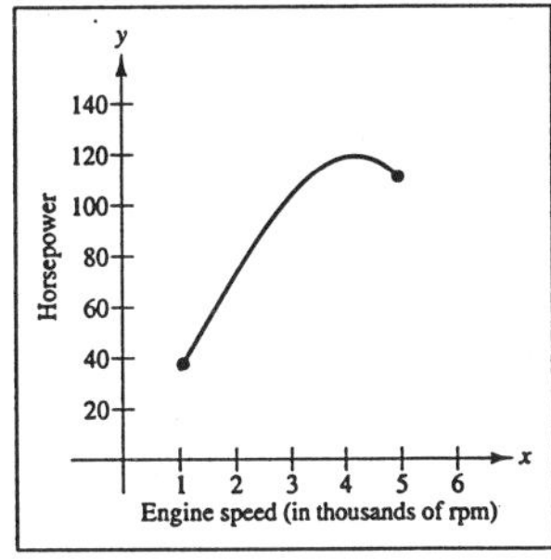

Solution:

Since $y = 110$ when $x = 5$ we have $x = 5$ as a zero to the equation $y - 110$.

$$\begin{aligned} f(x) &= y - 110, \quad 1 \le x \le 5 \\ &= -1.42x^3 + 5.04x^2 + 32.45x - 110.75 \end{aligned}$$

$$\begin{array}{r|rrrr} 5 & -1.42 & 5.04 & 32.45 & -110.75 \\ & & -7.10 & -10.30 & 110.75 \\ \hline & -1.42 & -2.06 & 22.15 & 0 \end{array}$$

Thus, $f(x) = (x - 5)(-1.42x^2 - 2.06x + 22.15)$. Using the Quadratic Formula to find the other two zeros yields

$$\begin{aligned} x &= \frac{-(-2.06) \pm \sqrt{(-2.06)^2 - 4(-1.42)(22.15)}}{2(-1.42)} \\ &= \frac{2.06 \pm \sqrt{130.0556}}{-2.84} \end{aligned}$$

$x \approx -4.74$ or $x \approx 3.29$.

Since $1 \le x \le 5$, we choose $x = 3.29$ which corresponds to 3290 rpm.

SECTION 3.4

Real Zeros of Polynomial Functions

- You should know Descartes's Rule of Signs.
  (a) The number of positive real zeros of f is either equal to the number of variations of sign of f or is less than that number by an even integer.
  (b) The number of negative real zeros of f is either equal to the number of variations in sign of $f(-x)$ or is less than that number by an even integer.
  (c) When there is only one variation in sign, there is exactly one positive (or negative) real zero.

- You should know the Rational Zero Test.

- You should know shortcuts for the Rational Zero Test.
  (a) Use a programmable calculator.
  (b) Sketch a graph.
  (c) After finding a root, use synthetic division to reduce the degree of the polynomial.

- You should be able to observe the last row obtained from synthetic division in order to determine upper or lower bounds.
  (a) If the test value is positive and all of the entries in the last row are positive or zero, then the test value is an upper bound.
  (b) If the test value is negative and the entries in the last row alternate from positive to negative, then the test value is a lower bound. (Zero entries count as positive or negative.)

Solutions to Selected Exercises

5. Use Descartes's Rule of Signs to determine the possible number of positive and negative zeros of $g(x) = 2x^3 - 3x^2 - 3$.

Solution:

$$g(x) = 2x^3 - 3x^2 - 3$$
$$g(-x) = -2x^3 - 3x^2 - 3$$

Since $g(x)$ has one variation in sign, g has exactly one positive real zero. Since $g(-x)$ has no variations in sign, there are no negative real zeros.

9. Use Descartes's Rule of Signs to determine the possible number of positive and negative real zeros of $h(x) = 4x^2 - 8x + 3$.

Solution:

$$h(x) = 4x^2 - 8x + 3$$
$$h(-x) = 4x^2 + 8x + 3$$

Since $h(x)$ has two variations in sign, h has either two or zero positive real zeros. Since $h(-x)$ has no variations in sign, there are no negative real zeros.

13. Use the Rational Zero Test to list all the possible rational zeros of $f(x) = -4x^3 + 15x^2 - 8x - 3$ and verify that the zeros of f shown on the graph are contained in the list.

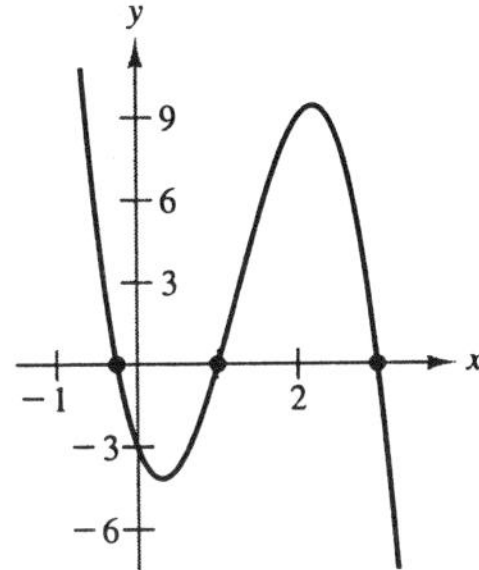

Solution:

Since the leading coefficient is -4 and the constant term is -3, the possible rational zeros of f are

$$\frac{\text{factors of } -3}{\text{factors of } -4} = \frac{\pm 1, \pm 3}{\pm 1, \pm 2, \pm 4} = \pm 1, \pm 3, \pm\frac{1}{2}, \pm\frac{3}{2}, \pm\frac{1}{4}, \pm\frac{3}{4}$$

The zeros shown on the graph are $-\frac{1}{4}$, 1 and 3 and are contained in the list.

17. Use synthetic division to determine if the given x-value is an upper bound or lower bound of the zeros of $f(x) = x^4 - 4x^3 + 15$.

(a) $x = 4$ (b) $x = -1$ (c) $x = 3$

Solution:

(a) 4	1	−4	0	0	15
		4	0	0	0
	1	0	0	0	15

Since the test value is positive and all the entries in the last row are positive, $x = 4$ is an upper bound.

(b) −1	1	−4	0	0	15
		−1	5	−5	5
	1	−5	5	−5	20

Since the test value is negative and the entries in the last row alternate in sign, $x = -1$ is a lower bound.

(c) 3	1	−4	0	0	15
		3	−3	−9	−27
	1	−1	−3	−9	−12

$x = 3$ is neither an upper nor a lower bound.

21. Find the real zeros of $f(x) = x^3 - 6x^2 + 11x - 6$.

Solution:

Possible rational zeros: $\pm1, \pm2, \pm3, \pm6$

1	1	−6	11	−6
		1	−5	6
	1	−5	6	0

$$x^3 - 6x^2 + 11x - 6 = (x-1)(x^2 - 5x + 6) = (x-1)(x-2)(x-3)$$

The zeros are 1, 2, and 3.

25. Find the real zeros of $h(t) = t^3 + 12t^2 + 21t + 10$.

Solution:

Possible rational zeros: $\pm 1, \ \pm 2, \ \pm 5, \ \pm 10$

$$\begin{array}{r|rrrr} -1 & 1 & 12 & 21 & 10 \\ & & -1 & -11 & -10 \\ \hline & 1 & 11 & 10 & 0 \end{array}$$

$$t^3 + 12t^2 + 21t + 10 = (t+1)(t^2 + 11t + 10) = (t+1)(t+1)(t+10)$$

Thus, the zeros are -1 and -10.

27. Find the real zeros of $f(x) = x^3 - 4x^2 + 5x - 2$.

Solution:

Possible rational zeros: $\pm 1, \ \pm 2$

$$\begin{array}{r|rrrr} 1 & 1 & -4 & 5 & -2 \\ & & 1 & -3 & 2 \\ \hline & 1 & -3 & 2 & 0 \end{array}$$

$$x^3 - 4x^2 + 5x - 2 = (x-1)(x^2 - 3x + 2) = (x-1)(x-1)(x-2)$$

Thus, the zeros are 1 and 2.

31. Find the real zeros of $f(x) = 4x^3 - 3x - 1$.

Solution:

Possible rational zeros: $\pm 1, \ \pm \frac{1}{2}, \ \pm \frac{1}{4}$

$$\begin{array}{r|rrrr} 1 & 4 & 0 & -3 & -1 \\ & & 4 & 4 & 1 \\ \hline & 4 & 4 & 1 & 0 \end{array}$$

$$4x^3 - 3x - 1 = (x-1)(4x^2 + 4x + 1) = (x-1)(2x+1)^2$$

Thus, the zeros are 1 and $-\frac{1}{2}$.

33. Find the real zeros of $f(y) = 4y^3 + 3y^2 + 8y + 6$.

Solution:

Possible rational zeros: $\pm 1, \pm 2, \pm 3, \pm 6, \pm\frac{1}{2}, \pm\frac{3}{2}, \pm\frac{1}{4}, \pm\frac{3}{4}$

$$\begin{array}{r|rrrr} -\frac{3}{4} & 4 & 3 & 8 & 6 \\ & & -3 & 0 & -6 \\ \hline & 4 & 0 & 8 & 0 \end{array}$$

$$4y^3 + 3y^2 + 8y + 6 = (y + \tfrac{3}{4})(4y^2 + 8) = (y + \tfrac{3}{4})4(y^2 + 2) = (4y + 3)(y^2 + 2)$$

Thus, the only zero is $-\frac{3}{4}$.

35. Find the real zeros of $f(x) = x^4 - 3x^2 + 2$.

Solution:

$$\begin{aligned} f(x) &= x^4 - 3x^2 + 2 \\ &= (x^2 - 1)(x^2 - 2) \\ &= (x + 1)(x - 1)(x + \sqrt{2})(x - \sqrt{2}) \end{aligned}$$

Thus, the zeros are ± 1 and $\pm\sqrt{2}$.

39. Find all the real solutions of $x^4 - 13x^2 - 12x = 0$.

Solution:

$$f(x) = x^4 - 13x^2 - 12x = x(x^3 - 13x - 12)$$

0 is a zero.

Possible rational zeros: $\pm 1, \pm 2, \pm 3, \pm 4, \pm 6, \pm 12$

$$\begin{array}{r|rrrr} -1 & 1 & 0 & -13 & -12 \\ & & -1 & 1 & 12 \\ \hline & 1 & -1 & -12 & 0 \end{array}$$

$$x^4 - 13x^2 - 12x = x(x^3 - 13x - 12) = x(x + 1)(x^2 - x - 12) = x(x + 1)(x + 3)(x - 4)$$

Thus, the zeros are $0, -1, -3,$ and 4.

43. Find all the real solutions of $x^5 - 7x^4 + 10x^3 + 14x^2 - 24x = 0$.

Solution:

$$f(x) = x^5 - 7x^4 + 10x^3 + 14x^2 - 24x = x(x^4 - 7x^3 + 10x^2 + 14x - 24)$$

0 is a zero.
Possible rational zeros: $\pm 1, \pm 2, \pm 3, \pm 4, \pm 6, \pm 8, \pm 12, \pm 24$

$$\begin{array}{r|rrrrr} 4 & 1 & -7 & 10 & 14 & -24 \\ & & 4 & -12 & -8 & 24 \\ \hline \end{array}$$

$$\begin{array}{r|rrrrr} 3 & 1 & -3 & -2 & 6 & 0 \\ & & 3 & 0 & -6 & \\ \hline & 1 & 0 & -2 & 0 & \end{array}$$

$$\begin{aligned} x^5 - 7x^4 + 10x^3 + 14x^2 - 24x &= x(x^4 - 7x^3 + 10x^2 + 14x - 24) \\ &= x(x-4)(x-3)(x^2-2) \\ &= x(x-4)(x-3)(x+\sqrt{2})(x-\sqrt{2}) \end{aligned}$$

Thus, the zeros are 0, 4, 3, and $\pm\sqrt{2}$.

47. For $f(x) = 4x^3 + 7x^2 - 11x - 18$:

(a) list all the possible rational zeros of f,
(b) sketch the graph of f so that some of the possible zeros in part (a) can be disregarded,
(c) then determine all the real zeros of f.

Solution:

(a) Possible rational roots: $\pm 1, \pm 2, \pm 3, \pm 6, \pm 9, \pm 18, \pm\frac{1}{2}, \pm\frac{3}{2}, \pm\frac{9}{2}, \pm\frac{1}{4}, \pm\frac{3}{4}, \pm\frac{9}{4}$

(b)

x	0	1	-1	$\frac{1}{2}$	-2	-3	$-\frac{3}{2}$
$f(x)$	-18	-18	-4	-21.25	0	-30	0.75

By testing values using synthetic division, we find that 2 is an upper bound and -3 is a lower bound. This eliminates 3, ± 6, ± 9, ± 18, $\pm\frac{9}{2}$, and $\frac{9}{4}$ as possible zeros.

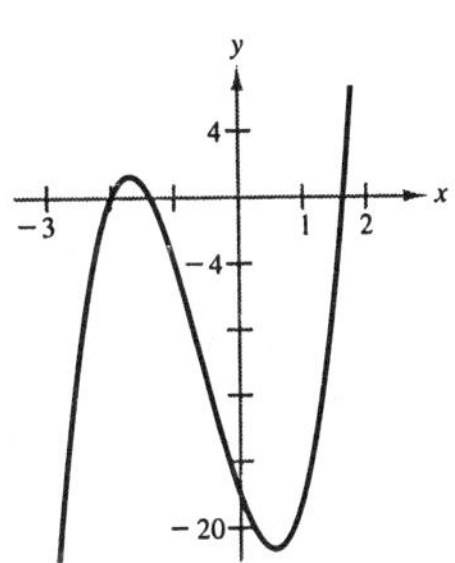

–CONTINUED ON NEXT PAGE–

47. –CONTINUED–

(c) -2 is a zero.

$4x^3 + 7x^2 - 11x - 18 = (x+2)(4x^2 - x - 9) = 0$

$4x^2 - x - 9$ does not factor, so by the Quadratic Formula

$x = \dfrac{1 \pm \sqrt{145}}{8}$ are also zeros.

51. Find all the rational zeros of $f(x) = x^3 - \frac{1}{4}x^2 - x + \frac{1}{4}$.

Solution:

$f(x) = x^3 - \frac{1}{4}x^2 - x + \frac{1}{4} = \frac{1}{4}(4x^3 - x^2 - 4x + 1)$

Possible rational zeros: $\pm 1,\ \pm\frac{1}{4},\ \pm\frac{1}{2}$

By testing these values, we see that $x = \pm 1$ and $x = \frac{1}{4}$ work.

55. Match the cubic equation $f(x) = x^3 - x$ with the number of rational and irrational zeros (a), (b), (c) or (d).

(a) Rational zeros: 0
Irrational zeros: 1

(b) Rational zeros: 3
Irrational zeros: 0

(c) Rational zeros: 1
Irrational zeros: 2

(d) Rational zeros: 1
Irrational zeros: 0

Solution:

$f(x) = x^3 - x = x(x^2 - 1) = x(x+1)(x-1)$

Zeros: $0,\ \pm 1$

Three rational zeros/no irrational zeros

Matches (b)

59. A rectangular package to be sent by a postal service can have a maximum combined length and girth (perimeter of a cross section) of 108 inches. (see figure) Find the dimensions of the package, given that the volume is to be 11,664 cubic inches.

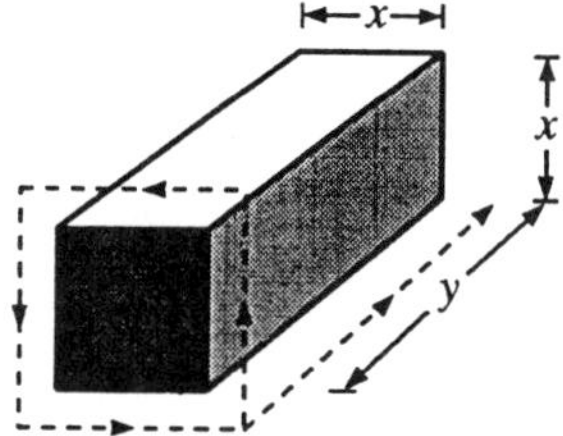

Solution:

The combined length and girth is

$$4x + y = 108$$
$$y = 108 - 4x.$$

The volume is

$$V = x^2 y = x^2(108 - 4x).$$

Since the volume is 11,664 cubic inches, we have

$$11{,}664 = x^2(108 - 4x)$$
$$11{,}664 = 108x^2 - 4x^3$$
$$4x^3 - 108x^2 + 11{,}664 = 0$$
$$4(x^3 - 27x^2 + 2916) = 0 \quad \text{Using the methods of this section yields}$$
$$4(x - 18)^2(x + 2) = 0 \quad \text{Thus, } x = 18 \text{ in. and } y = 36 \text{ in.}$$

The dimensions of the package are 18 inches × 18 inches × 36 inches.

SECTION 3.5

Complex Numbers

- You should know how to work with complex numbers.
- Operations on Complex Numbers

  (a) Addition: $(a+bi)+(c+di)=(a+c)+(b+d)i$
  (b) Subtraction: $(a+bi)-(c+di)=(a-c)+(b-d)i$
  (c) Multiplication: $(a+bi)(c+di)=(ac-bd)+(ad+bc)i$
  (d) Division: $\dfrac{a+bi}{c+di}=\dfrac{a+bi}{c+di}\cdot\dfrac{c-di}{c-di}=\dfrac{ac+bd}{c^2+d^2}+\dfrac{bc-ad}{c^2+d^2}i$
- The complex conjugate of $a+bi$ is $a-bi$:

  $$(a+bi)(a-bi)=a^2+b^2$$
- The additive inverse of $a+bi$ is $-a-bi$.
- The multiplicative inverse of $a+bi$ is

  $$\frac{a-bi}{a^2+b^2}.$$
- $\sqrt{-a}=\sqrt{a}\,i$ for $a>0$.

Solutions to Selected Exercises

3. Find real numbers a and b so that the equation $(a-1)+(b+3)i=5+8i$ is true.

Solution:

$$(a-1)+(b+3)i=5+8i$$

$$a-1=5 \quad\Rightarrow\quad a=6$$

$$b+3=8 \quad\Rightarrow\quad b=5$$

7. Write $2-\sqrt{-27}$ in standard form.

Solution:

$$2-\sqrt{-27}=2-\sqrt{27}\,i=2-3\sqrt{3}\,i$$

11. Write $-6i+i^2$ in standard form.

Solution:

$$-6i+i^2=-6i+(-1)=-1-6i$$

15. Write $\sqrt{-0.09}$ in standard form.

Solution:

$$\sqrt{-0.09} = \sqrt{0.09}\,i = 0.3i$$

19. Perform the indicated operation and write the result in standard form.

$$(8 - i) - (4 - i)$$

Solution:

$$(8 - i) - (4 - i) = 8 - i - 4 + i = 4$$

21. Perform the indicated operation and write the result in standard form.

$$(-2 + \sqrt{-8}\,) + (5 - \sqrt{-50}\,)$$

Solution:

$$(-2 + \sqrt{-8}\,) + (5 - \sqrt{-50}\,) = -2 + 2\sqrt{2}\,i + 5 - 5\sqrt{2}\,i = 3 - 3\sqrt{2}\,i$$

25. Perform the indicated operations and write the result in standard form.

$$-\left(\tfrac{3}{2} + \tfrac{5}{2}i\right) + \left(\tfrac{5}{3} + \tfrac{11}{3}i\right)$$

Solution:

$$\begin{aligned} -\left(\tfrac{3}{2} + \tfrac{5}{2}i\right) + \left(\tfrac{5}{3} + \tfrac{11}{3}i\right) &= -\tfrac{3}{2} - \tfrac{5}{2}i + \tfrac{5}{3} + \tfrac{11}{3}i \\ &= -\tfrac{9}{6} - \tfrac{15}{6}i + \tfrac{10}{6} + \tfrac{22}{6}i \\ &= \tfrac{1}{6} + \tfrac{7}{6}i \end{aligned}$$

29. Write the conjugate of $-2 - \sqrt{5}\,i$ and find the product of the number and its conjugate.

Solution:

The complex conjugate of $-2 - \sqrt{5}\,i$ is $-2 + \sqrt{5}\,i$.

$$(-2 - \sqrt{5}\,i)(-2 + \sqrt{5}\,i) = 4 - 5i^2 = 4 + 5 = 9.$$

31. Write the conjugate of $20i$ and find the product of the number and its conjugate.

Solution:

The complex conjugate of $20i$ is $-20i$.

$$(20i)(-20i) = -400i^2 = 400.$$

35. Perform the specified operation and write the result in standard form.

$\sqrt{-6}\sqrt{-2}$

Solution:

$$\sqrt{-6}\sqrt{-2} = (\sqrt{6}\,i)(\sqrt{2}\,i) = \sqrt{12}\,i^2 = 2\sqrt{3}(-1) = -2\sqrt{3}$$

Note: $\sqrt{-6}\sqrt{-2} \neq \sqrt{12}$

39. Perform the specified operation and write the result in standard form.

$(1+i)(3-2i)$

Solution:

$$(1+i)(3-2i) = 3 - 2i + 3i - 2i^2 = 3 + i + 2 = 5 + i$$

41. Perform the specified operation and write the result in standard form.

$6i(5-2i)$

Solution:

$$6i(5-2i) = 30i - 12i^2 = 12 + 30i$$

43. Perform the specified operation and write the result in standard form.

$(\sqrt{14} + \sqrt{10}\,i)(\sqrt{14} - \sqrt{10}\,i)$

Solution:

$$(\sqrt{14} + \sqrt{10}\,i)(\sqrt{14} - \sqrt{10}\,i) = 14 - 10i^2 = 14 + 10 = 24$$

47. Perform the specified operation and write the result in standard form.

$(2+3i)^2 + (2-3i)^2$

Solution:

$$\begin{aligned}(2+3i)^2 + (2-3i)^2 &= (4 + 12i + 9i^2) + (4 - 12i + 9i^2)\\ &= 8 + 18i^2\\ &= 8 - 18\\ &= -10\end{aligned}$$

51. Perform the specified operation and write the result in standard form.

$\dfrac{2+i}{2-i}$

Solution:

$$\frac{2+i}{2-i} = \frac{2+i}{2-i} \cdot \frac{2+i}{2+i} = \frac{4 + 4i + i^2}{4+1} = \frac{3+4i}{5} = \frac{3}{5} + \frac{4}{5}i$$

55. Perform the specified operation and write the result in standard form.

$$\frac{1}{(4-5i)^2}$$

Solution:

$$\frac{1}{(4-5i)^2} = \frac{1}{16-40i+25i^2} = \frac{1}{-9-40i} \bullet \frac{-9+40i}{-9+40i}$$

$$= \frac{-9+40i}{81+1600} = \frac{-9+40i}{1681} = -\frac{9}{1681} + \frac{40}{1681}i$$

59. Use the Quadratic Formula to solve $4x^2 + 16x + 17 = 0$.

Solution:

$4x^2 + 16x + 17 = 0;\ a = 4,\ b = 16,\ c = 17$

$$x = \frac{-16 \pm \sqrt{(16)^2 - 4(4)(17)}}{2(4)}$$

$$= \frac{-16 \pm \sqrt{-16}}{8} = \frac{-16 \pm 4i}{8}$$

$$= -2 \pm \frac{1}{2}i$$

63. Use the Quadratic Formula to solve $16t^2 - 4t + 3 = 0$.

Solution:

$16t^2 - 4t + 3 = 0;\ a = 16,\ b = -4,\ c = 3$

$$t = \frac{-(-4) \pm \sqrt{(-4)^2 - 4(16)(3)}}{2(16)}$$

$$= \frac{4 \pm \sqrt{-176}}{32} = \frac{4 \pm 4\sqrt{11}i}{32}$$

$$= \frac{1}{8} \pm \frac{\sqrt{11}}{8}i$$

65. Write out the first 16 positive powers of i and express each as i, $-i$, 1, or -1.

Solution:

$i = i$	$i^5 = i$	$i^9 = i$	$i^{13} = i$
$i^2 = -1$	$i^6 = -1$	$i^{10} = -1$	$i^{14} = -1$
$i^3 = -i$	$i^7 = -i$	$i^{11} = -i$	$i^{15} = -i$
$i^4 = 1$	$i^8 = 1$	$i^{12} = 1$	$i^{16} = 1$

69. Simplify $-5i^5$ and write it in standard form.

Solution:

$$-5i^5 = -5i$$

73. Simplify $\frac{1}{i^3}$ and write it in standard form.

Solution:

$$\frac{1}{i^3} = \frac{1}{-i} = \frac{1}{-i} \cdot \frac{i}{i} = \frac{i}{-i^2} = \frac{i}{1} = i$$

77. Prove that the sum of a complex number $a + bi$ and its conjugate is a real number.

Solution:

$$\begin{aligned}(a + bi) + (a - bi) &= (a + a) + (b - b)i \\ &= 2a + 0i = 2a \qquad \text{which is a real number.}\end{aligned}$$

81. Prove that the conjugate of the sum of two complex numbers $a_1 + b_1 i$ and $a_2 + b_2 i$ is the sum of their conjugates.

Solution:

$$(a_1 + b_1 i) + (a_2 + b_2 i) = (a_1 + a_2) + (b_1 + b_2)i$$

The complex conjugate of the sum is $(a_1 + a_2) - (b_1 + b_2)i$, and the sum of the conjugates is

$$\begin{aligned}(a_1 - b_1 i) + (a_2 - b_2 i) &= (a_1 + a_2) + (-b_1 - b_2)i \\ &= (a_1 + a_2) - (b_1 + b_2)i\end{aligned}$$

Thus, the conjugate of the sum is the sum of the conjugates.

SECTION 3.6

The Fundamental Theorem of Algebra

- You should know that if f is a polynomial of degree $n > 0$, then f has exactly n zeros (roots) in the complex number system.
- You should know that if $a+bi$ is a complex zero of a polynomial f, with real coefficients, then $a-bi$ is also a complex zero of f.
- You should know the difference between a factor that is irreducible over the rationals (such as x^2-7) and a factor that is irreducible over the reals (such as x^2+9).

Solutions to Selected Exercises

3. Find all the zeros of $h(x) = x^2 - 4x + 1$ and write the polynomial as a product of linear factors.

Solution:

h has no rational zeros. By the Quadratic Formula, the zeros are

$$x = \frac{4 \pm \sqrt{16-4}}{2} = 2 \pm \sqrt{3}.$$

$$h(x) = [x-(2+\sqrt{3})][x-(2-\sqrt{3})] = (x-2-\sqrt{3})(x-2+\sqrt{3})$$

7. Find all the zeros of $f(z) = z^2 - 2z + 2$ and write the polynomial as a product of linear factors.

Solution:

f has no rational zeros. By the Quadratic Formula, the zeros are

$$z = \frac{2 \pm \sqrt{4-8}}{2} = 1 \pm i.$$

$$f(z) = [z-(1+i)][z-(1-i)] = (z-1-i)(z-1+i)$$

9. Find all the zeros of $g(x) = x^3 - 6x^2 + 13x - 10$ and write the polynomial as a product of linear factors.

Solution:

Possible rational zeros: $\pm 1, \pm 2, \pm 5, \pm 10$

$$\begin{array}{r|rrrr} 2 & 1 & -6 & 13 & -10 \\ & & 2 & -8 & 10 \\ \hline & 1 & -4 & 5 & 0 \end{array}$$

$g(x) = (x - 2)(x^2 - 4x + 5)$

$x = 2$ is a zero, and by the Quadratic Formula $x = \dfrac{4 \pm \sqrt{16 - 20}}{2} = 2 \pm i$ are also zeros.

$g(x) = (x - 2)[x - (2 + i)][x - (2 - i)] = (x - 2)(x - 2 - i)(x - 2 + i)$

15. Find all the zeros of $f(x) = 16x^3 - 20x^2 - 4x + 15$ and write the polynomial as a product of linear factors.

Solution:

Possible rational zeros: $\pm 1, \pm 3, \pm 5, \pm 15, \pm\frac{1}{2}, \pm\frac{3}{2}, \pm\frac{5}{2}, \pm\frac{15}{2}, \pm\frac{1}{4}, \pm\frac{3}{4}, \pm\frac{5}{4}, \pm\frac{15}{4}, \pm\frac{1}{8}, \pm\frac{3}{8}, \pm\frac{5}{8}, \pm\frac{15}{8}, \pm\frac{1}{16}, \pm\frac{3}{16}, \pm\frac{5}{16}, \pm\frac{15}{16}$

$$\begin{array}{r|rrrr} -\frac{3}{4} & 16 & -20 & -4 & 15 \\ & & -12 & 24 & -15 \\ \hline & 16 & -32 & 20 & 0 \end{array}$$

$$f(x) = \left(x + \frac{3}{4}\right)(16x^2 - 32x + 20) = 4\left(x + \frac{3}{4}\right)(4x^2 - 8x + 5) = (4x + 3)(4x^2 - 8x + 5)$$

$x = -\dfrac{3}{4}$ is a zero, and by the Quadratic Formula $x = \dfrac{8 \pm \sqrt{64 - 80}}{8} = 1 \pm \dfrac{1}{2}i$ are also zeros.

$$f(x) = 16\left(x + \frac{3}{4}\right)\left[x - \left(1 + \frac{1}{2}i\right)\right]\left[x - \left(1 - \frac{1}{2}i\right)\right] = (4x + 3)(2x - 2 - i)(2x - 2 + i)$$

21. Find all the zeros of $g(x) = x^4 - 4x^3 + 8x^2 - 16x + 16$ and write the polynomial as a product of linear factors.

Solution:

Possible rational zeros: ±1, ±2, ±4, ±8, ±16

$$\begin{array}{r|rrrrr} 2 & 1 & -4 & 8 & -16 & 16 \\ & & 2 & -4 & 8 & -16 \\ \hline \end{array}$$

$$\begin{array}{r|rrrrr} 2 & 1 & -2 & 4 & -8 & 0 \\ & & 2 & 0 & 8 & \\ \hline & 1 & 0 & 4 & 0 & \end{array}$$

$$g(x) = (x-2)(x-2)(x^2+4) = (x-2)^2(x+2i)(x-2i)$$

The zeros of g are 2 and $\pm 2i$.

27. Find a polynomial with integer coefficients that has the zeros 1, $5i$, and $-5i$.

Solution:

$$\begin{aligned} f(x) &= (x-1)(x-5i)(x+5i) \\ &= (x-1)(x^2+25) \\ &= x^3 - x^2 + 25x - 25 \end{aligned}$$

Note: $f(x) = a(x^3 - x^2 + 25x - 25)$, where a is any real number, has the zeros 1 and $\pm 5i$.

31. Find a polynomial with integer coefficients that has the zeros i, $-i$, $6i$, and $-6i$.

Solution:

$$\begin{aligned} f(x) &= (x-i)(x+i)(x-6i)(x+6i) \\ &= (x^2+1)(x^2+36) \\ &= x^4 + 37x^2 + 36 \end{aligned}$$

Note: $f(x) = a(x^4 + 37x^2 + 36)$, where a is any real number, has the zeros $\pm i$, and $\pm 6i$.

35. Find a polynomial with integer coefficients that has the zeros $\frac{3}{4}$, -2, and $-\frac{1}{2}+i$.

Solution:

Since $-\frac{1}{2}+i$ is a zero, so is $-\frac{1}{2}-i$.

$$\begin{aligned} f(x) &= 16\left(x-\tfrac{3}{4}\right)(x+2)\left[x-\left(-\tfrac{1}{2}+i\right)\right]\left[x-\left(-\tfrac{1}{2}-i\right)\right] \quad \text{Multiply by 16 to clear all fractions.} \\ &= 4(4x-3)(x+2)\left[x^2+x+\left(\tfrac{1}{4}+1\right)\right] \\ &= (4x^2+5x-6)(4x^2+4x+5) \\ &= 16x^4+36x^3+16x^2+x-30 \end{aligned}$$

Note: $f(x)=a(16x^4+36x^3+16x^2+x-30)$, where a is any real number, has the zeros $\frac{3}{4}$, -2, and $-\frac{1}{2}\pm i$.

39. Write $f(x)=x^4-4x^3+5x^2-2x-6$

(a) as the product of factors that are irreducible over the rationals,

(b) as the product of linear and quadratic factors that are irreducible over the reals, and

(c) in completely factored form. [*Hint:* One factor is x^2-2x-2.]

Solution:

$$\begin{array}{r|l} & x^2 - 2x + 3 \\ \hline x^2-2x-2 & x^4-4x^3+5x^2-2x-6 \\ & x^4-2x^3-2x^2 \\ \hline & -2x^3+7x^2-2x \\ & -2x^3+4x^2+4x \\ \hline & 3x^2-6x-6 \\ & 3x^2-6x-6 \\ \hline & 0 \end{array}$$

$$f(x)=(x^2-2x+3)(x^2-2x-2)$$

(a) $f(x)=(x^2-2x+3)(x^2-2x-2)$

(b) $f(x)=(x^2-2x+3)(x-1+\sqrt{3})(x-1-\sqrt{3})$

(c) $f(x)=(x-1+\sqrt{2}\,i)(x-1-\sqrt{2}\,i)(x-1+\sqrt{3})(x-1-\sqrt{3})$

Note: Use the Quadratic Formula for (b) and (c).

43. Use the zero, $r = 2i$, to find all the zeros of $f(x) = 2x^4 - x^3 + 7x^2 - 4x - 4$.

Solution:

Since $2i$ is a zero of f, so is $-2i$.

$$\begin{array}{r|rrrrr} 2i & 2 & -1 & 7 & -4 & -4 \\ & & 0+4i & -8-2i & 4-2i & 4 \\ \hline \end{array}$$

$$\begin{array}{r|rrrr} -2i & 2 & -1+4i & -1-2i & -2i & 0 \\ & & 0-4i & 0+2i & 2i & \\ \hline & 2 & -1 & -1 & 0 & \end{array}$$

$$f(x) = (x - 2i)(x + 2i)(2x^2 - x - 1) = (x - 2i)(x + 2i)(2x + 1)(x - 1)$$

The zeros of f are $\pm 2i$, $-\frac{1}{2}$, and 1.

49. Use the zero $r = (1 - \sqrt{5}\,i)/2$ to find all the zeros of $h(x) = 8x^3 - 14x^2 + 18x - 9$.

Solution:

Since $(1 - \sqrt{5}\,i)/2$ is a zero, so is $(1 + \sqrt{5}\,i)/2$.

$$\begin{array}{r|rrrr} \frac{1-\sqrt{5}\,i}{2} & 8 & -14 & 18 & -9 \\ & & 4-4\sqrt{5}\,i & -15+3\sqrt{5}\,i & 9 \\ \hline \end{array}$$

$$\begin{array}{r|rrr} \frac{1+\sqrt{5}\,i}{2} & 8 & -10-4\sqrt{5}\,i & 3+3\sqrt{5}\,i & 0 \\ & & 4+4\sqrt{5}\,i & -3-3\sqrt{5}\,i & \\ \hline & 8 & -6 & 0 & \end{array}$$

$$f(x) = \left[x - \left(\frac{1}{2} - \frac{\sqrt{5}}{2}i\right)\right]\left[x - \left(\frac{1}{2} + \frac{\sqrt{5}}{2}i\right)\right](8x - 6)$$

The zeros of f are: $\frac{3}{4}$, $\frac{1}{2} \pm \frac{\sqrt{5}}{2}i$.

53. Find a quadratic function f (with integer coefficients) that has $\pm\sqrt{b}\,i$ as zeros. Assume that b is a positive integer.

Solution:

$$f(x) = (x - \sqrt{b}\,i)(x + \sqrt{b}\,i) = x^2 + b$$

Note: $f(x) = a(x^2 + b)$, where a is any real number, has the zeros $\pm\sqrt{b}\,i$.

SECTION 3.7

Rational Functions

- You should know the following basic facts about rational functions.
  - (a) A function of the form $f(x) = P(x)/Q(x)$, $Q(x) \neq 0$, where $P(x)$ and $Q(x)$ are polynomials, is called a rational function.
  - (b) The domain of a rational function is the set of all real numbers except those which make the denominator zero.
  - (c) If $f(x) = P(x)/Q(x)$ is in reduced form, and a is a value such that $Q(a) = 0$, then the line $x = a$ is a vertical asymptote of the graph of f.
  - (d) The line $y = b$ is a horizontal asymptote of the graph of f if $f(x) \to b$ as $x \to \infty$ or $x \to -\infty$.
  - (e) If $f(x) = P(x)/Q(x) = mx + b + R(x)/Q(x)$, then the line $y = mx + b$ is a slant asymptote of the graph of f.
- Be able to graph rational functions.

Solutions to Selected Exercises

3. Match $f(x) = (x + 1)/x$ with its graph.

Solution:

$$f(x) = \frac{x+1}{x}$$

Vertical asymptote: $x = 0$
Horizontal asymptote: $y = 1$
x-intercept: $(-1,\ 0)$
Matches graph (a)

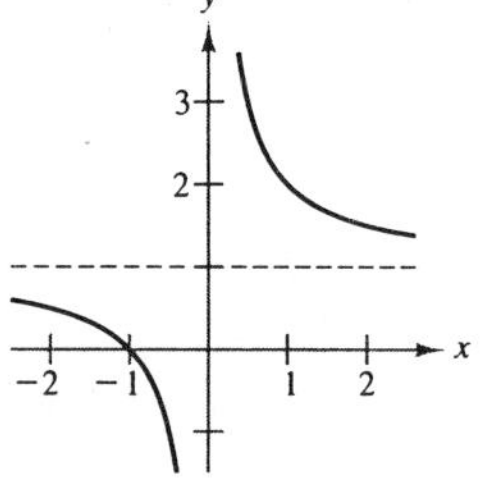

7. Match $f(x) = (x^2 + 1)/x$ with its graph.

Solution:

$$f(x) = \frac{x^2+1}{x} = x + \frac{1}{x}$$

Vertical asymptote: $x = 0$
Slant asymptote: $y = x$
No intercepts
Matches graph (h)

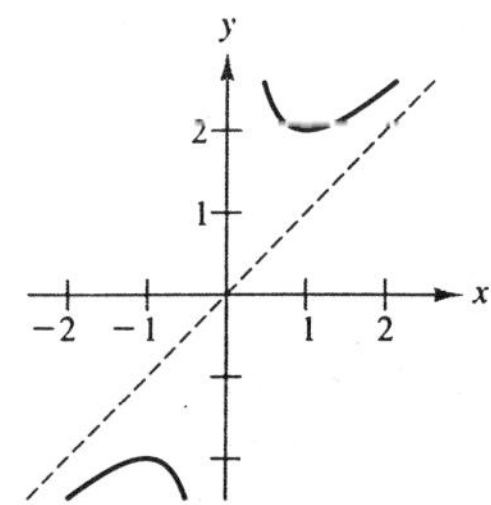

11. Find the domain of the following function and identify any horizontal, vertical, or slant asymptotes.

$$f(x) = \frac{2+x}{2-x}$$

Solution:

$$f(x) = \frac{2+x}{2-x} = \frac{x+2}{-x+2}$$

Domain: all real numbers except 2

Vertical asymptote: $x = 2$

Horizontal asymptote: $y = -1$

[degree of $p(x)$ = degree of $q(x)$]

15. Find the domain of the following function and identify any horizontal, vertical, or slant asymptotes.

$$f(x) = \frac{3x^2+1}{x^2+9}$$

Solution:

$$f(x) = \frac{3x^2+1}{x^2+9}$$

Domain: all real numbers

Horizontal asymptotes: $y = 3$

[degree of $p(x)$ = degree of $q(x)$]

19. Use the graph of $y = 1/x$ to sketch the graph of

$$y = f(x) + 1 = \frac{1}{x} + 1.$$

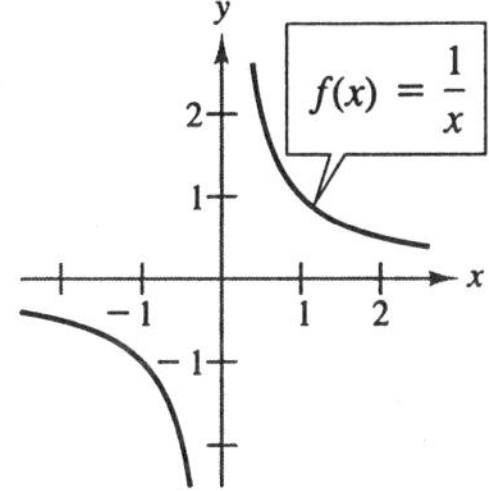

Solution:

$$y = f(x) + 1 = \frac{1}{x} + 1$$

Vertical shift one unit upward

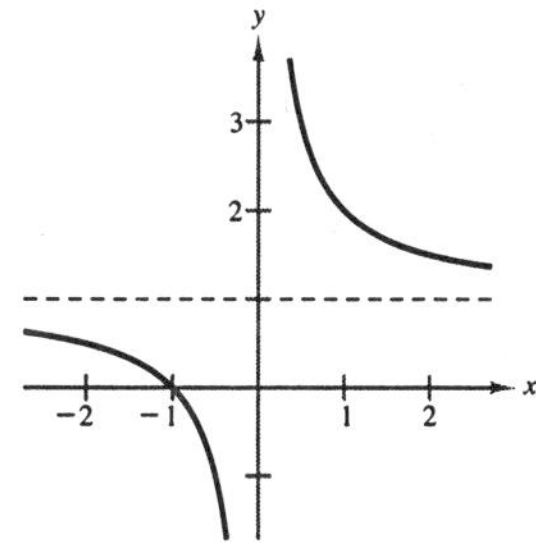

23. Use the graph of $f(x) = 4/x^2$ to sketch the graph of

$$y = f(x) - 2 = \frac{4}{x^2} - 2.$$

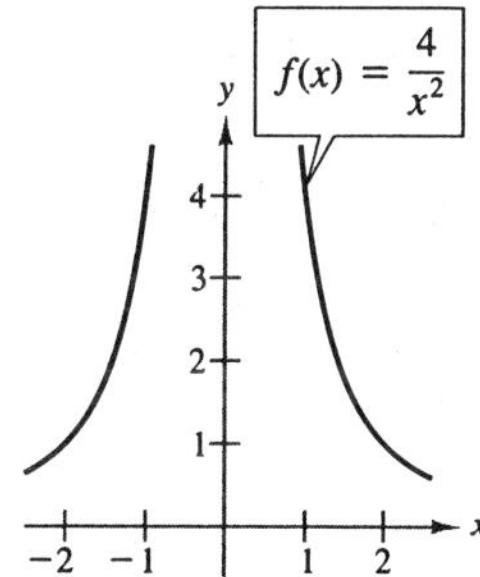

Solution:

$$y = f(x) - 2 = \frac{4}{x^2} - 2$$

Vertical shift two units downward

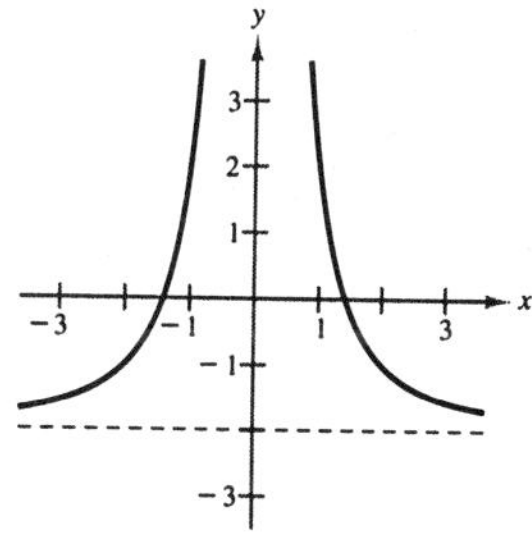

27. Use the graph of $g(x) = 8/x^3$ to sketch the graph of

$$y = g(x+2) = \frac{8}{(x+2)^3}.$$

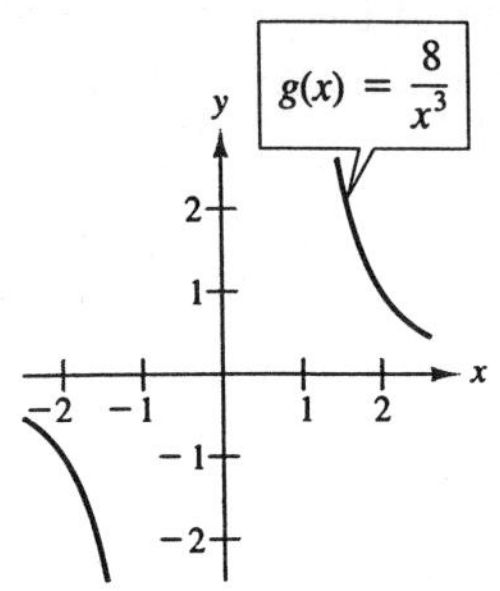

Solution:

$$y = g(x+2) = \frac{8}{(x+2)^3}$$

Horizontal shift two units to the left

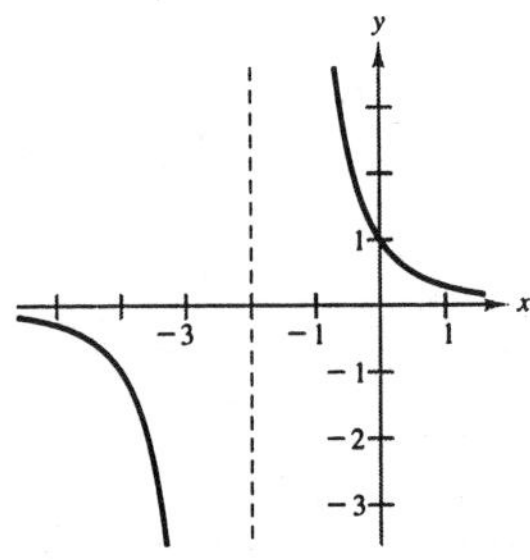

33. Sketch the graph of the following rational function. As sketching aids, check for intercepts, symmetry, vertical asymptotes, and horizontal asymptotes.

$$h(x) = \frac{-1}{x+2}$$

Solution:

Vertical asymptote: $x = -2$

Horizontal asymptote: $y = 0$

[degree $p(x) <$ degree $q(x)$]

y-intercept: $(0, -\frac{1}{2})$

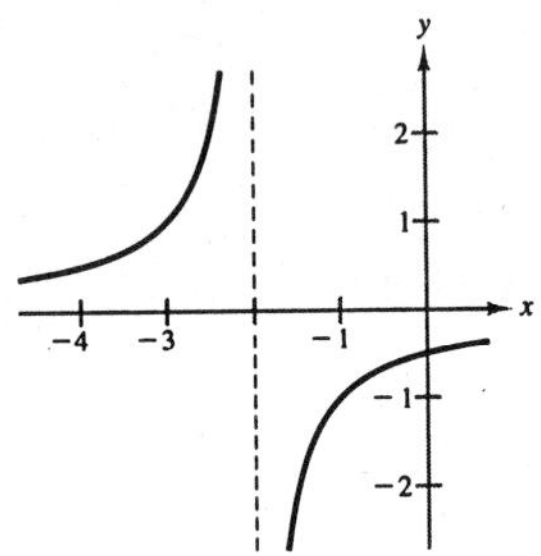

x	-4	-3	-1	0
y	$\frac{1}{2}$	1	-1	$-\frac{1}{2}$

35. Sketch the graph of the following rational function. As sketching aids, check for intercepts, symmetry, vertical asymptotes, and horizontal asymptotes.

$$f(x) = \frac{x+1}{x+2}$$

Solution:

$$f(x) = \frac{x+1}{x+2}$$

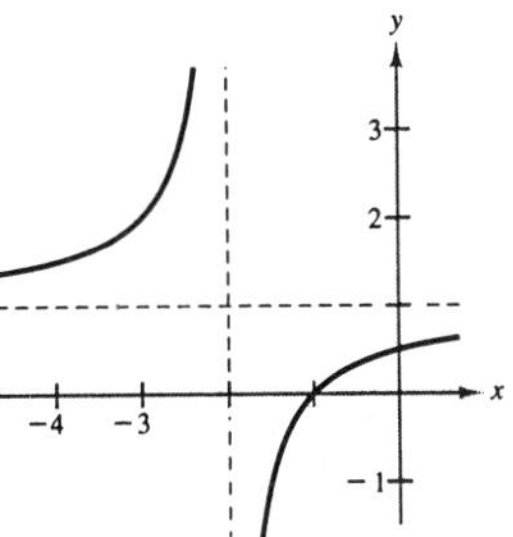

Vertical asymptote: $x = -2$

Horizontal asymptote: $y = 1$

[degree $p(x)$ = degree $q(x)$]

x-intercept: $(-1,\ 0)$

y-intercept: $(0,\ \frac{1}{2})$

x	-4	-3	-1	0	1
y	$\frac{3}{2}$	2	0	$\frac{1}{2}$	$\frac{2}{3}$

39. Sketch the graph of the following rational function. As sketching aids, check for intercepts, symmetry, vertical asymptotes, and horizontal asymptotes.

$$f(t) = \frac{3t+1}{t}$$

Solution:

$$f(t) = \frac{3t+1}{t}$$

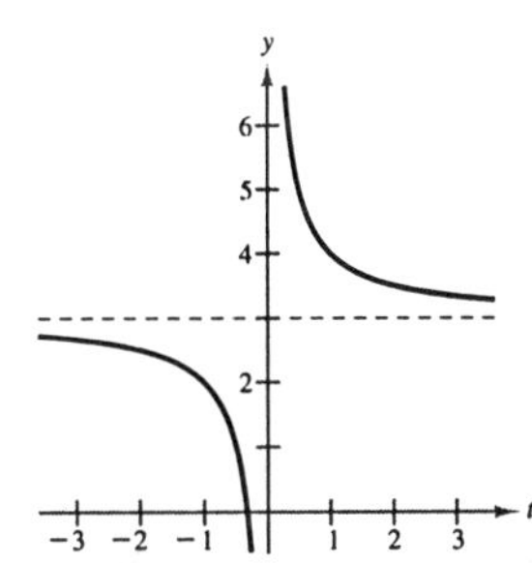

Vertical asymptote: $t = 0$

Horizontal asymptote: $y = 3$

[degree $p(x)$ = degree $q(x)$]

t-intercept: $(-\frac{1}{3},\ 0)$

x	-2	-1	1	2
y	$\frac{5}{2}$	2	4	$\frac{7}{2}$

43. Sketch the graph of the following rational function. As sketching aids, check for intercepts, symmetry, vertical asymptotes, and horizontal asymptotes.

$$C(x) = \frac{5+2x}{1+x}$$

Solution:

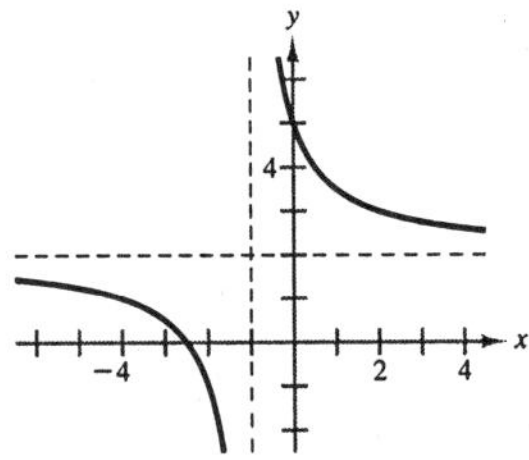

$$C(x) = \frac{5+2x}{1+x} = \frac{2x+5}{x+1}$$

Vertical asymptote: $x = -1$

Horizontal asymptote: $y = 2$

[degree $p(x)$ = degree $q(x)$]

x-intercept: $(-\frac{5}{2},\ 0)$

y-intercept: $(0,\ 5)$

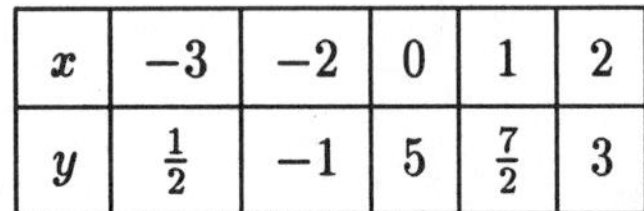

x	-3	-2	0	1	2
y	$\frac{1}{2}$	-1	5	$\frac{7}{2}$	3

47. Sketch the graph of the following rational function. As sketching aids, check for intercepts, symmetry, vertical asymptotes, and horizontal asymptotes.

$$h(x) = \frac{x^2}{x^2-9}$$

Solution:

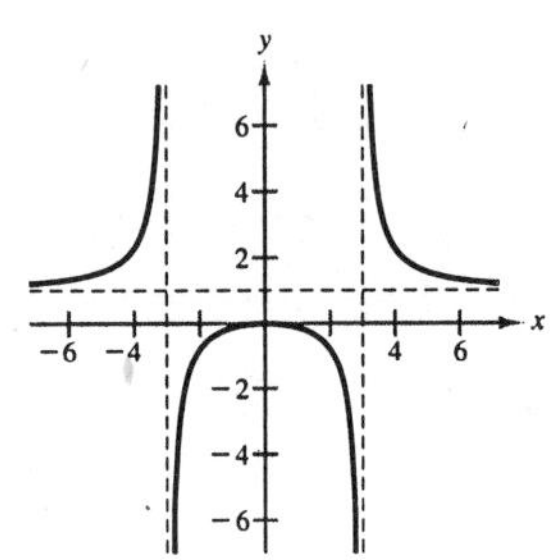

$$h(x) = \frac{x^2}{x^2-9}$$

Vertical asymptotes: $x = \pm 3$

Horizontal asymptote: $y = 1$

[degree $p(x)$ = degree $q(x)$]

Intercept: $(0,\ 0)$

y-axis symmetry

x	± 5	± 4	± 2	± 1	0
y	$\frac{25}{16}$	$\frac{16}{7}$	$-\frac{4}{5}$	$-\frac{1}{8}$	0

51. Sketch the graph of the following rational function. As sketching aids, check for intercepts, symmetry, vertical asymptotes, and horizontal asymptotes.

$$f(x) = -\frac{1}{(x-2)^2} + 3$$

Solution:

$$f(x) = \frac{-1 + 3(x-2)^2}{(x-2)^2}$$

$$= \frac{3x^2 - 12x + 11}{x^2 - 4x + 4}$$

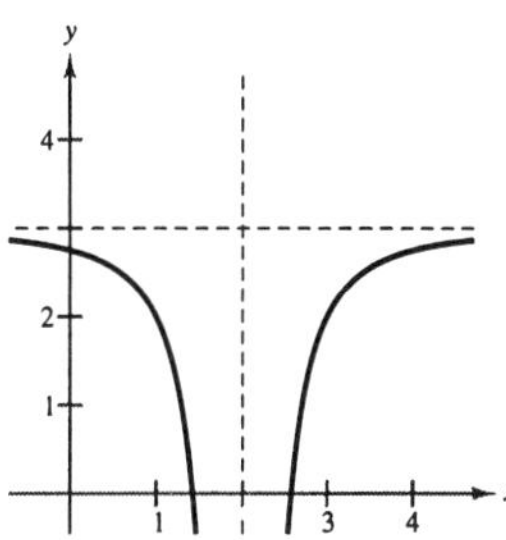

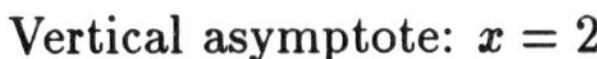

Vertical asymptote: $x = 2$

Horizontal asymptote: $y = 3$

[degree $p(x)$ = degree $q(x)$]

y-intercept: $(0, \frac{11}{4})$

x	-1	0	1	3	4
y	$\frac{26}{9}$	$\frac{11}{4}$	2	2	$\frac{11}{4}$

55. Sketch the graph of the following rational function. As sketching aids, check for intercepts, symmetry, vertical asymptotes, and slant asymptotes.

$$f(x) = \frac{2x^2 + 1}{x}$$

Solution:

$$f(x) = \frac{2x^2 + 1}{x} = 2x + \frac{1}{x}$$

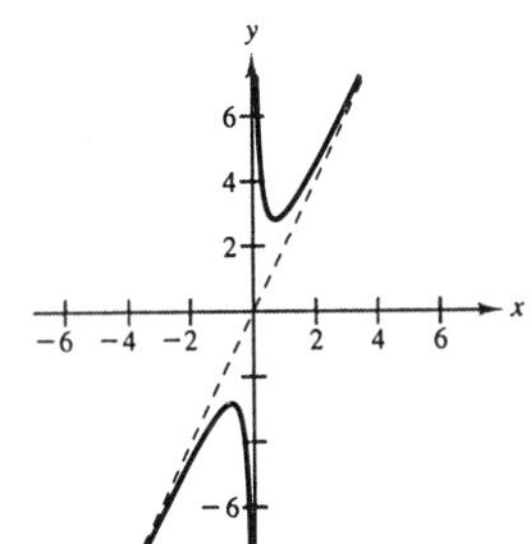

Vertical asymptote: $x = 0$

Slant asymptote: $y = 2x$

Origin symmetry

x	-3	-2	-1	$-\frac{1}{2}$	$\frac{1}{2}$	1	2	3
y	$-\frac{19}{3}$	$-\frac{9}{2}$	-3	-3	3	3	$\frac{9}{2}$	$\frac{19}{3}$

59. Sketch the graph of the following rational function. As sketching aids, check for intercepts, symmetry, vertical asymptotes, and slant asymptotes.

$$f(x) = \frac{x^3}{x^2 - 1} \quad \text{[Long Division of Polynomials]}$$

Solution:

$$f(x) = \frac{x^3}{x^2 - 1} = x + \frac{x}{x^2 - 1}$$

Vertical asymptotes: $x = \pm 1$

Slant asymptote: $y = x$

Intercept: (0, 0)

Origin symmetry

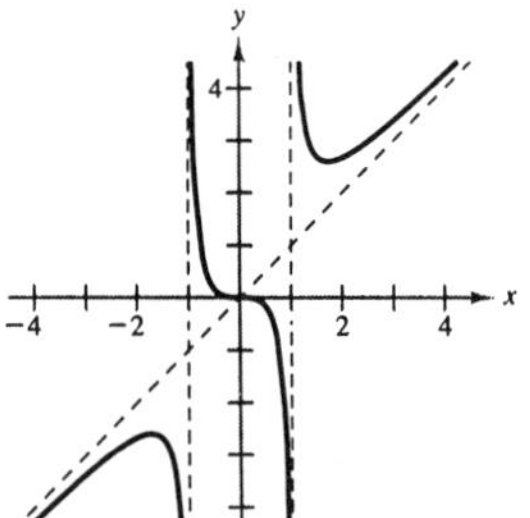

x	-3	-2	$-\frac{1}{2}$	0	$\frac{1}{2}$	2	3
y	$-\frac{27}{8}$	$-\frac{8}{3}$	$\frac{1}{6}$	0	$-\frac{1}{6}$	$\frac{8}{3}$	$\frac{27}{8}$

61. Sketch the graph of the following rational function. As sketching aids, check for intercepts, symmetry, vertical asymptotes, and slant asymptotes.

$$f(x) = \frac{x^2 - x + 1}{x - 1} \quad \text{[Long Division of Polynomials]}$$

Solution:

$$f(x) = \frac{x^2 - x + 1}{x - 1} = x + \frac{1}{x - 1}$$

Vertical asymptote: $x = 1$

Slant asymptote: $y = x$

y-intercept: $(0, -1)$

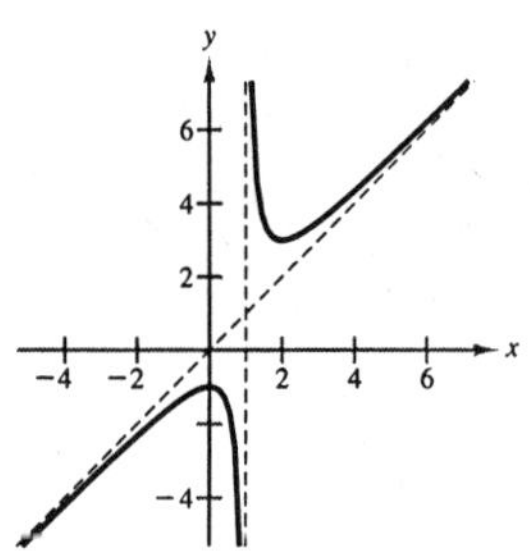

x	-2	-1	0	2	3	4
y	$-\frac{7}{3}$	$-\frac{3}{2}$	-1	3	$\frac{7}{2}$	$\frac{13}{3}$

67. The game commission introduces 50 deer into newly acquired state game lands. The population of the herd is

$$N = \frac{10(5+3t)}{1+0.04t}, \quad 0 \le t$$

where t is time in years (see figure).

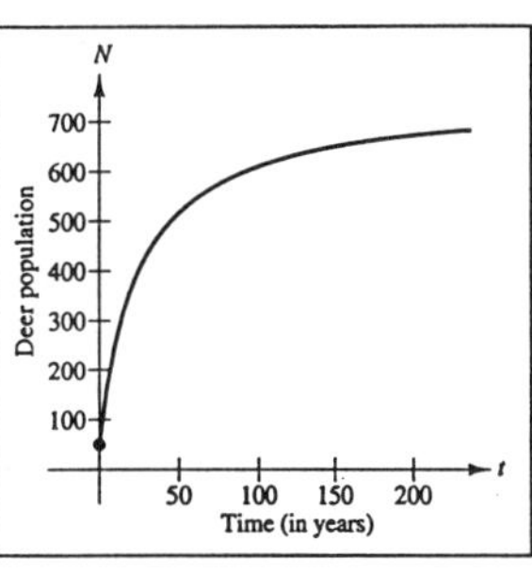

(a) Find the population when t is 5, 10, and 25.

(b) What is the limiting size of the herd as time increases?

Solution:

$$N = \frac{10(5+3t)}{1+0.04t} = \frac{30t+50}{0.04t+1}$$

(a) $N(5) \approx 167$ deer

$N(10) = 250$ deer

$N(25) = 400$ deer

(b) The herd is limited by a horizontal asymptote:

$$N = \frac{30}{0.04} = 750 \text{ deer.}$$

71. A right triangle is formed in the first quadrant by the x-axis, the y-axis, and a line segment through the point (2, 3), as shown in the figure.

(a) Show that an equation of the line segment is

$$y = \frac{3(x-a)}{2-a}, \quad 0 \le x \le a.$$

(b) Show that the area of the triangle is

$$A = \frac{-3a^2}{2(2-a)}.$$

(c) Sketch the graph of the area function of part (b), and from the graph estimate the value of a that yields a minimum area.

–CONTINUED ON NEXT PAGE–

71. -CONTINUED-

Solution:

(a) The line passes through the points $(a,\ 0)$ and $(2,\ 3)$ and has a slope of

$$m = \frac{3-0}{2-a} = \frac{3}{2-a}$$

$$y - 0 = \frac{3}{2-a}(x-a) \qquad \text{By the point slope equation}$$

$$y = \frac{3(x-a)}{2-a}$$

(b) The area of a triangle is $A = \frac{1}{2}bh$.

$b = a$

$h = y$ when $x = 0$, so $h = \dfrac{3(0-a)}{2-a} = \dfrac{-3a}{2-a}$.

$$A = \left(\frac{1}{2}a\right)\left(\frac{-3a}{2-a}\right) = \frac{-3a^2}{2(2-a)}$$

(c)
$$A = \frac{-3a^2}{2(2-a)}$$
$$= \frac{-3a^2}{-2a+4}$$
$$= \frac{3}{2}a + 3 + \frac{12}{2a-4}$$
$$= \frac{3}{2}a + 3 + \frac{6}{a-2}, \quad a > 2$$

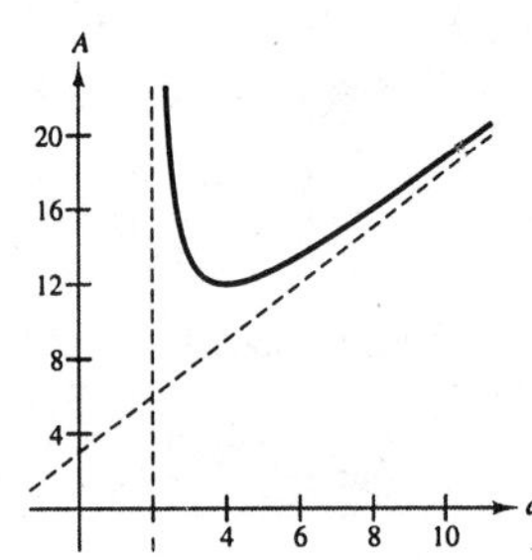

Vertical asymptote: $a = 2$

Slant asymptote: $A = \frac{3}{2}a + 3$

A is minimum when $a \approx 4$.

SECTION 3.8

Partial Fractions

- You should know how to decompose a rational function $N(x)/D(x)$ into partial fractions.
  (a) If the fraction is improper, divide to obtain
  $$\frac{N(x)}{D(x)} = p(x) + \frac{N_1(x)}{D(x)}$$
  where $p(x)$ is a polynomial.
  (b) Factor the denominator completely into linear and irreducible (over the reals) quadratic factors.
  (c) For each factor of the form $(px+q)^m$, the partial fraction decomposition includes the terms
  $$\frac{A_1}{(px+q)} + \frac{A_2}{(px+q)^2} + \cdots + \frac{A_m}{(px+q)^m}.$$
  (d) For each factor of the form $(ax^2+bx+c)^n$, the partial fraction decomposition includes the terms
  $$\frac{B_1x+C_1}{ax^2+bx+c} + \frac{B_2x+C_2}{(ax^2+bx+c)^2} + \cdots + \frac{B_nx+C_n}{(ax^2+bx+c)^n}.$$
- You should know how to determine the values of the constants in the numerators.
  (a) Set $\dfrac{N_1(x)}{D(x)} =$ partial fraction decomposition.
  (b) Multiply both sides by $D(x)$. This is called the basic equation.
  (c) For distinct linear factors, substitute the roots of the distinct linear factors into the basic equation.
  (d) For repeated linear factors, use the coefficients found in part (c) to rewrite the basic equation. Then use other values of x to solve for the remaining coefficients.
  (e) For quadratic factors, expand the basic equation, collect like terms, and then equate the coefficients of like powers.

Solutions to Selected Exercises

3. Write the partial fraction decomposition for the rational expression

$$\frac{1}{x^2+x}.$$

Solution:

Since $x^2 + x = x(x+1)$,

$$\frac{1}{x^2+x} = \frac{A}{x} + \frac{B}{x+1}$$

$$1 = A(x+1) + Bx. \qquad \text{Basic equation}$$

Let $x = 0$: $\quad 1 = A$

Let $x = -1$: $\quad 1 = -B \Rightarrow B = -1$

Thus, $\dfrac{1}{x^2+x} = \dfrac{1}{x} - \dfrac{1}{x+1}$.

7. Write the partial fraction decomposition for the rational expression

$$\frac{3}{x^2+x-2}.$$

Solution:

Since $x^2 + x - 2 = (x-1)(x+2)$,

$$\frac{3}{x^2+x-2} = \frac{A}{x-1} + \frac{B}{x+2}$$

$$3 = A(x+2) + B(x-1). \qquad \text{Basic equation}$$

Let $x = 1$: $\quad 3 = 3A$

$1 = A$

Let $x = -2$: $\quad 3 = -3B$

$-1 = B$

Thus, $\dfrac{3}{x^2+x-2} = \dfrac{1}{x-1} - \dfrac{1}{x+2}$.

11. Write the partial fraction decomposition for the rational expression $\dfrac{x^2+12x+12}{x^3-4x}$.

Solution:

Since $x^3 - 4x = x(x+2)(x-2)$,

$$\frac{x^2+12x+12}{x^3-4x} = \frac{A}{x} + \frac{B}{x+2} + \frac{C}{x-2}$$

$$x^2+12x+12 = A(x+2)(x-2) + Bx(x-2) + Cx(x+2). \quad \text{Basic equation}$$

Let $x = 0$: $\quad 12 = -4A$

$-3 = A$

Let $x = -2$: $\quad -8 = 8B$

$-1 = B$

Let $x = 2$: $\quad 40 = 8C$

$5 = C$

Thus, $\dfrac{x^2+12x+12}{x^3-4x} = -\dfrac{3}{x} - \dfrac{1}{x+2} + \dfrac{5}{x-2}$.

13. Write the partial fraction decomposition for the rational expression $\dfrac{4x^2+2x-1}{x^2(x+1)}$.

Solution:

$$\frac{4x^2+2x-1}{x^2(x+1)} = \frac{A}{x} + \frac{B}{x^2} + \frac{C}{x+1}$$

$$4x^2+2x-1 = Ax(x+1) + B(x+1) + Cx^2 \quad \text{Basic equation}$$

Let $x = 0$: $\quad -1 = B$

Let $x = -1$: $\quad 1 = C$

Let $x = 1$: $\quad 5 = 2A + 2B + C$

$5 = 2A - 2 + 1$

$6 = 2A$

$3 = A$

Thus, $\dfrac{4x^2+2x-1}{x^2(x+1)} = \dfrac{3}{x} - \dfrac{1}{x^2} + \dfrac{1}{x+1}$.

Note: $x^2 = (x-0)^2$ and is a linear factor squared. It is not an irreducible quadratic factor. Do not write $(Bx+C)/x^2$ in the partial fraction decomposition.

19. Write the partial fraction decomposition for the rational expression

$$\frac{x^2-1}{x(x^2+1)}.$$

Solution:

$$\frac{x^2-1}{x(x^2+1)} = \frac{A}{x} + \frac{Bx+C}{x^2+1}$$

$$x^2-1 = A(x^2+1)+(Bx+C)x \qquad \text{Basic equation}$$

Let $x=0$: $\quad -1 = A$

$$x^2-1 = Ax^2+A+Bx^2+Cx = -x^2-1+Bx^2+Cx = x^2(B-1)+Cx-1$$

Equating coefficients of like powers,

$1 = B-1$

$2 = B$ and $0 = C$

Thus, $\dfrac{x^2-1}{x(x^2+1)} = -\dfrac{1}{x} + \dfrac{2x}{x^2+1}$.

23. Write the partial fraction decomposition for the rational expression $\dfrac{x}{16x^4-1}$.

Solution:

Since $16x^4-1 = (4x^2+1)(2x+1)(2x-1)$,

$$\frac{x}{16x^4-1} = \frac{A}{2x+1} + \frac{B}{2x-1} + \frac{Cx+D}{4x^2+1}$$

$$x = A(2x-1)(4x^2+1) + B(2x+1)(4x^2+1) + (Cx+D)(2x+1)(2x-1). \qquad \text{Basic equation}$$

Let $x=-\frac{1}{2}$: $\quad -\frac{1}{2} = -4A$

$\frac{1}{8} = A$

Let $x=\frac{1}{2}$: $\quad \frac{1}{2} = 4B$

$\frac{1}{8} = B$

Let $x=0$: $\quad 0 = -A+B-D$

$0 = -\frac{1}{8}+\frac{1}{8}-D$

$0 = D$

–CONTINUED ON NEXT PAGE–

23. –CONTINUED–

Let $x = 1$: $\quad 1 = 5A + 15B + 3C + 3D$

$$1 = \tfrac{5}{8} + \tfrac{15}{8} + 3C + 0$$

$$1 = \tfrac{20}{8} + 3C$$

$$-\tfrac{3}{2} = 3C$$

$$-\tfrac{1}{2} = C$$

Thus, $\dfrac{x}{16x^4 - 1} = \dfrac{1/8}{2x+1} + \dfrac{1/8}{2x-1} - \dfrac{x/2}{4x^2+1} = \dfrac{1}{8}\left[\dfrac{1}{2x+1} + \dfrac{1}{2x-1} - \dfrac{4x}{4x^2+1}\right].$

27. Write the partial fraction decomposition for the rational expression

$$\frac{x^2+5}{(x+1)(x^2-2x+3)}.$$

Solution:

$$\frac{x^2+5}{(x+1)(x^2-2x+3)} = \frac{A}{x+1} + \frac{Bx+C}{x^2-2x+3}$$

$$x^2 + 5 = A(x^2 - 2x + 3) + (Bx + C)(x + 1) \qquad \text{Basic equation}$$

Let $x = -1$: $\quad 6 = 6A$

$$1 = A$$

$$x^2 + 5 = x^2 - 2x + 3 + Bx^2 + Bx + Cx + C$$

$$= x^2(1 + B) + x(-2 + B + C) + (3 + C)$$

Equating coefficients of like powers:

$1 = 1 + B,\quad 0 = -2 + B + C,\quad$ and $\quad 5 = 3 + C$

$0 = B \qquad 0 = -2 + 0 + C \qquad 2 = C$

$2 = C$

Thus, $\dfrac{x^2+5}{(x+1)(x^2-2x+3)} = \dfrac{1}{x+1} + \dfrac{2}{x^2-2x+3}.$

31. Write the partial fraction decomposition for the rational expression $\dfrac{x^4}{(x-1)^3}$.

Solution:

$$\begin{aligned}\frac{x^4}{(x-1)^3} &= \frac{x^4}{x^3-3x^2+3x-1}\\ &= x+3+\frac{6x^2-8x+3}{(x-1)^3} \qquad \text{By long division of polynomials}\end{aligned}$$

$$\begin{aligned}\frac{6x^2-8x+3}{(x-1)^3} &= \frac{A}{x-1}+\frac{B}{(x-1)^2}+\frac{C}{(x-1)^3}\\ 6x^2-8x+3 &= A(x-1)^2+B(x-1)+C \qquad \text{Basic equation}\end{aligned}$$

Let $x=1$: $\quad 1 = C$

$$6x^2-8x+3 = Ax^2-2Ax+A+Bx-B+1$$

$$6x^2-8x+3 = Ax^2+(-2A+B)x+(A-B+1)$$

Equating coefficients of like powers:

$$\begin{aligned}6 = A \quad -8 &= -2A+B \quad \text{and} \quad 3 = A-B+1\\ -8 &= -12+B \qquad\qquad 3 = 6-B+1\\ 4 &= B \qquad\qquad\qquad\ \ 4 = B\end{aligned}$$

Thus, $\dfrac{x^4}{(x-1)^3} = x+3+\dfrac{6}{x-1}+\dfrac{4}{(x-1)^2}+\dfrac{1}{(x-1)^3}$.

35. Write the partial fraction decomposition for the rational expression $1/[y(L-y)]$, L is a constant.

Solution:

$$\begin{aligned}\frac{1}{y(L-y)} &= \frac{A}{y}+\frac{B}{L-y}\\ 1 &= A(L-y)+By \qquad \text{Basic equation}\end{aligned}$$

Let $y=0$: $\quad 1 = LA$

$\quad 1/L = A$

Let $y=L$: $\quad 1 = LB$

$\quad 1/L = B$

Thus, $\dfrac{1}{y(L-y)} = \dfrac{1/L}{y}+\dfrac{1/L}{L-y} = \dfrac{1}{L}\left(\dfrac{1}{y}+\dfrac{1}{L-y}\right)$.

REVIEW EXERCISES FOR CHAPTER 3

Solutions to Selected Exercises

3. Sketch the graph of the quadratic function $f(x) = \frac{1}{3}(x^2 + 5x - 4)$. Identify the vertex and the intercepts.

Solution:

x-intercepts: Let $y = 0$.

$$0 = \frac{1}{3}(x^2 + 5x - 4)$$

$$0 = x^2 + 5x - 4 \qquad \text{Multiply both sides by 3.}$$

$$x = \frac{-5 \pm \sqrt{5^2 - 4(1)(-4)}}{2(1)} \qquad \text{Use the Quadratic Formula.}$$

$$= \frac{-5 \pm \sqrt{41}}{2}$$

$\left(\frac{-5 + \sqrt{41}}{2}, 0\right)$, $\left(\frac{-5 - \sqrt{41}}{2}, 0\right)$ are the x-intercepts.

y-intercept: Let $x = 0$.

$$y = \frac{1}{3}(0^2 + 5(0) - 4) = -\frac{4}{3}$$

$\left(0, -\frac{4}{3}\right)$ is the y-intercept.

Vertex: Complete the square.

$$f(x) = \frac{1}{3}(x^2 + 5x - 4)$$

$$= \frac{1}{3}\left(x^2 + 5x + \frac{25}{4} - \frac{25}{4} - 4\right)$$

$$= \frac{1}{3}\left[\left(x^2 + 5x + \frac{25}{4}\right) - \left(\frac{25}{4} + \frac{16}{4}\right)\right]$$

$$= \frac{1}{3}\left(x + \frac{5}{2}\right)^2 - \frac{1}{3}\left(\frac{41}{4}\right) = \frac{1}{3}\left(x + \frac{5}{2}\right)^2 - \frac{41}{12}$$

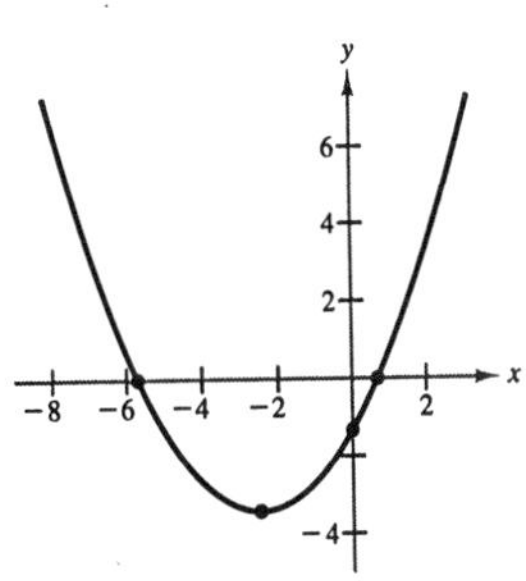

$\left(-\frac{5}{2}, -\frac{41}{12}\right)$ is the vertex.

9. Find the maximum or minimum value of $g(x) = x^2 - 2x$.

Solution:

$$\begin{aligned} g(x) &= x^2 - 2x \qquad \text{has a minimum since } a > 0. \\ &= x^2 - 2x + 1 - 1 \\ &= (x-1)^2 - 1 \end{aligned}$$

The minimum occurs at the vertex $(1, -1)$.

13. Find the maximum or minimum value of $f(t) = -2t^2 + 4t + 1$.

Solution:

$$\begin{aligned} f(t) &= -2t^2 + 4t + 1 \qquad \text{has a maximum since } a < 0. \\ &= -2(t^2 - 2t - \tfrac{1}{2}) \\ &= -2(t^2 - 2t + 1 - 1 - \tfrac{1}{2}) \\ &= -2[(t-1)^2 - \tfrac{3}{2}] \\ &= -2(t-1)^2 + 3 \end{aligned}$$

The maximum occurs at the vertex $(1, 3)$.

15. A rectangle is inscribed in the region bounded by the x-axis, the y-axis, and the graph of $x + 2y - 6 = 0$ (see figure). Find the coordinates (x, y) that yield a maximum area for the rectangle.

Solution:

$$x + 2y - 6 = 0 \quad \Rightarrow \quad y = \frac{6-x}{2}$$

The area of the rectangle is

$$\begin{aligned} A = xy = x\left(\frac{6-x}{2}\right) &= -\frac{1}{2}x^2 + 3x \\ &= -\frac{1}{2}(x^2 - 6x + 9 - 9) \\ &= -\frac{1}{2}[(x-3)^2 - 9] \\ &= -\frac{1}{2}(x-3)^2 + \frac{9}{2} \end{aligned}$$

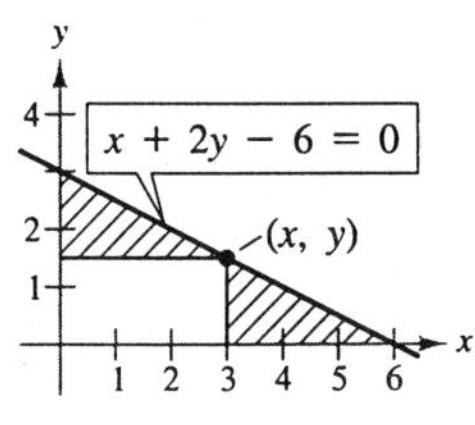

The maximum value of the area is $9/2$ square units and this occurs when $x = 3$ and $y = (6-3)/2 = 3/2$, which corresponds to the coordinates $(3, 3/2)$.

17. Find the number of units x that produce a maximum revenue R, where $R = 900x - 0.1x^2$.

Solution:

$$R = 900x - 0.1x^2$$
$$= -0.1(x^2 - 9000x + 4500^2) + 2{,}025{,}000$$
$$= 2{,}025{,}000 - 0.1(x - 4500)^2$$
$$x = 4500 \text{ units}$$

The revenue is maximized when 4500 units are produced.

21. Determine the right-hand and left-hand behavior of the graph of the polynomial function $f(x) = -x^2 + 6x + 9$.

Solution:

$$f(x) = -x^2 + 6x + 9$$

Since the degree is even and the leading coefficient is negative, the graph falls to the left and to the right.

27. Sketch the graph of $g(x) = x^4 - x^3 - 2x^2$.

Solution:

$$g(x) = x^4 - x^3 - 2x^2$$
$$= x^2(x^2 - x - 2)$$
$$= x^2(x + 1)(x - 2)$$

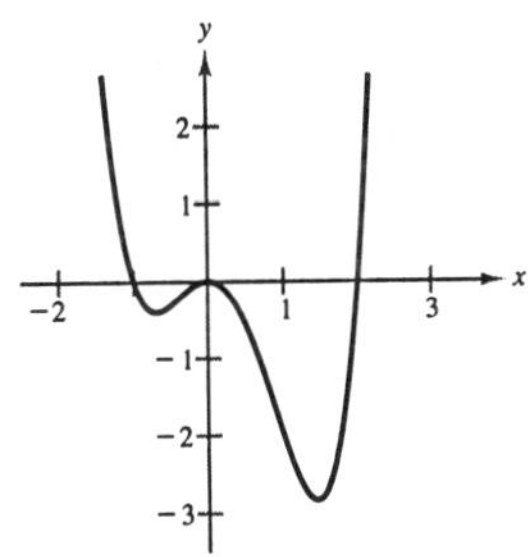

The zeros of g of 0, -1, and 2, so the graph of g crosses the x-axis at these points. Since the degree of g is even and the leading coefficient is positive, the graph moves up to the right and left.

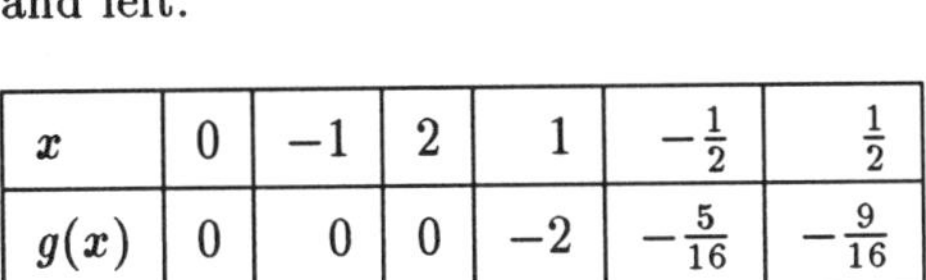

x	0	-1	2	1	$-\frac{1}{2}$	$\frac{1}{2}$
$g(x)$	0	0	0	-2	$-\frac{5}{16}$	$-\frac{9}{16}$

31. Sketch the graph of $f(x) = x(x+3)^2$.

Solution:

$$f(x) = x(x+3)^2 = x^3 + 6x^2 + 9x$$

The zeros of f and 0 and -3, so the graph of f crosses the x-axis at these points. Since the degree of f is odd and the leading coefficient is positive, the graph falls to the left and rises to the right.

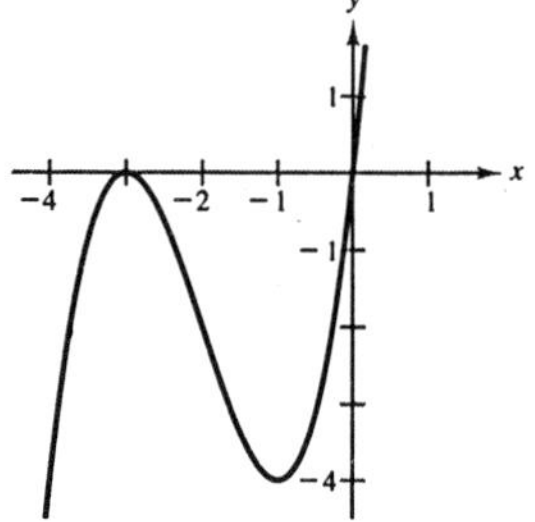

x	-4	-3	-2	-1	0	1
$f(x)$	-4	0	-2	-4	0	16

35. Perform the indicated division.

$$\frac{x^4 + x^3 - x^2 + 2x}{x^2 + 2x}$$

Solution:

$$\begin{array}{r l}
 & x^2 - x + 1 \\
x^2 + 2x \,) & x^4 + x^3 - x^2 + 2x \\
 & -(x^4 + 2x^3) \\
 & \quad -x^3 - x^2 \\
 & \quad -(-x^3 - 2x^2) \\
 & \qquad x^2 + 2x \\
 & \qquad -(x^2 + 2x) \\
 & \qquad\qquad 0
\end{array}$$

Thus, $\dfrac{x^4 + x^3 - x^2 + 2x}{x^2 + 2x} = x^2 - x + 1.$

41. Perform the indicated operations and write the result in standard form.

$$\left(\frac{\sqrt{2}}{2} - \frac{\sqrt{2}}{2}i\right) - \left(\frac{\sqrt{2}}{2} + \frac{\sqrt{2}}{2}i\right)$$

Solution:

$$\left(\frac{\sqrt{2}}{2} - \frac{\sqrt{2}}{2}i\right) - \left(\frac{\sqrt{2}}{2} + \frac{\sqrt{2}}{2}i\right) = \left(\frac{\sqrt{2}}{2} - \frac{\sqrt{2}}{2}\right) + \left(-\frac{\sqrt{2}}{2} - \frac{\sqrt{2}}{2}\right)i = 0 - \sqrt{2}\,i = -\sqrt{2}\,i$$

45. Perform the indicated operations and write the result in standard form.

$(10 - 8i)(2 - 3i)$

Solution:

$$(10 - 8i)(2 - 3i) = 20 - 30i - 16i + 24i^2 = -4 - 46i$$

49. Perform the indicated operations and write the result in standard form.

$$\frac{4}{-3i}$$

Solution:

$$\frac{4}{-3i} = \frac{4}{-3i} \bullet \frac{3i}{3i} = \frac{12i}{9} = \frac{4}{3}i$$

53. Use synthetic division to perform the indicated division.

$$\frac{6x^4 - 4x^3 - 27x^2 + 18x}{x - (2/3)}$$

Solution:

$\frac{2}{3}$	6	−4	−27	18	0
		4	0	−18	0
	6	0	−27	0	0

Thus,

$$\frac{6x^4 - 4x^3 - 27x^2 + 18x}{x - (2/3)} = 6x^3 - 27x.$$

59. Use synthetic division to determine whether the given values of x are zeros of

$f(x) = 2x^3 + 7x^2 - 18x - 30$.

(a) $x = 1$ (b) $x = \frac{5}{2}$ (c) $x = -3 + \sqrt{3}$ (d) $x = 0$

Solution:

(a)

1	2	7	−18	−30
		2	9	−9
	2	9	−9	−39

$x = 1$ is *not* a zero of f.

(b)

$\frac{5}{2}$	2	7	−18	−30
		5	30	30
	2	12	12	0

$x = \frac{5}{2}$ *is* a zero of f.

(c)

$-3+\sqrt{3}$	2	7	−18	−30
		$-6+2\sqrt{3}$	$3-5\sqrt{3}$	30
	2	$1+2\sqrt{3}$	$-15-5\sqrt{3}$	0

$x = -3 + \sqrt{3}$ *is* a zero of f.

(d)

0	2	7	−18	−30
		0	0	0
	2	7	−18	−30

$x = 0$ is *not* a zero of f.

63. Use synthetic division to find the specified value of $f(x) = x^4 + 10x^3 - 24x^2 + 20x + 44$.

(a) $f(-3)$ (b) $f(\sqrt{2}\,i)$

Solution:

(a)

-3	1	10	-24	20	44
		-3	-21	135	-465
	1	7	-45	155	-421

Thus, $f(-3) = -421$.

(b)

$\sqrt{2}\,i$	1	10	-24	20	44
		$0 + \sqrt{2}\,i$	$-2 + 10\sqrt{2}\,i$	$-20 - 26\sqrt{2}\,i$	52
	1	$10 + \sqrt{2}\,i$	$-26 + 10\sqrt{2}\,i$	$-26\sqrt{2}\,i$	96

Thus, $f(\sqrt{2}\,i) = 96$.

65. Find a polynomial with integer coefficients that has the zeros -1, -1, $\frac{1}{3}$, and $-\frac{1}{2}$.

Solution:

$$f(x) = 6(x+1)^2\left(x - \tfrac{1}{3}\right)\left(x + \tfrac{1}{2}\right)$$ Multiply by 6 to clear the fractions.

$$= (x+1)^2 3\left(x - \tfrac{1}{3}\right) 2\left(x + \tfrac{1}{2}\right)$$

$$= (x^2 + 2x + 1)(3x - 1)(2x + 1)$$

$$= (x^2 + 2x + 1)(6x^2 + x - 1)$$

$$= 6x^4 + 13x^3 + 7x^2 - x - 1$$

71. Find all the zeros of $f(x) = 6x^3 - 5x^2 + 24x - 20$.

Solution:

Possible rational zeros:

$\pm1,\ \pm2,\ \pm4,\ \pm5,\ \pm10,\ \pm20,\ \pm\frac{1}{2},\ \pm\frac{5}{2},\ \pm\frac{1}{3},\ \pm\frac{2}{3},\ \pm\frac{4}{3},\ \pm\frac{5}{3},\ \pm\frac{10}{3},\ \pm\frac{20}{3},\pm\frac{1}{6},\ \pm\frac{5}{6}$

$$\begin{array}{r|rrrr} \frac{5}{6} & 6 & -5 & 24 & -20 \\ & & 5 & 0 & 20 \\ \hline & 6 & 0 & 24 & 0 \end{array}$$

$$f(x) = \left(x - \tfrac{5}{6}\right)(6x^2 + 24) = 6\left(x - \tfrac{5}{6}\right)(x^2 + 4) = 6\left(x - \tfrac{5}{6}\right)(x + 2i)(x - 2i)$$

The zeros of f are $\frac{5}{6}$ and $\pm 2i$. This problem can also be solved by factoring by grouping.

$$\begin{aligned} f(x) &= 6x^3 - 5x^2 + 24x - 20 \\ &= x^2(6x - 5) + 4(6x - 5) = (6x - 5)(x^2 + 4) = (6x - 5)(x + 2i)(x - 2i) \end{aligned}$$

The zeros of f are $\frac{5}{6}$ and $\pm 2i$.

79. Sketch the graph of the rational function. As sketching aids, check for intercept, symmetry, vertical asymptotes, horizontal asymptotes, and slant asymptotes.

$$P(x) = \frac{x^2}{x^2 + 1}$$

Solution:

$$P(x) = \frac{x^2}{x^2 + 1}$$

Intercept: $(0, 0)$

y-axis symmetry

Horizontal asymptote: $y = 1$ (degree $p(x)$ = degree $q(x)$)

x	±2	±1	0
$P(x)$	$\frac{4}{5}$	$\frac{1}{2}$	0

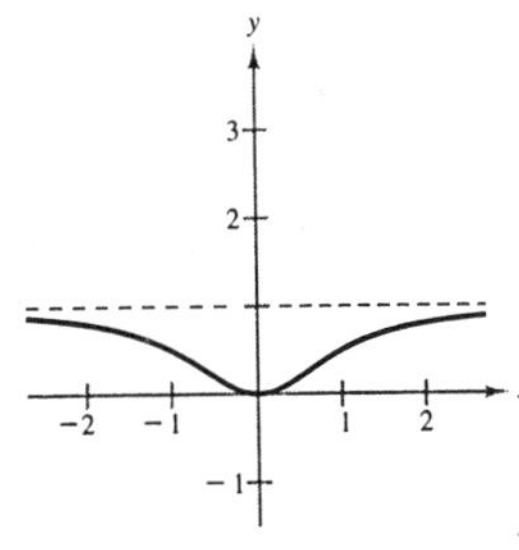

83. Sketch the graph of the rational function. As sketching aids, check for intercept, symmetry, vertical asymptotes, horizontal asymptotes, and slant asymptotes.

$$y = \frac{2x^2}{x^2 - 4}$$

Solution:

$$y = \frac{2x^2}{x^2 - 4}$$

Intercept: (0, 0)
y-axis symmetry
Vertical asymptotes: $x = 2$ and $x = -2$
Horizontal asymptote: $y = 2$ (degree $p(x)$ = degree $q(x)$)

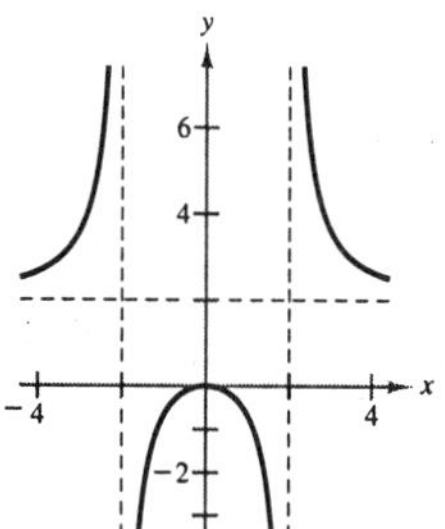

x	± 4	± 3	± 1	0
y	$\frac{8}{3}$	$\frac{18}{5}$	$-\frac{2}{3}$	0

85. The cost in millions of dollars for the federal government to seize $p\%$ of an illegal drug as it enters the country is given by

$$C = \frac{528p}{100 - p}, \quad 0 \le p < 100.$$

(a) Find the cost of seizing 25%.
(b) Find the cost of seizing 50%.
(c) Find the cost of seizing 75%.
(d) According to this model, would it be possible to seize 100% of the drug?

Solution:

(a) When $p = 25$, $C = \dfrac{528(25)}{100 - 25} = \$176$ million.

(b) When $p = 50$, $C = \dfrac{528(50)}{100 - 50} = \$528$ million.

(c) When $p = 75$, $C = \dfrac{528(75)}{100 - 75} = \$1584$ million.

(d) As $p \to 100$, $C \to \infty$. This tells us that it is not possible to seize 100% of an illegal drug as it enters the country.

89. Write the partial fraction decomposition for $\dfrac{4-x}{x^2+6x+8}$.

Solution:

$$\frac{4-x}{x^2+6x+8} = \frac{4-x}{(x+2)(x+4)}$$

$$\frac{4-x}{(x+2)(x+4)} = \frac{A}{x+2} + \frac{B}{x+4}$$

$$4-x = A(x+4) + B(x+2)$$

Let $x = -2$: $6 = 2A \Rightarrow A = 3$

Let $x = -4$: $8 = -2B \Rightarrow B = -4$

$$\frac{4-x}{x^2+6x+8} = \frac{3}{x+2} - \frac{4}{x+4}$$

91. Write the partial fraction decomposition for $\dfrac{x^2+2x}{x^3-x^2+x-1}$.

Solution:

$$\frac{x^2+2x}{x^3-x^2+x-1} = \frac{x^2+2x}{(x-1)(x^2+1)}$$

Factor the denominator by grouping.

$$\frac{x^2+2x}{(x-1)(x^2+1)} = \frac{A}{x-1} + \frac{Bx+C}{x^2+1}$$

$$x^2+2x = A(x^2+1) + (Bx+C)(x-1)$$

Let $x = 1$: $3 = 2A$, $A = 3/2$

Let $x = 0$: $0 = A - C$, $C = 3/2$

Let $x = 2$: $8 = 5A + 2B + C$

$$8 = (15/2) + 2B + (3/2),\ B = -1/2$$

$$\frac{3/2}{x-1} + \frac{-(1/2)x+3/2}{x^2+1} = \frac{1}{2}\left(\frac{3}{x-1} - \frac{x-3}{x^2+1}\right)$$

Practice Test for Chapter 3

1. Sketch the graph of $f(x) = x^2 - 9$. Identify the vertex and intercepts.

2. Sketch the graph of $g(x) = 3x^2 - 5x - 28$. Identify the vertex and intercepts.

3. Find the quadratic function that has a vertex at $(-2,\ 1)$ and passes through the point $(3,\ 4)$.

4. Find all the real zeros of $f(x) = x^4 - 5x^2 + 4$.

5. Find a polynomial that has the zeros -1, $\sqrt{3}$, and $-\sqrt{3}$.

6. Sketch the graph of $f(x) = x^3 - 9x$.

7. Use long division to divide $6x^3 - 5x + 1$ by $x^2 - 2$.

8. Use synthetic division to divide $x^4 + 3x^3 - 9x^2 + 1$ by $x + 3$.

9. Use synthetic division to find $f(-2)$ for $f(x) = x^6 - 3x^3 + x - 5$.

10. Find all real zeros of $p(x) = x^3 - 2x^2 - 5x + 6$.

11. Find all real zeros of $x^3 + x^2 - 5x - 6 = 0$.

12. List all the possible rational zeros of $p(x) = 4x^3 - 7x^2 + x - 20$.

13. Write $5i^3 - (\sqrt{-9})^2$ in standard form.

14. Multiply $(6 + 5i)(-2 + 9i)$ and write the result in standard form.

15. Divide $\dfrac{3 + 2i}{4 - 7i}$ and write the result in standard form.

16. Find all the zeros of $f(x) = x^2 - 3x + 5$.

17. Find all the zeros of $p(x) = x^5 + x^3 - 8x^2 - 8$.

18. Find a polynomial with integer coefficients that has the zeros 0, 4, $2+i$.

19. Find any horizontal or vertical asymptotes of $h(x) = \dfrac{x^2+1}{x^2-16}$.

20. Graph $f(x) = \dfrac{x+3}{x-2}$.

21. Graph $g(x) = \dfrac{x}{x^2-1}$.

22. Find any slant asymptotes of $h(x) = \dfrac{x^2+6x-7}{x+3}$.

23. Write the partial fraction decomposition for $\dfrac{8x-2}{x^2-2x-8}$.

24. Write the partial fraction decomposition for $\dfrac{3x^2+15}{x^3+3x}$.

25. Write the partial fraction decomposition for $\dfrac{2x^3+8x^2+9x-1}{x^2+4x+4}$.

CHAPTER 4

Exponential and Logarithmic Functions

Section 4.1 Exponential Functions **179**

Section 4.2 Logarithmic Functions **185**

Section 4.3 Properties of Logarithms **189**

Section 4.4 Solving Exponential and Logarithmic Equations **193**

Section 4.5 Exponential and Logarithmic Applications **197**

Review Exercises **205**

Practice Test **210**

SECTION 4.1

Exponential Functions

- You should know that a function of the form $y = a^x$, where $a > 0$, $a \neq 1$, is called an exponential function with base a.
- You should be able to graph exponential functions.
- You should know some properties of exponential functions where $a > 0$ and $a \neq 1$.
  (a) If $a^x = a^y$, then $x = y$.
  (b) If $a^x = b^x$ and $x \neq 0$, then $a = b$.
- You should know formulas for compound interest.
  (a) For n compoundings per year: $A = P\left(1 + \frac{r}{n}\right)^{nt}$.
  (b) For continuous compoundings: $A = Pe^{rt}$.

Solutions to Selected Exercises

3. Use a calculator to evaluate $1000(1.06)^{-5}$. Round your answer to three decimal places.

Solution:

$$1000(1.06)^{-5} \approx 747.258$$

1.06 $\boxed{y^x}$ 5 $\boxed{+/-}$ $\boxed{\times}$ 1000 $\boxed{=}$

7. Use a calculator to evaluate $8^{2\pi}$. Round your answer to three decimal places.

Solution:

$$8^{2\pi} \approx 472{,}369.379$$

8 $\boxed{y^x}$ $\boxed{(}$ 2 $\boxed{\times}$ π $\boxed{)}$ $\boxed{=}$

11. Use a calculator to evaluate e^2. Round your answer to three decimal places.

Solution:

$e^2 \approx 7.389$

2 $\boxed{e^x}$ or $\boxed{\text{INV}}$ $\boxed{\ln x}$

15. Match $f(x) = 3^x$ with its graph.

Solution:

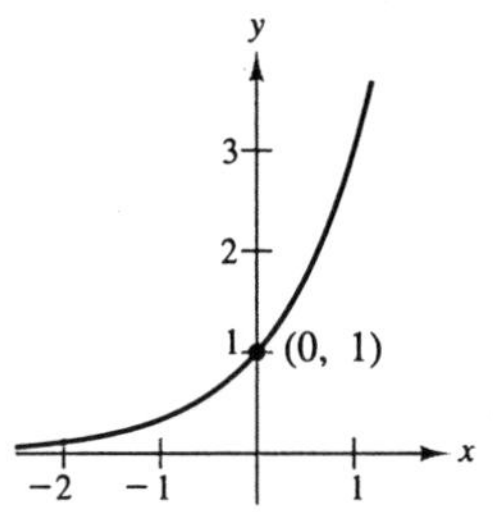

$f(x) = 3^x$

y-intercept: (0, 1)

3^x increases as x increases

Matches graph (g)

19. Match $f(x) = 3^x - 4$ with its graph.

Solution:

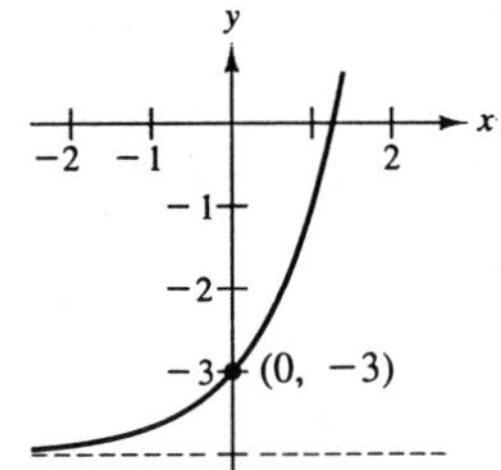

$f(x) = 3^x - 4$

y-intercept: (0, −3)

$3^x - 4$ increases as x increases

Matches graph (d)

23. Sketch the graph of $g(x) = 5^x$.

Solution:

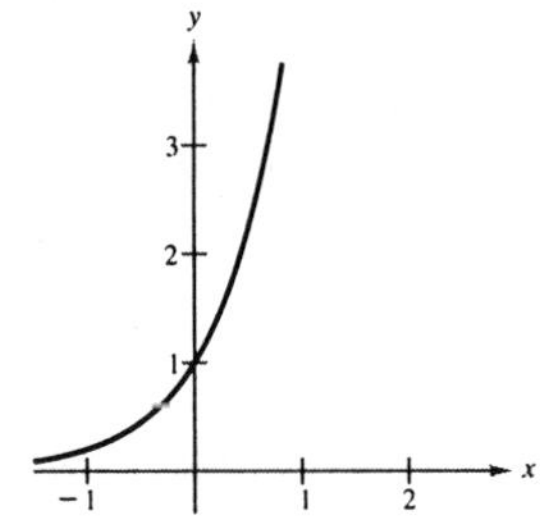

$g(x) = 5^x$

x	−2	−1	0	1	2
$g(x)$	$\frac{1}{25}$	$\frac{1}{5}$	1	5	25

27. Sketch the graph of $h(x) = 5^{x-2}$.

Solution:

$h(x) = 5^{x-2}$

x	-1	0	1	2	3
$h(x)$	$\frac{1}{125}$	$\frac{1}{25}$	$\frac{1}{5}$	1	5

31. Sketch the graph of $f(x) = 3^{x-2} + 1$.

Solution:

$f(x) = 3^{x-2} + 1$

x	-1	0	1	2	3	4
y	$1\frac{1}{27}$	$1\frac{1}{9}$	$1\frac{1}{3}$	2	4	10

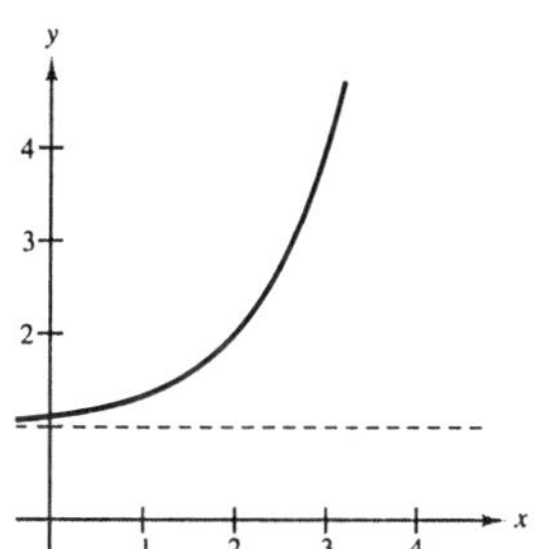

37. Sketch the graph of $f(x) = e^{2x}$.

Solution:

$f(x) = e^{2x}$

x	0	1	2	-1	-2
$f(x)$	1	7.39	54.60	0.135	0.02

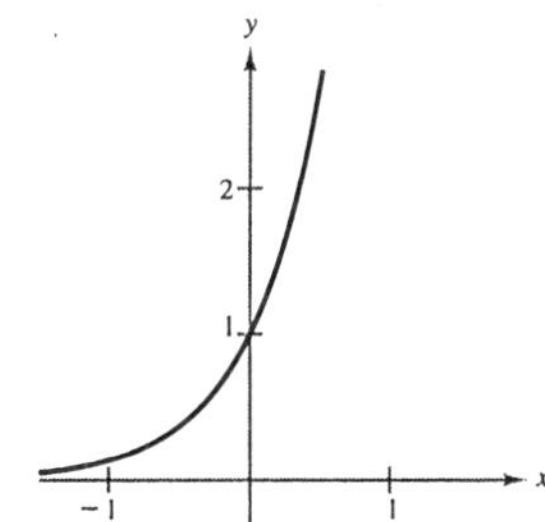

43. Use the table in the textbook to determine the balance A for \$2500 invested at 12% for 20 years and compounded n times per year.

Solution:

$$A = P\left(1 + \frac{r}{n}\right)^{nt}$$

$$P = 2500, \quad r = 0.12, \quad t = 20$$

–CONTINUED ON NEXT PAGE–

43. –CONTINUED–

When $n = 1$, $A = 2500\left(1 + \frac{0.12}{1}\right)^{(1)(20)} \approx \$24{,}115.73$

When $n = 2$, $A = 2500\left(1 + \frac{0.12}{2}\right)^{(2)(20)} \approx \$25{,}714.29$

When $n = 4$, $A = 2500\left(1 + \frac{0.12}{4}\right)^{(4)(20)} \approx \$26{,}602.23$

When $n = 12$, $A = 2500\left(1 + \frac{0.12}{12}\right)^{(12)(20)} \approx \$27{,}231.38$

When $n = 365$, $A = 2500\left(1 + \frac{0.12}{365}\right)^{(365)(20)} \approx \$27{,}547.07$

For continuous compounding, $A = Pe^{rt}$, $A = 2500e^{(0.12)(20)} \approx \$27{,}557.94$

n	1	2	4	12	365	Continuous compounding
A	\$24,115.73	\$25,714.29	\$26,602.23	\$27,231.38	\$27,547.07	\$27,557.94

45. Use the table in the textbook to determine the amount of money P that should be invested at 9% compounded continuously to produce a balance of \$100,000 in t years.

Solution:

$$A = Pe^{rt}$$

$$100{,}000 = Pe^{0.09t}$$

$$\frac{100{,}000}{e^{0.09t}} = P$$

$$P = 100{,}000e^{-0.09t}$$

–CONTINUED ON NEXT PAGE–

45. –CONTINUED–

When $t = 1$, $P = 100{,}000e^{-0.09(1)} \approx \$91{,}393.12$

When $t = 10$, $P = 100{,}000e^{-0.09(10)} \approx \$40{,}656.97$

When $t = 20$, $P = 100{,}000e^{-0.09(20)} \approx \$16{,}529.89$

When $t = 30$, $P = 100{,}000e^{-0.09(30)} \approx \$6{,}720.55$

When $t = 40$, $P = 100{,}000e^{-0.09(40)} \approx \$2{,}732.37$

When $t = 50$, $P = 100{,}000e^{-0.09(50)} \approx \$1{,}110.90$

t	1	10	20	30	40	50
P	\$91,393.12	\$40,656.97	\$16,529.89	\$6,720.55	\$2,732.37	\$1,110.90

49. On the day of your grandchild's birth, you deposited \$25,000 in a trust fund that pays 8.75% interest, compounded continuously. Determine the balance in this account on your grandchild's 25th birthday.

Solution:

$$A = 25{,}000e^{(0.0875)(25)} \approx \$222{,}822.57$$

51. The demand equation for a certain product is given by $p = 500 - 0.5e^{0.004x}$. Find the price p for a demand of (a) $x = 1000$ units and (b) $x = 1500$ units.

Solution:

(a) $x = 1000$

$$p = 500 - 0.5e^4 \approx \$472.70$$

(b) $x = 1500$

$$p = 500 - 0.5e^6 \approx \$298.29$$

55. Let Q represent the mass of radium (Ra^{226}) whose half-life is 1620 years. The quantity of radium present after t years is given by

$$Q = 25\left(\frac{1}{2}\right)^{t/1620}$$

(a) Determine the initial quantity (when $t = 0$).

(b) Determine the quantity present after 1000 years.

(c) Sketch the graph of this function over the interval $t = 0$ to $t = 5000$.

–CONTINUED ON NEXT PAGE–

55. –CONTINUED–

Solution:

(a) When $t = 0$, $Q = 25\left(\frac{1}{2}\right)^{0/1620} = 25(1) = 25$ units

(b) When $t = 1000$, $Q = 25\left(\frac{1}{2}\right)^{1000/1620} \approx 16.297$ units

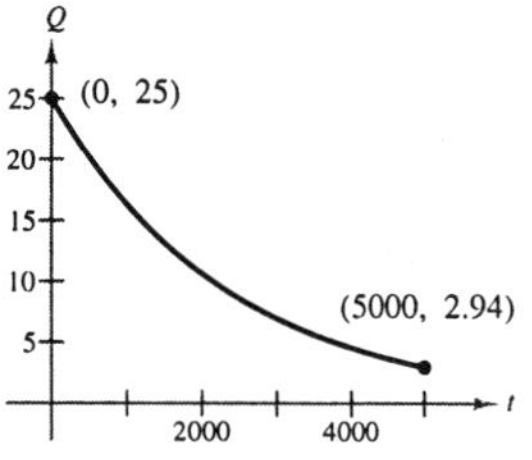

(c)

t	0	1000	2000	3000	4000	5000
Q	25	16.297	10.624	6.926	4.515	2.943

59. After t years, the value of a car that cost you \$20,000 is given by

$$V(t) = 20{,}000\left(\frac{3}{4}\right)^t.$$

Sketch a graph of the function and determine the value of the car two years after it was purchased.

Solution:

t	0	1	2
V	\$20,000	\$15,000	\$11,250

t	3	4	5
V	\$8,437.50	\$6,328.13	\$4,746.09

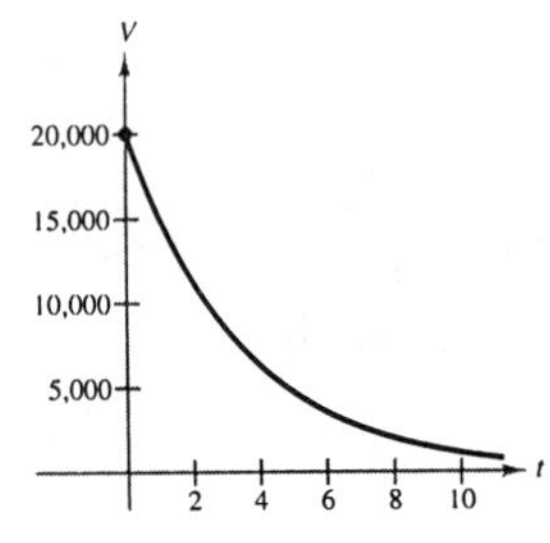

When $t = 2$, $V = \$11{,}250$.

61. Given the exponential function $f(x) = a^x$, show that (a) $f(u+v) = f(u) \cdot f(v)$ and (b) $f(2x) = [f(x)]^2$.

Solution:

(a) $f(u+v) = a^{u+v}$

$= a^u \cdot a^v = f(u) \cdot f(v)$

(b) $f(2x) = a^{2x}$

$= (a^x)^2 = [f(x)]^2$

SECTION 4.2

Logarithmic Functions

- You should know that a function of the form $y = \log_b M$, where $b > 0$, $b \neq 1$, and $M > 0$, is called a logarithm of M to base b.
- You should be able to convert from logarithmic form to exponential form and vice versa.

  $y = \log_b M \Longleftrightarrow b^y = M$
- You should know the following properties of logarithms.

  (a) $\log_a 1 = 0$ (b) $\log_a a = 1$ (c) $\log_a a^x = x$
- You should know the definition of the natural logarithmic function.

  $\log_e x = \ln x, \quad x > 0$
- You should know the properties of the natural logarithmic function.

  (a) $\ln 1 = 0$ (b) $\ln e = 1$ (c) $\ln e^x = x$
- You should be able to graph logarithmic functions.

Solutions to Selected Exercises

5. Evaluate $\log_{16} 4$ without using a calculator.

Solution:

$$\log_{16} 4 = \log_{16} \sqrt{16}$$
$$= \log_{16} 16^{1/2} = \tfrac{1}{2}$$

9. Evaluate $\log_{10} 0.01$ without using a calculator.

Solution:

$$\log_{10} 0.01 = \log_{10} \tfrac{1}{100}$$
$$= \log_{10} 10^{-2} = -2$$

13. Evaluate $\ln e^{-2}$ without using a calculator.

Solution:

$$\ln e^{-2} = -2$$

17. Use the definition of a logarithm to write $5^3 = 125$ in logarithmic form.

Solution:

$$5^3 = 125$$

$$\log_5 125 = 3$$

23. Use the definition of a logarithm to write $e^3 = 20.0855\ldots$ in logarithmic form.

Solution:

$$e^3 = 20.0855\ldots$$

$$\log_e 20.0855\ldots = 3$$

$$\ln 20.0855\ldots = 3$$

27. Use a calculator to evaluate $\log_{10} 345$. Round your answer to three decimal places.

Solution:

$$\log_{10} 345 = 2.537819095\ldots \approx 2.538$$

33. Use a calculator to evaluate $\ln(1+\sqrt{3})$. Round your answer to three decimal places.

Solution:

$$\ln(1+\sqrt{3}) = 1.005052539\ldots \approx 1.005$$

37. Demonstrate that $f(x) = e^x$ and $g(x) = \ln x$ are inverses of each other by sketching their graphs on the same coordinate plane.

Solution:

x	-2	-1	0	1	2	3
$f(x)$	0.135	0.368	1	2.718	7.389	20.086
$g(x)$	—	—	—	0	0.693	1.097

The graph of g is obtained by reflecting the graph of f about the line $y = x$.

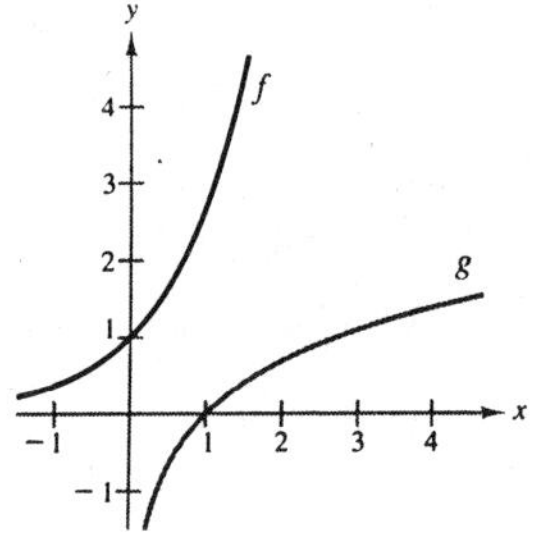

41. Use the graph of $y = \ln x$ to match $f(x) = -\ln(x+2)$ to its graph.

Solution:

$$f(x) = -\ln(x+2)$$

Reflection and horizontal shift two units to the left
Vertical asymptote: $x = -2$
x-intercept: $(-1, 0)$
Matches graph (a)

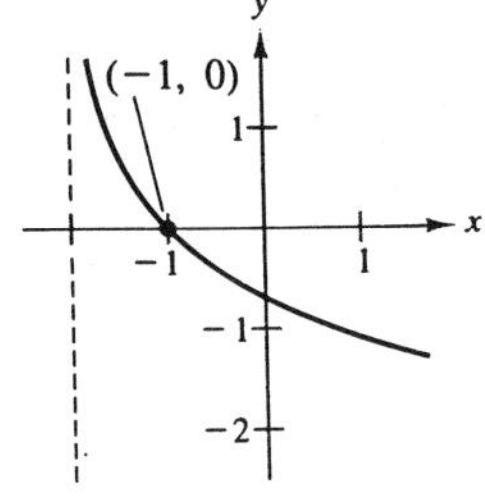

47. Find the domain, vertical asymptote, and x-intercept of $h(x) = \log_4(x-3)$, and sketch its graph.

Solution:

$h(x) = \log_4(x-3)$

Domain: $x - 3 > 0 \Rightarrow x > 3$

The domain is $(3, \infty)$.

Vertical asymptote: $x - 3 = 0 \Rightarrow x = 3$

The vertical asymptote is the line $x = 3$.

x-intercept: $\log_4(x-3) = 0$

$x - 3 = 4^0$

$x - 3 = 1 \Rightarrow x = 4$

The x-intercept is $(4, 0)$.

x	3.5	4	5	7
$h(x)$	−0.5	0	0.5	1

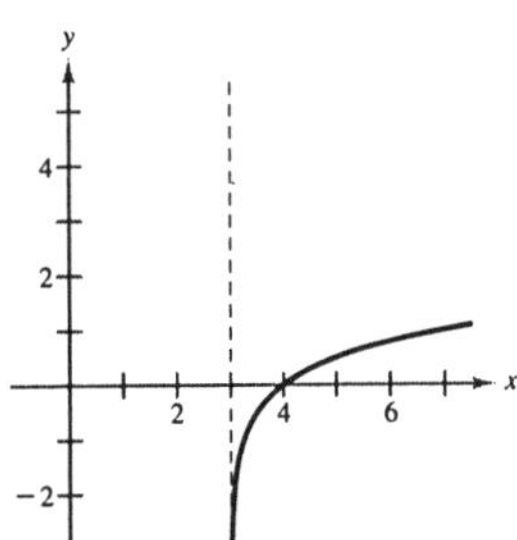

51. Find the domain, vertical asymptote, and x-intercept of $y = \log_{10}(x/5)$, and sketch its graph.

Solution:

$y = \log_{10}\left(\frac{x}{5}\right)$

Domain: $\frac{x}{5} > 0 \Rightarrow x > 0$

The domain is $(0, \infty)$.

Vertical asymptote: $\frac{x}{5} = 0 \Rightarrow x = 0$

The vertical asymptote is the y-axis.

x-intercept: $\log_{10}\left(\frac{x}{5}\right) = 0$

$\frac{x}{5} = 10^0$

$\frac{x}{5} = 1 \Rightarrow x = 5$

The x-intercept is $(5, 0)$.

x	1	2	3	4
y	−0.699	−0.398	−0.222	−0.097

x	5	6	7
y	0	0.079	0.146

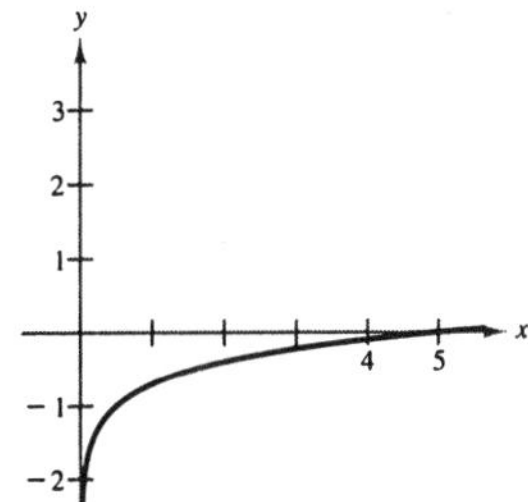

57. Students in a mathematics class were given an exam and then retested monthly with an equivalent exam. The average score for the class was given by the human memory model $f(t) = 80 - 17\log_{10}(t+1)$, $0 \le t \le 12$, where t is the time in months.

(a) What was the average score on the original exam ($t = 0$)?

(b) What was the average score after four months?

(c) What was the average score after ten months?

Solution:

(a) $f(0) = 80 - 17\log_{10} 1 = 80.0$

(b) $f(4) = 80 - 17\log_{10} 5 \approx 68.1$

(c) $f(10) = 80 - 17\log_{10} 11 \approx 62.3$

61. Use the model $y = 80.4 - 11\ln x$ which approximates the minimum required ventilation rate in terms of the air space per child in a public school classroom. In the model, x is the air space per child in cubic feet and y is the ventilation rate in cubic feet per minute. Use the model to approximate the required ventilation rate if there are 300 cubic feet of air space per child.

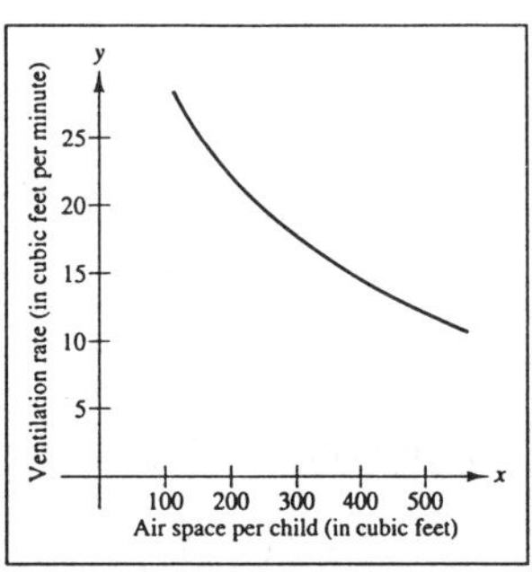

Solution:

$$y = 80.4 - 11\ln 300 \approx 17.658 \text{ cubic feet per minute}$$

67. The work (in foot-pounds) done in compressing an initial volume of 9 cubic feet at a pressure of 15 pounds per square inch to a volume of 3 cubic feet is $W = 19{,}440(\ln 9 - \ln 3)$. Find W.

Solution:

$$W = 19{,}440(\ln 9 - \ln 3) \approx 21{,}357.023 \text{ foot-pounds}$$

69. (a) Use a calculator to complete the table (shown in the textbook) for the function

$$f(x) = \frac{\ln x}{x}.$$

(b) Use the table in part (a) to determine what $f(x)$ approaches as x increases without bound.

Solution:

(a)

x	1	5	10	10^2	10^4	10^6
$f(x)$	0	0.322	0.230	0.046	0.00092	0.0000138

(b) As $x \to \infty$, $f(x) \to 0$.

SECTION 4.3

Properties of Logarithms

- You should know the following properties of logarithms.
  (a) $\log_a(uv) = \log_a u + \log_a v$
  (b) $\log_a(u/v) = \log_a u - \log_a v$
  (c) $\log_a u^n = n \log_a u$
  (d) $\log_a{}^x = \dfrac{\log_b{}^x}{\log_b{}^a}$
- You should be able to rewrite logarithmic expressions.

Solutions to Selected Exercises

1. Use the change of base formula to write $\log_3 5$ as a multiple of a common logarithm.

Solution:

$$\log_3 5 = \frac{\log_{10} 5}{\log_{10} 3}$$

5. Use the change of base formula to write $\log_3 5$ as a multiple of a natural logarithm.

Solution:

$$\log_3 5 = \frac{\ln 5}{\ln 3}$$

9. Evaluate $\log_3 7$ using the change of base formula. Do the problem twice; once with common logarithms and once with natural logarithms. Round to three decimal places.

Solution:

$$\log_3 7 = \frac{\log_{10} 7}{\log_{10} 3} \approx 1.771$$

$$\log_3 7 = \frac{\ln 7}{\ln 3} \approx 1.771$$

13. Evaluate $\log_9(0.4)$ using the change of base formula. Do the problem twice; once with common logarithms and once with natural logarithms. Round to three decimal places.

Solution:

$$\log_9(0.4) = \frac{\log_{10} 0.4}{\log_{10} 9} \approx -0.417$$

$$\log_9(0.4) = \frac{\ln 0.4}{\ln 9} \approx -0.417$$

17. Use the properties of logarithms to write $\log_{10} 5x$ as a sum, difference and/or multiple of logarithms.

Solution:

$$\log_{10} 5x = \log_{10} 5 + \log_{10} x$$

21. Use the properties of logarithms to write $\log_8 x^4$ as a sum, difference, and/or multiple of logarithms.

Solution:

$$\log_8 x^4 = 4 \log_8 x$$

25. Use the properties of logarithms to write $\ln xyz$ as a sum, difference, and/or multiple of logarithms.

Solution:

$$\begin{aligned}\ln xyz &= \ln[x(yz)]\\ &= \ln x + \ln yz\\ &= \ln x + \ln y + \ln z\end{aligned}$$

29. Use the properties of logarithms to write $\ln z(z-1)^2$ as a sum, difference and/or multiple of logarithms.

Solution:

$$\begin{aligned}\ln z(z-1)^2 &= \ln z + \ln(z-1)^2\\ &= \ln z + 2\ln(z-1)\end{aligned}$$

33. Use the properties of logarithms to write the following expression as a sum, difference, and/or multiple of logarithms.

$$\ln \frac{x^4\sqrt{y}}{z^5}$$

Solution:

$$\begin{aligned}\ln \frac{x^4\sqrt{y}}{z^5} &= \ln x^4\sqrt{y} - \ln z^5\\ &= \ln x^4 + \ln \sqrt{y} - \ln z^5\\ &= 4\ln x + \frac{1}{2}\ln y - 5\ln z\end{aligned}$$

35. Use the properties of logarithms to write the following expression as a sum, difference, and/or multiple of logarithms.

$$\log_b \frac{x^2}{y^2z^3}$$

Solution:

$$\begin{aligned}\log_b \frac{x^2}{y^2z^3} &= \log_b x^2 - \log_b y^2z^3\\ &= \log_b x^2 - [\log_b y^2 + \log_b z^3]\\ &= 2\log_b x - 2\log_b y - 3\log_b z\end{aligned}$$

39. Write $\log_4 z - \log_4 y$ as the logarithm of a single quantity.

Solution:

$$\log_4 z - \log_4 y = \log_4 \frac{z}{y}$$

45. Write $\ln x - 3\ln(x+1)$ as the logarithm of a single quantity.

Solution:

$$\begin{aligned}\ln x - 3\ln(x+1) &= \ln x - \ln(x+1)^3\\ &= \ln \frac{x}{(x+1)^3}\end{aligned}$$

49. Write $\ln x - 2[\ln(x+2) + \ln(x-2)]$ as the logarithm of a single quantity.

Solution:

$$\begin{aligned}\ln x - 2[\ln(x+2) + \ln(x-2)] &= \ln x - 2\ln(x+2)(x-2)\\ &= \ln x - 2\ln(x^2-4)\\ &= \ln x - \ln(x^2-4)^2\\ &= \ln \frac{x}{(x^2-4)^2}\end{aligned}$$

53. Write $\frac{1}{3}[\ln y + 2\ln(y+4)] - \ln(y-1)$ as the logarithm of a single quantity.

Solution:

$$\begin{aligned}\frac{1}{3}[\ln y + 2\ln(y+4)] - \ln(y-1) &= \frac{1}{3}[\ln y + \ln(y+4)^2] - \ln(y-1)\\ &= \frac{1}{3}\ln[y(y+4)^2] - \ln(y-1)\\ &= \ln\sqrt[3]{y(y+4)^2} - \ln(y-1)\\ &= \ln\frac{\sqrt[3]{y(y+4)^2}}{y-1}\end{aligned}$$

57. Approximate $\log_b 6$ using the properties of logarithms, given $\log_b 2 \approx 0.3562$ and $\log_b 3 \approx 0.5646$.

Solution:

$$\log_b 6 = \log_b(2 \cdot 3) = \log_b 2 + \log_b 3 \approx 0.3562 + 0.5646 = 0.9208$$

63. Approximate $\log_b \sqrt{2}$ using the properties of logarithms, given $\log_b 2 \approx 0.3562$.

Solution:

$$\log_b \sqrt{2} = \log_b(2^{1/2}) = \tfrac{1}{2}\log_b 2 \approx \tfrac{1}{2}(0.3562) = 0.1781$$

65. Approximate $\log_b \frac{1}{4}$ using the properties of logarithms, given $\log_b 2 \approx 0.3562$.

Solution:

$$\log_b \tfrac{1}{4} = \log_b 1 - \log_b 4 = 0 - \log_b 2^2 = -2\log_b 2 \approx -2(0.3562) = -0.7124$$

69. Approximate the following using the properties of logarithms, given $\log_b 2 \approx 0.3562$ and $\log_b 3 \approx 0.5646$.

$$\log_b\left[\frac{(4.5)^3}{\sqrt{3}}\right]$$

Solution:

$$\begin{aligned}\log_b\left[\frac{(4.5)^3}{\sqrt{3}}\right] &= 3\log_b 4.5 - \frac{1}{2}\log_b 3\\ &= 3\log_b \frac{9}{2} - \frac{1}{2}\log_b 3\\ &= 3[\log_b 9 - \log_b 2] - \frac{1}{2}\log_b 3\\ &= 3[2\log_b 3 - \log_b 2] - \frac{1}{2}\log_b 3\\ &\approx 3[2(0.5646) - 0.3562] - \frac{1}{2}(0.5646) = 2.0367\end{aligned}$$

73. Find the exact value of $\log_4 16^{1.2}$.

Solution:

$$\log_4 16^{1.2} = 1.2\log_4 16 = 1.2\log_4 4^2 = (1.2)(2)\log_4 4 = (2.4)(1) = 2.4$$

77. Use the properties of logarithms to simplify $\log_4 8$.

Solution:

$$\log_4 8 = \log_4 2^3 = 3\log_4 2 = 3\log_4 \sqrt{4} = 3\log_4 4^{1/2} = 3\left(\tfrac{1}{2}\right)\log_4 4 = \tfrac{3}{2}$$

81. Use the properties of logarithms to simplify $\log_5 \frac{1}{250}$.

Solution:

$$\begin{aligned}\log_5 \tfrac{1}{250} &= \log_5 1 - \log_5 250 = 0 - \log_5(125 \cdot 2)\\ &= -\log_5(5^3 \cdot 2) = -[\log_5 5^3 + \log_5 2]\\ &= -[3\log_5 5 + \log_5 2] = -3 - \log_5 2\end{aligned}$$

85. The relationship between the number of decibels β and the intensity of a sound I in watts per meter squared is given by

$$\beta = 10\log_{10}\left(\frac{I}{10^{-16}}\right).$$

Use properties of logarithms to write the formula in simpler form, and determine the number of decibels of a sound with an intensity of 10^{-10} watts per meter squared.

Solution:

$$\begin{aligned}\beta &= 10\log_{10}\left(\frac{I}{10^{-16}}\right)\\ &= 10\left[\log_{10} I - \log_{10} 10^{-16}\right]\\ &= 10[\log_{10} I + 16]\\ &= 160 + 10\log_{10} I\end{aligned}$$

When $I = 10^{-10}$, we have

$$\begin{aligned}\beta &= 160 + 10\log_{10} 10^{-10}\\ &= 160 + 10(-10)\\ &= 60 \text{ decibels.}\end{aligned}$$

87. Prove that $\log_b \dfrac{u}{v} = \log_b u - \log_b v$.

Solution:

Let $x = \log_b u$ and $y = \log_b v$, then $b^x = u$ and $b^y = v$.

$$\begin{aligned}\frac{u}{v} &= \frac{b^x}{b^y} = b^{x-y}\\ \log_b\left(\frac{u}{v}\right) &= \log_b(b^{x-y})\\ &= x - y\\ &= \log_b u - \log_b v\end{aligned}$$

SECTION 4.4

Solving Exponential and Logarithmic Equations

- You should be able to solve exponential and logarithmic equations.
- To solve an exponential equation, take the logarithm of both sides.
- To solve a logarithmic equation, rewrite it in exponential form.

Solutions to Selected Exercises

5. Solve $\left(\frac{3}{4}\right)^x = \frac{27}{64}$ for x without a calculator.

Solution:

$$\left(\frac{3}{4}\right)^x = \frac{27}{64}$$

$$\left(\frac{3}{4}\right)^x = \left(\frac{3}{4}\right)^3$$

$$x = 3$$

9. Solve $\log_{10} x = -1$ for x without a calculator.

Solution:

$$\log_{10} x = -1$$

$$x = 10^{-1} = \tfrac{1}{10}$$

13. Apply the inverse properties of $\ln x$ and e^x to simplify $e^{\ln(5x+2)}$.

Solution:

$$e^{\ln(5x+2)} = 5x + 2$$

17. Solve $e^x = 10$. Round to three decimal places.

Solution:

$$e^x = 10$$

$$x = \ln 10 \approx 2.303$$

21. Solve $e^x - 5 = 10$. Round to three decimal places.

Solution:

$$e^x - 5 = 10$$

$$e^x = 15$$

$$x = \ln 15 \approx 2.708$$

27. Solve $500e^{-x} = 300$. Round to three decimal places.

Solution:

$$500e^{-x} = 300$$

$$e^{-x} = \tfrac{3}{5}$$

$$-x = \ln \tfrac{3}{5}$$

$$x = -\ln \tfrac{3}{5} = \ln \tfrac{5}{3} \approx 0.511$$

31. Solve $e^{2x} - 4e^x - 5 = 0$. Round to three decimal places.

Solution:

$$e^{2x} - 4e^x - 5 = 0$$

$$(e^x + 1)(e^x - 5) = 0$$

$$e^x + 1 = 0 \quad \text{or} \quad e^x - 5 = 0$$

$$e^x = -1 \qquad\qquad e^x = 5$$

No solution $\qquad x = \ln 5 \approx 1.609$

33. Solve $3(1 + e^{2x}) = 4$. Round to three decimal places.

Solution:

$$3(1 + e^{2x}) = 4$$

$$1 + e^{2x} = \tfrac{4}{3}$$

$$e^{2x} = \tfrac{1}{3}$$

$$2x = \ln \tfrac{1}{3}$$

$$x = \tfrac{1}{2} \ln \tfrac{1}{3} \approx -0.549$$

37. Solve $10^x = 42$. Round to three decimal places.

Solution:

$$10^x = 42$$

$$x = \log_{10} 42 \approx 1.623$$

41. Solve $5^{-t/2} = 0.20$. Round to three decimal places.

Solution:

$$5^{-t/2} = 0.20$$

$$5^{-t/2} = \frac{1}{5}$$

$$5^{-t/2} = 5^{-1}$$

$$-\frac{t}{2} = -1$$

$$t = 2$$

45. Solve $3\left(5^{x-1}\right) = 21$. Round to three decimal places.

Solution:

$$3\left(5^{x-1}\right) = 21$$

$$5^{x-1} = 7$$

$$\ln 5^{x-1} = \ln 7$$

$$(x - 1) \ln 5 = \ln 7$$

$$x - 1 = \frac{\ln 7}{\ln 5}$$

$$x = 1 + \frac{\ln 7}{\ln 5} \approx 2.209$$

49. Solve $\left(1 + \frac{0.10}{12}\right)^{12t} = 2$. Round to three decimal places.

Solution:

$$\left(1 + \frac{0.10}{12}\right)^{12t} = 2$$

$$\ln\left(1 + \frac{0.10}{12}\right)^{12t} = \ln 2$$

$$12t \ln\left(1 + \frac{0.10}{12}\right) = \ln 2$$

$$t = \frac{\ln 2}{12 \ln\left(1 + \frac{0.10}{12}\right)}$$

$$\approx 6.960$$

53. Solve $\ln 2x = 2.4$. Round to three decimal places.

Solution:

$$\ln 2x = 2.4$$
$$2x = e^{2.4}$$
$$x = \frac{e^{2.4}}{2} \approx 5.512$$

57. Solve $\ln \sqrt{x+2} = 1$. Round to three decimal places.

Solution:

$$\ln \sqrt{x+2} = 1$$
$$\sqrt{x+2} = e^1$$
$$x + 2 = e^2$$
$$x = -2 + e^2 \approx 5.389$$

61. Solve $\log_{10}(z-3) = 2$. Round to three decimal places.

Solution:

$$\log_{10}(z-3) = 2$$
$$z - 3 = 10^2$$
$$z = 10^2 + 3 = 103$$

63. Solve $\log_{10}(x+4) - \log_{10} x = \log_{10}(x+2)$. Round to three decimal places.

Solution:

$$\log_{10}(x+4) - \log_{10} x = \log_{10}(x+2)$$
$$\log_{10}\left(\frac{x+4}{x}\right) = \log_{10}(x+2)$$
$$\frac{x+4}{x} = x + 2$$
$$x + 4 = x^2 + 2x$$
$$0 = x^2 + x - 4$$
$$x = \frac{-1 \pm \sqrt{17}}{2} = -\frac{1}{2} \pm \frac{\sqrt{17}}{2} \qquad \text{Quadratic Formula}$$
$$x = -\frac{1}{2} + \frac{\sqrt{17}}{2} \qquad \left(\text{since } -\tfrac{1}{2} - \tfrac{\sqrt{17}}{2} \text{ is not in the domain}\right)$$

Choosing the positive value of x (the negative value is extraneous), we have $-\frac{1}{2} + \frac{\sqrt{17}}{2} \approx 1.562$.

67. Solve $\ln(x+5) = \ln(x-1) - \ln(x+1)$. Round to three decimal places.

Solution:

$$\ln(x+5) = \ln(x-1) - \ln(x+1)$$

$$\ln(x+5) = \ln\left(\frac{x-1}{x+1}\right)$$

$$x+5 = \frac{x-1}{x+1}$$

$$(x+5)(x+1) = x-1$$

$$x^2 + 6x + 5 = x - 1$$

$$x^2 + 5x + 6 = 0$$

$$(x+2)(x+3) = 0$$

$$x = -2 \text{ or } x = -3$$

Both of these solutions are extraneous, so the equation has no solution.

71. Find the time required for a \$1000 investment to double at an interest rate of $r = 0.085$ compounded continuously.

Solution:

$$A = Pe^{rt}$$

$$2000 = 1000e^{0.085t}$$

$$2 = e^{0.085t}$$

$$\ln 2 = 0.085t$$

$$\frac{\ln 2}{0.085} = t$$

$$t \approx 8.2 \text{ years}$$

75. The demand equation for a certain product is given by $p = 500 - 0.5(e^{0.004x})$. Find the demand x for a price of (a) $p = \$350$ and (b) $p = \$300$.

Solution:

(a)

$$350 = 500 - 0.5(e^{0.004x})$$

$$-150 = -0.5(e^{0.004x})$$

$$300 = e^{0.004x}$$

$$0.004x = \ln 300$$

$$x = \frac{\ln 300}{0.004} \approx 1426 \text{ units}$$

(b)

$$300 = 500 - 0.5(e^{0.004x})$$

$$-200 = -0.5(e^{0.004x})$$

$$400 = e^{0.004x}$$

$$0.004x = \ln 400$$

$$x = \frac{\ln 400}{0.004} \approx 1498 \text{ units}$$

SECTION 4.5

Exponential and Logarithmic Applications

- You should be able to solve compound interest problems.

  (a) Compound interest formulas:

  1. $A = P\left(1 + \frac{r}{n}\right)^{nt}$
  2. $A = Pe^{rt}$

  (b) Doubling time:

  1. $t = \frac{\ln 2}{n \ln[1 + (r/n)]}$, n compoundings per year
  2. $t = \frac{\ln 2}{r}$, continuous compounding

  (c) Effective yield:

  1. Effective yield $= \left(1 + \frac{r}{n}\right)^n - 1$, n compoundings per year
  2. Effective yield $= e^r - 1$, continuous compounding

- You should be able to solve growth and decay problems.

  $Q(t) = Ce^{kt}$

  (a) If $k > 0$, the population grows.

  (b) If $k < 0$, the population decays.

  (c) Ratio of Carbon 14 to Carbon 12 is $R(t) = \frac{1}{10^{12}} e^{-t/8223}$

- You should be able to solve logistics model problems.

  $y = \frac{a}{1 + be^{-(x-c)/d}}$

- You should be able to solve intensity model problems.

  $S = k \log_{10} \frac{I}{I_0}$

Solutions to Selected Exercises

5. Five hundred dollars is deposited into an account with continuously compounded interest. If the balance is \$1292.85 after 10 years, find the annual percentage rate, the effective yield, and the time to double.

Solution:

$P = 500,\ A = 1292.85,\ t = 10$

$A = Pe^{rt}$

$$1292.85 = 500e^{10r}$$

$$\frac{1292.85}{500} = e^{10r}$$

$$10r = \ln\left(\frac{1292.85}{500}\right)$$

$$r = \frac{1}{10}\ln\left(\frac{1292.85}{100}\right) \approx 0.095 = 9.5\%$$

Effective yield $= e^{0.095} - 1 \approx 0.09966 \approx 9.97\%$

Time to double: $1000 = 500e^{0.095t}$

$$2 = e^{0.095t}$$

$$0.095t = \ln 2$$

$$t = \frac{\ln 2}{0.095} \approx 7.30 \text{ years}$$

9. Five thousand dollars is deposited into an account with continuously compounded interest. If the effective yield is 8.33%, find the annual percentage rate, the time to double, and the amount after 10 years.

Solution:

$P = 5000$

Effective yield $= 8.33\%$

$0.0833 = e^r - 1$

$$r = \ln 1.0833 \approx 0.0800 = 8\%$$

Time to double: $10{,}000 = 5000e^{0.08t}$

$$t = \frac{\ln 2}{0.08} \approx 8.66 \text{ years}$$

After 10 years: $A = 5000e^{0.08(10)} = \$11{,}127.70$

13. Determine the time necessary for \$1000 to double if it is invested at 11% compounded (a) annually, (b) monthly, (c) daily, and (d) continuously.

Solution:

$P = 1000, \ r = 11\%$

(a) $n = 1$

$$t = \frac{\ln 2}{\ln(1 + 0.11)} \approx 6.642 \text{ years}$$

(b) $n = 12$

$$t = \frac{\ln 2}{12 \ln\left(1 + \frac{0.11}{12}\right)} \approx 6.330 \text{ years}$$

(c) $n = 365$

$$t = \frac{\ln 2}{365 \ln\left(1 + \frac{0.11}{365}\right)} \approx 6.302 \text{ years}$$

(d) Continuously

$$t = \frac{\ln 2}{0.11} \approx 6.301 \text{ years}$$

17. The half-life of the isotope Ra^{226} is 1620 years. If the initial quantity is 10 grams, how much will remain after 1000 years, and after 10,000 years?

Solution:

$$Q(t) = Ce^{kt}$$

$$Q = 10 \quad \text{when } t = 0 \Rightarrow 10 = Ce^{0} \Rightarrow 10 = C$$

$$Q(t) = 10e^{kt}$$

$$Q = 5 \quad \text{when } t = 1620$$

$$5 = 10e^{1620k}$$

$$k = \tfrac{1}{1620} \ln\left(\tfrac{1}{2}\right)$$

$$Q(t) = 10e^{[\ln(1/2)/1620]t}$$

When $t = 1000$, $Q(t) = 10e^{[\ln(1/2)/1620](1000)} \approx 6.52$ grams.
When $t = 10{,}000$, $Q(t) = 10e^{[\ln(1/2)/1620](10000)} \approx 0.14$ gram.

21. The half-life of the isotope Pu^{230} is 24,360 years. If 2.1 grams remain after 1000 years, what is the initial quantity and how much will remain after 10,000 years?

Solution:

$$y = Ce^{[\ln(1/2)/24360]t}$$

$$2.1 = Ce^{[\ln(1/2)/24360](1000)}$$

$$C \approx 2.16$$

The initial quantity is 2.16 grams.
When $t = 10{,}000$, $y = 2.16e^{[\ln(1/2)/24360](10000)} \approx 1.63$ grams.

23. Find the constant k such that the exponential function $y = Ce^{kt}$ passes through the points (0, 1) and (4, 10).

Solution:

$$y = Ce^{kt}$$

Since the graph passes through the point (0, 1), we have

$$1 = Ce^{k(0)},$$

$$1 = C.$$

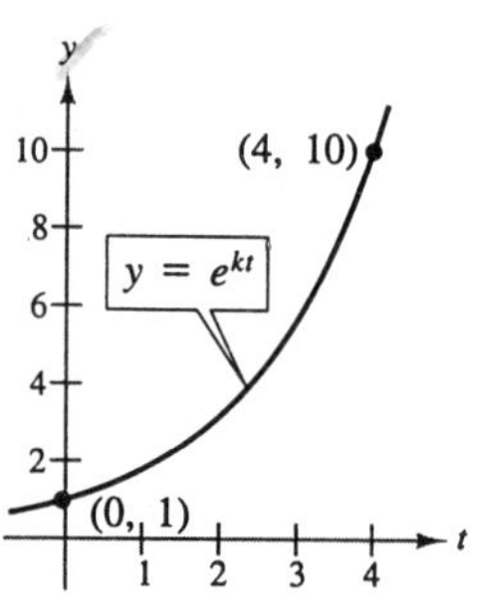

$$y = e^{kt}$$

Since the graph passes through the point (4, 10), we have

$$10 = e^{4k},$$

$$4k = \ln 10$$

$$k = \frac{\ln 10}{4} \approx 0.5756.$$

27. The population P of a city is given by $P = 105{,}300e^{0.015t}$ where t is the time in years with $t = 0$ corresponding to 1990. According to this model, in what year will the city have a population of 150,000?

Solution:

$$150{,}000 = 105{,}300e^{0.015t}$$

$$0.015t = \ln\left(\frac{150{,}000}{105{,}300}\right)$$

$$t = \frac{1}{0.015}\ln\left(\frac{150{,}000}{105{,}300}\right) \approx 23.588 \text{ years}$$

$$1990 + 23 = 2013$$

The city will have a population of 150,000 in the year 2013.

31. The population of Dhaka, Bangladesh was 4.22 million in 1990, and its projected population for the year 2000 is 6.49 million. (*Source:* U.S. Bureau of Census) Find the exponential growth model $y = Ce^{kt}$ for the population growth of Dhaka by letting $t = 0$ correspond to 1990. Use the model to predict the population of the city in 2010.

–CONTINUED ON NEXT PAGE–

31. –CONTINUED–

Solution:

$$P = Ce^{kt}$$

$$4.22 = Ce^{k(0)} \Rightarrow C = 4.22$$

$$6.49 = 4.22e^{10k}$$

$$\frac{6.49}{4.22} = e^{10k}$$

$$k = \frac{1}{10}\ln\left(\frac{6.49}{4.22}\right) \approx 0.0430$$

$$P = 4.22e^{0.0430t}$$

When $t = 20$: $P \approx 9.97$ million.

35. The half-life of radioactive radium (Ra^{226}) is 1620 years. What percentage of a present amount of radioactive radium will remain after 100 years?

Solution:

$$Q(t) = Ce^{kt}$$

$$\tfrac{1}{2}C = Ce^{k(1620)}$$

$$\tfrac{1}{2} = e^{1620k}$$

$$k = \tfrac{1}{1620}\ln(\tfrac{1}{2}) \approx -0.0004$$

$$Q(t) = Ce^{-0.0004t}$$

When $t = 100$: $Q(100) \approx 0.961C = 96.1\%$ of C.

39. The sales S (in thousands of units) of a new product after it has been on the market t years are given by $S(t) = 100(1 - e^{kt})$.

(a) Find S as a function of t if 15,000 units have been sold after one year.

(b) How many units will be sold after five years?

Solution:

(a)
$$S(t) = 100(1 - e^{kt})$$
$$S = 15 \text{ when } t = 1$$
$$15 = 100(1 - e^{k})$$
$$0.15 = 1 - e^{k}$$
$$e^{k} = 0.85$$
$$k = \ln 0.85 \approx -0.1625$$
$$S(t) = 100[1 - e^{-0.1625t}]$$

(b)
$$S(5) = 100[1 - e^{-0.1625t}]$$
$$\approx 55.625 \text{ thousands of units}$$
$$= 55{,}625 \text{ units}$$

41. A certain lake was stocked with 500 fish and the fish population increased according to the logistics curve

$$p(t) = \frac{10{,}000}{1 + 19e^{-t/5}}$$

where t is measured in months (see figure).

(a) Estimate the fish population after 5 months.

(b) After how many months will the fish population be 2000?

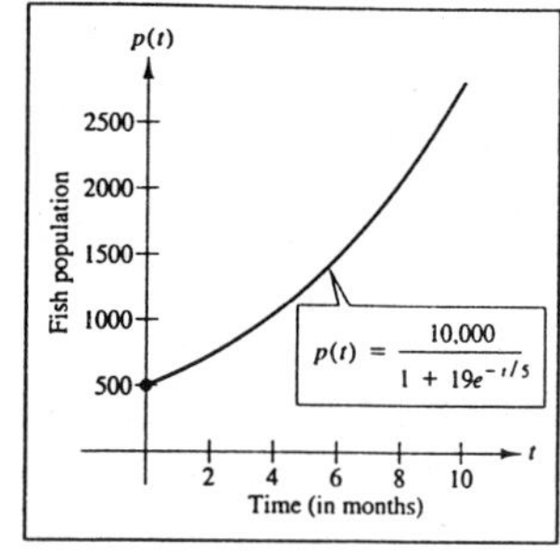

Solution:

$$p(t) = \frac{10{,}000}{1 + 19e^{-t/5}}$$

(a)
$$p(5) = \frac{10{,}000}{1 + 19e^{-1}} \approx 1252 \text{ fish}$$

(b)
$$2000 = \frac{10{,}000}{1 + 19e^{-t/5}}$$
$$1 + 19e^{-t/5} = 5$$
$$e^{-t/5} = \frac{4}{19}$$
$$-\frac{t}{5} = \ln\left(\frac{4}{19}\right)$$
$$t = -5\ln\left(\frac{4}{19}\right) \approx 7.8 \text{ months}$$

45. Find the magnitude R of an earthquake of intensity I (let $I_0 = 1$).

(a) $I = 80{,}500{,}000$ (b) $I = 48{,}275{,}000$

Solution:

$$R = \log_{10} \frac{I}{I_0} = \log_{10} I \text{ since } I_0 = 1$$

(a) $R = \log_{10} 80{,}500{,}000 \approx 7.9$ (b) $R = \log_{10} 48{,}275{,}000 \approx 7.7$

47. The intensity level β, in decibels, of a sound wave is defined by

$$\beta(I) = 10 \log_{10} \frac{I}{I_0}$$

where I_0 is an intensity of 10^{-16} watts per square centimeter, corresponding roughly to the faintest sound that can be heard. Determine $\beta(I)$ for the following conditions.

(a) $I = 10^{-14}$ watts per square centimeter (whisper)

(b) $I = 10^{-9}$ watts per square centimeter (busy street corner)

(c) $I = 10^{-6.5}$ watts per square centimeter (air hammer)

(d) $I = 10^{-4}$ watts per square centimeter (threshold of pain)

Solution:

$$\beta(I) = 10 \log_{10} \frac{I}{I_0} \text{ where } I_0 = 10^{-16} \text{ watt/cm}^2$$

(a) $\beta(10^{-14}) = 10 \log_{10} \dfrac{10^{-14}}{10^{-16}} = 10 \log_{10} 10^2 = 20$ decibels

(b) $\beta(10^{-9}) = 10 \log_{10} \dfrac{10^{-9}}{10^{-16}} = 10 \log_{10} 10^7 = 70$ decibels

(c) $\beta(10^{-6.5}) = 10 \log_{10} \dfrac{10^{-6.5}}{10^{-16}} = 10 \log_{10} 10^{9.5} = 95$ decibels

(d) $\beta(10^{-4}) = 10 \log_{10} \dfrac{10^{-4}}{10^{-16}} = 10 \log_{10} 10^{12} = 120$ decibels

51. Use the acidity model $\text{pH} = -\log_{10}[\text{H}^+]$, where acidity (pH) is a measure of the hydrogen ion concentration $[\text{H}^+]$ (measured in moles of hydrogen per liter) of a solution. Find the pH if $[\text{H}^+] = 2.3 \times 10^{-5}$.

Solution:

$$\text{pH} = -\log_{10}[\text{H}^+] = -\log_{10}[2.3 \times 10^{-5}] \approx 4.64$$

57. At 8:30 A.M. a coroner was called to the home of a person who had died during the night. In order to estimate the time of death, the coroner took the person's temperature twice. At 9:00 A.M. the temperature was 85.7° and at 9:30 A.M. the temperature was 82.8°. From these two temperatures the coroner was able to determine that the time elapsed since death and the body temperature were related by the formula

$$t = -2.5 \ln \frac{T - 70}{98.6 - 70}$$

where t is the time in hours that has elapsed since the person died and T is the temperature (in degrees Fahrenheit) of the person's body at 9:00 A.M. Assume that the person had a normal body temperature of 98.6° at death, and that the room temperature was a constant 70°. (This formula is derived from a general cooling principle called Newton's Law of Cooling.) Use this formula to estimate the time of death of the person.

Solution:

$$t = -2.5 \ln \left(\frac{T - 70}{98.6 - 70}\right)$$

At 9:00 A.M. we have: $t = -2.5 \ln \left(\frac{85.7 - 70}{98.6 - 70}\right) \approx 1.5$ hours. From this we can conclude that the person died at 7:30 A.M.

REVIEW EXERCISES FOR CHAPTER 4

Solutions to Selected Exercises

5. Match $f(x) = \log_2 x$ with the sketch of its graph.

Solution:

$f(x) = \log_2 x$
Vertical asymptote: $x = 0$
Intercept: $(1,\ 0)$
Matches (c)

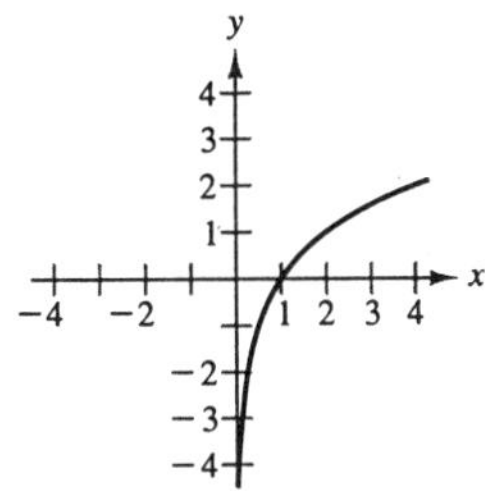

9. Sketch the graph of $g(x) = 6^{-x}$.

Solution:

$$g(x) = 6^{-x} = \left(\frac{1}{6}\right)^x$$

x	0	1	-1
$g(x)$	1	$\frac{1}{6}$	6

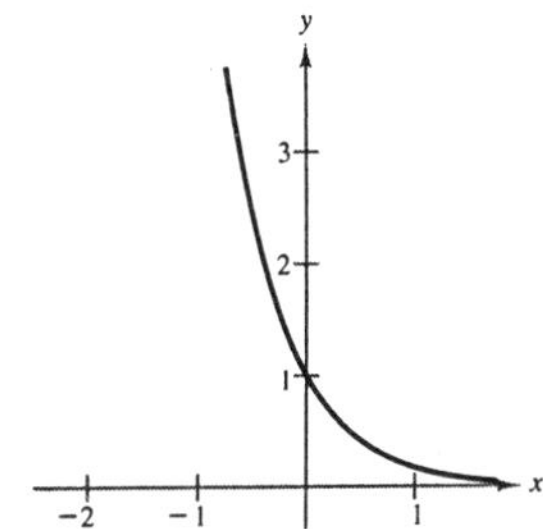

15. Use the table in the textbook to determine the balance A for \$3500 invested at 10.5% for 10 years and compounded n times per year.

Solution:

$$n = 1: \quad A = 3500(1 + 0.105)^{10} \approx \$9{,}499.28$$

$$n = 2: \quad A = 3500\left(1 + \frac{0.105}{2}\right)^{20} \approx \$9{,}738.91$$

$$n = 4: \quad A = 3500\left(1 + \frac{0.105}{4}\right)^{40} \approx \$9{,}867.22$$

$$n = 12: \quad A = 3500\left(1 + \frac{0.105}{12}\right)^{120} \approx \$9{,}956.20$$

$$n = 365: \quad A = 3500\left(1 + \frac{0.105}{365}\right)^{3650} \approx \$10{,}000.27$$

Continuous: $A = 3500e^{(0.105)(10)} \approx \$10{,}001.78$

n	1	2	4	12	365	Continuous compounding
A	\$9,499.28	\$9,738.91	\$9,867.22	\$9,956.20	\$10,000.27	\$10,001.78

17. Use the table in the textbook to determine the amount of money P that should be invested at 8% compounded continuously to produce a final balance of \$200,000 in t years.

Solution:

$$A = Pe^{rt}$$
$$P = Ae^{-rt} = 200{,}000e^{-0.08t}$$

t	1	10	20	30	40	50
P	\$184,623.27	\$89,865.79	\$40,379.30	\$18,143.59	\$8,152.44	\$3,663.13

21. A solution of a certain drug contained 500 units per milliliter when prepared. It was analyzed after 40 days and found to contain 300 units per milliliter. Assuming that the rate of decomposition is proportional to the amount present, the equation giving the amount A after t days is $A = 500e^{-0.013t}$. Use this model to find A when $t = 60$.

Solution:

$$A = 500e^{-0.013(60)} \approx 229.2 \text{ units per milliliter}$$

23. A certain automobile gets 28 miles per gallon of gasoline for speeds up to 50 miles per hour. Over 50 miles per hour, the number of miles per gallon drops at the rate of 12% for each 10 miles per hour. If s is the speed and y is the number of miles per gallon, then

$$y = 28e^{0.6-0.012s}, \quad s \geq 50.$$

Use this function to complete the table shown in the textbook.

Solution:

When $s = 50$, $y = 28e^{0.6-0.012(50)} = 28$ miles per gallon

When $s = 55$, $y = 28e^{0.6-0.012(55)} \approx 26.4$ miles per gallon

When $s = 60$, $y = 28e^{0.6-0.012(60)} \approx 24.8$ miles per gallon

When $s = 65$, $y = 28e^{0.6-0.012(65)} \approx 23.4$ miles per gallon

When $s = 70$, $y = 28e^{0.6-0.012(70)} \approx 22.0$ miles per gallon

Speed	50	55	60	65	70
Miles per gallon	28	26.4	24.8	23.4	22.0

27. Sketch the graph of $f(x) = \ln x + 3$.

Solution:

$f(x) = \ln x + 3$
Domain: $(0, \infty)$

x	1	2	3	$\frac{1}{2}$	$\frac{1}{4}$
$f(x)$	3	3.69	4.10	2.31	1.61

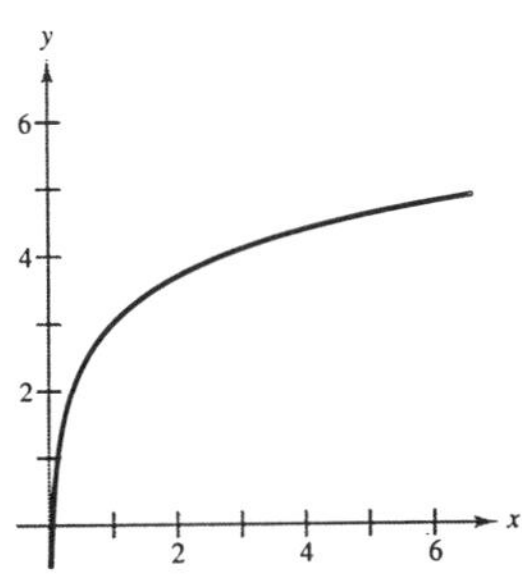

29. Sketch the graph of $h(x) = \ln(e^{x-1})$.

Solution:

$$h(x) = \ln(e^{x-1}) = (x-1)\ln e = x - 1$$

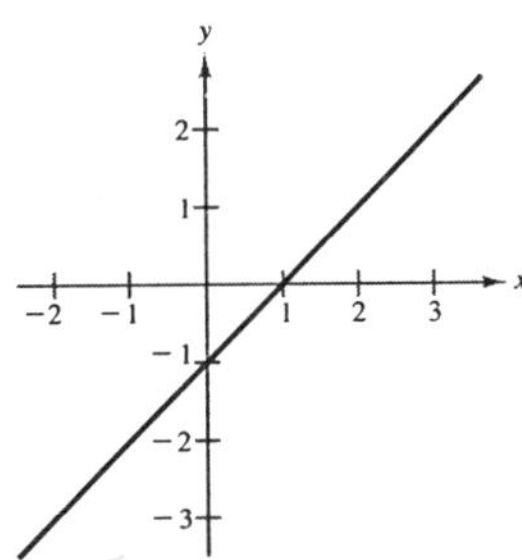

33. Evaluate $\log_{10} 1000$ without using a calculator.

Solution:

$$\log_{10} 1000 = \log_{10} 10^3 = 3$$

37. Evaluate $\ln e^7$ without using a calculator.

Solution:

$$\ln e^7 = 7 \ln e = 7(1) = 7$$

41. Evaluate $\log_4 9$ using the change of base formula. Do the problem twice; once with common logarithms and once with natural logarithms. Round to three decimal places.

Solution:

$$\log_4 9 = \frac{\log_{10} 9}{\log_{10} 4} \approx 1.585$$

$$\log_4 9 = \frac{\ln 9}{\ln 4} \approx 1.585$$

47. Use the properties of logarithms to write the following expression as a sum, difference, and/or multiple of logarithms.

$$\log_{10} \frac{5\sqrt{y}}{x^2}$$

Solution:

$$\log_{10} \frac{5\sqrt{y}}{x^2} = \log_{10} 5\sqrt{y} - \log_{10} x^2$$
$$= \log_{10} 5 + \log_{10} \sqrt{y} - \log_{10} x^2$$
$$= \log_{10} 5 + \frac{1}{2} \log_{10} y - 2 \log_{10} x$$

49. Use the properties of logarithms to write the following expression as a sum, difference, and/or multiple of logarithms.

$$\ln[(x^2+1)(x-1)]$$

Solution:

$$\ln[(x^2+1)(x-1)] = \ln(x^2+1) + \ln(x-1)$$

53. Write $\frac{1}{2}\ln|2x-1| - 2\ln|x+1|$ as the logarithm of a single quantity.

Solution:

$$\frac{1}{2}\ln|2x-1| - 2\ln|x+1| = \ln\sqrt{|2x-1|} - \ln(x+1)^2 = \ln\frac{\sqrt{|2x-1|}}{(x+1)^2}$$

55. Write $\ln 3 + \frac{1}{3}\ln(4-x^2) - \ln x$ as the logarithm of a single quantity.

Solution:

$$\ln 3 + \frac{1}{3}\ln(4-x^2) - \ln x = \ln 3 + \ln\sqrt[3]{4-x^2} - \ln x = \ln\left(3\sqrt[3]{4-x^2}\right) - \ln x = \ln\frac{3\sqrt[3]{4-x^2}}{x}$$

59. Determine whether the equation $\ln(x+y) = \ln x + \ln y$ is true or false.

Solution:

False, since $\ln x + \ln y = \ln(xy)$

63. Approximate $\log_b\sqrt{3}$ using the properties of logarithms given $\log_b 3 \approx 0.5646$.

Solution:

$$\log_b\sqrt{3} = \tfrac{1}{2}\log_b 3 \approx \tfrac{1}{2}(0.5646) = 0.2823$$

67. Solve $e^x = 12$. Round to three decimal places.

Solution:

$$e^x = 12$$

$$x = \ln 12 \approx 2.485$$

71. Solve $e^{2x} - 7e^x + 10 = 0$. Round to three decimal places.

Solution:

$$e^{2x} - 7e^x + 10 = 0$$

$$(e^x - 2)(e^x - 5) = 0$$

$e^x = 2$ or $e^x = 5$

$x = \ln 2$ $\quad$ $x = \ln 5$

$x \approx 0.693$ $\quad$ $x \approx 1.609$

75. Solve $\ln x - \ln 3 = 2$. Round to three decimal places.

Solution:

$$\ln x - \ln 3 = 2$$
$$\ln\left(\frac{x}{3}\right) = 2$$
$$\frac{x}{3} = e^2$$
$$x = 3e^2 \approx 22.167$$

79. Find the exponential function $y = Ce^{kt}$ that passes through the points (0, 4) and $(5, \frac{1}{2})$.

Solution:

Since the graph passes through the point (0, 4), we have

$$4 = Ce^{k(0)}$$
$$4 = C(1) \quad \text{so} \quad y = 4e^{kt}.$$

Since the graph passes through the point $(5, \frac{1}{2})$, we have

$$\tfrac{1}{2} = 4e^{5k}$$
$$\tfrac{1}{8} = e^{5k}$$
$$5k = \ln\tfrac{1}{8}$$
$$k \approx -0.4159.$$

Thus, $y = 4e^{-0.4159t}$.

81. The demand equation for a certain product is given by $p = 500 - 0.5e^{0.004x}$. Find the demand x for a price of (a) $p = \$450$ and (b) $p = \$400$.

Solution:

(a) $p = 450$

$$450 = 500 - 0.5e^{0.004x}$$
$$0.5e^{0.004x} = 50$$
$$e^{0.004x} = 100$$
$$0.004x = \ln 100$$
$$x \approx 1151 \text{ units}$$

(b) $p = 400$

$$400 = 500 - 0.5e^{0.004x}$$
$$0.5e^{0.004x} = 100$$
$$e^{0.004x} = 200$$
$$0.004x = \ln 200$$
$$x \approx 1325 \text{ units}$$

Practice Test for Chapter 4

1. Solve for x: $x^{3/5} = 8$.

2. Solve for x: $3^{x-1} = \frac{1}{81}$.

3. Graph $f(x) = 2^{-x}$.

4. Graph $g(x) = e^x + 1$.

5. If \$5000 is invested at 9% interest, find the amount after three years if the interest is compounded (a) monthly, (b) quarterly, and (c) continuously.

6. Write the equation in logarithmic form: $7^{-2} = \frac{1}{49}$.

7. Solve for x: $x - 4 = \log_2 \frac{1}{64}$.

8. Given $\log_b 2 = 0.3562$ and $\log_b 5 = 0.8271$, evaluate $\log_b \sqrt[4]{8/25}$.

9. Write $5 \ln x - \frac{1}{2} \ln y + 6 \ln z$ as a single logarithm.

10. Using your calculator and the change of base formula, evaluate $\log_9 28$.

11. Use your calculator to solve for N: $\log_{10} N = 0.6646$.

12. Graph $y = \log_4 x$.

13. Determine the domain of $f(x) = \log_3(x^2 - 9)$.

14. Graph $y = \ln(x - 2)$.

15. True or false: $\dfrac{\ln x}{\ln y} = \ln(x - y)$

16. Solve for x: $5^x = 41$.

17. Solve for x: $x - x^2 = \log_5 \frac{1}{25}$.

18. Solve for x: $\log_2 x + \log_2(x - 3) = 2$.

19. Solve for x: $\dfrac{e^x + e^{-x}}{3} = 4$.

20. Six thousand dollars is deposited into a fund at an annual percentage rate of 13%. Find the time required for the investment to double if the interest is compounded continuously.

CHAPTER 5

Trigonometry

Section 5.1 Radian and Degree Measure 212

Section 5.2 The Trigonometric Functions and the Unit Circle 217

Section 5.3 Trigonometric Functions and Right Triangles 220

Section 5.4 Trigonometric Functions of Any Angle 226

Section 5.5 Graphs of Sine and Cosine Functions 234

Section 5.6 Graphs of Other Trigonometric Functions 241

Section 5.7 Other Graphing Techniques 245

Section 5.8 Inverse Trigonometric Functions 248

Section 5.9 Applications of Trigonometry 252

Review Exercises 256

Practice Test 261

SECTION 5.1

Radian and Degree Measure

- If two angles have the same initial and terminal sides, they are coterminal angles.
- The radian measure of a central angle θ is found by taking the arc length s and dividing it by the radius r.

$$\theta = \frac{s}{r}$$

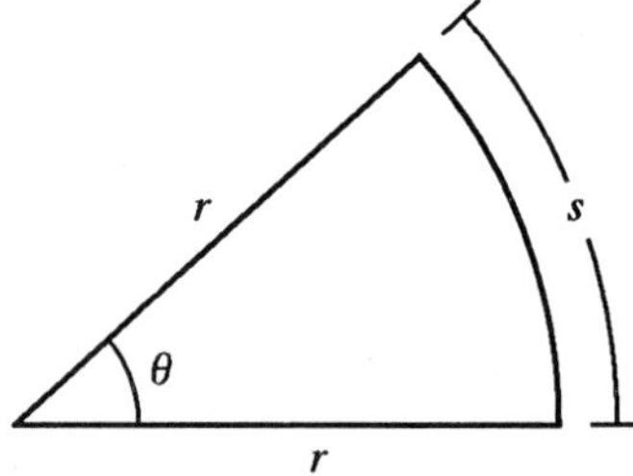

- You should know the following about angles:

  (a) θ is acute if $0 < \theta < \pi/2$.
  (b) θ is a right angle if $\theta = \pi/2$.
  (c) θ is obtuse if $\pi/2 < \theta < \pi$.
  (d) α and β are complementary if $\alpha + \beta = \pi/2$.
  (e) α and β are supplementary if $\alpha + \beta = \pi$.

- To convert degrees to radians, multiply by $\pi/180$.
- To convert radians to degrees, multiply by $180/\pi$.
- You should be able to convert angles to degrees, minutes, and seconds.

  (a) One minute: $1' = \frac{1}{60}(1^\circ)$

  (b) One second: $1'' = \frac{1}{60}(1') = \frac{1}{3600}(1^\circ)$

- Speed $= \frac{\text{distance}}{\text{time}} = \frac{s}{t}$
- Angular speed $= \frac{\theta}{t}$

Solutions to Selected Exercises

3. Determine the quadrant in which the terminal side of the angle lies. The angle is given in radians.

(a) -1 (b) -2

Solution:

(a) Since $-\pi/2 \approx -1.57$ and $-\pi/2 < -1 < 0$, the terminal side of $\theta = -1$ lies in Quadrant IV.

(b) Since $-\pi/2 \approx -1.57$, $-\pi \approx -3.14$ and $-\pi < -2 < -\pi/2$, the terminal side of $\theta = -2$ lies in Quadrant III.

7. Sketch the angle in standard position.

(a) $\dfrac{5\pi}{4}$ (b) $\dfrac{2\pi}{3}$

Solution:

(a) $\dfrac{5\pi}{4} = \pi + \dfrac{\pi}{4}$

The terminal side lies in Quadrant III.

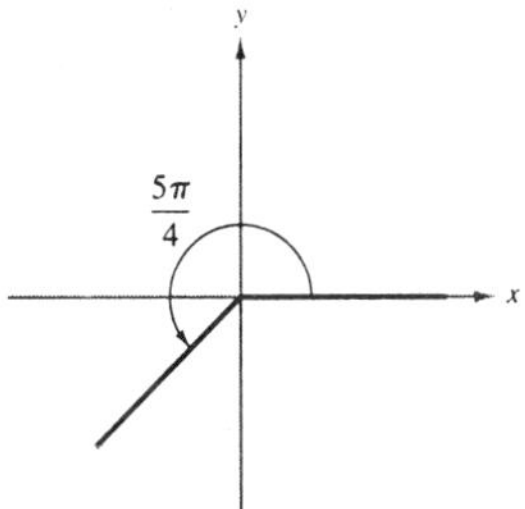

(b) 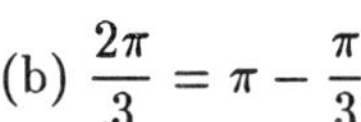
$\dfrac{2\pi}{3} = \pi - \dfrac{\pi}{3}$

The terminal side lies in Quadrant II.

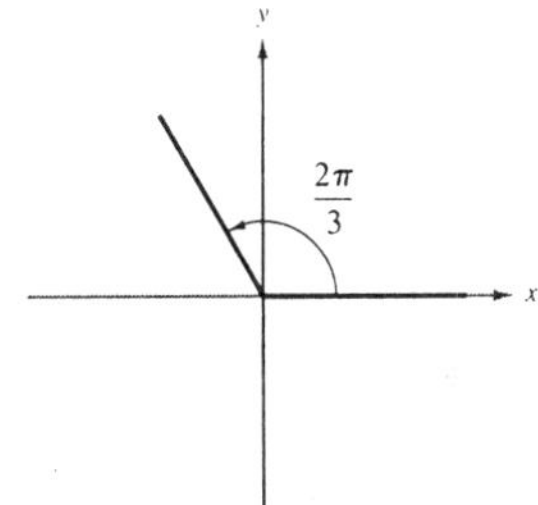

11. Determine two coterminal angles (one positive and one negative) for (a) $\theta = \pi/9$ and (b) $\theta = 4\pi/3$. Give the answers in radians.

Solution:

(a) Coterminal angles for $\dfrac{\pi}{9}$:

$$\frac{\pi}{9} + 2\pi = \frac{19\pi}{9}$$
$$\frac{\pi}{9} - 2\pi = -\frac{17\pi}{9}$$

(b) Coterminal angles for $\dfrac{4\pi}{3}$:

$$\frac{4\pi}{3} + 2\pi = \frac{10\pi}{3}$$
$$\frac{4\pi}{3} - 2\pi = -\frac{2\pi}{3}$$

13. Determine two coterminal angles (one positive and one negative) for (a) $\theta = 36°$ and (b) $\theta = -45°$. Give the answers in degrees.

Solution:

a) Coterminal angles for 36°:

$$36° + 360° = 396°$$
$$36° - 360° = -324°$$

(b) Coterminal angles for −45°:

$$-45° + 360° = 315°$$
$$-45° - 360° = -405°$$

17. Find (if possible) the positive angle complement and the positive angle supplement of the angles (a) $\pi/3$ and (b) $3\pi/4$.

Solution:

(a) Complement: $\dfrac{\pi}{2} - \dfrac{\pi}{3} = \dfrac{\pi}{6}$

Supplement: $\pi - \dfrac{\pi}{3} = \dfrac{2\pi}{3}$

(b) Complement: Not possible since $\dfrac{3\pi}{4} > \dfrac{\pi}{2}$.

Supplement: $\pi - \dfrac{3\pi}{4} = \dfrac{\pi}{4}$

23. Express (a) $7\pi/3$ and (b) $-11\pi/30$ in degree measure. (Do not use a calculator.)

Solution:

(a) $\dfrac{7\pi}{3} = \dfrac{7\pi}{3}\left(\dfrac{180}{\pi}\right) = 420°$

(b) $-\dfrac{11\pi}{30} = -\dfrac{11\pi}{30}\left(\dfrac{180}{\pi}\right) = -66°$

27. Express (a) −20° and (b) −240° in radian measure as a multiple of π. (Do not use a calculator.)

Solution:

(a) $-20° = -20\left(\dfrac{\pi}{180}\right) = -\dfrac{\pi}{9}$

(b) $-240° = -240\left(\dfrac{\pi}{180}\right) = -\dfrac{4\pi}{3}$

31. Convert (a) 532° and (b) 0.54° to radian measure. List your answer to three decimal places.

Solution:

(a) $532° = 532\left(\dfrac{\pi}{180}\right) \approx 9.285$ radians

(b) $0.54° = 0.54\left(\dfrac{\pi}{180}\right) \approx 0.009$ radian

33. Convert (a) $\pi/7$ and (b) $5\pi/11$ from radian to degree measure. List your answer to three decimal places.

Solution:

(a) $\dfrac{\pi}{7} = \dfrac{\pi}{7}\left(\dfrac{180}{\pi}\right) \approx 25.714°$

(b) $\dfrac{5\pi}{11} = \dfrac{5\pi}{11}\left(\dfrac{180}{\pi}\right) \approx 81.818°$

35. Convert (a) -4.2π and (b) 4.8 from radian to degree measure. List your answer to three decimal places.

Solution:

(a) $-4.2\pi = -4.2\pi\left(\dfrac{180}{\pi}\right) = -756°$

(b) $4.8 = 4.8\left(\dfrac{180}{\pi}\right) \approx 275.020°$

39. Convert (a) 240.6° and (b) −145.8° to D° M′ S″ form.

Solution:

(a) $240.6° = 240 + 0.6(60) = 240°\ 36'$

(b) $-145.8° = -[145 + 0.8(60)] = -145°\ 48'$

43. Find the radian measure of the central angle of a circle with radius $r = 15$ inches which intercepts an arc of length $s = 4$ inches.

Solution:

$$\theta = \frac{s}{r} = \frac{4 \text{ inches}}{15 \text{ inches}} = \frac{4}{15} \text{ radian}$$

49. Find the length of the arc intercepted by the central angle $\theta = 2$ radians on the circle of radius $r = 6$ meters.

Solution:

$$s = r\theta$$

$$s = (6)(2) = 12 \text{ m}$$

53. Find the distance between the two cities. Assume that earth is a sphere of radius 4000 miles and that the cities are on the same meridian (one city is due north of the other).

City	Latitude
Miami	25°46′37″ N
Erie	42°7′15″ N

Solution:

The central angle is the difference in latitudes.

$$\theta = 42°7'15'' - 25°46'37''$$

$$= \left(42 + \tfrac{7}{60} + \tfrac{15}{3600}\right) - \left(25 + \tfrac{46}{60} + \tfrac{37}{3600}\right)$$

$$= 16.343\overline{8}° \approx 0.28525 \text{ radian}$$

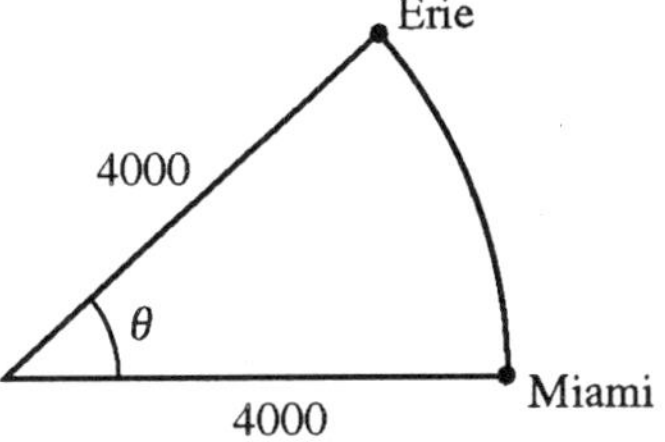

The distance between the cities is found by calculating the arc length.

$$s = r\theta = 4000(0.28525) = 1141 \text{ miles}$$

The cities are approximately 1141 miles apart.

55. Assuming that the earth is a sphere of radius 4000 miles, what is the difference in latitude of two cities, one of which is 325 miles due north of the other?

Solution:

Use the formula $\theta = s/r$.

$$\theta = \frac{325}{4000} \cdot \frac{180}{\pi} \approx 4.655°$$

The difference in latitude is approximately 4.655°.

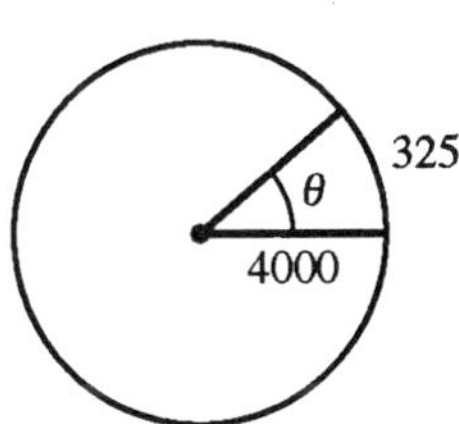

59. A car is moving at the rate of 50 miles per hour, and the diameter of each wheel is 2.5 feet.

(a) Find the number of revolutions per minute of the rotating wheels.

(b) Find the *angular speed* of the wheels in radians per minute.

Solution:

(a) 50 miles per hour $= 50(5280)/60 = 4400$ feet per minute
The circumference of the tire is $C = 2.5\pi$ feet.
The number of revolutions per minute is $r = 4400/2.5\pi \approx 560.2$

(b) The angular speed is θ/t.

$$\theta = \frac{4400}{2.5\pi}(2\pi) = 3520 \text{ radians}$$

$$\text{Angular speed} = \frac{3520 \text{ radians}}{1 \text{ minute}} = 3520 \text{ rad/min}$$

SECTION 5.2

The Trigonometric Functions and the Unit Circle

- You should know the definition of the trigonometric functions. Let t be a real number and $(x,\ y)$ the point on the unit circle corresponding to t.

  (a) $\sin t = y$ (b) $\cos t = x$
  (c) $\tan t = y/x,\quad x \neq 0$ (d) $\cot t = x/y,\quad y \neq 0$
  (e) $\sec t = 1/x,\quad x \neq 0$ (f) $\csc t = 1/y,\quad y \neq 0$

- The cosine and secant functions are even.

  (a) $\cos(-t) = \cos t$ (b) $\sec(-t) = \sec t$

- The other four trigonometric functions are odd.

  (a) $\sin(-t) = -\sin t$ (b) $\tan(-t) = -\tan t$
  (c) $\cot(-t) = -\cot t$ (d) $\csc(-t) = -\csc t$

- You should be able to evaluate trigonometric functions with a calculator.

Solutions to Selected Exercises

3. Find the point $(x,\ y)$ on the unit circle that corresponds to $t = 5\pi/6$.

Solution:

Using Figure 5.19 in the text, move counterclockwise to obtain the second quadrant point

$$(x,\ y) = \left(-\frac{\sqrt{3}}{2},\ \frac{1}{2}\right).$$

5. Find the point $(x,\ y)$ on the unit circle that corresponds to $\theta = 4\pi/3$.

Solution:

Using Figure 5.19 again and the fact that $4\pi/3 = 8\pi/6$, we obtain the point

$$(x,\ y) = \left(-\frac{1}{2},\ -\frac{\sqrt{3}}{2}\right).$$

9. Evaluate the sine, cosine, and tangent of $t = \pi/4$.

Solution:

$t = \dfrac{\pi}{4}$ corresponds to the point $\left(\dfrac{\sqrt{2}}{2}, \dfrac{\sqrt{2}}{2}\right)$.

$$\sin\frac{\pi}{4} = y = \frac{\sqrt{2}}{2}$$

$$\cos\frac{\pi}{4} = x = \frac{\sqrt{2}}{2}$$

$$\tan\frac{\pi}{4} = \frac{y}{x} = \frac{\sqrt{2}/2}{\sqrt{2}/2} = 1$$

13. Evaluate the sine, cosine, and tangent of $t = 11\pi/6$.

Solution:

$t = \dfrac{11\pi}{6}$ corresponds to the point $\left(\dfrac{\sqrt{3}}{2}, -\dfrac{1}{2}\right)$.

$$\sin\frac{11\pi}{6} = y = -\frac{1}{2}$$

$$\cos\frac{11\pi}{6} = x = \frac{\sqrt{3}}{2}$$

$$\tan\frac{11\pi}{6} = \frac{y}{x} = \frac{-1/2}{\sqrt{3}/2} = -\frac{1}{\sqrt{3}} = -\frac{\sqrt{3}}{3}$$

17. Evaluate the six trigonometric functions of $t = 3\pi/4$.

Solution:

$t = \dfrac{3\pi}{4}$ corresponds to the point $\left(-\dfrac{\sqrt{2}}{2}, \dfrac{\sqrt{2}}{2}\right)$.

$$\sin\frac{3\pi}{4} = y = \frac{\sqrt{2}}{2} \qquad \csc\frac{3\pi}{4} = \frac{1}{y} = \frac{1}{\sqrt{2}/2} = \frac{2}{\sqrt{2}} = \sqrt{2}$$

$$\cos\frac{3\pi}{4} = x = -\frac{\sqrt{2}}{2} \qquad \sec\frac{3\pi}{4} = \frac{1}{x} = \frac{1}{-\sqrt{2}/2} = -\frac{2}{\sqrt{2}} = -\sqrt{2}$$

$$\tan\frac{3\pi}{4} = \frac{y}{x} = \frac{\sqrt{2}/2}{-\sqrt{2}/2} = -1 \qquad \cot\frac{3\pi}{4} = \frac{x}{y} = \frac{-\sqrt{2}/2}{\sqrt{2}/2} = -1$$

23. Evaluate $\sin 3\pi$ using its period as an aid.

Solution:

$$\begin{aligned}\sin 3\pi &= \sin(2\pi + \pi)\\ &= \sin\pi\\ &= 0\end{aligned}$$

27. Evaluate $\cos(19\pi)/6$ using its period as an aid.

Solution:

$$\begin{aligned}\cos\frac{19\pi}{6} &= \cos\left(2\pi + \frac{7\pi}{6}\right)\\ &= \cos\frac{7\pi}{6}\\ &= -\frac{\sqrt{3}}{2}\end{aligned}$$

31. Use $\sin t = \frac{1}{3}$ to evaluate the functions (a) $\sin(-t)$ and (b) $\csc(-t)$.

Solution:

(a) $\sin(-t) = -\sin t = -\dfrac{1}{3}$

(b) $\csc(-t) = -\csc t$

$$= -\frac{1}{\sin t} = -\frac{1}{1/3} = -3$$

35. Use $\sin t = \frac{4}{5}$ to evaluate the functions (a) $\sin(\pi - t)$ and (b) $\sin(t + \pi)$.

Solution:

(a) $\sin(\pi - t) = \sin t = \frac{4}{5}$

If t is in Quadrant I, $\pi - t$ is in Quadrant II and they both have the same y-value on the unit circle. If t is in Quadrant II, $\pi - t$ is in Quadrant I and they both have the same y-value on the unit circle. (t cannot be in Quadrant III or Quadrant IV since $\sin t = \frac{4}{5}$ is positive.)

(b) $\sin(t + \pi) = -\sin t = -\frac{4}{5}$

If t is in Quadrant I, $\pi + t$ is in Quadrant III and has a negative y-value on the unit circle. If t is in Quadrant II, $\pi + t$ is in Quadrant IV and has a negative y-value on the unit circle.

39. Use a calculator to evaluate $\cos(-3)$. [Set your calculator in radian mode and round your answer to four decimal places.]

Solution:

$$\cos(-3) = \cos 3 \approx -0.9899925 \approx -0.9900$$

43. Use a calculator to evaluate $\csc 0.8$. [Set your calculator in radian mode and round your answer to four decimal places.]

Solution:

$$\csc 0.8 = \frac{1}{\sin 0.8} \approx 1.3940078 \approx 1.3940$$

49. The displacement from equilibrium of an oscillating weight suspended by a spring is $y(t) = \frac{1}{4}\cos 6t$ where y is the displacement in feet and t is the time in seconds. Find the displacement when (a) $t = 0$, (b) $t = \frac{1}{4}$, and (c) $t = \frac{1}{2}$.

Solution:

(a) $y(0) = \frac{1}{4}\cos[6(0)] = \frac{1}{4}\cos 0 = \frac{1}{4}(1) = 0.2500$ foot

(b) $y\left(\frac{1}{4}\right) = \frac{1}{4}\cos\left[6\left(\frac{1}{4}\right)\right] = \frac{1}{4}\cos\frac{3}{2} \approx 0.0177$ foot

(c) $y\left(\frac{1}{2}\right) = \frac{1}{4}\cos\left[6\left(\frac{1}{2}\right)\right] = \frac{1}{4}\cos 3 \approx -0.2475$ foot

SECTION 5.3

Trigonometric Functions and Right Triangles

- You should know the right triangle definition of trigonometric functions.

  (a) $\sin\theta = \dfrac{\text{opp}}{\text{hyp}}$ (b) $\cos\theta = \dfrac{\text{adj}}{\text{hyp}}$

  (c) $\tan\theta = \dfrac{\text{opp}}{\text{adj}}$ (d) $\cot\theta = \dfrac{\text{adj}}{\text{opp}}$

  (e) $\sec\theta = \dfrac{\text{hyp}}{\text{adj}}$ (f) $\csc\theta = \dfrac{\text{hyp}}{\text{opp}}$

- You should know the sine, cosine, and tangent of the special angles 30°, 45°, and 60°.

  (a) For 45°, use the triangle

  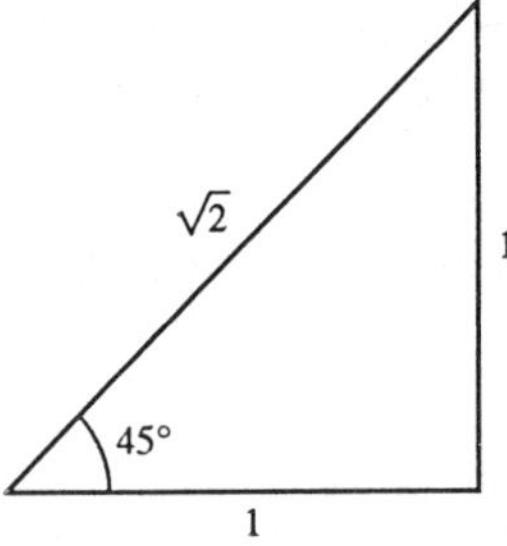

  (b) For 30° and 60°, use the triangle

  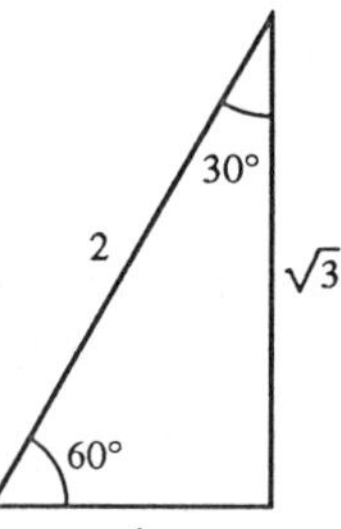

- You should know the fundamental trigonometric identities.

  (a) $\sin\theta = \dfrac{1}{\csc\theta}$ (b) $\cos\theta = \dfrac{1}{\sec\theta}$

  (c) $\tan\theta = \dfrac{1}{\cot\theta}$ (d) $\cot\theta = \dfrac{1}{\tan\theta}$

  (e) $\sec\theta = \dfrac{1}{\cos\theta}$ (f) $\csc\theta = \dfrac{1}{\sin\theta}$

  (g) $\tan\theta = \dfrac{\sin\theta}{\cos\theta}$ (h) $\cot\theta = \dfrac{\cos\theta}{\sin\theta}$

  (i) $\sin^2\theta + \cos^2\theta = 1$ (j) $1 + \tan^2\theta = \sec^2\theta$

  (k) $1 + \cot^2\theta = \csc^2\theta$

Solutions to Selected Exercises

5. Sketch a right triangle corresponding to $\sin\theta = \frac{2}{3}$. Use the Pythagorean Theorem to determine the third side and then find the other five trigonometric functions of θ.

Solution:

$\sin\theta = \dfrac{2}{3}$

$2^2 + (\text{adj})^2 = 3^2$

Adjacent side $= \sqrt{9-4} = \sqrt{5}$

$\cos\theta = \dfrac{\sqrt{5}}{3}$ $\qquad \sec\theta = \dfrac{3}{\sqrt{5}} = \dfrac{3\sqrt{5}}{5}$

$\tan\theta = \dfrac{2}{\sqrt{5}} = \dfrac{2\sqrt{5}}{5}$ $\qquad \csc\theta = \dfrac{3}{2}$

$\cot\theta = \dfrac{\sqrt{5}}{2}$

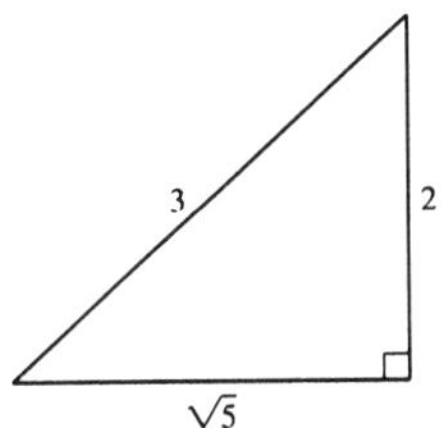

7. Sketch a right triangle corresponding to $\tan\theta = 3$. Use the Pythagorean Theorem to determine the third side and then find the other five trigonometric functions of θ.

Solution:

$\tan\theta = 3 = \dfrac{3}{1}$

$3^2 + 1^2 = (\text{hyp})^2$

Hypotenuse $= \sqrt{1+9} = \sqrt{10}$

$\sin\theta = \dfrac{3}{\sqrt{10}} = \dfrac{3\sqrt{10}}{10}$ $\qquad \csc\theta = \dfrac{\sqrt{10}}{3}$

$\cos\theta = \dfrac{1}{\sqrt{10}} = \dfrac{\sqrt{10}}{10}$ $\qquad \sec\theta = \sqrt{10}$

$\cot\theta = \dfrac{1}{3}$

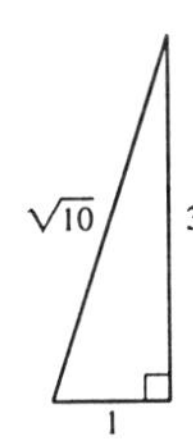

11. Use $\csc\theta = 3$ and $\sec\theta = \frac{3\sqrt{2}}{4}$ and trigonometric identities to find the following trigonometric functions.

(a) $\sin\theta$ (b) $\cos\theta$ (c) $\tan\theta$ (d) $\sec(90^\circ - \theta)$

Solution:

(a) $\sin\theta = \dfrac{1}{\csc\theta} = \dfrac{1}{3}$

(b) $\cos\theta = \dfrac{1}{\sec\theta} = \dfrac{4}{3\sqrt{2}} = \dfrac{2\sqrt{2}}{3}$

(c) $\tan\theta = \dfrac{\sin\theta}{\cos\theta}$

$$= \frac{1/3}{(2\sqrt{2})/3} = \frac{1}{2\sqrt{2}} = \frac{\sqrt{2}}{4}$$

(d) $\sec(90^\circ - \theta) = \csc\theta = 3$

13. Evaluate (a) $\cos 60^\circ$ and (b) $\tan \pi/6$ by memory or by constructing an appropriate triangle.

Solution:

(a) $\cos 60^\circ = \dfrac{1}{2}$

(b) $\tan\dfrac{\pi}{6} = \dfrac{1}{\sqrt{3}} = \dfrac{\sqrt{3}}{3}$

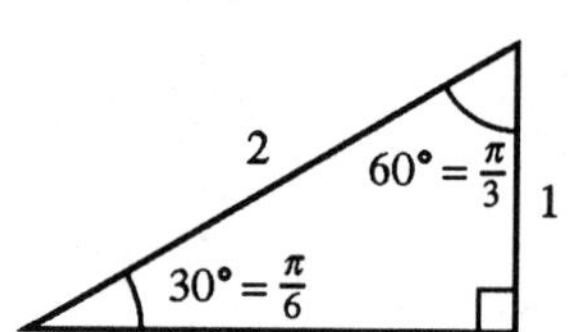

19. Use a calculator to evaluate (a) $\cot(\pi/16)$ and (b) $\tan(\pi/16)$. Round your answers to four decimal places.

Solution:

Make sure that your calculator is in radian mode.

(a) $\cot\dfrac{\pi}{16} = \dfrac{1}{\tan(\pi/16)} \approx 5.0273$

(b) $\tan\dfrac{\pi}{16} \approx 0.1989$

23. Find the value of θ in degrees ($0° < \theta < 90°$) and radians ($0 < \theta < \pi/2$) for (a) $\sin\theta = 1/2$ and (b) $\csc\theta = 2$ without a calculator.

Solution:

(a) $\sin\theta = \dfrac{1}{2}$

$\theta = 30° = \dfrac{\pi}{6}$

(b) $\csc\theta = 2 \Rightarrow \sin\theta = \dfrac{1}{2}$ Same as (a)

$\theta = 30° = \dfrac{\pi}{6}$

27. Find the value of θ in degrees ($0° < \theta < 90°$) and radians ($0 < \theta < \pi/2$) for (a) $\csc\theta = (2\sqrt{3})/3$ and (b) $\sin\theta = \sqrt{2}/2$ without a calculator.

Solution:

(a) $\csc\theta = \dfrac{2\sqrt{3}}{3} = \dfrac{2}{\sqrt{3}}$

$\theta = 60° = \dfrac{\pi}{3}$

(b) $\sin\theta = \dfrac{\sqrt{2}}{2} = \dfrac{1}{\sqrt{2}}$

$\theta = 45° = \dfrac{\pi}{4}$

31. Find the value of θ in degrees ($0° < \theta < 90°$) and radians ($0 < \theta < \pi/2$) by using the inverse key on a calculator for (a) $\tan\theta = 1.1920$ and (b) $\tan\theta = 0.4663$.

Solution:

(a) $\tan\theta = 1.1920$

$\theta = \tan^{-1} 1.1920$

$\theta \approx 50.01° \approx 0.873$ radian

(b) $\tan\theta = 0.4663$

$\theta = \tan^{-1} 0.4663$

$\theta \approx 25° \approx 0.436$ radian

35. Solve for x.

Solution:

$\tan 60° = \dfrac{25}{x}$

$\sqrt{3} = \dfrac{25}{x}$

$x = \dfrac{25}{\sqrt{3}} = \dfrac{25\sqrt{3}}{3} \approx 14.43$

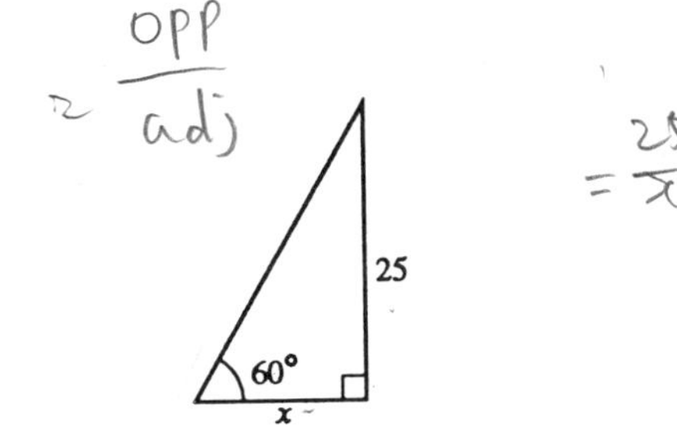

39. Solve for y.

Solution:

$$\sin 50° = \frac{y}{12}$$

$$0.7660 \approx \frac{y}{12}$$

$$y \approx 9.19$$

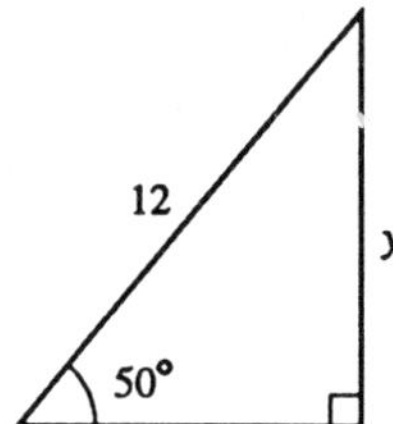

41. A six-foot person standing 12 feet from a streetlight casts an eight-foot shadow, as shown in the figure. What is the height of the streetlight?

Solution:

$$\frac{h}{6} = \frac{20}{8}$$

$$h = \frac{120}{8} = 15 \text{ ft}$$

The height of the streetlight is 15 feet.

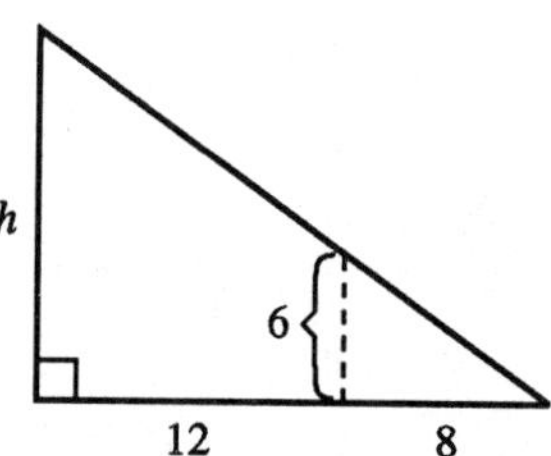

45. From a 150-foot observation tower on the coast, a Coast Guard officer sights a boat in difficulty. The angle of depression of the boat is 4°, as shown in the figure. How far is the boat from the shoreline?

Solution:

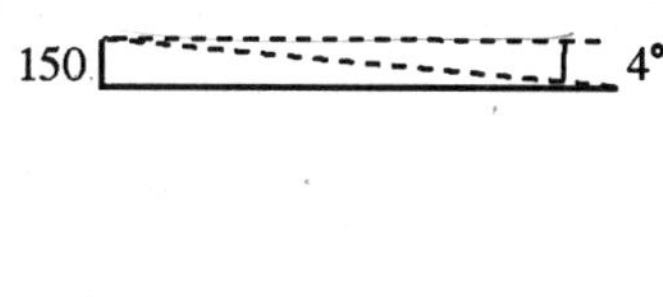

Let $x =$ distance from the boat to the shoreline.

$$\tan 4° = \frac{150}{x}$$

$$x = \frac{150}{\tan 4°} \approx 2145 \text{ ft}$$

The boat is approximately 2145 feet from shore.

53. Determine whether the statement is true or false, and give a reason for your answer.

$\sin 45° + \cos 45° = 1$

Solution:

False; $\sin 45° + \cos 45° = \dfrac{\sqrt{2}}{2} + \dfrac{\sqrt{2}}{2} = \sqrt{2} \neq 1$

55. Determine whether the statement is true or false, and give a reason for your answer.

$\dfrac{\sin 60°}{\sin 30°} = \sin 2°$

Solution:

False; $\dfrac{\sin 60°}{\sin 30°} = \dfrac{\sqrt{3}/2}{1/2} = \sqrt{3} \approx 1.732$ and $\sin 2° \approx 0.035$

Thus, $\dfrac{\sin 60°}{\sin 30°} \neq \sin 2°$.

SECTION 5.4

Trigonometric Functions of Any Angle

- You should know the trigonometric functions of any angle θ in standard position with $(x,\ y)$ on the terminal side of θ and $r = \sqrt{x^2 + y^2}$.

  (a) $\sin\theta = \dfrac{y}{r}$ (b) $\cos\theta = \dfrac{x}{r}$ (c) $\tan\theta = \dfrac{y}{x},\quad x \neq 0$

  (d) $\cot\theta = \dfrac{x}{y},\quad y \neq 0$ (e) $\sec\theta = \dfrac{r}{x},\quad x \neq 0$ (f) $\csc\theta = \dfrac{r}{y},\quad y \neq 0$

- You should know the signs of the trigonometric functions in the four quadrants.
- You should be able to find the trigonometric functions of the quadrant angles (if they exist).

  (a) For 0, use (1, 0). (b) For $\dfrac{\pi}{2}$, use (0, 1).

  (c) For π, use $(-1,\ 0)$. (d) For $\dfrac{3\pi}{2}$, use $(0,\ -1)$.

- You should be able to use reference angles with the special angles to find trigonometric values.

Solutions to Selected Exercises

3. Determine the exact value of the six trigonometric functions of the given angle θ.

(a)

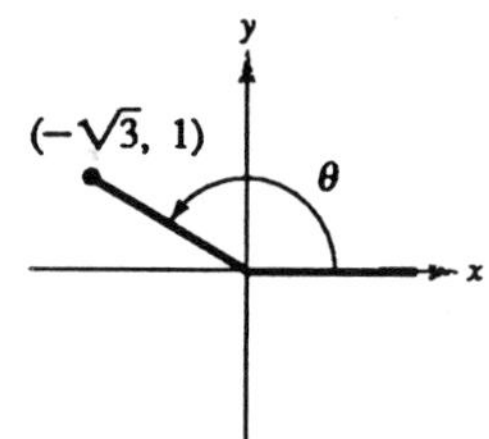

(b)

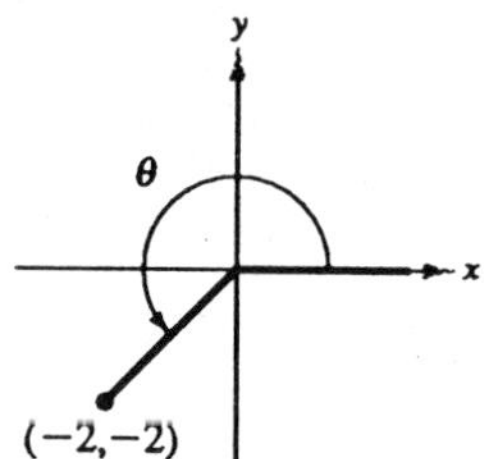

–CONTINUED ON NEXT PAGE–

3. –CONTINUED–

Solution:

(a) $x = -\sqrt{3},\ y = 1,\ r = \sqrt{3+1} = 2$

$\sin\theta = \frac{y}{r} = \frac{1}{2}$ $\qquad$ $\csc\theta = \frac{r}{y} = 2$

$\cos\theta = \frac{x}{r} = -\frac{\sqrt{3}}{2}$ $\qquad$ $\sec\theta = \frac{r}{x} = -\frac{2}{\sqrt{3}} = -\frac{2\sqrt{3}}{3}$

$\tan\theta = \frac{y}{x} = -\frac{1}{\sqrt{3}} = -\frac{\sqrt{3}}{3}$ $\qquad$ $\cot\theta = \frac{x}{y} = -\sqrt{3}$

(b) $x = -2,\ y = -2,\ r = \sqrt{4+4} = 2\sqrt{2}$

$\sin\theta = \frac{y}{r} = -\frac{2}{2\sqrt{2}} = -\frac{\sqrt{2}}{2}$ $\qquad$ $\csc\theta = \frac{r}{y} = \frac{2\sqrt{2}}{-2} = -\sqrt{2}$

$\cos\theta = \frac{x}{r} = -\frac{2}{2\sqrt{2}} = -\frac{\sqrt{2}}{2}$ $\qquad$ $\sec\theta = \frac{r}{x} = \frac{2\sqrt{2}}{-2} = -\sqrt{2}$

$\tan\theta = \frac{y}{x} = \frac{-2}{-2} = 1$ $\qquad$ $\cot\theta = \frac{x}{y} = \frac{-2}{-2} = 1$

7. The point is on the terminal side of an angle in standard position. Determine the exact value of the six trigonometric functions of the angle.

(a) $(-4,\ 10)$ $\qquad$ (b) $(3,\ -5)$

Solution:

(a) $x = -4,\ y = 10,\ r = \sqrt{16+100} = 2\sqrt{29}$

$\sin\theta = \frac{10}{2\sqrt{29}} = \frac{5\sqrt{29}}{29}$ $\qquad$ $\csc\theta = \frac{2\sqrt{29}}{10} = \frac{\sqrt{29}}{5}$

$\cos\theta = \frac{-4}{2\sqrt{29}} = -\frac{2\sqrt{29}}{29}$ $\qquad$ $\sec\theta = \frac{2\sqrt{29}}{-4} = -\frac{\sqrt{29}}{2}$

$\tan\theta = \frac{10}{-4} = -\frac{5}{2}$ $\qquad$ $\cot\theta = \frac{-4}{10} = -\frac{2}{5}$

(b) $x = 3,\ y = -5,\ r = \sqrt{9+25} = \sqrt{34}$

$\sin\theta = \frac{-5}{\sqrt{34}} = -\frac{5\sqrt{34}}{34}$ $\qquad$ $\csc\theta = -\frac{\sqrt{34}}{5}$

$\cos\theta = \frac{3}{\sqrt{34}} = \frac{3\sqrt{34}}{34}$ $\qquad$ $\sec\theta = \frac{\sqrt{34}}{3}$

$\tan\theta = -\frac{5}{3}$ $\qquad$ $\cot\theta = -\frac{3}{5}$

9. Use the two similar triangles in the figure to find (a) the unknown sides of the triangles and (b) the six trigonometric functions of the angles α_1 and α_2.

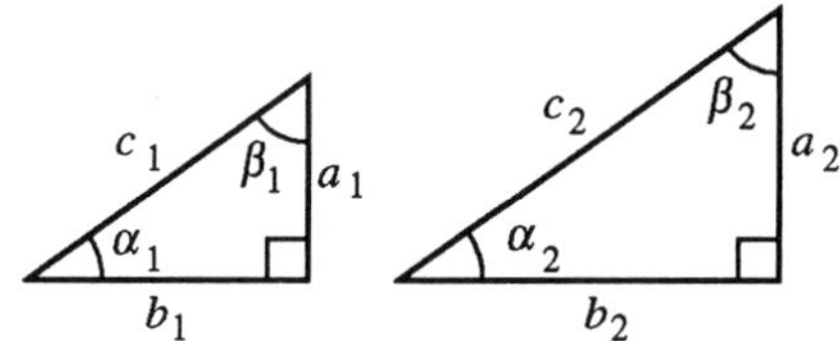

Solution:

Given $a_1 = 3,\ b_1 = 4,\ a_2 = 9,$

(a) $c_1 = \sqrt{a_1{}^2 + b_1{}^2} = \sqrt{9+16} = 5$

$$\frac{a_1}{a_2} = \frac{b_1}{b_2}$$

$$\frac{3}{9} = \frac{4}{b_2}$$

$$b_2 = 12$$

$$c_2 = \sqrt{a_2{}^2 + b_2{}^2}$$

$$= \sqrt{81+144} = 15$$

(b) $\sin\alpha_1 = \sin\alpha_2 = \frac{3}{5}$ $\qquad$ $\csc\alpha_1 = \csc\alpha_2 = \frac{5}{3}$

$\cos\alpha_1 = \cos\alpha_2 = \frac{4}{5}$ $\qquad$ $\sec\alpha_1 = \sec\alpha_2 = \frac{5}{4}$

$\tan\alpha_1 = \tan\alpha_2 = \frac{3}{4}$ $\qquad$ $\cot\alpha_1 = \cot\alpha_2 = \frac{4}{3}$

13. Determine the quadrant in which θ lies.

(a) $\sin\theta < 0$ and $\cos\theta < 0$

(b) $\sin\theta > 0$ and $\cos\theta < 0$

Solution:

(a) $\sin\theta < 0 \Rightarrow \theta$ lies in Quadrant III or in Quadrant IV.
$\cos\theta < 0 \Rightarrow \theta$ lies in Quadrant II or in Quadrant III.
$\sin\theta < 0$ *and* $\cos\theta < 0 \Rightarrow \theta$ lies in Quadrant III.

(b) $\sin\theta > 0 \Rightarrow \theta$ lies in Quadrant I or in Quadrant II.
$\cos\theta < 0 \Rightarrow \theta$ lies in Quadrant II or in Quadrant III.
$\sin\theta > 0$ *and* $\cos\theta < 0 \Rightarrow \theta$ lies in Quadrant II.

17. Find the values of the six trigonometric functions of θ, given θ lies in Quadrant II and $\sin\theta = \frac{3}{5}$.

Solution:

$y = 3, \; r = 5, \; x = -\sqrt{25-9} = -4$

x is negative since θ lies in Quadrant II.

$\sin\theta = \frac{3}{5}$ $\qquad \csc\theta = \frac{5}{3}$

$\cos\theta = -\frac{4}{5}$ $\qquad \sec\theta = -\frac{5}{4}$

$\tan\theta = -\frac{3}{4}$ $\qquad \cot\theta = -\frac{4}{3}$

21. Find the values of the six trigonometric functions of θ, given $\sin\theta > 0$ and $\sec\theta = -2$.

Solution:

θ is in Quadrant II.

$\sec\theta = \dfrac{2}{-1}, \; r = 2,$

$x = -1, \; y = \sqrt{4-1} = \sqrt{3}$

$\sin\theta = \dfrac{\sqrt{3}}{2}$ $\qquad \csc\theta = \dfrac{2\sqrt{3}}{3}$

$\cos\theta = -\dfrac{1}{2}$ $\qquad \sec\theta = -2$

$\tan\theta = -\sqrt{3}$ $\qquad \cot\theta = -\dfrac{\sqrt{3}}{3}$

25. Find the values of the six trigonometric functions of θ, given the terminal side of θ is in Quadrant III and lies on the line $y = 2x$.

Solution:

To find a point on the terminal side of θ, use any point on the line $y = 2x$ that lies in Quadrant III. $(-1, -2)$ is one such point.

$x = -1, \; y = -2, \; r = \sqrt{5}$

$\sin\theta = -\dfrac{2}{\sqrt{5}} = -\dfrac{2\sqrt{5}}{5}$ $\qquad \csc\theta = \dfrac{\sqrt{5}}{-2} = -\dfrac{\sqrt{5}}{2}$

$\cos\theta = -\dfrac{1}{\sqrt{5}} = -\dfrac{\sqrt{5}}{5}$ $\qquad \sec\theta = \dfrac{\sqrt{5}}{-1} = -\sqrt{5}$

$\tan\theta = \dfrac{-2}{-1} = 2$ $\qquad \cot\theta = \dfrac{-1}{-2} = \dfrac{1}{2}$

29. Find the reference angle θ', and draw a sketch for (a) $\theta = -245°$ and (b) $\theta = -72°$.

Solution:

(a) $\theta = -245°$

$\theta' = 245° - 180°$

$\theta' = 65°$

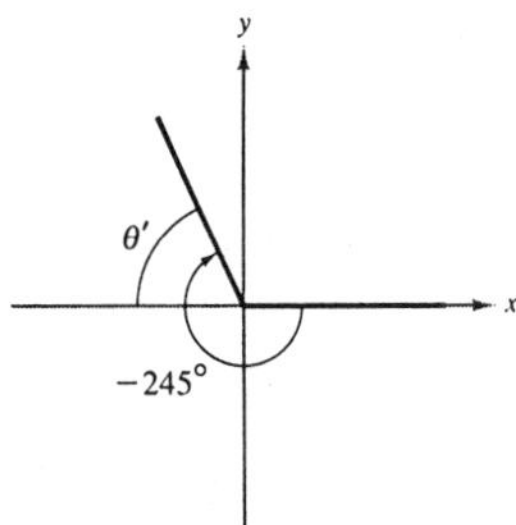

(b) $\theta = -72°$

$\theta' = |-72°| = 72°$

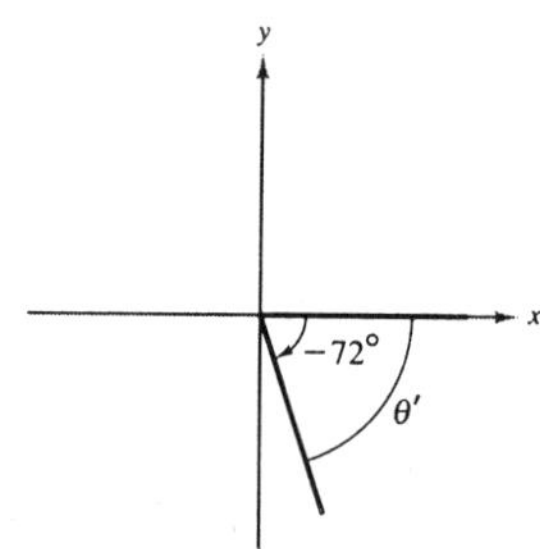

33. Find the reference angle θ' and draw a sketch for (a) $\theta = 3.5$ and (b) $\theta = 5.8$.

Solution:

(a) $\theta = 3.5$

$\theta' = 3.5 - \pi$

$\theta' \approx 0.3584$

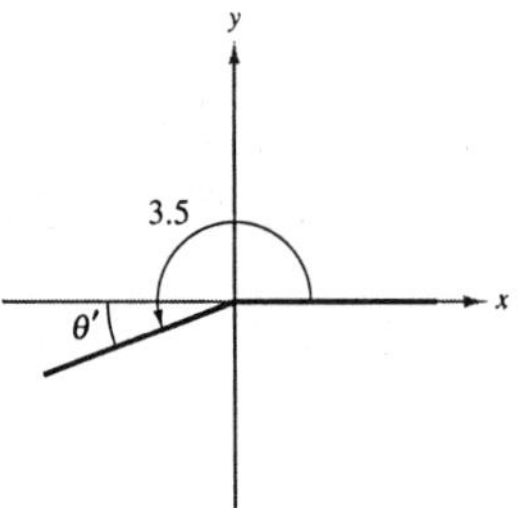

(b) $\theta = 5.8$

$\theta' = 2\pi - 5.8$

$\theta' \approx 0.4832$

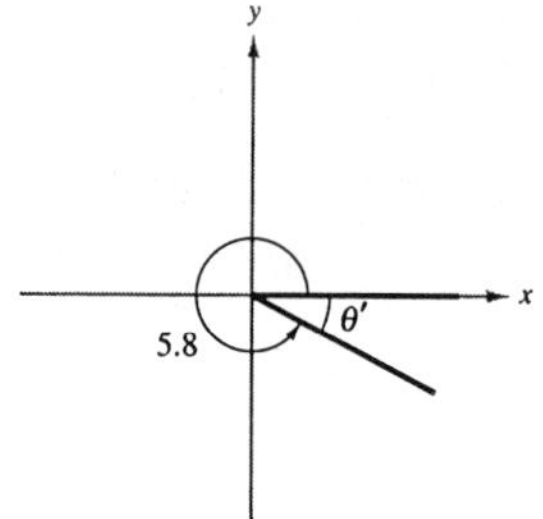

35. Evaluate the sine, cosine, and tangent of the angles without using a calculator.

(a) $225°$ (b) $-225°$

Solution:

(a) The reference angle of $225°$ is $45°$, and $225°$ lies in Quadrant III.

$$\sin 225° = -\sin 45° = -\frac{\sqrt{2}}{2}$$

$$\cos 225° = -\cos 45° = -\frac{\sqrt{2}}{2}$$

$$\tan 225° = \tan 45° = 1$$

(b) The reference angle is $45°$, and $-225°$ lies in Quadrant II.

$$\sin(-225°) = \sin 45° = \frac{\sqrt{2}}{2}$$

$$\cos(-225°) = -\cos 45° = -\frac{\sqrt{2}}{2}$$

$$\tan(-225°) = -\tan 45° = -1$$

39. Evaluate the sine, cosine, and tangent of the angles without using a calculator.

(a) $\frac{4\pi}{3}$ (b) $\frac{2\pi}{3}$

Solution:

(a) The reference angle of $4\pi/3$ is $\pi/3$, and $4\pi/3$ lies in Quadrant III.

$$\sin\frac{4\pi}{3} = -\sin\frac{\pi}{3} = -\frac{\sqrt{3}}{2}$$

$$\cos\frac{4\pi}{3} = -\cos\frac{\pi}{3} = -\frac{1}{2}$$

$$\tan\frac{4\pi}{3} = \tan\frac{\pi}{3} = \sqrt{3}$$

(b) The reference angle of $2\pi/3$ is $\pi/3$, and $2\pi/3$ lies in Quadrant II.

$$\sin\frac{2\pi}{3} = \sin\frac{\pi}{3} = \frac{\sqrt{3}}{2}$$

$$\cos\frac{2\pi}{3} = -\cos\frac{\pi}{3} = -\frac{1}{2}$$

$$\tan\frac{2\pi}{3} = -\tan\frac{\pi}{3} = -\sqrt{3}$$

45. Use a calculator to evaluate (a) $\sin 10°$ and (b) $\csc 10°$ to four decimal places. (Be sure the calculator is set in the correct mode.)

Solution:

(a) $\sin 10° \approx 0.1736$

(b) $\csc 10° = \dfrac{1}{\sin 10°} \approx 5.7588$

47. Use a calculator to evaluate (a) $\cos(-110°)$ and (b) $\cos 250°$ to four decimal places. (Be sure the calculator is set in the correct mode.)

Solution:

(a) $\cos(-110°) = \cos 110° \approx -0.3420$

(b) $\cos 250° \approx -0.3420$

53. Find two values of θ that satisfy (a) $\sin\theta = \frac{1}{2}$ and (b) $\sin\theta = -\frac{1}{2}$. List your answers in degrees $(0° \le \theta < 360°)$ and radians $(0 \le \theta < 2\pi)$. Do not use a calculator.

Solution:

(a) $\sin\theta = \dfrac{1}{2} > 0$

θ is in either Quadrant I or Quadrant II.

$\theta = 30° = \dfrac{\pi}{6}$ or $\theta = 150° = \dfrac{5\pi}{6}$

(b) $\sin\theta = -\dfrac{1}{2} < 0$

θ is in either Quadrant III or Quadrant IV.

$\theta = 210° = \dfrac{7\pi}{6}$ or $\theta = 330° = \dfrac{11\pi}{6}$

57. Find two values of θ that satisfy (a) $\tan\theta = 1$ and (b) $\cot\theta = -\sqrt{3}$. List your answers in degrees $(0° \le \theta < 360°)$ and radians $(0 \le \theta < 2\pi)$. Do not use a calculator.

Solution:

(a) $\tan\theta = 1$, θ lies in either Quadrant I or Quadrant III.

$\theta = 45° = \dfrac{\pi}{4}$ or $\theta = 225° = \dfrac{5\pi}{4}$

(b) $\cot\theta = -\sqrt{3}$, θ lies in either Quadrant II or Quadrant IV.

$\theta = 150° = \dfrac{5\pi}{6}$ or $\theta = 330° = \dfrac{11\pi}{6}$

61. Use a calculator to approximate two values of θ $(0 \le \theta < 2\pi)$ that satisfy (a) $\cos\theta = 0.9848$ and (b) $\cos\theta = -0.5890$. Round to three decimal places.

Solution:

Make sure the calculator is in radian mode.

(a)	$\cos\theta = 0.9848$	θ lies in Quadrant I or Quadrant IV.
	Quadrant I:	$\theta = \cos^{-1} 0.9848 \approx 0.175$
	Quadrant IV:	$\theta = 2\pi - \cos^{-1} 0.9848 \approx 6.109$
(b)	$\cos\theta = -0.5890$	θ lies in Quadrant II or Quadrant III.
	Quadrant II:	$\theta = \cos^{-1}(-0.5890) \approx 2.201$
	Quadrant III:	$\theta = \pi + \cos^{-1} 0.5890 \approx 4.083$

65. Use $\sin\theta = -\frac{3}{5}$, θ in Quadrant IV, and trigonometric identities to find $\cos\theta$.

Solution:

Since $\sin^2\theta + \cos^2\theta = 1$, we have:

$$\cos^2\theta = 1 - \sin^2\theta$$

$$\cos\theta = \pm\sqrt{1 - \sin^2\theta}$$

Also, since θ lies in Quadrant IV, we know that $\cos\theta > 0$. Thus,

$$\begin{aligned}\cos\theta &= +\sqrt{1 - \sin^2\theta}\\ &= \sqrt{1 - \left(-\tfrac{3}{5}\right)^2}\\ &= \sqrt{1 - \left(\tfrac{9}{25}\right)}\\ &= \sqrt{\tfrac{16}{25}}\\ &= \tfrac{4}{5}.\end{aligned}$$

71. The average daily temperature (in degrees Fahrenheit) for a city is

$$T = 45 - 23\cos\left[\frac{2\pi}{365}(t - 32)\right]$$

where t is the time in days with $t = 1$ corresponding to January 1. Find the average temperature on (a) January 1, (b) July 4 ($t = 185$), and (c) October 18 ($t = 291$).

Solution:

(a) $t = 1$

$$T = 45 - 23\cos\left[\frac{2\pi}{365}(1 - 32)\right] \approx 25.2°\text{F}$$

(b) $t = 185$

$$T = 45 - 23\cos\left[\frac{2\pi}{365}(185 - 32)\right] \approx 65.1°\text{F}$$

(c) $t = 291$

$$T = 45 - 23\cos\left[\frac{2\pi}{365}(291 - 32)\right] \approx 50.8°\text{F}$$

SECTION 5.5

Graphs of Sine and Cosine Functions

- You should be able to graph $y = a\sin(bx - c)$ and $y = a\cos(bx - c)$.
- Amplitude: $|a|$
- Period: $\dfrac{2\pi}{|b|}$
- Shift: Solve $bx - c = 0$ and $bx - c = 2\pi$.
- Key Increments: $\dfrac{1}{4}$ (period)

Solutions to Selected Exercises

5. Determine the period and amplitude of $y = \frac{1}{2}\sin \pi x$.

Solution:

$$y = \frac{1}{2}\sin \pi x; \quad a = \frac{1}{2}, \ b = \pi, \ c = 0$$

Period: $\dfrac{2\pi}{|b|} = \dfrac{2\pi}{\pi} = 2$

Amplitude: $|a| = \left|\dfrac{1}{2}\right| = \dfrac{1}{2}$

11. Determine the period and amplitude of

$$y = \frac{1}{2}\cos\frac{2x}{3}.$$

Solution:

$$y = \frac{1}{2}\cos\frac{2x}{3}; \quad a = \frac{1}{2}, \ b = \frac{2}{3}, \ c = 0$$

Period: $\dfrac{2\pi}{|b|} = \dfrac{2\pi}{2/3} = 3\pi$

Amplitude: $|a| = \left|\dfrac{1}{2}\right| = \dfrac{1}{2}$

15. Describe the relationship between the graphs of $f(x) = \sin x$ and $g(x) = \sin(x - \pi)$.

Solution:

$f(x) = \sin x$ and $g(x) = \sin(x - \pi)$ both have a period of 2π and an amplitude of 1. However, the graph of $g(x) = \sin(x - \pi)$ is the graph of $f(x) = \sin x$ shifted to the right π units.

19. Describe the relationship between the graphs of $f(x) = \cos x$ and $g(x) = \cos 2x$.

Solution:

$f(x) = \cos x$ and $g(x) = \cos 2x$ both have an amplitude of 1. However, $f(x) = \cos x$ has a period of 2π, whereas $g(x) = \cos 2x$ has a period of π.

23. Sketch the graphs of $f(x) = -2\sin x$ and $g(x) = 4\sin x$ on the same coordinate plane. (Include two full periods.)

Solution:

$f(x) = -2\sin x$

Period: 2π

Amplitude: 2

$g(x) = 4\sin x$

Period: 2π

Amplitude: 4

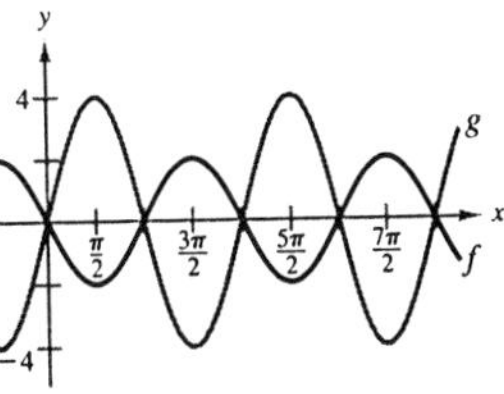

27. Sketch the graphs of the following on the same coordinate plane. (Include two full periods.)

$$f(x) = -\frac{1}{2}\sin\frac{x}{2} \quad \text{and} \quad g(x) = 3 - \frac{1}{2}\sin\frac{x}{2}$$

Solution:

$f(x) = -\dfrac{1}{2}\sin\dfrac{x}{2}$

Period: 4π

Amplitude: $\dfrac{1}{2}$

$g(x) = 3 - \dfrac{1}{2}\sin\dfrac{x}{2}$ is the graph of $f(x)$ shifted vertically three units upward.

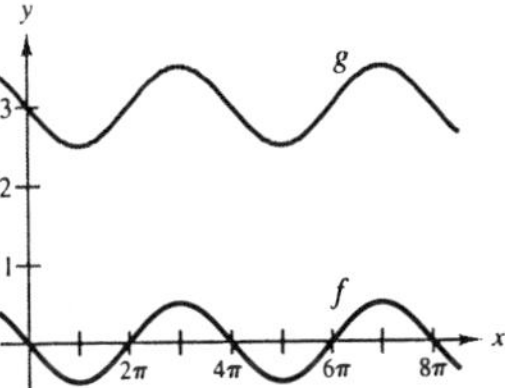

31. Sketch f and g on the same coordinate axes and show that $f(x) = g(x)$ for all x. (Include two full periods.)

$$f(x) = \sin x \quad \text{and} \quad g(x) = \cos\left(x - \frac{\pi}{2}\right)$$

Solution:

Since sine and cosine are cofunctions and x and $x - (\pi/2)$ are complementary, we have

$$\sin x = \cos\left(x - \frac{\pi}{2}\right).$$

Period: 2π

Amplitude: 1

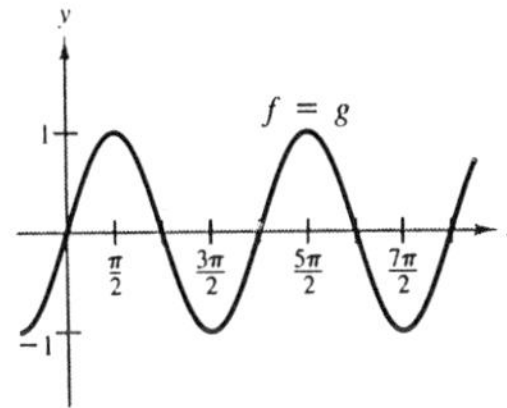

35. Sketch the graph of $y = -2\sin 6x$. (Include two full periods.)

Solution:

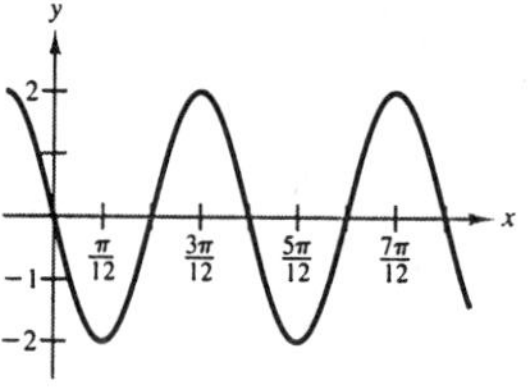

$y = -2\sin 6x$; $a = -2$, $b = 6$, $c = 0$

Period: $\dfrac{2\pi}{6} = \dfrac{\pi}{3}$

Amplitude: $|-2| = 2$

Key points: $(0, 0)$, $\left(\frac{\pi}{12}, -2\right)$, $\left(\frac{\pi}{6}, 0\right)$, $\left(\frac{\pi}{4}, 2\right)$, $\left(\frac{\pi}{3}, 0\right)$

39. Sketch the graph of the following. (Include two full periods.)

$$y = -\sin \frac{2\pi x}{3}$$

Solution:

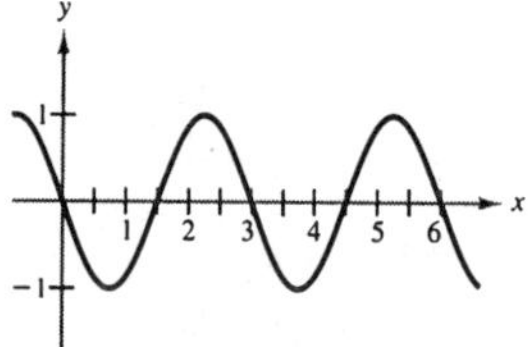

$y = -\sin \dfrac{2\pi x}{3}$; $a = -1$, $b = \dfrac{2\pi}{3}$, $c = 0$

Period: $\dfrac{2\pi}{2\pi/3} = 3$

Amplitude: 1

Key points: $(0, 0)$, $\left(\frac{3}{4}, -1\right)$, $\left(\frac{3}{2}, 0\right)$, $\left(\frac{9}{4}, 1\right)$, $(3, 0)$

43. Sketch the graph of the following. (Include two full periods.)

$$y = \sin\left(x - \frac{\pi}{4}\right)$$

Solution:

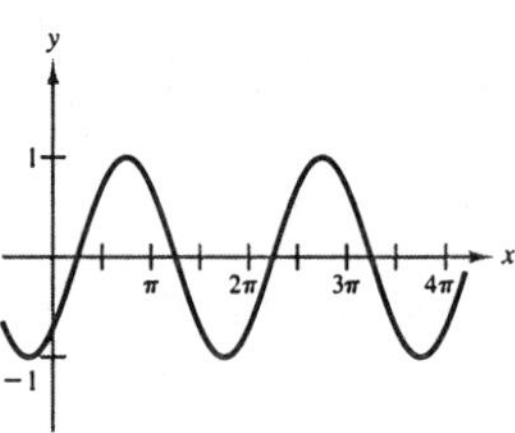

$y = \sin\left(x - \dfrac{\pi}{4}\right)$; $a = 1$, $b = 1$, $c = \dfrac{\pi}{4}$

Period: 2π

Amplitude: 1

Shift: Set $x - \dfrac{\pi}{4} = 0$ and $x - \dfrac{\pi}{4} = 2\pi$

$x = \dfrac{\pi}{4}$ $\qquad$ $x = \dfrac{9\pi}{4}$

Key points: $\left(\frac{\pi}{4}, 0\right)$, $\left(\frac{3\pi}{4}, 1\right)$, $\left(\frac{5\pi}{4}, 0\right)$, $\left(\frac{7\pi}{4}, -1\right)$, $\left(\frac{9\pi}{4}, 0\right)$

49. Sketch the graph of the following. (Include two full periods.)

$$y = \frac{2}{3}\cos\left(\frac{x}{2} - \frac{\pi}{4}\right)$$

Solution:

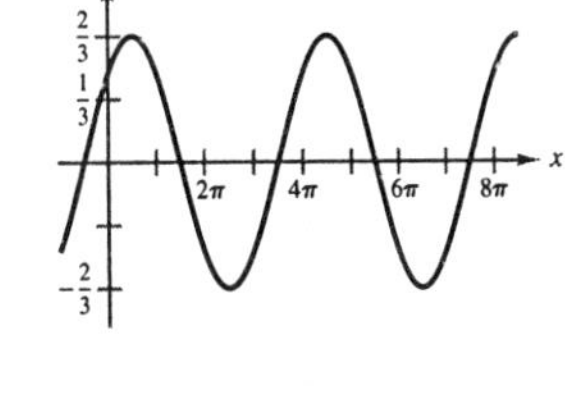

$y = \frac{2}{3}\cos\left(\frac{x}{2} - \frac{\pi}{4}\right);\quad a = \frac{2}{3},\ b = \frac{1}{2},\ c = \frac{\pi}{4}$

Period: 4π

Amplitude: $\frac{2}{3}$

Shift: Set $\frac{x}{2} - \frac{\pi}{4} = 0$ and $\frac{x}{2} - \frac{\pi}{4} = 2\pi$

$x = \frac{\pi}{2}$ $\qquad x = \frac{9\pi}{2}$

Key points: $\left(\frac{\pi}{2}, \frac{2}{3}\right), \left(\frac{3\pi}{2}, 0\right), \left(\frac{5\pi}{2}, \frac{-2}{3}\right), \left(\frac{7\pi}{2}, 0\right), \left(\frac{9\pi}{2}, \frac{2}{3}\right)$

53. Sketch the graph of the following. (Include two full periods.)

$$y = \cos\left(2\pi x - \frac{\pi}{2}\right) + 1$$

Solution:

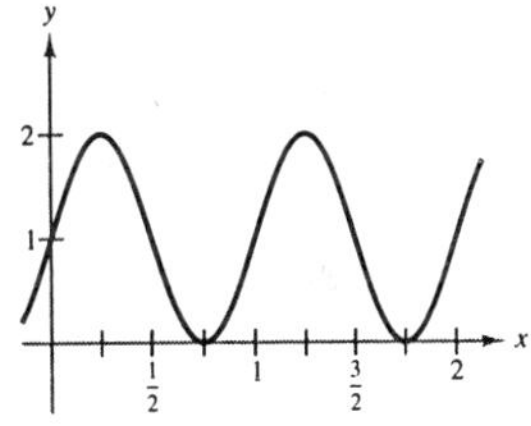

$y = \cos\left(2\pi x - \frac{\pi}{2}\right) + 1;\quad a = 1,\ b = 2\pi,\ c = \frac{\pi}{2}$

Period: 1

Amplitude: 1

Shift: Set $2\pi x - \frac{\pi}{2} = 0$ and $2\pi x - \frac{\pi}{2} = 2\pi$

$x = \frac{1}{4}$ $\qquad x = \frac{5}{4}$

Key points: $\left(\frac{1}{4}, 2\right), \left(\frac{1}{2}, 1\right), \left(\frac{3}{4}, 0\right), (1, 1), \left(\frac{5}{4}, 2\right)$

Vertical shift: One unit upward

55. Sketch the graph of the following. (Include two full periods.)

$$y = -0.1 \sin\left(\frac{\pi x}{10} + \pi\right)$$

Solution:

$y = -0.1 \sin\left(\frac{\pi x}{10} + \pi\right)$; $a = -0.1$, $b = \frac{\pi}{10}$, $c = -\pi$

Period: $\frac{2\pi}{\pi/10} = 20$

Amplitude: $|-0.1| = 0.1$

Shift: Set $\frac{\pi x}{10} + \pi = 0$ and $\frac{\pi x}{10} + \pi = 2\pi$

$x = -10$ $\qquad$ $x = 10$

Key points: $(-10, 0)$, $(-5, -0.1)$, $(0, 0)$, $(5, 0.1)$, $(10, 0)$

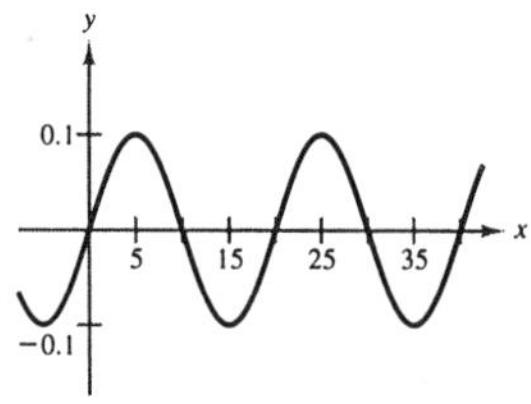

59. Sketch the graph of $y = \frac{1}{10} \cos 60\pi x$. (Include two full periods.)

Solution:

$y = \frac{1}{10} \cos(60\pi x)$; $a = \frac{1}{10}$, $b = 60\pi$, $c = 0$

Period: $\frac{2\pi}{60\pi} = \frac{1}{30}$

Amplitude: $\frac{1}{10}$

Key points: $\left(0, \frac{1}{10}\right)$, $\left(\frac{1}{120}, 0\right)$, $\left(\frac{1}{60}, -\frac{1}{10}\right)$,
$\left(\frac{1}{40}, 0\right)$, $\left(\frac{1}{30}, \frac{1}{10}\right)$

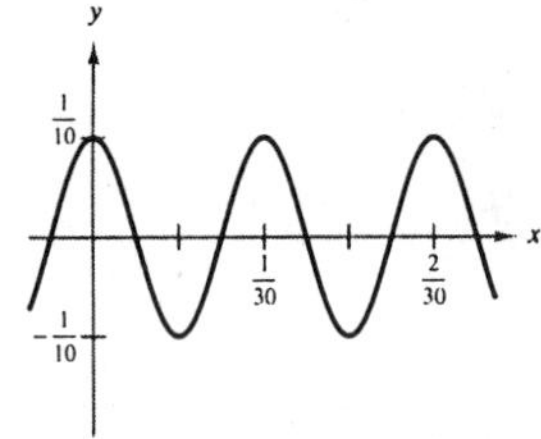

63. Use the graph of $\cos x$ to find all real numbers x in the interval $[-2\pi, 2\pi]$ that gives the functional value of $\sqrt{2}/2$.

Solution:

$$\cos x = \frac{\sqrt{2}}{2}$$

$$x = \pm\frac{\pi}{4}, \pm\frac{7\pi}{4}$$

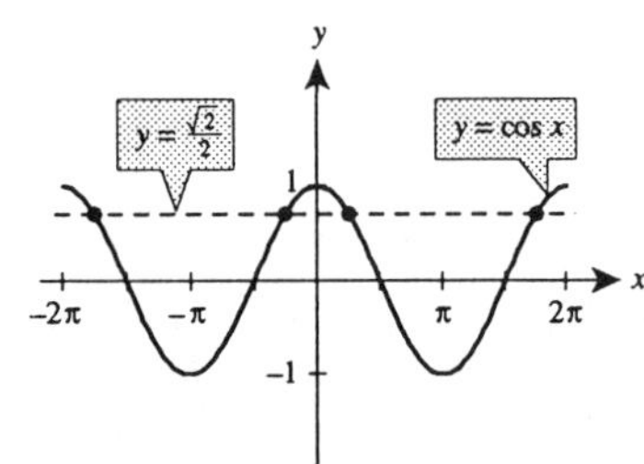

67. Find a, b, and c so that the graph of the function matches the graph in the figure.

Solution:

$y = a\cos(bx - c)$

Amplitude: $1 \Rightarrow a = 1$

Period: $\dfrac{2\pi}{b} = \pi \Rightarrow b = 2$

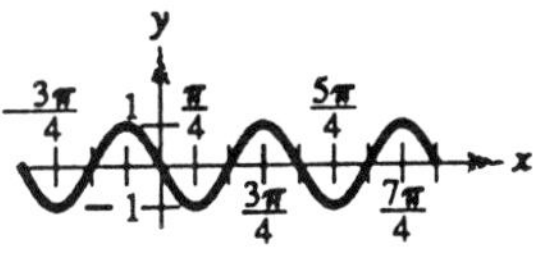

Shift: The graph begins at $-\dfrac{\pi}{4}$.

$$2\left(-\frac{\pi}{4}\right) - c = 0 \Rightarrow c = -\frac{\pi}{2}$$

Thus, $y = \cos\left(2x + \dfrac{\pi}{2}\right)$.

69. For a person at rest, the velocity v (in liters per second) of air flow during a respiratory cycle is

$$v = 0.85\sin\frac{\pi t}{3}$$

where t is the time in seconds. (Inhalation occurs when $v > 0$, and exhalation occurs when $v < 0$.)

(a) Find the time for one full respiratory cycle.

(b) Find the number of cycles per minute.

(c) Sketch the graph of the velocity function.

Solution:

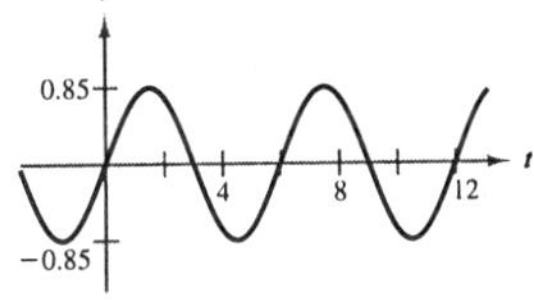

(a) Time for one cycle = one period = $\dfrac{2\pi}{\pi/3} = 6$ sec

(b) Cycles per min = $\dfrac{60}{6} = 10$ cycles per min

(c) Amplitude: 0.85

Period: 6

Key points: $(0, 0)$, $\left(\dfrac{3}{2}, 0.85\right)$, $(3, 0)$, $\left(\dfrac{9}{2}, -0.85\right)$, $(6, 0)$

71. When tuning a piano, a technician strikes a tuning fork for the A above middle C and sets up wave motion that can be approximated by

$$y = 0.001 \sin 880\pi t$$

where t is the time in seconds.

(a) What is the period p of this function?

(b) The frequency f is given by $f = 1/p$. What is the frequency of this note?

(c) Sketch the graph of this function.

Solution:

(a) Period: $\dfrac{2\pi}{880\pi} = \dfrac{1}{440}$

(b) $f = \dfrac{1}{p} = 440$

(c) Amplitude: 0.001

Period: $\dfrac{1}{440}$

Key points: $(0, 0)$, $\left(\dfrac{1}{1760}, 0.001\right)$, $\left(\dfrac{1}{880}, 0\right)$, $\left(\dfrac{3}{1760}, -0.001\right)$, $\left(\dfrac{1}{440}, 0\right)$

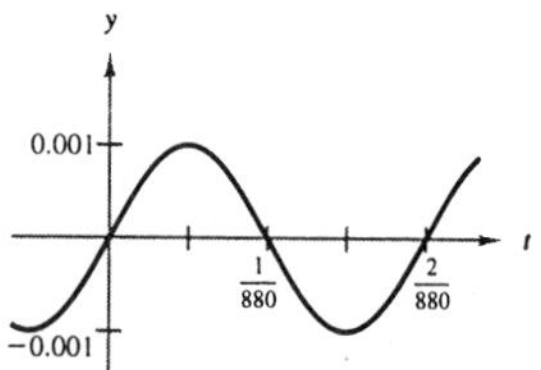

75. Determine the relationship between the graphs of the functions f and g.

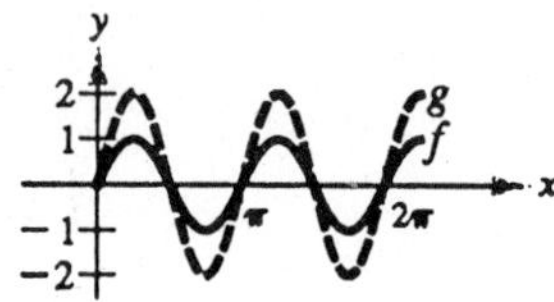

Solution:

The graphs have the same period of π, and the amplitude of g is twice the amplitude of f. Thus, $g = 2f$. On set of possible functions is

$$f(x) = \sin(2x)$$

$$g(x) = 2\sin(2x).$$

SECTION 5.6

Graphs of Other Trigonometric Functions

- You should be able to graph

  $y = a\tan(bx - c) \quad y = a\cot(bx - c)$

  $y = a\sec(bx - c) \quad y = a\csc(bx - c)$

- When graphing $y = a\sec(bx - c)$ or $y = a\csc(bx - c)$ you should know to first graph $y = a\cos(bx - c)$ or $y = a\sin(bx - c)$ since

  (a) The intercepts of sine and cosine are vertical asymptotes of cosecant and secant.

  (b) The maximums of sine and cosine are local minimums of cosecant and secant.

  (c) The minimums of sine and cosine are local maximums of cosecant and secant.

Solutions to Selected Exercises

5. Match $y = \cot \pi x$ with the correct graph and give the period of the function.

Solution:

Period: $\dfrac{\pi}{\pi} = 1$

Matches graph (d)

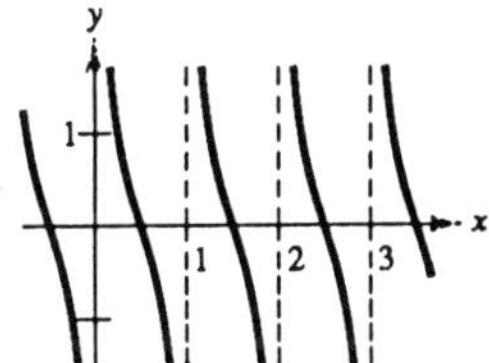

11. Sketch the graph of $y = \tan 2x$ through two periods.

Solution:

Period: $\dfrac{\pi}{2}$

One cycle: $-\dfrac{\pi}{4}$ to $\dfrac{\pi}{4}$

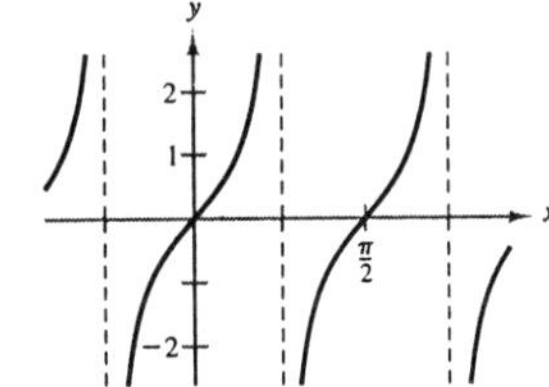

17. Sketch the graph of $y = -2 \sec 4x$ through two periods.

Solution:

Graph $y = -2 \cos 4x$ first.

Period: $\dfrac{2\pi}{4} = \dfrac{\pi}{2}$

One cycle: 0 to $\dfrac{\pi}{2}$

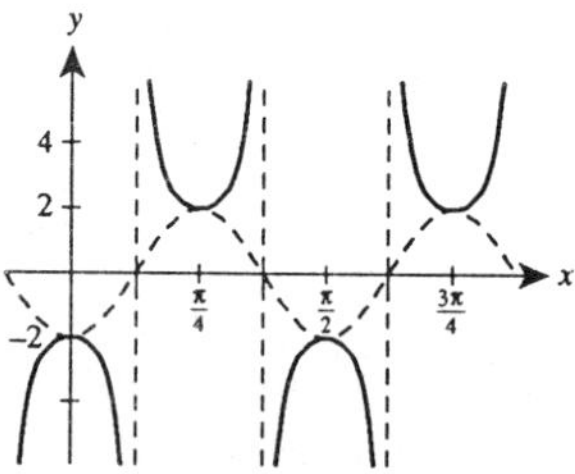

23. Sketch the graph of the following through two periods.

$$y = \csc \frac{x}{2}$$

Solution:

Graph $y = \sin \dfrac{x}{2}$ first.

Period: $\dfrac{2\pi}{1/2} = 4\pi$

One cycle: 0 to 4π

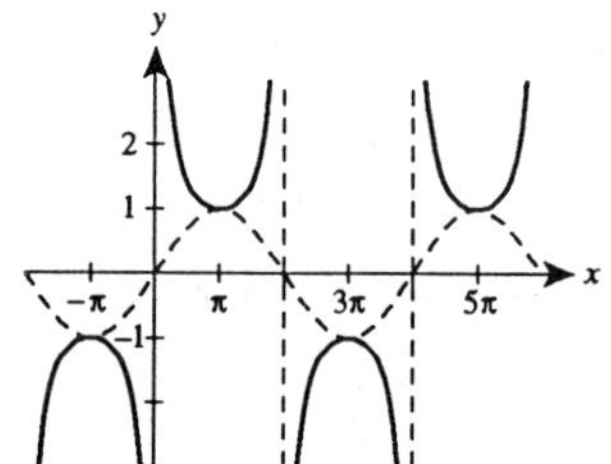

25. Sketch the graph of the following through two periods.

$$y = \cot \frac{x}{2}$$

Solution:

Period: $\dfrac{\pi}{1/2} = 2\pi$

One cycle: 0 to 2π

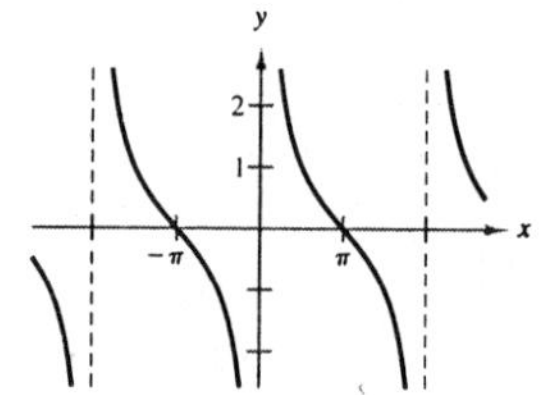

29. Sketch the graph of the following through two periods.

$$y = \tan\left(x - \frac{\pi}{4}\right)$$

Solution:

Period: π

Shift: Set $x - \dfrac{\pi}{4} = -\dfrac{\pi}{2}$ and $x - \dfrac{\pi}{4} = \dfrac{\pi}{2}$

$x = -\dfrac{\pi}{4}$ to $x = \dfrac{3\pi}{4}$

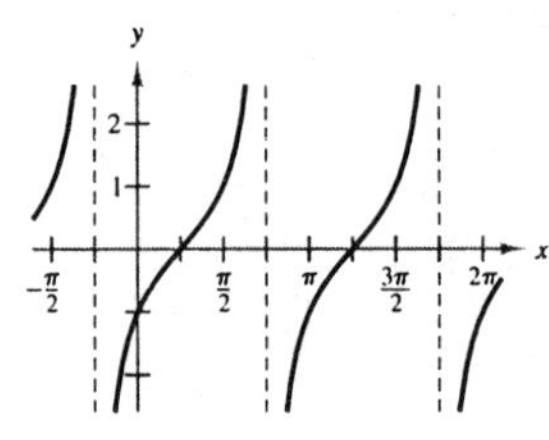

33. Sketch the graph of the following through two periods.

$$y = \frac{1}{4}\cot\left(x - \frac{\pi}{2}\right)$$

Solution:

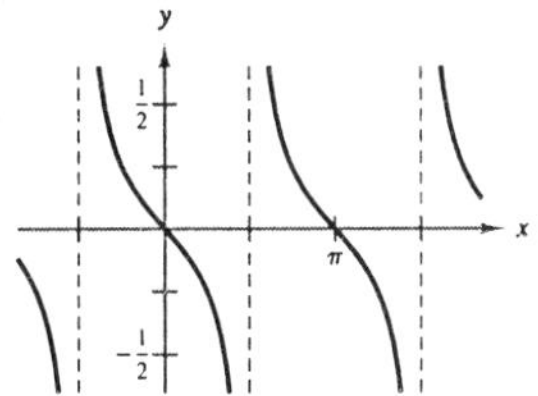

Period: π

Shift: Set $x - \dfrac{\pi}{2} = 0$ and $x - \dfrac{\pi}{2} = \pi$

$x = \dfrac{\pi}{2}$ to $x = \dfrac{3\pi}{2}$

35. Sketch the graph of $y = 2\sec(2x - \pi)$ through two periods.

Solution:

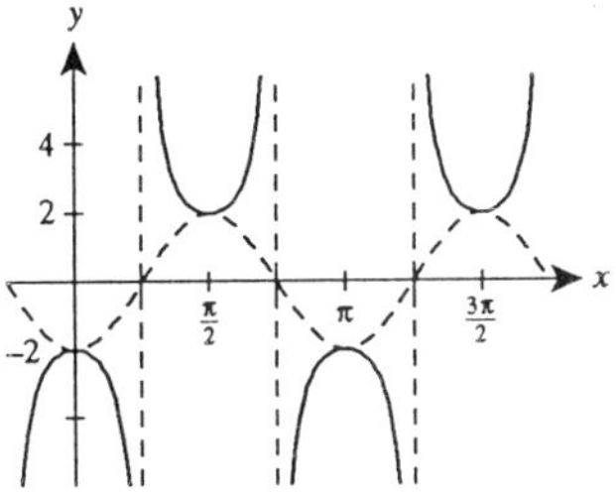

Graph $y = 2\cos(2x - \pi)$ first.

Period: π

Shift: Set $2x - \pi = 0$ and $2x - \pi = 2\pi$

$x = \dfrac{\pi}{2}$ to $x = \dfrac{3\pi}{2}$

39. Sketch the graph of $y = \csc(\pi - x)$ through two periods.

Solution:

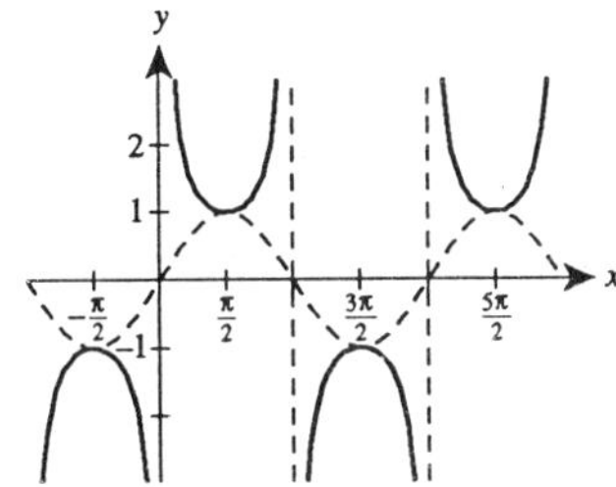

Graph $y = \sin(\pi - x)$ first.

Period: 2π

Shift: Set $\pi - x = 0$ and $\pi - x = 2\pi$

$x = \pi$ to $x = -\pi$

43. Use the graph of sec x to find all real numbers x in the interval $[-2\pi, 2\pi]$ that gives the functional value of -2.

Solution:

$$\sec x = -2$$

$$x = \pm\frac{2\pi}{3}, \pm\frac{4\pi}{3}$$

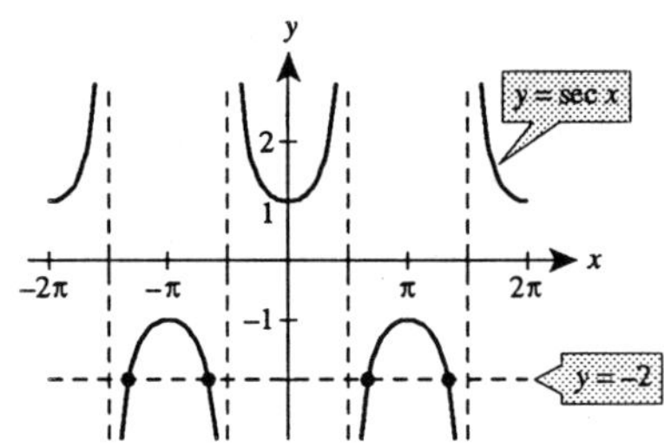

45. A plane flying at an altitude of 6 miles over level ground will pass directly over a radar antenna, as shown in the figure. Let d be the ground distance from the antenna to the point directly under the plane, and let x be the angle of elevation to the plane from the antenna. Write d as a function of x, and sketch the graph of the function over the interval $0 < x < \pi$.

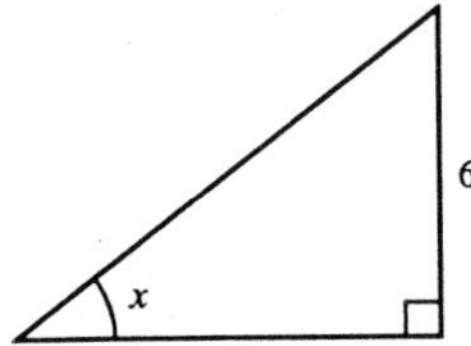

Solution:

$$\tan x = \frac{6}{d}$$

$$d = \frac{6}{\tan x} = 6 \cot x$$

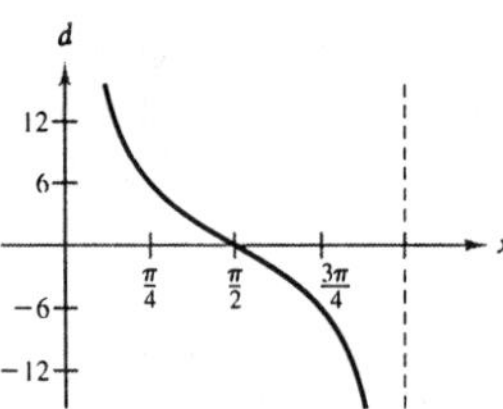

SECTION 5.7

Other Graphing Techniques

- You should be able to graph by addition of ordinates.
- You should be able to graph vertical translations.
- You should be able to graph using a damping factor.

Solutions to Selected Exercises

1. Use addition of ordinates to sketch the graph of

$$y = 2 - 2\sin\frac{x}{2}.$$

Solution:

Vertical translation of the graph of $y = -2\sin(x/2)$ by two units

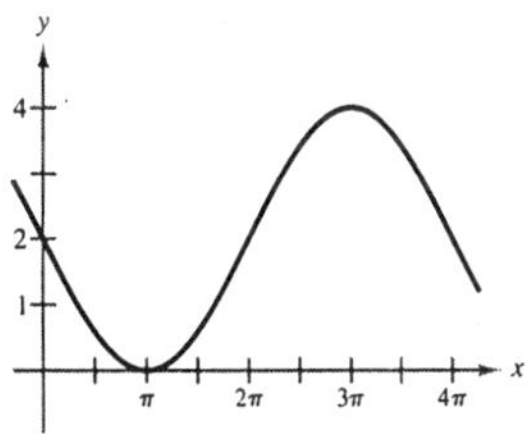

7. Use addition of ordinates to sketch the graph of $y = 1 + \csc x$.

Solution:

Vertical translation of the graph of $y = 1 + \csc x$ by one unit

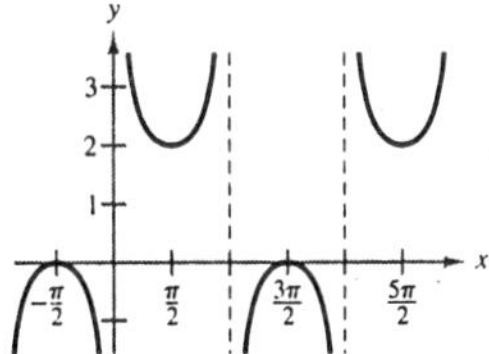

11. Use addition of ordinates to sketch the graph of $y = \frac{1}{2}x - 2\cos x$.

Solution:

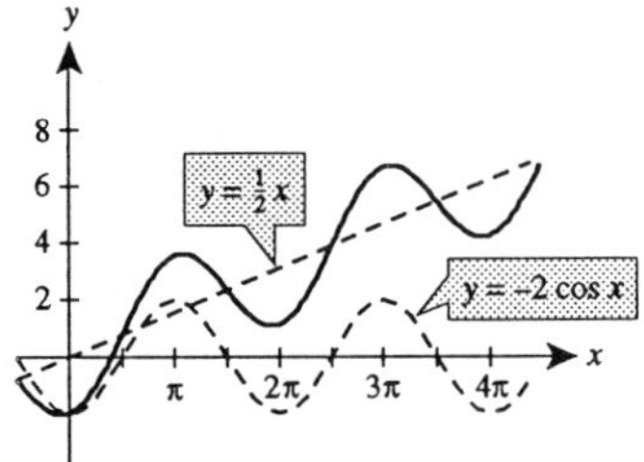

15. Use addition of ordinates to sketch the graph of $y = 2\sin x + \sin 2x$.

Solution:

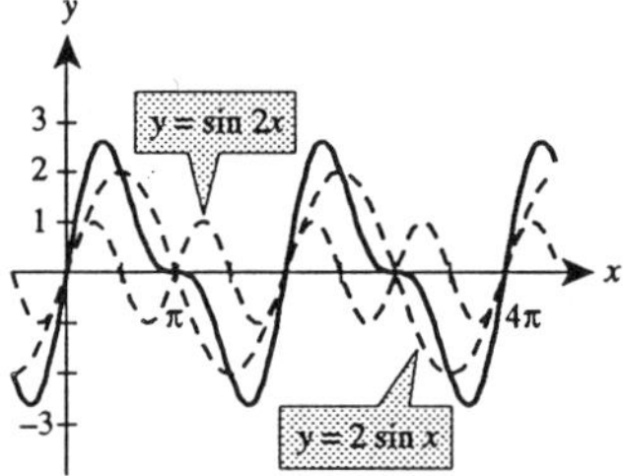

21. Use addition of ordinates to sketch the graph of $y = -3 + \cos x + 2\sin 2x$.

Solution:

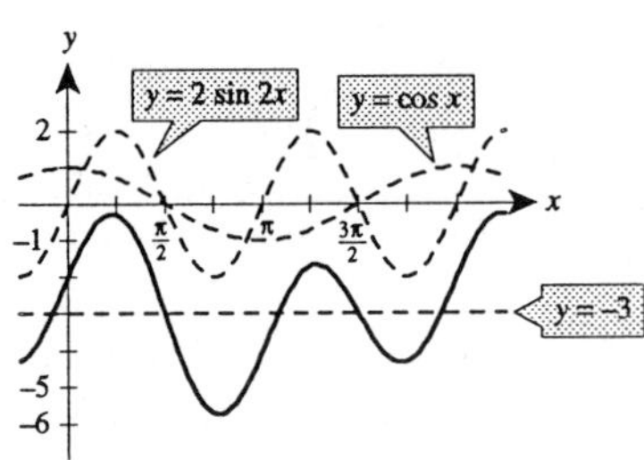

23. Sketch the graph of $y = x\cos x$.

Solution:

$$y = x\cos x$$

$$|x\cos x| = |x||\cos x| \le |x|$$

Thus, the graph lies between the lines $y = -x$ and $y = x$.

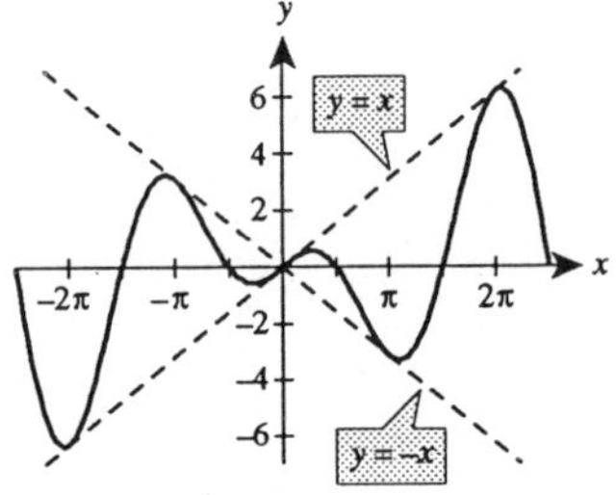

27. Sketch the graph of $y = e^{-x^2/2}\sin x$.

Solution:

$$y = e^{-x^2/2}\sin x$$

$$|e^{-x^2/2}\sin x| = |e^{-x^2/2}||\sin x| \le |e^{-x^2/2}|$$

Thus, $-e^{-x^2/2} \le y \le e^{-x^2/2}$.

31. The projected monthly sales S (in thousands of units) of a seasonal product is modeled by

$$S = 74 + 3t + 40 \sin \frac{\pi t}{6}$$

where t is the time in months, with $t = 1$ corresponding to January. Sketch the graph of this sales function over one year.

Solution:

t	0	1	2	3	4	5	6
S	74	97	114.64	123	120.64	109	92

t	7	8	9	10	11	12
S	75	63.36	61	69.36	87	110

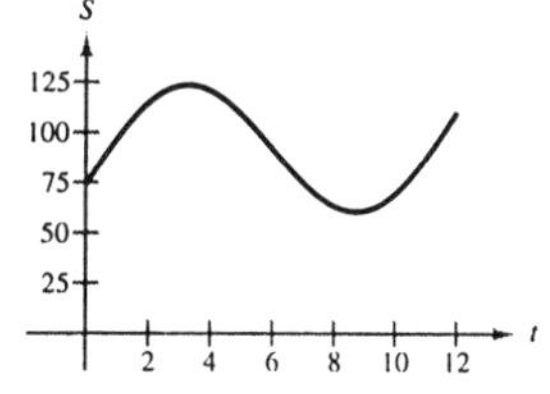

35. An object weighing W pounds is suspended from the ceiling by a steel spring (see figure). The weight is pulled downward (positive direction) from its equilibrium position and released. The resulting motion of the weight is described by the function

$$y = \tfrac{1}{2}e^{-t/4} \cos 4t, \quad t > 0$$

where y is the distance in feet and t is the time in seconds. Sketch the graph of this function.

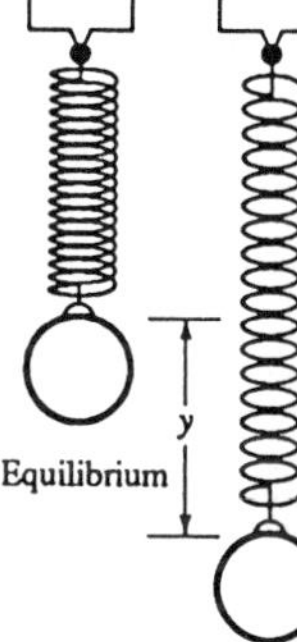

Solution:

$$y = \tfrac{1}{2}e^{-t/4} \cos 4t, \quad t > 0$$

$$\left|\tfrac{1}{2}e^{-t/4} \cos 4t\right| = \tfrac{1}{2}e^{-t/4}|\cos 4t|$$

$$\leq \tfrac{1}{2}e^{-t/4}$$

Thus, $-\frac{1}{2}e^{-t/4} \leq y \leq \frac{1}{2}e^{-t/4}$.

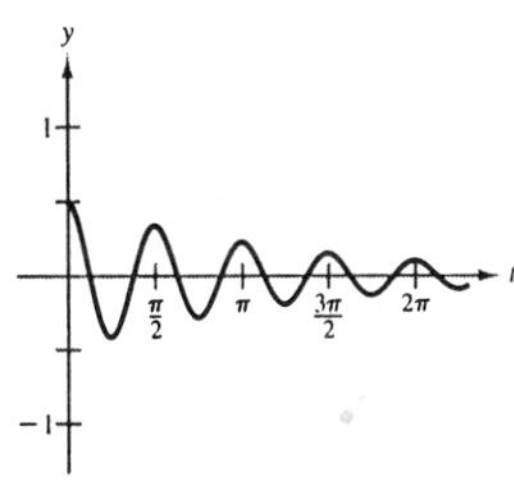

SECTION 5.8

Inverse Trigonometric Functions

- You should know the definitions, domains, and ranges of $y = \arcsin x$, $y = \arccos x$, and $y = \arctan x$.
- You should know the inverse properties of the inverse trigonometric functions.
- You should be able to use the triangle technique to convert trigonometric expressions into algebraic expressions.

Solutions to Selected Exercises

3. Evaluate $\arccos \frac{1}{2}$ without using a calculator.

Solution:

$$\arccos \frac{1}{2} = \theta$$

$$\cos \theta = \frac{1}{2}$$

$$\theta = \frac{\pi}{3}$$

7. Evaluate $\arccos\left(-\frac{\sqrt{3}}{2}\right)$ without using a calculator.

Solution:

$$\arccos\left(-\frac{\sqrt{3}}{2}\right) = \theta$$

$$\cos \theta = -\frac{\sqrt{3}}{2}, \quad \frac{\pi}{2} < \theta < \pi$$

$$\theta = \frac{5\pi}{6}$$

9. Evaluate $\arctan(-\sqrt{3})$ without using a calculator.

Solution:

$$\arctan(-\sqrt{3}) = \theta$$

$$\tan \theta = -\sqrt{3}, \quad -\frac{\pi}{2} < \theta < 0$$

$$\theta = -\frac{\pi}{3}$$

13. Evaluate $\arcsin \frac{\sqrt{3}}{2}$ without using a calculator.

Solution:

$$\arcsin \frac{\sqrt{3}}{2} = \theta$$

$$\sin \theta = \frac{\sqrt{3}}{2}$$

$$\theta = \frac{\pi}{3}$$

17. Use a calculator to approximate $\arccos 0.28$. (Round to two decimal places.)

Solution:

Make sure that your calculator is in radian mode.

$$\arccos 0.28 \approx 1.29$$

21. Use a calculator to approximate arctan(−2). (Round to two decimal places.)

Solution:

$\arctan(-2) \approx -1.11$

27. Use a calculator to approximate arctan 0.92. (Round to two decimal places.)

Solution:

$\arctan 0.92 \approx 0.74$

31. Use the properties of inverse trigonometric functions to evaluate cos[arccos(−0.1)].

Solution:

$\cos[\arccos(-0.1)] = -0.1$

35. Find the exact value of $\sin(\arctan \frac{3}{4})$ without using a calculator. [*Hint:* Make a sketch of a right triangle, as illustrated in Example 6.]

Solution:

Let $y = \arctan \frac{3}{4}$. Then,

$$\tan y = \frac{3}{4}, \quad 0 < y < \frac{\pi}{2}$$

and $\sin y = \frac{3}{5}$.

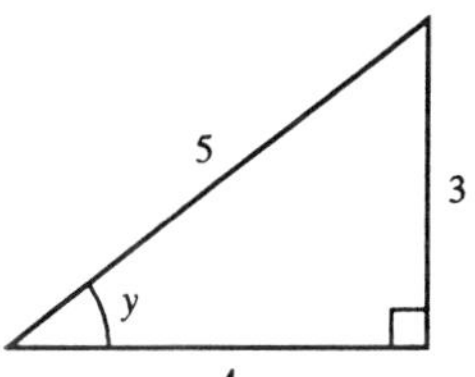

41. Find the exact value of $\sec[\arctan(-\frac{3}{5})]$ without using a calculator. [*Hint:* Make a sketch of a right triangle, as illustrated in Example 6.]

Solution:

Let $y = \arctan\left(-\frac{3}{5}\right)$. Then,

$$\tan y = -\frac{3}{5}, \quad -\frac{\pi}{2} < y < 0$$

and $\sec y = \sqrt{34}/5$.

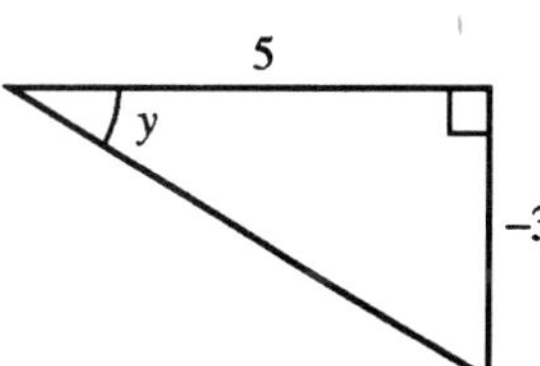

45. Write an algebraic expression that is equivalent to $\cos(\arcsin 2x)$. [*Hint:* Sketch a right triangle, as demonstrated in Example 7.]

Solution:

Let $y = \arcsin(2x)$. Then,

$$\sin y = 2x = \frac{2x}{1}$$

and $\cos y = \sqrt{1 - 4x^2}$.

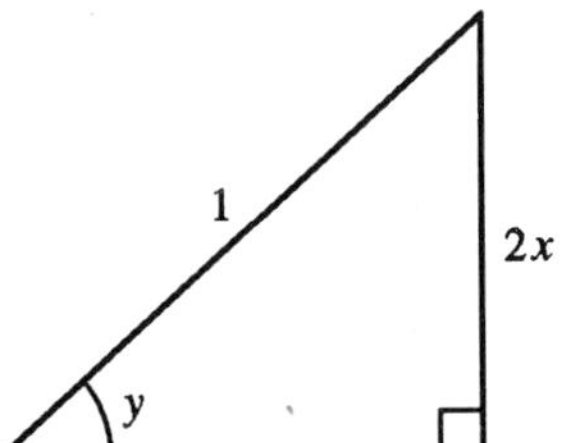

49. Write an algebraic expression that is equivalent to $\tan[\arccos(x/3)]$. [*Hint:* Sketch a right triangle, as demonstrated in Example 7.]

Solution:

Let $y = \arccos(x/3)$. Then,

$$\cos y = \frac{x}{3} \text{ and } \tan y = \frac{\sqrt{9 - x^2}}{x}.$$

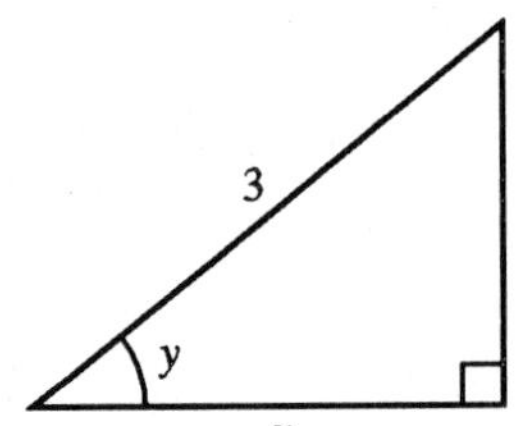

53. Fill in the blank.

$$\arctan \frac{9}{x} = \arcsin (\_\_)$$

Solution:

$$\arctan \frac{9}{x} = \arcsin \frac{9}{\sqrt{x^2 + 81}}$$

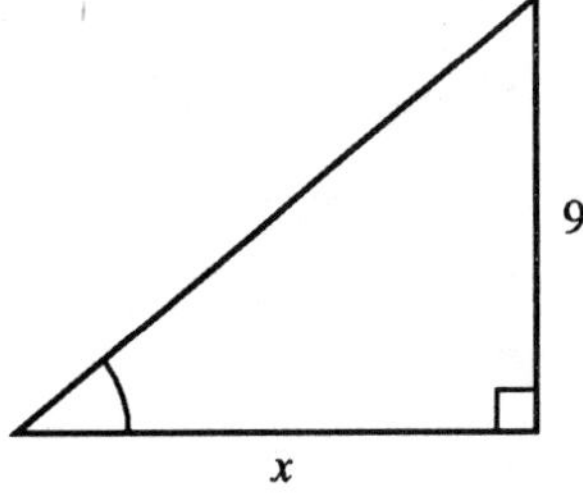

57. Sketch the graph of

$$f(x) = \arcsin(x - 1).$$

Solution:

The graph of $f(x) = \arcsin(x - 1)$ is a horizontal translation of the graph of $y = \arcsin x$ by one unit.

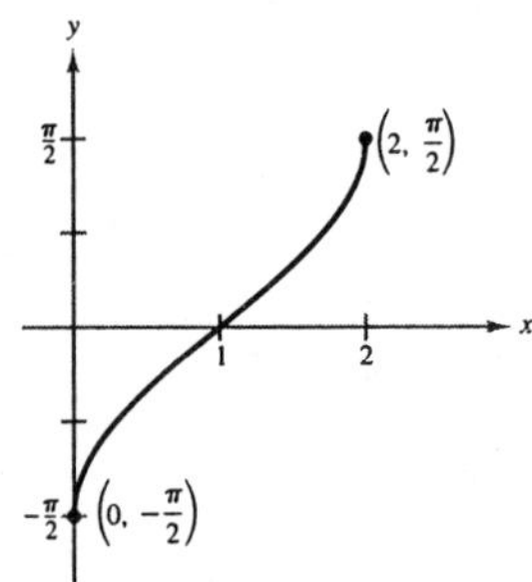

61. A photographer is taking a picture of a four-foot square painting hung in an art gallery. The camera lens is one foot below the lower edge of the painting, as shown in the figure. The angle β subtended by the camera lens x feet from the painting is given by

$$\beta = \arctan \frac{4x}{x^2+5}.$$

Find β when (a) $x = 3$ feet and (b) $x = 6$ feet.

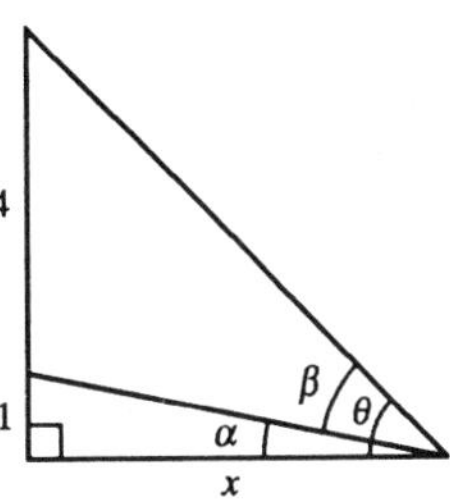

Solution:

(a) When $x = 3$, $\beta = \arctan\left(\frac{12}{9+5}\right) \approx 0.7086$ radians or $40.6°$.

(b) When $x = 6$, $\beta = \arctan\left(\frac{24}{36+5}\right) \approx 0.5296$ radians or $30.3°$.

65. Define the inverse cotangent function by restricting the domain of the cotangent to the interval $(0, \pi)$ and sketch its graph.

Solution:

$y = \operatorname{arccot} x$ if and only if $\cot y = x$

Domain: $-\infty < x < \infty$

Range: $0 < x < \pi$

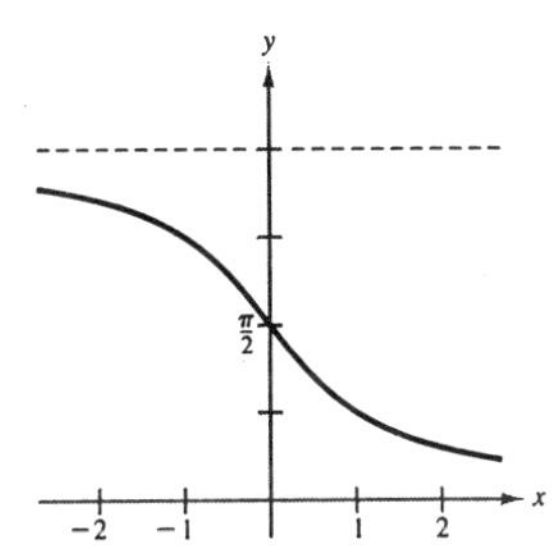

69. Prove the identity of $\arcsin(-x) = -\arcsin x$.

Solution:

Let $y = \arcsin(-x)$. Then,

$$\begin{aligned}
\sin y &= -x \\
-\sin y &= x \\
\sin(-y) &= x \\
-y &= \arcsin x \\
y &= -\arcsin x.
\end{aligned}$$

Therefore, $\arcsin(-x) = -\arcsin x$.

73. Prove $\arcsin x + \arccos x = \pi/2$.

Solution:

Let $\alpha = \arcsin x$ and $\beta = \arccos x$, then, $\sin \alpha = x$ and $\cos \beta = x$. Thus, $\sin \alpha = \cos \beta$ which implies that α and β are complementary angles and we have

$$\begin{aligned}
\alpha + \beta &= \frac{\pi}{2} \\
\arcsin x + \arccos x &= \frac{\pi}{2}.
\end{aligned}$$

SECTION 5.9

Applications of Trigonometry

- You should be able to solve right triangles.
- You should be able to solve right triangle applications.
- You should be able to solve applications of simple harmonic motion.

Solutions to Selected Exercises

5. Solve the right triangle, given $A = 12°15'$ and $c = 430.5$. (Round to two decimal places.)

Solution:

$A = 12°15', \quad c = 430.5$

$B = 90° - 12°15' = 77°45'$

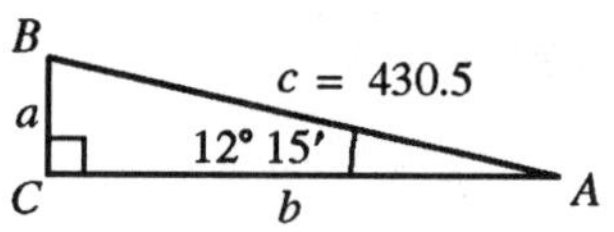

$$\sin 12°15' = \frac{a}{430.5}$$

$$a = 430.5 \sin 12°15' \approx 91.34$$

$$\cos 12°15' = \frac{b}{430.5}$$

$$b = 430.5 \cos 12°15' \approx 420.70$$

9. Solve the right triangle, given $b = 16$ and $c = 52$. (Round to two decimal places.)

Solution:

$b = 16, \quad c = 52$

$$a = \sqrt{52^2 - 16^2}$$

$$= \sqrt{2448} = 12\sqrt{17} \approx 49.48$$

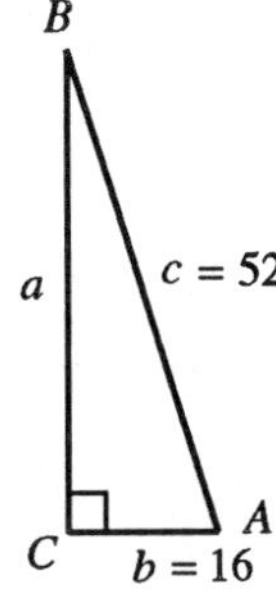

$$\cos A = \tfrac{16}{52}$$

$$A = \arccos \tfrac{16}{52} \approx 72.08°$$

$$B = 90 - 72.08 \approx 17.92°$$

11. An isosceles triangle has two angles of 52°, as shown in the figure. The base of the triangle is 4 inches. Find the altitude of the triangle.

Solution:

Divide the triangle in half. Then

$$\tan 52^\circ = \frac{h}{2}$$

$$h = 2\tan 52^\circ$$

$$\approx 2.56 \text{ inches.}$$

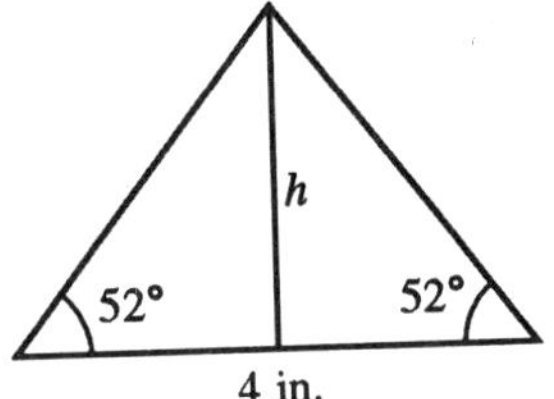

15. A ladder of length 16 feet leans against the side of a house (see figure). Find the height h of the top of the ladder if the angle of elevation of the ladder is 74°.

Solution:

$$\sin 74^\circ = \frac{h}{16}$$

$$16\sin 74^\circ = h$$

$$h \approx 15.4 \text{ feet}$$

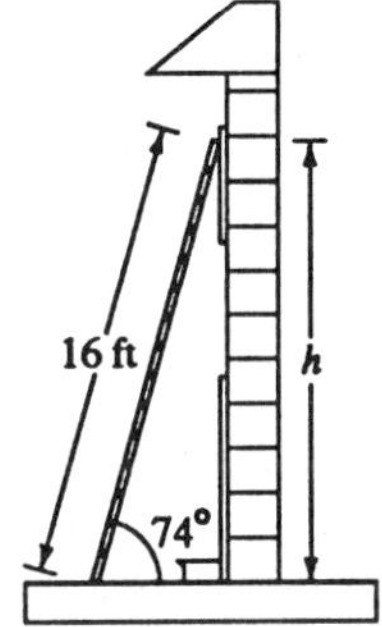

17. An amateur radio operator erects a 75-foot vertical tower for his antenna. Find the angle of elevation to the top of the tower at a point on level ground 50 feet from the base.

Solution:

$$\tan\theta = \tfrac{75}{50}$$

$$\theta = \arctan\tfrac{3}{2} \approx 56.3^\circ$$

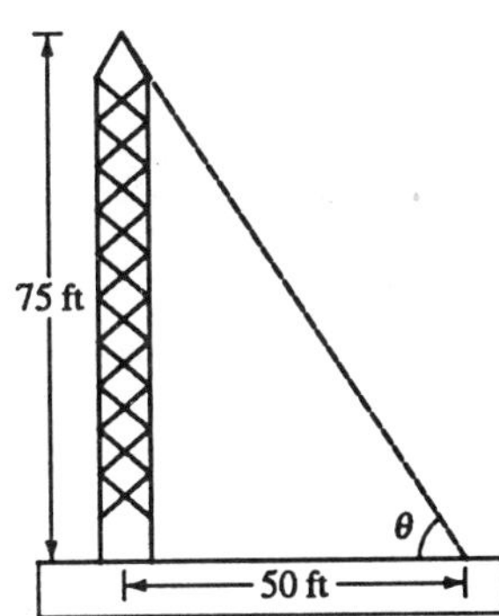

23. From a point 50 feet in front of a church, the angles of elevation to the base of the steeple and the top of the steeple are 35° and 47°40′, respectively, as shown in the figure. Find the height of the steeple.

Solution:

Let the height of the church $= x$ and the height of the church and steeple $= y$. Then,

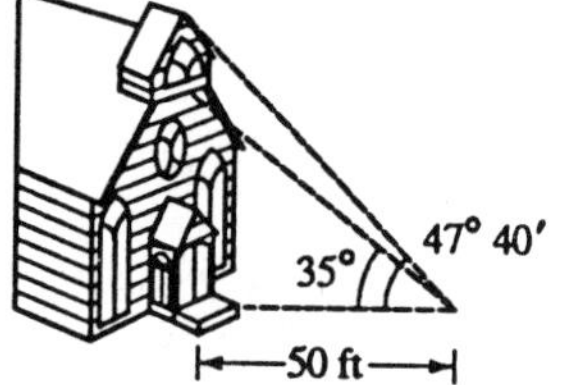

$$\tan 35° = \frac{x}{50} \qquad \text{and} \qquad \tan 47°40' = \frac{y}{50}$$

$$x \approx 35.01 \qquad\qquad y \approx 54.88$$

Height of the steeple $= y - x = 19.9$ feet

27. A ship is 45 miles east and 30 miles south of port. If the captain wants to travel directly to port, what bearing should be taken?

Solution:

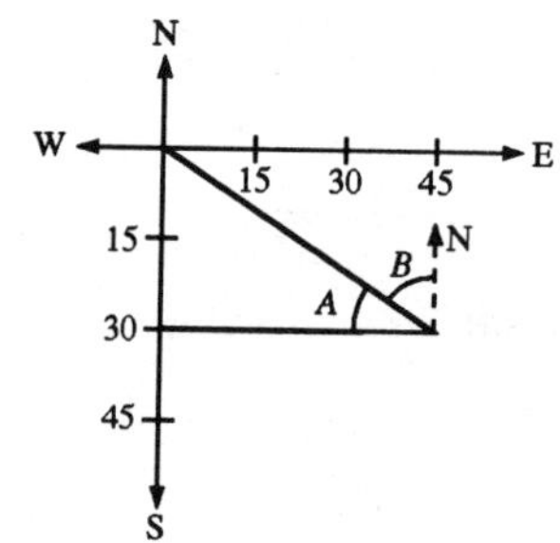

$$\tan A = \tfrac{30}{45}$$

$$A = \arctan \tfrac{2}{3}$$

$$A \approx 33.69°$$

Bearing $= 90 - 33.69 = 56.3° = \text{N } 56.3° \text{ W}$

31. An observer in a lighthouse 300 feet above sea level spots two ships directly offshore. The angles of depression to the ships are 4° and 6.5°, as shown in the figure. How far apart are the ships?

Solution:

$$\tan 4° = \frac{300}{y}$$

$$y = \frac{300}{\tan 4°} \approx 4290$$

$$\tan 6.5° = \frac{300}{x}$$

$$x = \frac{300}{\tan 6.5} \approx 2633$$

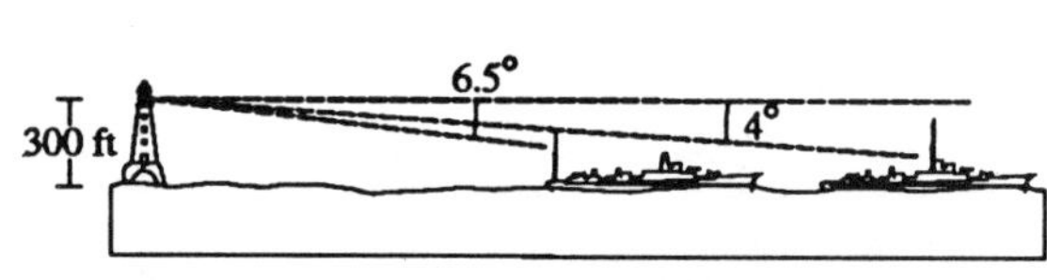

Distance between ships $= 1657$ feet

37. Use the figure to find the distance y across the flat sides of the hexagonal nut as a function of r.

Solution:

$$\sin 60° = \frac{\text{opp.}}{\text{hyp.}} = \frac{\frac{1}{2}y}{r}$$

$$\frac{\sqrt{3}}{2} = \frac{y}{2r}$$

$$y = \sqrt{3}r$$

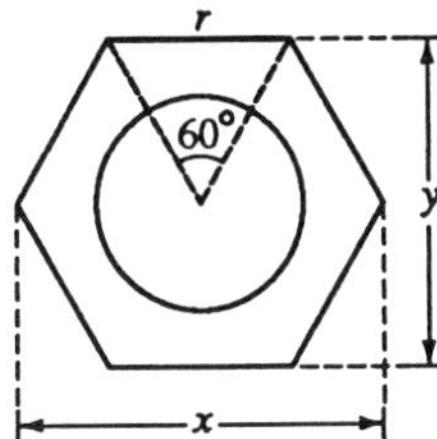

41. For the simple harmonic motion described by $d = 4\cos 8\pi t$, find (a) the maximum displacement, (b) the frequency, and (c) the least positive value of t for which $d = 0$.

Solution:

(a) Maximum displacement $=$ amplitude $= 4$

(b) Frequency $= \dfrac{\omega}{\text{period}} = \dfrac{8\pi}{2\pi} = 4$ cycles per unit of time

(c) $d = 0$ when $8\pi t = \dfrac{\pi}{2}$ or $t = \dfrac{1}{16}$

45. A point on the end of a tuning fork moves in simple harmonic motion described by $d = a\sin\omega t$. Find ω given that the tuning fork for middle C has a frequency of 264 vibrations per second.

Solution:

$$\text{Frequency} = \frac{\omega}{2\pi} = 264$$

$$\omega = 528\pi$$

REVIEW EXERCISES FOR CHAPTER 5

Solutions to Selected Exercises

3. Sketch the angle $-110°$ in standard position, and list one positive and one negative coterminal angle.

Solution:

$\theta = -110°$

Coterminal angles:

$\theta_1 = 360° + (-110°) = 250°$

$\theta_2 = -360° + (-110°) = -470°$

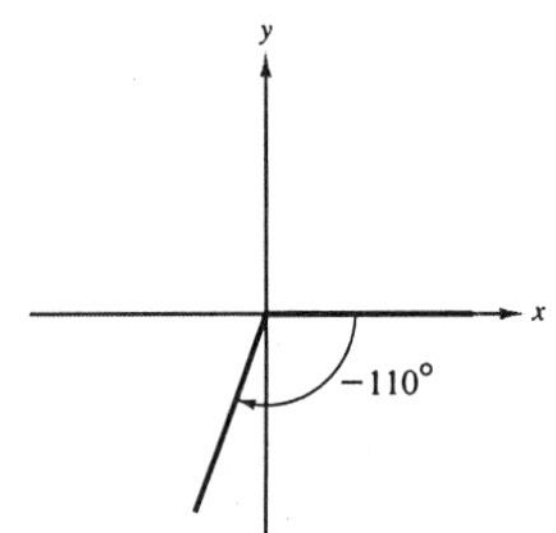

7. Convert $5°22'53''$ to decimal form. Round to two decimal places.

Solution:

$$5°22'53'' = \left(5 + \tfrac{22}{60} + \tfrac{53}{3600}\right)^{\circ} \approx 5.38°$$

11. Convert $-85.15°$ to D° M′ S″ form.

Solution:

$$-85.15° = -[85° + .15(60)'] = -85°9'$$

15. Convert -3.5 from radians to degrees. Round to two decimal places.

Solution:

$$-3.5 = -3.5\left(\frac{180}{\pi}\right)^{\circ} \approx -200.54°$$

19. Convert $-33°45'$ from degrees to radians. Round to four decimal places.

Solution:

$$\begin{aligned} -33°45' &= -\left[33 + \frac{45}{60}\right]^{\circ} \\ &= -33.75° \\ &= -33.75\left(\frac{\pi}{180}\right) \\ &\approx -0.5890 \text{ radians} \end{aligned}$$

21. Find the reference angle for $252°$.

Solution:

$252°$ is in Quadrant III. The reference angle is $\theta = 252° - 180° = 72°$.

25. Find the six trigonometric functions of the angle θ (in standard position) whose terminal side passes through the point (12, 16).

Solution:

$x = 12,\ y = 16,\ r = \sqrt{(12)^2 + (16)^2} = 20$

$\sin\theta = \frac{y}{r} = \frac{16}{20} = \frac{4}{5}$ $\qquad \csc\theta = \frac{r}{y} = \frac{20}{16} = \frac{5}{4}$

$\cos\theta = \frac{x}{r} = \frac{12}{20} = \frac{3}{5}$ $\qquad \sec\theta = \frac{r}{x} = \frac{20}{12} = \frac{5}{3}$

$\tan\theta = \frac{y}{x} = \frac{16}{12} = \frac{4}{3}$ $\qquad \cot\theta = \frac{x}{y} = \frac{12}{16} = \frac{3}{4}$

29. Find the six trigonometric functions of the angle θ (in standard position) whose terminal side passes through the point (−4, −6).

Solution:

$x = -4,\ y = -6,\ r = \sqrt{(-4)^2 + (-6)^2} = 2\sqrt{13}$

$\sin\theta = \frac{y}{r} = \frac{-6}{2\sqrt{13}} = -\frac{3\sqrt{13}}{13}$ $\qquad \csc\theta = \frac{r}{y} = \frac{2\sqrt{13}}{-6} = -\frac{\sqrt{13}}{3}$

$\cos\theta = \frac{x}{r} = \frac{-4}{2\sqrt{13}} = -\frac{2\sqrt{13}}{13}$ $\qquad \sec\theta = \frac{r}{x} = \frac{2\sqrt{13}}{-4} = -\frac{\sqrt{13}}{2}$

$\tan\theta = \frac{y}{x} = \frac{-6}{-4} = \frac{3}{2}$ $\qquad \cot\theta = \frac{x}{y} = \frac{-4}{-6} = \frac{2}{3}$

33. Find the remaining five trigonometric functions of θ, given $\sin\theta = \frac{3}{8}$ and $\cos\theta < 0$. [*Hint:* Sketch a right triangle.]

Solution:

$\sin\theta = \frac{3}{8},\ \cos\theta < 0,\ \theta$ is in Quadrant II.

$y = 3,\ r = 8,\ x = -\sqrt{64 - 9} = -\sqrt{55}$

$\cos\theta = -\frac{\sqrt{55}}{8}$ $\qquad \sec\theta = \frac{8}{-\sqrt{55}} = -\frac{8\sqrt{55}}{55}$

$\tan\theta = \frac{3}{-\sqrt{55}} = -\frac{3\sqrt{55}}{55}$ $\qquad \cot\theta = -\frac{\sqrt{55}}{3}$

$\csc\theta = \frac{8}{3}$

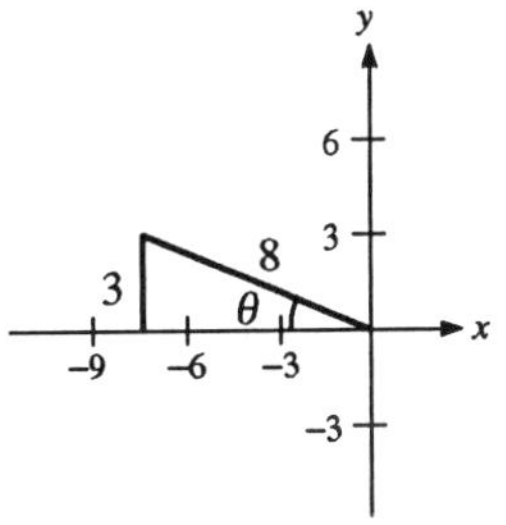

39. Evaluate $\cos 495°$ without the aid of a calculator.

Solution:

The reference angle for 495° is 45° and 495° is in Quadrant II. Therefore,

$$\cos 495° = -\cos 45° = -\frac{\sqrt{2}}{2}.$$

43. Use a calculator to evaluate $\sec(12\pi/5)$. Round your answer to two decimal places.

Solution:

Make sure that your calculator is in radian mode.

$$\sec\left(\frac{12\pi}{5}\right) = \frac{1}{\cos(12\pi/5)} \approx 3.24$$

47. Given $\csc\theta = -2$, find two values of θ in degrees $(0° \le \theta < 360°)$ and in radians $(0 \le \theta < 2\pi)$ without using a calculator.

Solution:

Since $\csc\theta < 0$, we know that θ is in either Quadrant III or in Quadrant IV. Also, since $\csc 30° = 2$, we know that 30° is the reference angle.

In Quadrant III: $\theta = 180° + 30° = 210° = \dfrac{7\pi}{6}$

In Quadrant IV: $\theta = 360° - 30° = 330° = \dfrac{11\pi}{6}$

51. Given $\sec\theta = -1.0353$, find two values of θ in degrees $(0° \le \theta < 360°)$ and in radians $(0 \le \theta < 2\pi)$ by using a calculator.

Solution:

Since $\sec\theta < 0$, we know that θ is in either Quadrant II or in Quadrant III. To find the reference angle θ', use

$$\cos\theta' = \frac{1}{\sec\theta'} = \frac{1}{1.0353}$$

$$\theta' = \cos^{-1}\left(\frac{1}{1.0353}\right).$$

The reference angle θ' is 15°.

In Quadrant II: $\theta = 180° - 15° = 165° \approx 2.8798$

In Quadrant III: $\theta = 180° + 15° = 195° \approx 3.4034$

55. Write an algebraic expression for

$$\sin\left(\arccos\frac{x^2}{4-x^2}\right).$$

Solution:

Use a right triangle to find $\sin\left(\arccos\dfrac{x^2}{4-x^2}\right)$.

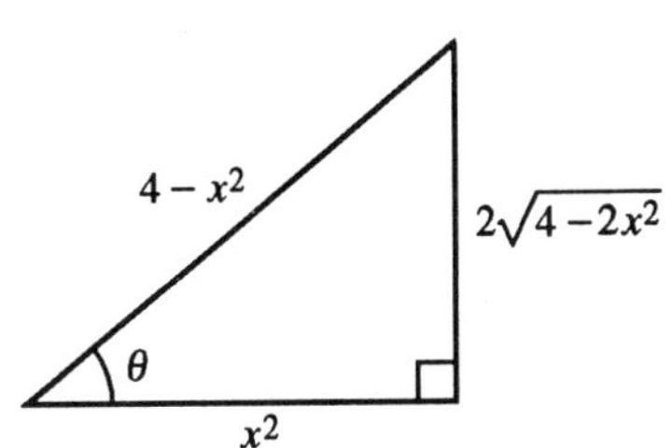

Let $\theta = \arccos\dfrac{x^2}{4-x^2}$, then, $\cos\theta = \dfrac{x^2}{4-x^2}$.

The opposite side is

$$\sqrt{(4-x^2)^2-(x^2)^2} = \sqrt{16-8x^2}$$
$$= 2\sqrt{4-2x^2}.$$

Thus, $\sin\theta = \dfrac{2\sqrt{4-2x^2}}{4-x^2}$.

61. Sketch the graph of

$$f(x) = -\frac{1}{4}\cos\frac{\pi x}{4}.$$

Solution:

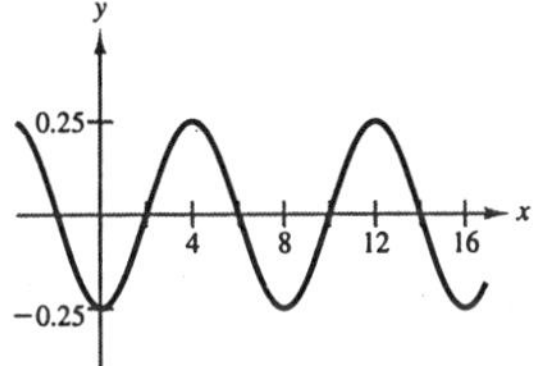

Amplitude: $\left|-\dfrac{1}{4}\right| = \dfrac{1}{4}$

Period: $\dfrac{2\pi}{\pi/4} = 8$

67. Sketch the graph of

$$f(t) = \csc\left(3t-\frac{\pi}{2}\right).$$

Solution:

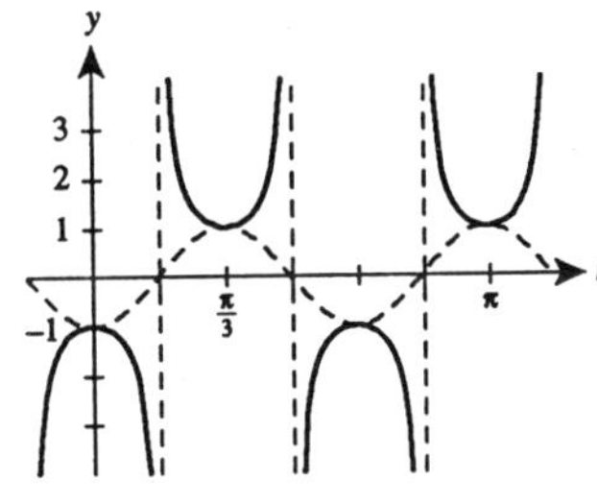

Period: $\dfrac{2\pi}{3}$

Shift: $3t-\dfrac{\pi}{2}=0$ and $3t-\dfrac{\pi}{2}=2\pi$

$t=\dfrac{\pi}{6}$ $\qquad$ $t=\dfrac{5\pi}{6}$

71. Sketch the graph of

$$f(x) = \frac{x}{4} - \sin x.$$

Solution:

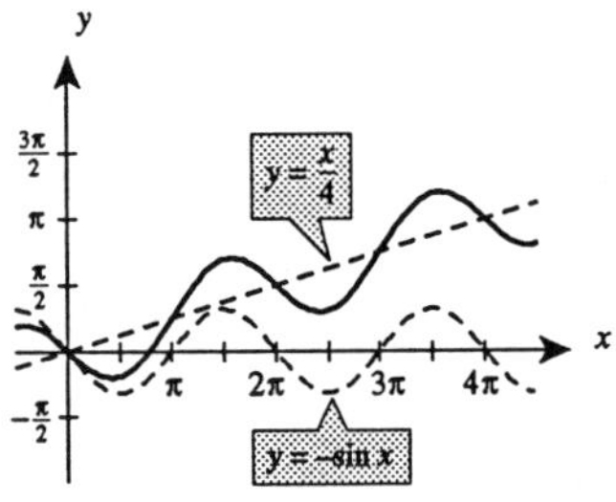

77. Sketch the graph of

$$y = \arcsin \frac{x}{2}.$$

Solution:

Domain: $-2 \leq x \leq 2$

Range: $-\frac{\pi}{2} \leq y \leq \frac{\pi}{2}$

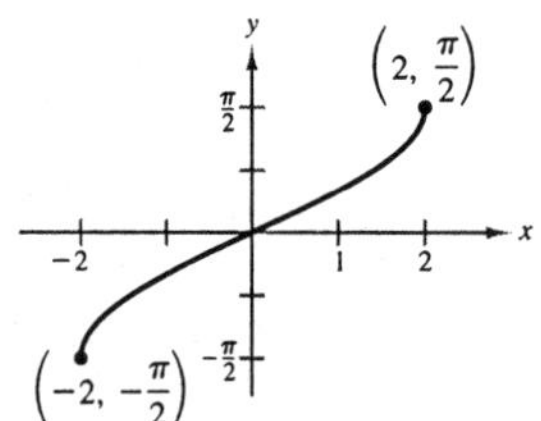

83. An observer 2.5 miles from the launch pad of a space shuttle measures the angle of elevation to the base of the vehicle to be 28° soon after liftoff (see figure). How high is the shuttle at that instant? Assume that the shuttle is still moving vertically.

Solution:

$$\tan 28° = \frac{x}{2.5}$$

$$x = 2.5 \tan 28° \approx 1.33 \text{ miles}$$

The shuttle is approximately 1.33 miles off the ground.

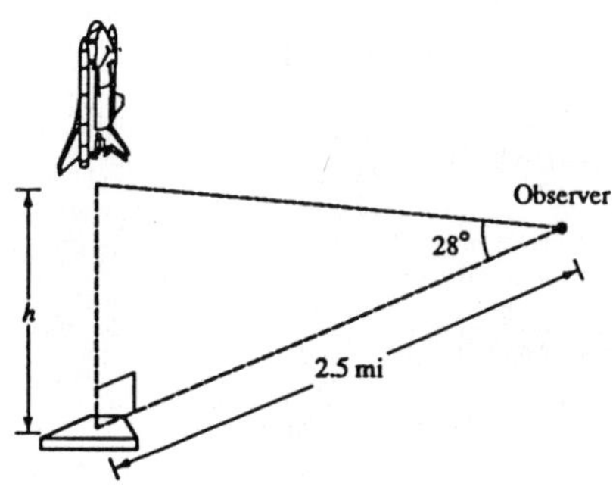

Practice Test for Chapter 5

1. (a) Express $350°$ in radian measure.

(b) Express $\frac{5\pi}{9}$ in degree measure.

2. (a) Convert $135°14'12''$ to decimal form.

(b) Convert $-22.569°$ to D° M′ S″ form.

3. Use the unit circle to evaluate the following.

(a) $\sin \frac{5\pi}{6}$

(b) $\tan \frac{5\pi}{4}$

4. Use the unit circle and the periodic nature of sine and cosine to evaluate the following.

(a) $\sin 7\pi$

(b) $\cos\left(-\frac{13\pi}{3}\right)$

5. If $\cos\theta = \frac{2}{3}$, use the trigonometric identities to find $\tan\theta$ $(0° \le \theta \le 90°)$.

6. Find θ given $\sin\theta = 0.9063$ $(0° \le \theta \le 90°)$.

7. Solve for x.

35

20°

x

8. Find the magnitude of the reference angle for $\theta = \frac{6\pi}{5}$.

9. Evaluate $\csc 3.92$.

10. Find $\sec\theta$ given that θ lies in Quadrant III and $\tan\theta = 6$.

11. Graph $y = 3\sin\frac{x}{2}$.

12. Graph $y = -2\cos(x - \pi)$.

13. Graph $y = \tan 2x$.

14. Graph $y = -\csc\left(x + \frac{\pi}{4}\right)$.

15. Graph $y = 2x + \sin x$.

16. Graph $y = 3x \cos x$.

17. Evaluate $\arcsin 1$.

18. Evaluate $\arctan(-3)$.

19. Evaluate $\sin\left(\arccos \dfrac{4}{\sqrt{35}}\right)$.

20. Write an algebraic expression for $\cos\left(\arcsin \dfrac{x}{4}\right)$.

For Exercises 21–23, solve the right triangle.

21. $A = 40°, \; c = 12$

22. $B = 6.84°, \; a = 21.3$

23. $a = 5, \; b = 9$

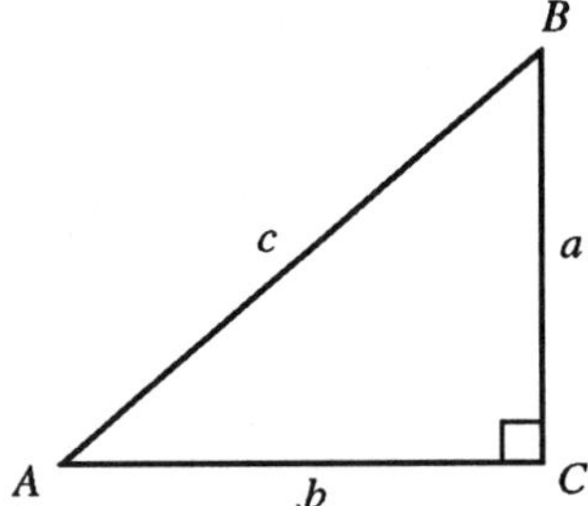

24. A 20-foot ladder leans against the side of a barn. Find the height of the top of the ladder if the angle of elevation of the ladder is 67°.

25. An observer in a lighthouse 250 feet above sea level spots a ship off the shore. If the angle of depression to the ship is 5°, how far out is the ship?

CHAPTER 6

Analytic Trigonometry

Section 6.1 Applications of Fundamental Identities **264**

Section 6.2 Verifying Trigonometric Identities **270**

Section 6.3 Solving Trigonometric Equations **273**

Section 6.4 Sum and Difference Formulas **278**

Section 6.5 Multiple-Angle Formulas and Product-Sum Formulas **283**

Review Exercises **291**

Practice Test **295**

SECTION 6.1

Applications of Fundamental Identities

- You should know the fundamental trigonometric identities.

  (a) Reciprocal Identities

  $$\sin u = \frac{1}{\csc u} \qquad \csc u = \frac{1}{\sin u}$$

  $$\cos u = \frac{1}{\sec u} \qquad \sec u = \frac{1}{\cos u}$$

  $$\tan u = \frac{1}{\cot u} = \frac{\sin u}{\cos u} \qquad \cot u = \frac{1}{\tan u} = \frac{\cos u}{\sin u}$$

  (b) Pythagorean Identities

  $$\sin^2 u + \cos^2 u = 1$$

  $$1 + \tan^2 u = \sec^2 u$$

  $$1 + \cot^2 u = \csc^2 u$$

  (c) Cofunction Identities

  $$\sin\left(\frac{\pi}{2} - u\right) = \cos u \qquad \cos\left(\frac{\pi}{2} - u\right) = \sin u$$

  $$\tan\left(\frac{\pi}{2} - u\right) = \cot u \qquad \cot\left(\frac{\pi}{2} - u\right) = \tan u$$

  $$\sec\left(\frac{\pi}{2} - u\right) = \csc u \qquad \csc\left(\frac{\pi}{2} - u\right) = \sec u$$

  (d) Negative Angle Identities

  $$\sin(-x) = -\sin x \qquad \csc(-x) = -\csc x$$

  $$\cos(-x) = \cos x \qquad \sec(-x) = \sec x$$

  $$\tan(-x) = -\tan x \qquad \cot(-x) = -\cot x$$

- You should be able to use these fundamental identities to find function values.

- You should be able to convert trigonometric expressions to equivalent forms by using the fundamental identities.

Solutions to Selected Exercises

3. Given $\sec\theta = \sqrt{2}$ and $\sin\theta = -\sqrt{2}/2$, use the fundamental identities to find the values of the other trigonometric functions.

Solution:

$$\cos\theta = \frac{1}{\sec\theta} = \frac{1}{\sqrt{2}} = \frac{\sqrt{2}}{2} \qquad \cot\theta = \frac{1}{\tan\theta} = \frac{1}{-1} = -1$$

$$\tan\theta = \frac{\sin\theta}{\cos\theta} = \frac{-\sqrt{2}/2}{\sqrt{2}/2} = -1 \qquad \csc\theta = \frac{1}{\sin\theta} = \frac{1}{-\sqrt{2}/2} = -\sqrt{2}$$

7. Given $\sec\phi = -1$ and $\sin\phi = 0$, use the fundamental identities to find the values of the other trigonometric functions.

Solution:

$$\cos\phi = \frac{1}{\sec\phi} = \frac{1}{-1} = -1 \qquad \cot\phi = \frac{1}{\tan\phi} = \frac{1}{0}, \quad \text{which is undefined}$$

$$\tan\phi = \frac{\sin\phi}{\cos\phi} = \frac{0}{-1} = 0 \qquad \csc\phi = \frac{1}{\sin\phi} = \frac{1}{0}, \quad \text{which is undefined}$$

11. Given $\tan\theta = 2$ and $\sin\theta < 0$, use the fundamental identities to find the values of the other trigonometric functions.

Solution:

θ is in Quadrant III.

$$\sec\theta = -\sqrt{1+\tan^2\theta} = -\sqrt{1+(2)^2} = -\sqrt{5}$$

$$\cot\theta = \frac{1}{\tan\theta} = \frac{1}{2}$$

$$\csc\theta = -\sqrt{1+\cot^2\theta} = -\sqrt{1+(1/2)^2} = -\sqrt{1+(1/4)} = -\frac{\sqrt{5}}{2}$$

$$\sin\theta = \frac{1}{\csc\theta} = -\frac{2}{\sqrt{5}} = -\frac{2\sqrt{5}}{5}$$

$$\cos\theta = \frac{1}{\sec\theta} = -\frac{1}{\sqrt{5}} = -\frac{\sqrt{5}}{5}$$

17. Match $\tan^2 x - \sec^2 x$ with one of the following.

(a) -1 (b) $\cos x$ (c) $\cot x$
(d) 1 (e) $-\tan x$ (f) $\sin x$

Solution:

$$\tan^2 x - \sec^2 x = (\sec^2 x - 1) - \sec^2 x = -1$$

Matches (a).

21. Match $\sin x \sec x$ with one of the following.

(a) $\csc x$ (b) $\tan x$ (c) $\sin^2 x$
(d) $\sin x \tan x$ (e) $\sec^2 x$ (f) $\sec^2 x + \tan^2 x$

Solution:

$$\sin x \sec x = \sin x\left(\frac{1}{\cos x}\right) = \frac{\sin x}{\cos x} = \tan x$$

Matches (b).

25. Match $\sec^4 x - \tan^4 x$ with one of the following.

(a) $\csc x$ (b) $\tan x$ (c) $\sin^2 x$
(d) $\sin x \tan x$ (e) $\sec^2 x$ (f) $\sec^2 x + \tan^2 x$

Solution:

$$\begin{aligned}\sec^4 x - \tan^4 x &= (\sec^2 x + \tan^2 x)(\sec^2 x - \tan^2 x)\\ &= (\sec^2 x + \tan^2 x)[(1 + \tan^2 x) - \tan^2 x]\\ &= \sec^2 x + \tan^2 x\end{aligned}$$

Matches (f).

29. Use the fundamental identities to simplify $\cos\beta\tan\beta$.

Solution:

$$\cos\beta\tan\beta = \cos\beta\left(\frac{\sin\beta}{\cos\beta}\right) = \sin\beta$$

33. Use the fundamental identities to simplify $\sec^2 x(1 - \sin^2 x)$.

Solution:

$$\sec^2 x(1 - \sin^2 x) = \sec^2 x(\cos^2 x) = \frac{1}{\cos^2 x}(\cos^2 x) = 1$$

39. Use the fundamental identities to simplify

$$\frac{\cos^2 y}{1 - \sin y}.$$

Solution:

$$\frac{\cos^2 y}{1 - \sin y} = \frac{1 - \sin^2 y}{1 - \sin y}$$

$$= \frac{(1 + \sin y)(1 - \sin y)}{1 - \sin y}$$

$$= 1 + \sin y$$

43. Factor $\sin^2 x \sec^2 x - \sin^2 x$ and use the fundamental identities to simplify the result.

Solution:

$$\sin^2 x \sec^2 x - \sin^2 x = \sin^2 x(\sec^2 x - 1)$$

$$= \sin^2 x \tan^2 x$$

47. Factor $\sin^4 x - \cos^4 x$ and use the fundamental identities to simplify the result.

Solution:

$$\sin^4 x - \cos^4 x = (\sin^2 x + \cos^2 x)(\sin^2 x - \cos^2 x)$$

$$= 1(\sin^2 x - \cos^2 x)$$

$$= \sin^2 x - \cos^2 x$$

51. Multiply $(\sec x + 1)(\sec x - 1)$ and use the fundamental identities to simplify the result.

Solution:

$$(\sec x + 1)(\sec x - 1) = \sec^2 x - 1 = \tan^2 x$$

55. Add the following and use the fundamental identities to simplify the result.

$$\frac{\cos x}{1 + \sin x} + \frac{1 + \sin x}{\cos x}$$

Solution:

$$\frac{\cos x}{1 + \sin x} + \frac{1 + \sin x}{\cos x} = \frac{(\cos x)(\cos x) + (1 + \sin x)(1 + \sin x)}{\cos x(1 + \sin x)}$$

$$= \frac{\cos^2 x + 1 + 2\sin x + \sin^2 x}{\cos x(1 + \sin x)}$$

$$= \frac{(\sin^2 x + \cos^2 x) + 1 + 2\sin x}{\cos x(1 + \sin x)}$$

$$= \frac{1 + 1 + 2\sin x}{\cos x(1 + \sin x)}$$

$$= \frac{2(1 + \sin x)}{\cos x(1 + \sin x)}$$

$$= \frac{2}{\cos x} = 2\left(\frac{1}{\cos x}\right) = 2\sec x$$

59. Rewrite the following expression so that it is *not* in fractional form.

$$\frac{3}{\sec x - \tan x}$$

Solution:

$$\begin{aligned}\frac{3}{\sec x - \tan x} &= \frac{3}{\sec x - \tan x} \cdot \frac{\sec x + \tan x}{\sec x + \tan x}\\ &= \frac{3(\sec x + \tan x)}{\sec^2 x - \tan^2 x}\\ &= \frac{3(\sec x + \tan x)}{(1 + \tan^2 x) - \tan^2 x}\\ &= 3(\sec x + \tan x)\end{aligned}$$

63. Use $x = 3\sec\theta$ to write $\sqrt{x^2 - 9}$ as a trigonometric function of θ, where $0 < \theta < \pi/2$.

Solution:

Since $x = 3\sec\theta$,

$$\begin{aligned}\sqrt{x^2 - 9} &= \sqrt{(3\sec\theta)^2 - 9}\\ &= \sqrt{9\sec^2\theta - 9}\\ &= \sqrt{9(\sec^2\theta - 1)}\\ &= \sqrt{9\tan^2\theta}\\ &= 3\tan\theta.\end{aligned}$$

69. Use $x = 3\tan\theta$ to write $\sqrt{(9 + x^2)^3}$ as a trigonometric function of θ, where $0 < \theta < \pi/2$.

Solution:

Since $x = 3\tan\theta$,

$$\begin{aligned}\sqrt{(9 + x^2)^3} &= \sqrt{[9 + (3\tan\theta)^2]^3}\\ &= \sqrt{(9 + 9\tan^2\theta)^3}\\ &= \sqrt{[9(1 + \tan^2\theta)]^3}\\ &= \sqrt{(9\sec^2\theta)^3}\\ &= (\sqrt{9\sec^2\theta})^3\\ &= (3\sec\theta)^3\\ &= 27\sec^3\theta.\end{aligned}$$

73. Determine whether the equation is an identity, and give a reason for your answer.

$$\frac{\sin k\theta}{\cos k\theta} = \tan\theta, \quad k \text{ is constant}$$

Solution:

$\frac{\sin k\theta}{\cos k\theta} = \tan\theta$ is not an identity, since $\frac{\sin k\theta}{\cos k\theta} = \tan k\theta$.

77. Use a calculator to demonstrate that $\csc^2\theta - \cot^2\theta = 1$ is true for the following.

(a) $\theta = 132°$

(b) $\theta = \frac{2\pi}{7}$

Solution:

(a) $\theta = 132°$

$\csc 132° \approx 1.34563$

$\cot 132° \approx -0.90040$

Therefore,

$$\csc^2 132° - \cot^2 132° \approx 1.81072 - 0.81072 = 1.$$

(b) $\theta = \frac{2\pi}{7}$

$\csc\frac{2\pi}{7} \approx 1.2790$

$\cot\frac{2\pi}{7} \approx 0.7975$

Therefore,

$$\csc^2\frac{2\pi}{7} - \cot^2\frac{2\pi}{7} \approx 1.6360 - 0.6360 = 1.$$

81. Express each of the other trigonometric functions of θ in terms of $\sin\theta$.

Solution:

Since $\sin^2\theta + \cos^2\theta = 1$ and $\cos^2\theta = 1 - \sin^2\theta$:

$$\cos\theta = \pm\sqrt{1-\sin^2\theta}$$

$$\tan\theta = \frac{\sin\theta}{\cos\theta} = \pm\frac{\sin\theta}{\sqrt{1-\sin^2\theta}}$$

$$\cot\theta = \frac{1}{\tan\theta} = \pm\frac{\sqrt{1-\sin^2\theta}}{\sin\theta}$$

$$\sec\theta = \frac{1}{\cos\theta} = \pm\frac{1}{\sqrt{1-\sin^2\theta}}$$

$$\csc\theta = \frac{1}{\sin\theta}$$

SECTION 6.2

Verifying Trigonometric Identities

- You should know the difference between an expression, a conditional equation, and an identity.
- You should be able to solve trigonometric identities, using the following techniques.
  (a) Work with *one* side at a time. Do not "cross" the equal sign.
  (b) Use algebraic techniques such as combining fractions, factoring expressions, rationalizing denominators, and squaring binomials.
  (c) Use the fundamental identities.
  (d) Convert all the terms into sines and cosines.

Solutions to Selected Exercises

5. Verify $\cos^2\beta - \sin^2\beta = 1 - 2\sin^2\beta$.

Solution:

$$\begin{aligned}\cos^2\beta - \sin^2\beta &= (1 - \sin^2\beta) - \sin^2\beta\\ &= 1 - 2\sin^2\beta\end{aligned}$$

9. Verify $\sin^2\alpha - \sin^4\alpha = \cos^2\alpha - \cos^4\alpha$.

Solution:

$$\begin{aligned}\sin^2\alpha - \sin^4\alpha &= \sin^2\alpha(1 - \sin^2\alpha)\\ &= (1 - \cos^2\alpha)(\cos^2\alpha)\\ &= \cos^2\alpha - \cos^4\alpha\end{aligned}$$

13. Verify $\dfrac{\cot^2 t}{\csc t} = \csc t - \sin t$.

Solution:

$$\begin{aligned}\frac{\cot^2 t}{\csc t} &= \frac{\csc^2 t - 1}{\csc t}\\ &= \frac{\csc^2 t}{\csc t} - \frac{1}{\csc t}\\ &= \csc t - \sin t\end{aligned}$$

17. Verify $\dfrac{1}{\sec x \tan x} = \csc x - \sin x$.

Solution:

$$\begin{aligned}\frac{1}{\sec x \tan x} &= \frac{1}{\sec x} \cdot \frac{1}{\tan x}\\ &= \cos x \cot x\\ &= \cos x\left(\frac{\cos x}{\sin x}\right)\\ &= \frac{\cos^2 x}{\sin x} = \frac{1 - \sin^2 x}{\sin x}\\ &= \frac{1}{\sin x} - \frac{\sin^2 x}{\sin x}\\ &= \csc x - \sin x\end{aligned}$$

19. Verify $\csc x - \sin x = \cos x \cot x$.

Solution:

$$\csc x - \sin x = \frac{1}{\sin x} - \frac{\sin^2 x}{\sin x}$$

$$= \frac{1 - \sin^2 x}{\sin x} = \frac{\cos^2 x}{\sin x} = \cos x \left(\frac{\cos x}{\sin x}\right) = \cos x \cot x$$

25. Verify $\dfrac{\cos\theta \cot\theta}{1 - \sin\theta} - 1 = \csc\theta$.

Solution:

$$\frac{\cos\theta \cot\theta}{1 - \sin\theta} - 1 = \frac{\cos\theta\left(\dfrac{\cos\theta}{\sin\theta}\right)}{1 - \sin\theta} \bullet \frac{1 + \sin\theta}{1 + \sin\theta} - 1$$

$$= \frac{\cos^2\theta(1 + \sin\theta)}{\sin\theta(1 - \sin^2\theta)} - 1$$

$$= \frac{\cos^2\theta(1 + \sin\theta)}{\sin\theta\cos^2\theta} - 1$$

$$= \frac{1 + \sin\theta}{\sin\theta} - 1$$

$$= \frac{1}{\sin\theta} + \frac{\sin\theta}{\sin\theta} - 1$$

$$= \csc\theta + 1 - 1$$

$$= \csc\theta$$

29. Verify $2\sec^2 x - 2\sec^2 x \sin^2 x - \sin^2 x - \cos^2 x = 1$.

Solution:

$$2\sec^2 x - 2\sec^2 x \sin^2 x - \sin^2 x - \cos^2 x = 2\sec^2 x(1 - \sin^2 x) - (\sin^2 x + \cos^2 x)$$

$$= 2\sec^2 x(\cos^2 x) - 1$$

$$= 2\left(\frac{1}{\cos^2 x}\right)(\cos^2 x) - 1$$

$$= 2 - 1 = 1$$

33. Verify $\csc^4 x - 2\csc^2 x + 1 = \cot^4 x$.

Solution:

$$\csc^4 x - 2\csc^2 x + 1 = (\csc^2 x - 1)^2 = (\cot^2 x)^2 = \cot^4 x$$

37. Verify $\dfrac{\sin\beta}{1-\cos\beta} = \dfrac{1+\cos\beta}{\sin\beta}$.

Solution:

$$\frac{\sin\beta}{1-\cos\beta} = \frac{\sin\beta}{1-\cos\beta}\cdot\frac{1+\cos\beta}{1+\cos\beta} = \frac{\sin\beta(1+\cos\beta)}{1-\cos^2\beta} = \frac{\sin\beta(1+\cos\beta)}{\sin^2\beta} = \frac{1+\cos\beta}{\sin\beta}$$

41. Verify $\cos\left(\dfrac{\pi}{2}-x\right)\csc x = 1$.

Solution:

$$\cos\left(\frac{\pi}{2}-x\right)\csc x = \sin x \csc x = \sin x\left(\frac{1}{\sin x}\right) = 1$$

45. Verify $\dfrac{\cos(-\theta)}{1+\sin(-\theta)} = \sec\theta + \tan\theta$.

Solution:

$$\begin{aligned}\frac{\cos(-\theta)}{1+\sin(-\theta)} &= \frac{\cos\theta}{1-\sin\theta}\\ &= \frac{\cos\theta}{1-\sin\theta}\cdot\frac{1+\sin\theta}{1+\sin\theta}\\ &= \frac{\cos\theta(1+\sin\theta)}{1-\sin^2\theta}\\ &= \frac{\cos\theta(1+\sin\theta)}{\cos^2\theta}\\ &= \frac{1+\sin\theta}{\cos\theta}\\ &= \frac{1}{\cos\theta}+\frac{\sin\theta}{\cos\theta}\\ &= \sec\theta+\tan\theta\end{aligned}$$

49. Verify $\dfrac{\tan x + \cot y}{\tan x \cot y} = \tan y + \cot x$.

Solution:

$$\begin{aligned}\frac{\tan x+\cot y}{\tan x\cot y} &= \frac{\tan x}{\tan x\cot y}+\frac{\cot y}{\tan x\cot y}\\ &= \frac{1}{\cot y}+\frac{1}{\tan x}\\ &= \tan y+\cot x\end{aligned}$$

53. Verify $\sin^2 x + \sin^2\left(\dfrac{\pi}{2}-x\right) = 1$.

Solution:

$$\begin{aligned}\sin^2 x+\sin^2\left(\frac{\pi}{2}-x\right) &= \sin^2 x+\cos^2 x\\ &= 1\end{aligned}$$

59. Explain why $\sqrt{\tan^2 x} = \tan x$ is *not* an identity and find one value of the variable for which the equation is not true.

Solution:

$$\sqrt{\tan^2 x} = |\tan x|$$

To show that $\sqrt{\tan^2 x} = \tan x$ is not true, pick any value of x whose tangent is negative. For example, $\sqrt{\tan^2 135°} = 1$, whereas, $\tan 135° = -1$.

SECTION 6.3

Solving Trigonometric Equations

- You should be able to identify and solve trigonometric equations.
- A trigonometric equation is a conditional equation. It is true for a specific set of values.
- To solve trigonometric equations, use algebraic techniques such as collecting like terms, taking square roots, factoring, squaring, converting to quadratic form, and using formulas.

Solutions to Selected Exercises

5. Verify that the given values of x are solutions of the equation $2\sin^2 x - \sin x - 1 = 0$.

(a) $x = \dfrac{\pi}{2}$ (b) $x = \dfrac{7\pi}{6}$

Solution:

(a)
$$x = \frac{\pi}{2}$$
$$\sin\frac{\pi}{2} = 1$$
$$2\sin^2\frac{\pi}{2} - \sin\frac{\pi}{2} - 1 = 2(1)^2 - 1 - 1 = 0$$

(b)
$$x = \frac{7\pi}{6}$$
$$\sin\frac{7\pi}{6} = -\frac{1}{2}$$
$$2\sin^2\frac{7\pi}{6} - \sin\frac{7\pi}{6} - 1 = 2\left(-\frac{1}{2}\right)^2 - \left(-\frac{1}{2}\right) - 1 = 2\left(\frac{1}{4}\right) + \frac{1}{2} - 1 = \frac{1}{2} + \frac{1}{2} - 1 = 0$$

9. Find all solutions of $\sqrt{3}\csc x - 2 = 0$. (Do not use a calculator.)

Solution:

$$\sqrt{3}\csc x - 2 = 0$$

$$\csc x = \frac{2}{\sqrt{3}}$$

$$x = \frac{\pi}{3} \text{ or } \frac{2\pi}{3} \text{ in } [0,\ 2\pi)$$

In general form, $x = (\pi/3) + 2n\pi$ or $x = (2\pi/3) + 2n\pi$ where n is an integer.

13. Find all solutions of $3\sec^2 x - 4 = 0$. (Do not use a calculator.)

Solution:

$$3\sec^2 x - 4 = 0$$

$$\sec^2 x = \frac{4}{3}$$

$$\sec x = \pm\sqrt{\frac{4}{3}} = \pm\frac{2}{\sqrt{3}} = \pm\frac{2\sqrt{3}}{3}$$

$$\sec x = \frac{2\sqrt{3}}{3} \quad \text{or} \quad \sec x = -\frac{2\sqrt{3}}{3}$$

$$x = \frac{\pi}{6}, \frac{11\pi}{6} \qquad x = \frac{5\pi}{6}, \frac{7\pi}{6} \text{ in } [0,\ 2\pi)$$

In general form, the solutions are $x = (\pi/6) + n\pi$ or $x = (5\pi/6) + n\pi$ where n is an integer.

17. Find all solutions of $\sin x(\sin x + 1) = 0$. (Do not use a calculator.)

Solution:

$$\sin x = 0 \quad \text{or} \quad \sin x = -1$$

$$x = 0,\ \pi \qquad x = \frac{3\pi}{2} \text{ in } [0,\ 2\pi)$$

In general form, the solutions are $x = n\pi$ and $x = (3\pi/2) + 2n\pi$, where n is an integer.

21. Find all solutions of $\sec x \csc x - 2\csc x = 0$ in the interval $[0, 2\pi)$. (Do not use a calculator.)

Solution:

$$\sec x \csc x - 2\csc x = 0$$
$$\csc x(\sec x - 2) = 0$$

$\csc x = 0$ or $\sec x = 2$

Not possible $\qquad x = \frac{\pi}{3}, \frac{5\pi}{3}$

25. Find all solutions of $\cos^3 x = \cos x$ in the interval $[0, 2\pi)$. (Do not use a calculator.)

Solution:

$$\cos^3 x = \cos x$$
$$\cos^3 x - \cos x = 0$$
$$\cos x(\cos^2 x - 1) = 0$$

$\cos x = 0$ or $\cos^2 x - 1 = 0$

$x = \frac{\pi}{2}, \frac{3\pi}{2} \qquad \cos x = \pm 1$

$x = 0, \pi$

$$x = 0, \frac{\pi}{2}, \pi, \frac{3\pi}{2}$$

29. Find all solutions of $2\sec^2 x + \tan^2 x - 3 = 0$ in the interval $[0, 2\pi)$. (Do not use a calculator.)

Solution:

$$2\sec^2 x + \tan^2 x - 3 = 0$$
$$2\sec^2 x + (\sec^2 x - 1) - 3 = 0$$
$$3\sec^2 x - 4 = 0$$
$$\sec^2 x = \frac{4}{3}$$
$$\sec x = \pm\frac{2}{\sqrt{3}}$$

$\sec x = \frac{2}{\sqrt{3}}$ or $\sec x = -\frac{2}{\sqrt{3}}$

$x = \frac{\pi}{6}, \frac{11\pi}{6} \qquad x = \frac{5\pi}{6}, \frac{7\pi}{6}$

$$x = \frac{\pi}{6}, \frac{5\pi}{6}, \frac{7\pi}{6}, \frac{11\pi}{6}$$

33. Find all solutions of $\sin 2x = -\sqrt{3}/2$ in the interval $[0,\ 2\pi)$. (Do not use a calculator.)

Solution:

$$\sin 2x = -\frac{\sqrt{3}}{2}$$

$$2x = \frac{4\pi}{3}, \quad 2x = \frac{5\pi}{3}, \quad 2x = \frac{10\pi}{3}, \quad 2x = \frac{11\pi}{3}$$

$$x = \frac{2\pi}{3}, \quad x = \frac{5\pi}{6}, \quad x = \frac{5\pi}{3}, \quad x = \frac{11\pi}{6}$$

39. Find all solutions of

$$\frac{1 + \sin x}{\cos x} + \frac{\cos x}{1 + \sin x} = 4$$

in the interval $[0,\ 2\pi)$. (Do not use a calculator.)

Solution:

$$\frac{1 + \sin x}{\cos x} + \frac{\cos x}{1 + \sin x} = 4$$

$$(1 + \sin x)^2 + (\cos x)^2 = 4 \cos x(1 + \sin x)$$

$$1 + 2 \sin x + \sin^2 x + \cos^2 x = 4 \cos x(1 + \sin x)$$

$$1 + 2 \sin x + 1 = 4 \cos x(1 + \sin x)$$

$$2(1 + \sin x) - 4 \cos(1 + \sin x) = 0$$

$$2(1 + \sin x)(1 - 2 \cos x) = 0$$

$\sin x = -1$ or $\cos x = \dfrac{1}{2}$

$x = \dfrac{3\pi}{2}$, extraneous $\qquad x = \dfrac{\pi}{3}, \dfrac{5\pi}{3}$

(Makes the denominator zero.)

The only solutions are $x = \pi/3,\ 5\pi/3$.

43. Solve $12y^2 - 13y + 3 = 0$ and $12\sin^2 x - 13\sin x + 3 = 0$. Restrict the solutions to the interval $[0,\ 2\pi)$.

Solution:

$$12y^2 - 12y + 3 = 0$$

$$(4y-3)(3y-1) = 0$$

$$y = \tfrac{3}{4} \quad \text{or} \quad y = \tfrac{1}{3}$$

$$12\sin^2 x - 13\sin x + 3 = 0$$

$$(4\sin x - 3)(3\sin x - 1) = 0$$

$$\sin x = \tfrac{3}{4} \qquad \text{or} \qquad \sin x = \tfrac{1}{3}$$

$$x \approx 0.8481,\ 2.2935 \qquad\qquad x \approx 0.3398,\ 2.8018$$

47. Solve $y^2 - 8y + 13 = 0$ and $\tan^2 x - 8\tan x + 13 = 0$. Restrict the solutions to the interval $[0,\ 2\pi)$.

Solution:

$$y^2 - 8y + 13 = 0$$

$$\begin{aligned} y &= \frac{8 \pm \sqrt{64-52}}{2} \\ &= \frac{8 \pm \sqrt{12}}{2} \\ &= \frac{8 \pm 2\sqrt{3}}{2} \\ &= 4 \pm \sqrt{3} \end{aligned}$$

$$\tan^2 x - 8\tan x + 13 = 0$$

$$\begin{aligned} \tan x &= \frac{8 \pm \sqrt{64-52}}{2} \\ &= \frac{8 \pm 2\sqrt{3}}{2} \end{aligned}$$

$$\tan x = 4 + \sqrt{3} \qquad \text{or} \qquad \tan x = 4 - \sqrt{3}$$

$$x \approx 1.3981,\ 4.5397 \qquad\qquad x \approx 1.1555,\ 4.2971$$

53. A 5-pound weight is oscillating on the end of a spring, and the position of the weight relative to the point of equilibrium is given by $y = \frac{1}{4}(\cos 8t - 3\sin 8t)$ where t is the time in seconds. Find the times when the weight is at the point of equilibrium $[y = 0]$ for $0 \le t \le 1$.

Solution:

$$\frac{1}{4}(\cos 8t - 3\sin 8t) = 0$$

$$\begin{aligned} \cos 8t &= 3\sin 8t \\ \frac{1}{3} &= \tan 8t \\ 8t &\approx 0.32175 + n\pi \\ t &\approx 0.04 + \frac{n\pi}{8} \end{aligned}$$

In the interval $0 \le t \le 1$, we have $t = 0.04,\ 0.43$, and 0.83.

SECTION 6.4

Sum and Difference Formulas

- You should memorize the sum and difference formulas.

$$\sin(u \pm v) = \sin u \cos v \pm \cos u \sin v$$

$$\cos(u \pm v) = \cos u \cos v \mp \sin u \sin v$$

$$\tan(u \pm v) = \frac{\tan u \pm \tan v}{1 \mp \tan u \tan v}$$

- You should be able to use these formulas to find the values of the trigonometric functions of angles whose sums or differences are special angles.
- You should be able to use these formulas to solve trigonometric equations.

Solutions to Selected Exercises

5. Determine the exact value of the sine, cosine, and tangent of the angle $195° = 225° - 30°$.

Solution:

$$\sin 195° = \sin(225° - 30°) = \sin 225° \cos 30° - \cos 225° \sin 30°$$

$$= (-\sin 45°)\cos 30° - (-\cos 45°)\sin 30° = -\frac{\sqrt{2}}{2} \cdot \frac{\sqrt{3}}{2} + \frac{\sqrt{2}}{2} \cdot \frac{1}{2} = \frac{\sqrt{2}}{4}(1 - \sqrt{3})$$

$$\cos 195° = \cos(225° - 30°) = \cos 225° \cos 30° + \sin 225° \sin 30°$$

$$= (-\cos 45°)\cos 30° + (-\sin 45°)\sin 30° = -\frac{\sqrt{2}}{2} \cdot \frac{\sqrt{3}}{2} - \frac{\sqrt{2}}{2} \cdot \frac{1}{2} = -\frac{\sqrt{2}}{4}(1 + \sqrt{3})$$

$$\tan 195° = \tan(225° - 30°) = \frac{\tan 225° - \tan 30°}{1 + \tan 225° \tan 30°} = \frac{\tan 45° - \tan 30°}{1 + \tan 45° \tan 30°}$$

$$= \frac{1 - \sqrt{3}/3}{1 + (1)(\sqrt{3}/3)} \cdot \frac{3}{3} = \frac{3 - \sqrt{3}}{3 + \sqrt{3}} \cdot \frac{3 - \sqrt{3}}{3 - \sqrt{3}}$$

$$= \frac{(3 - \sqrt{3})^2}{9 - 3} = \frac{9 - 6\sqrt{3} + 3}{6}$$

$$= \frac{12 - 6\sqrt{3}}{6} = \frac{6(2 - \sqrt{3})}{6} = 2 - \sqrt{3}$$

11. Simplify $\cos 25° \cos 15° - \sin 25° \sin 15°$.

Solution:

$$\cos 25° \cos 15° - \sin 25° \sin 15° = \cos(25° + 15°) = \cos 40°$$

15. Simplify $\dfrac{\tan 325° - \tan 86°}{1 + \tan 325° \tan 86°}$.

Solution:

$$\frac{\tan 325° - \tan 86°}{1 + \tan 325° \tan 86°} = \tan(325° - 86°)$$
$$= \tan 239°$$

19. Simplify $\dfrac{\tan 2x + \tan x}{1 - \tan 2x \tan x}$.

Solution:

$$\frac{\tan 2x + \tan x}{1 - \tan 2x \tan x} = \tan(2x + x)$$
$$= \tan 3x$$

23. Find the exact value of $\cos(v + u)$ given that $\sin u = 5/13$, $0 < u < \pi/2$, and $\cos v = -3/5$, $\pi/2 < v < \pi$.

Solution:

$$\sin u = \frac{5}{13},\ 0 < u < \frac{\pi}{2} \quad \Rightarrow \quad \cos u = \frac{12}{13}$$
$$\cos v = -\frac{3}{5},\ \frac{\pi}{2} < v < \pi \quad \Rightarrow \quad \sin v = \frac{4}{5}$$

$$\cos(v + u) = \cos v \cos u - \sin v \sin u = \left(-\frac{3}{5}\right)\left(\frac{12}{13}\right) - \left(\frac{4}{5}\right)\left(\frac{5}{13}\right) = -\frac{56}{65}$$

27. Find the exact value of $\sin(v - u)$ given that $\sin u = 7/25$, $\pi/2 < u < \pi$ and $\cos v = 4/5$, $3\pi/2 < v < 2\pi$.

Solution:

$$\sin u = \frac{7}{25},\ \frac{\pi}{2} < u < \pi \quad \Rightarrow \quad \cos u = -\frac{24}{25}$$
$$\cos v = \frac{4}{5},\ \frac{3\pi}{2} < v < 2\pi \quad \Rightarrow \quad \sin v = -\frac{3}{5}$$

$$\sin(v - u) = \sin v \cos u - \cos v \sin u = \left(-\frac{3}{5}\right)\left(-\frac{24}{25}\right) - \left(\frac{4}{5}\right)\left(\frac{7}{25}\right) = \frac{44}{125}$$

31. Verify $\cos\left(\dfrac{3\pi}{2} - x\right) = -\sin x$.

Solution:

$$\cos\left(\frac{3\pi}{2} - x\right) = \cos\frac{3\pi}{2}\cos x + \sin\frac{3\pi}{2}\sin x = (0)(\cos x) + (-1)(\sin x) = -\sin x$$

35. Verify $\cos(\pi - \theta) + \sin\left(\frac{\pi}{2} + \theta\right) = 0$.

Solution:

$$\begin{aligned}\cos(\pi - \theta) + \sin\left(\frac{\pi}{2} + \theta\right) &= (\cos\pi\cos\theta + \sin\pi\sin\theta) + \left(\sin\frac{\pi}{2}\cos\theta + \cos\frac{\pi}{2}\sin\theta\right)\\ &= (-1)\cos\theta + (0)\sin\theta + (1)\cos\theta + (0)\sin\theta\\ &= -\cos\theta + 0 + \cos\theta + 0\\ &= 0\end{aligned}$$

39. Verify $\cos(x + y)\cos(x - y) = \cos^2 x - \sin^2 y$.

Solution:

$$\begin{aligned}\cos(x + y)\cos(x - y) &= (\cos x\cos y - \sin x\sin y)(\cos x\cos y + \sin x\sin y)\\ &= \cos^2 x\cos^2 y - \sin^2 x\sin^2 y\\ &= \cos^2 x(1 - \sin^2 y) - (1 - \cos^2 x)\sin^2 y\\ &= \cos^2 x - \cos^2 x\sin^2 y - \sin^2 y + \cos^2 x\sin^2 y\\ &= \cos^2 x - \sin^2 y\end{aligned}$$

41. Verify $\sin(x + y) + \sin(x - y) = 2\sin x\cos y$.

Solution:

$$\begin{aligned}\sin(x + y) + \sin(x - y) &= (\sin x\cos y + \cos x\sin y) + (\sin x\cos y - \cos x\sin y)\\ &= 2\sin x\cos y\end{aligned}$$

45. Verify $a\sin B\theta + b\cos B\theta = \sqrt{a^2 + b^2}\sin(B\theta + C)$, where $C = \arctan(b/a)$.

Solution:

$$\begin{aligned}\sqrt{a^2 + b^2}\sin(B\theta + C) &= \sqrt{a^2 + b^2}(\sin B\theta\cos C + \cos B\theta\sin C)\\ &= \sqrt{a^2 + b^2}\left[\sin B\theta\left(\frac{a}{\sqrt{a^2 + b^2}}\right) + \cos B\theta\left(\frac{b}{\sqrt{a^2 + b^2}}\right)\right]\\ &= \frac{\sqrt{a^2 + b^2}}{\sqrt{a^2 + b^2}}(a\sin B\theta + b\cos B\theta)\\ &= a\sin B\theta + b\cos B\theta\end{aligned}$$

49. Use the formulas given in Exercises 45 and 46 to write $12\sin 3\theta + 5\cos 3\theta$ in the following forms.

(a) $\sqrt{a^2+b^2}\sin(B\theta + C)$ (b) $\sqrt{a^2+b^2}\cos(B\theta - C)$

Solution:

$a = 12,\ b = 5,\ B = 3,\ C = \arctan\frac{5}{12} \approx 0.3948$

$\sqrt{12^2+5^2} = \sqrt{169} = 13$

(a) Thus, $\sqrt{a^2+b^2}\sin(B\theta + C) = 13\sin(3\theta + 0.3948)$, and

(b) $\sqrt{a^2+b^2}\cos(B\theta - C) = 13\cos(3\theta - 0.3948)$.

53. Write $\sin(\arcsin x + \arccos x)$ as an algebraic expression.

Solution:

$$\begin{aligned}\sin(\arcsin x + \arccos x) &= \sin(\arcsin x)\cos(\arccos x) + \cos(\arcsin x)\sin(\arccos x)\\ &= (x)(x) + \frac{\sqrt{1-x^2}}{1}\bullet\frac{\sqrt{1-x^2}}{1}\\ &= x^2 + 1 - x^2 = 1\end{aligned}$$

57. Find all solutions in the interval $[0,\ 2\pi)$ of $\cos\left(x + \frac{\pi}{4}\right) + \cos\left(x - \frac{\pi}{4}\right) = 1$.

Solution:

$$\begin{aligned}\cos\left(x + \frac{\pi}{4}\right) + \cos\left(x - \frac{\pi}{4}\right) &= 1\\ \cos x\cos\frac{\pi}{4} - \sin x\sin\frac{\pi}{4} + \cos x\cos\frac{\pi}{4} + \sin x\sin\frac{\pi}{4} &= 1\\ \frac{\sqrt{2}}{2}\cos x + \frac{\sqrt{2}}{2}\cos x &= 1\\ \sqrt{2}\cos x &= 1\\ \cos x &= \frac{1}{\sqrt{2}}\\ x &= \frac{\pi}{4},\ \frac{7\pi}{4}\end{aligned}$$

61. The equation of a standing wave is obtained by adding the displacements of two waves traveling in opposite directions (see figure). Assume that each of the waves has amplitude A, period T, and wavelength λ. If the models for these waves are

$$y_1 = A\cos 2\pi\left(\frac{t}{T} - \frac{x}{\lambda}\right) \quad \text{and}$$

$$y_2 = A\cos 2\pi\left(\frac{t}{T} + \frac{x}{\lambda}\right)$$

show that

$$y_1 + y_2 = 2A\cos\frac{2\pi t}{T}\cos\frac{2\pi x}{\lambda}.$$

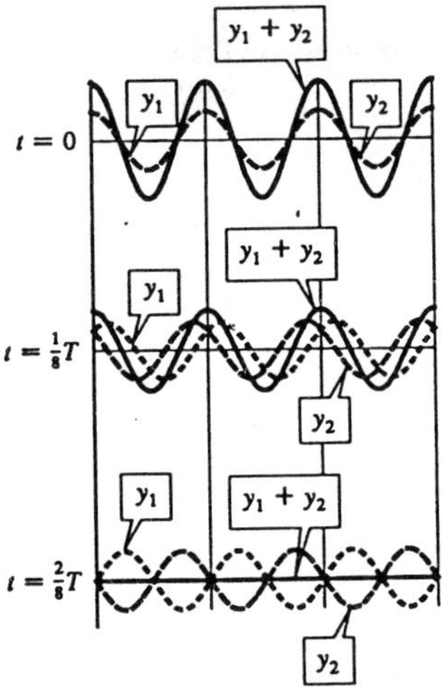

Solution:

$$\begin{aligned} y_1 + y_2 &= A\cos 2\pi\left(\frac{t}{T} - \frac{x}{\lambda}\right) + A\cos 2\pi\left(\frac{t}{T} + \frac{x}{\lambda}\right) \\ &= A\left[\cos\left(\frac{2\pi t}{T} - \frac{2\pi x}{\lambda}\right) + \cos\left(\frac{2\pi t}{T} + \frac{2\pi x}{\lambda}\right)\right] \\ &= A\left[\cos\frac{2\pi t}{T}\cos\frac{2\pi x}{\lambda} + \sin\frac{2\pi t}{T}\sin\frac{2\pi x}{\lambda} + \cos\frac{2\pi t}{T}\cos\frac{2\pi x}{\lambda} - \sin\frac{2\pi t}{T}\sin\frac{2\pi x}{\lambda}\right] \\ &= 2A\cos\frac{2\pi t}{T}\cos\frac{2\pi x}{\lambda} \end{aligned}$$

63. Verify that $\dfrac{\cos(x+h) - \cos x}{h} = \cos x\left(\dfrac{\cos h - 1}{h}\right) - \sin x\left(\dfrac{\sin h}{h}\right)$.

Solution:

$$\begin{aligned} \frac{\cos(x+h) - \cos x}{h} &= \frac{\cos x\cos h - \sin x\sin h - \cos x}{h} \\ &= \frac{\cos x(\cos h - 1) - \sin x\sin h}{h} \\ &= \frac{\cos x(\cos h - 1)}{h} - \frac{\sin x\sin h}{h} \\ &= \cos x\left(\frac{\cos h - 1}{h}\right) - \sin x\left(\frac{\sin h}{h}\right) \end{aligned}$$

SECTION 6.5

Multiple-Angle Formulas and Product-Sum Formulas

- You should know the following double-angle formulas.

  (a) $\sin 2u = 2\sin u \cos u$

  (b) $\cos 2u = \cos^2 u - \sin^2 u$

  $= 2\cos^2 u - 1$

  $= 1 - 2\sin^2 u$

  (c) $\tan 2u = \dfrac{2\tan u}{1 - \tan^2 u}$

- You should be able to reduce the power of a trigonometric function.

  (a) $\sin^2 u = \dfrac{1 - \cos 2u}{2}$

  (b) $\cos^2 u = \dfrac{1 + \cos 2u}{2}$

  (c) $\tan^2 u = \dfrac{1 - \cos 2u}{1 + \cos 2u}$

- You should be able to use the half-angle formulas.

  (a) $\sin \dfrac{u}{2} = \pm\sqrt{\dfrac{1 - \cos u}{2}}$

  (b) $\cos \dfrac{u}{2} = \pm\sqrt{\dfrac{1 + \cos u}{2}}$

  (c) $\tan \dfrac{u}{2} = \dfrac{1 - \cos u}{\sin u} = \dfrac{\sin u}{1 + \cos u}$

- You should be able to use the product-sum formulas.

  (a) $\sin u \sin v = \dfrac{1}{2}[\cos(u - v) - \cos(u + v)]$

  (b) $\cos u \cos v = \dfrac{1}{2}[\cos(u - v) + \cos(u + v)]$

  (c) $\sin u \cos v = \dfrac{1}{2}[\sin(u + v) + \sin(u - v)]$

  (d) $\cos u \sin v = \dfrac{1}{2}[\sin(u + v) - \sin(u - v)]$

■ You should be able to use the sum-product formulas.

(a) $\sin x + \sin y = 2\sin\left(\frac{x+y}{2}\right)\cos\left(\frac{x-y}{2}\right)$

(b) $\sin x - \sin y = 2\cos\left(\frac{x+y}{2}\right)\sin\left(\frac{x-y}{2}\right)$

(c) $\cos x + \cos y = 2\cos\left(\frac{x+y}{2}\right)\cos\left(\frac{x-y}{2}\right)$

(d) $\cos x - \cos y = -2\sin\left(\frac{x+y}{2}\right)\sin\left(\frac{x-y}{2}\right)$

Solutions to Selected Exercises

3. Find all the solutions of $4\sin x\cos x = 1$ in the interval $[0, 2\pi)$.

Solution:

$$\begin{aligned}4\sin x\cos x &= 1\\ 2[2\sin x\cos x] &= 1\\ 2\sin 2x &= 1\\ \sin 2x &= \frac{1}{2}\end{aligned}$$

$$2x = \frac{\pi}{6},\quad 2x = \frac{5\pi}{6},\quad 2x = \frac{13\pi}{6},\quad 2x = \frac{17\pi}{6}$$

$$x = \frac{\pi}{12},\quad x = \frac{5\pi}{12},\quad x = \frac{13\pi}{12},\quad x = \frac{17\pi}{12}$$

9. Find all solutions of $\sin 4x + 2\sin 2x = 0$ in the interval $[0, 2\pi)$.

Solution:

$$\begin{aligned}\sin 4x + 2\sin 2x &= 0\\ 2\sin 2x\cos 2x + 2\sin 2x &= 0\\ 2\sin 2x(\cos 2x + 1) &= 0\end{aligned}$$

$\sin 2x = 0$ or $\cos 2x = -1$

$2x = 0, \pi, 2\pi, 3\pi$ $\quad$ $2x = \pi, 3\pi$

$x = 0, \frac{\pi}{2}, \pi, \frac{3\pi}{2}$ $\quad$ $x = \frac{\pi}{2}, \frac{3\pi}{2}$

13. Use a double-angle identity to rewrite $g(x) = 4 - 8\sin^2 x$ and sketch its graph.

Solution:

$$g(x) = 4 - 8\sin^2 x$$
$$= 4(1 - 2\sin^2 x)$$
$$= 4\cos 2x$$

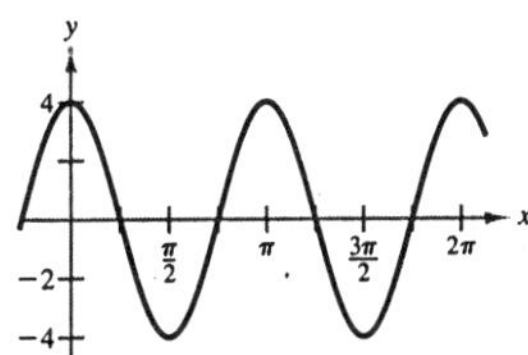

17. Find the exact values of $\sin 2u$, $\cos 2u$, and $\tan 2u$, given $\tan u = 1/2$, $\pi < u < 3\pi/2$.

Solution:

$\tan u = \frac{1}{2}$, u lies in Quadrant III.

$\sin u = -\frac{1}{\sqrt{5}}$ and $\cos u = -\frac{2}{\sqrt{5}}$

$$\sin 2u = 2\left(-\frac{1}{\sqrt{5}}\right)\left(-\frac{2}{\sqrt{5}}\right) = \frac{4}{5}$$

$$\cos 2u = \left(-\frac{2}{\sqrt{5}}\right)^2 - \left(-\frac{1}{\sqrt{5}}\right)^2 = \frac{3}{5}$$

$$\tan 2u = \frac{\sin 2u}{\cos 2u} = \frac{4}{3}$$

21. Use the power-reducing formulas to write $\cos^4 x$ in terms of the first power of the cosine.

Solution:

$$\cos^4 x = (\cos^2 x)^2$$
$$= \left(\frac{1 + \cos 2x}{2}\right)^2$$
$$= \frac{1}{4}(1 + 2\cos 2x + \cos^2 2x)$$
$$= \frac{1}{4}\left(1 + 2\cos 2x + \frac{1 + \cos 4x}{2}\right)$$
$$= \frac{1}{4}\left(\frac{3}{2} + 2\cos 2x + \frac{1}{2}\cos 4x\right)$$
$$= \frac{1}{8}(3 + 4\cos 2x + \cos 4x)$$

25. Use the power-reducing formulas to write $\sin^2 x \cos^4 x$ in terms of the first power of the cosine.

Solution:

$$
\begin{aligned}
\sin^2 x \cos^4 x &= \left(\frac{1-\cos 2x}{2}\right)\cos^4 x \\
&= \frac{1}{2}(1-\cos 2x)\frac{1}{8}(3+4\cos 2x+\cos 4x) \quad \text{from Exercise 21} \\
&= \frac{1}{16}(3+4\cos 2x+\cos 4x-3\cos 2x-4\cos^2 2x-\cos 4x\cos 2x) \\
&= \frac{1}{16}\left[3+\cos 2x+\cos 4x-4\left(\frac{1+\cos 4x}{2}\right)-\frac{1}{2}(\cos 2x+\cos 6x)\right] \\
&= \frac{1}{16}\left(3+\cos 2x+\cos 4x-2-2\cos 4x-\frac{1}{2}\cos 2x-\frac{1}{2}\cos 6x\right) \\
&= \frac{1}{16}\left(1+\frac{1}{2}\cos 2x-\cos 4x-\frac{1}{2}\cos 6x\right) \\
&= \frac{1}{32}(2+\cos 2x-2\cos 4x-\cos 6x)
\end{aligned}
$$

29. Use half-angle formulas to determine the exact values of the sine, cosine, and tangent of the angle $112°30'$.

Solution:

$$\sin 112°30' = +\sqrt{\frac{1-\cos 225°}{2}} = \sqrt{\frac{1-(-\sqrt{2}/2)}{2}} = \sqrt{\frac{2+\sqrt{2}}{4}} = \frac{\sqrt{2+\sqrt{2}}}{2}$$

$$\cos 112°30' = -\sqrt{\frac{1+\cos 225°}{2}} = -\sqrt{\frac{1+(-\sqrt{2}/2)}{2}} = -\sqrt{\frac{2-\sqrt{2}}{4}} = -\frac{\sqrt{2-\sqrt{2}}}{2}$$

$$
\begin{aligned}
\tan 112°30' &= -\sqrt{\frac{1-\cos 225°}{1+\cos 225°}} = -\sqrt{\frac{1+\sqrt{2}/2}{1-\sqrt{2}/2}} = -\sqrt{\frac{2+\sqrt{2}}{2-\sqrt{2}}} \cdot \sqrt{\frac{2+\sqrt{2}}{2+\sqrt{2}}} \\
&= -\sqrt{\frac{(2+\sqrt{2})^2}{4-2}} = -\frac{2+\sqrt{2}}{\sqrt{2}} = -(\sqrt{2}+1) = -1-\sqrt{2}
\end{aligned}
$$

35. Find the exact values of $\sin(u/2)$, $\cos(u/2)$, and $\tan(u/2)$ by using the half-angle formulas, given $\tan u = -5/8$, $3\pi/2 < u < 2\pi$.

Solution:

$\tan u = -\frac{5}{8}$, u lies in Quadrant IV; $\sin u = -\frac{5}{\sqrt{89}}$ and $\cos u = \frac{8}{\sqrt{89}}$

$$\sin\frac{u}{2} = \sqrt{\frac{1 - 8/\sqrt{89}}{2}} = \sqrt{\frac{\sqrt{89} - 8}{2\sqrt{89}}} = \sqrt{\frac{89 - 8\sqrt{89}}{178}}$$

$$\cos\frac{u}{2} = -\sqrt{\frac{1 + 8/\sqrt{89}}{2}} = -\sqrt{\frac{\sqrt{89} + 8}{2\sqrt{89}}} = \sqrt{\frac{89 + 8\sqrt{89}}{178}}$$

$$\tan\frac{u}{2} = -\sqrt{\frac{1 - 8/\sqrt{89}}{1 + 8/\sqrt{89}}} = -\sqrt{\frac{\sqrt{89} - 8}{\sqrt{89} + 8}} \cdot \sqrt{\frac{\sqrt{89} - 8}{\sqrt{89} - 8}} = \sqrt{\frac{(\sqrt{89} - 8)^2}{89 - 64}} = -\left(\frac{\sqrt{89} - 8}{5}\right)$$

$$= \frac{1}{5}(8 - \sqrt{89})$$

39. Use the half-angle formulas to simplify $\sqrt{\frac{1 - \cos 6x}{2}}$.

Solution:

$$\sqrt{\frac{1 - \cos 6x}{2}} = \sqrt{\frac{1 - \cos 2(3x)}{2}} = \sin 3x$$

43. Find all the solutions of $\sin(x/2) + \cos x = 0$ in the interval $[0,\ 2\pi)$.

Solution:

$$\sin\frac{x}{2} + \cos x = 0$$

$$\pm\sqrt{\frac{1 - \cos x}{2}} = -\cos x$$

$$\frac{1 - \cos x}{2} = \cos^2 x$$

$$0 = 2\cos^2 x + \cos x - 1 = (2\cos x - 1)(\cos x + 1)$$

$$\cos x = \frac{1}{2} \qquad \text{or} \qquad \cos x = -1$$

$$x = \frac{\pi}{3}, \frac{5\pi}{3} \qquad\qquad x = \pi$$

By checking these values in the original equation, we see that $x = \pi/3$ and $x = 5\pi/3$ are extraneous, and $x = \pi$ is the only solution.

49. Rewrite $\sin 5\theta \cos 3\theta$ as a sum.

Solution:

$$\sin 5\theta \cos 3\theta = \tfrac{1}{2}[\sin(5\theta + 3\theta) + \sin(5\theta - 3\theta)] = \tfrac{1}{2}(\sin 8\theta + \sin 2\theta)$$

53. Rewrite $\sin(x + y)\sin(x - y)$ as a sum.

Solution:

$$\begin{aligned}\sin(x + y)\sin(x - y) &= \tfrac{1}{2}\{\cos[(x + y) - (x - y)] - \cos[(x + y) + (x - y)]\} \\ &= \tfrac{1}{2}\{\cos 2y - \cos 2x\}\end{aligned}$$

57. Express $\sin 60° + \sin 30°$ as a product.

Solution:

$$\sin 60° + \sin 30° = 2 \sin\left(\frac{60° + 30°}{2}\right)\cos\left(\frac{60° - 30°}{2}\right) = 2 \sin 45° \cos 15°$$

61. Express $\cos 6x + \cos 2x$ as a product.

Solution:

$$\cos 6x + \cos 2x = 2 \cos\left(\frac{6x + 2x}{2}\right)\cos\left(\frac{6x - 2x}{2}\right) = 2 \cos 4x \cos 2x$$

65. Express $\cos(\phi + 2\pi) + \cos\phi$ as a product.

Solution:

$$\cos(\phi + 2\pi) + \cos\phi = 2 \cos\left(\frac{\phi + 2\pi + \phi}{2}\right)\cos\left(\frac{\phi + 2\pi - \phi}{2}\right) = 2 \cos(\phi + \pi) \cos \pi$$

67. Find all solutions of $\sin 6x + \sin 2x = 0$ in the interval $[0,\ 2\pi)$.

Solution:

$$\sin 6x + \sin 2x = 0$$

$$2\sin\left(\frac{6x+2x}{2}\right)\cos\left(\frac{6x-2x}{2}\right) = 0$$

$$2(\sin 4x)\cos 2x = 0$$

$$\sin 4x = 0 \qquad \text{or} \qquad \cos 2x = 0$$

$$4x = n\pi \qquad\qquad 2x = \frac{\pi}{2} + n\pi$$

$$x = \frac{n\pi}{4} \qquad\qquad x = \frac{\pi}{4} + \frac{n\pi}{2}$$

In the interval $[0,\ 2\pi)$, we have

$$x = 0,\ \frac{\pi}{4},\ \frac{\pi}{2},\ \frac{3\pi}{4},\ \pi,\ \frac{5\pi}{4},\ \frac{3\pi}{2},\ \frac{7\pi}{4}.$$

71. Verify $\csc 2\theta = \dfrac{\csc\theta}{2\cos\theta}$.

Solution:

$$\csc 2\theta = \frac{1}{\sin 2\theta} = \frac{1}{2\sin\theta(\cos\theta)} = \frac{1}{\sin\theta}\cdot\frac{1}{2\cos\theta} = \frac{\csc\theta}{2\cos\theta}$$

75. Verify $(\sin x + \cos x)^2 = 1 + \sin 2x$.

Solution:

$$\begin{aligned}(\sin x + \cos x)^2 &= \sin^2 x + 2\sin x\cos x + \cos^2 x\\ &= (\sin^2 x + \cos^2 x) + 2\sin x\cos x\\ &= 1 + \sin 2x\end{aligned}$$

79. Verify $1 + \cos 10y = 2\cos^2 5y$.

Solution:

$$2\cos^2 5y = 2\left[\frac{1+\cos 10y}{2}\right] = 1 + \cos 10y$$

85. Verify $\dfrac{\cos 4x - \cos 2x}{2\sin 3x} = -\sin x$.

Solution:

$$\frac{\cos 4x - \cos 2x}{2\sin 3x} = \frac{-2\sin\left(\dfrac{4x+2x}{2}\right)\sin\left(\dfrac{4x-2x}{2}\right)}{2\sin 3x} = \frac{-2\sin 3x \sin x}{2\sin 3x} = -\sin x$$

87. Verify $\dfrac{\cos t + \cos 3t}{\sin 3t - \sin t} = \cot t$.

Solution:

$$\frac{\cos t + \cos 3t}{\sin 3t - \sin t} = \frac{2\cos\left(\dfrac{t+3t}{2}\right)\cos\left(\dfrac{t-3t}{2}\right)}{2\cos\left(\dfrac{3t+t}{2}\right)\sin\left(\dfrac{3t-t}{2}\right)} = \frac{2\cos 2t\cos(-t)}{2\cos 2t \sin t} = \frac{\cos t}{\sin t} = \cot t$$

89. Sketch the graph of $f(x) = \sin^2 x$ by using the power-reducing formulas.

Solution:

$$\begin{aligned} f(x) &= \sin^2 x \\ &= \frac{1-\cos 2x}{2} \\ &= \frac{1}{2} - \frac{1}{2}\cos 2x \end{aligned}$$

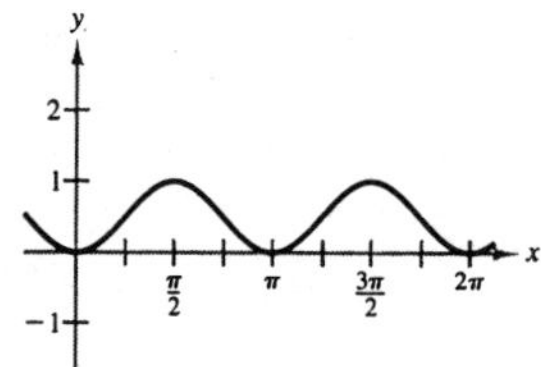

Period: π

x	$-\pi$	$-\dfrac{3\pi}{4}$	$-\dfrac{\pi}{2}$	$-\dfrac{\pi}{4}$	0	$\dfrac{\pi}{4}$	$\dfrac{\pi}{2}$	$\dfrac{3\pi}{4}$	π
$f(x)$	0	$\dfrac{1}{2}$	1	$\dfrac{1}{2}$	0	$\dfrac{1}{2}$	1	$\dfrac{1}{2}$	0

93. Prove $\cos u \sin v = \frac{1}{2}[\sin(u+v) - \sin(u-v)]$.

Solution:

$$\begin{aligned} \tfrac{1}{2}[\sin(u+v) - \sin(u-v)] &= \tfrac{1}{2}\{(\sin u\cos v + \cos u \sin v) - (\sin u \cos v - \cos u \sin v)\} \\ &= \tfrac{1}{2}\{2\cos u \sin v\} \\ &= \cos u \sin v \end{aligned}$$

REVIEW EXERCISES FOR CHAPTER 6

Solutions to Selected Exercises

3. Simplify $\dfrac{\sin^2\alpha - \cos^2\alpha}{\sin^2\alpha - \sin\alpha\cos\alpha}$.

Solution:

$$\begin{aligned}\frac{\sin^2\alpha - \cos^2\alpha}{\sin^2\alpha - \sin\alpha\cos\alpha} &= \frac{(\sin\alpha + \cos\alpha)(\sin\alpha - \cos\alpha)}{\sin\alpha(\sin\alpha - \cos\alpha)}\\ &= \frac{\sin\alpha + \cos\alpha}{\sin\alpha} = \frac{\sin\alpha}{\sin\alpha} + \frac{\cos\alpha}{\sin\alpha} = 1 + \cot\alpha\end{aligned}$$

7. Simplify $\tan^2\theta(\csc^2\theta - 1)$.

Solution:

$$\begin{aligned}\tan^2\theta(\csc^2\theta - 1) &= \tan^2\theta(\cot^2\theta)\\ &= \tan^2\theta\left(\frac{1}{\tan^2\theta}\right)\\ &= 1\end{aligned}$$

11. Verify $\tan x(1 - \sin^2 x) = \frac{1}{2}\sin 2x$.

Solution:

$$\begin{aligned}\tan x(1 - \sin^2 x) &= \frac{\sin x}{\cos x}(\cos^2 x)\\ &= (\sin x)\cos x\\ &= \frac{1}{2}(2(\sin x)\cos x)\\ &= \frac{1}{2}\sin 2x\end{aligned}$$

15. Verify $\sin^5 x\cos^2 x = (\cos^2 x - 2\cos^4 x + \cos^6 x)\sin x$.

Solution:

$$\begin{aligned}(\cos^2 x - 2\cos^4 x + \cos^6 x)\sin x &= \cos^2 x(1 - 2\cos^2 x + \cos^4 x)\sin x\\ &= \cos^2 x(1 - \cos^2 x)^2\sin x\\ &= \cos^2 x(\sin^2 x)^2\sin x\\ &= \cos^2 x\sin^4 x\sin x\\ &= \cos^2 x\sin^5 x\\ &= \sin^5 x\cos^2 x\end{aligned}$$

19. Verify $\sqrt{\dfrac{1-\sin\theta}{1+\sin\theta}} = \dfrac{1-\sin\theta}{|\cos\theta|}$.

Solution:

$$\sqrt{\frac{1-\sin\theta}{1+\sin\theta}} = \sqrt{\frac{1-\sin\theta}{1+\sin\theta} \cdot \frac{1-\sin\theta}{1-\sin\theta}}$$

$$= \sqrt{\frac{(1-\sin\theta)^2}{1-\sin^2\theta}} = \sqrt{\frac{(1-\sin\theta)^2}{\cos^2\theta}} = \frac{|1-\sin\theta|}{|\cos\theta|} = \frac{1-\sin\theta}{|\cos\theta|}$$

Note: We can drop the absolute value on $1-\sin\theta$ since it is always nonnegative.

23. Verify $\sin\left(x - \dfrac{3\pi}{2}\right) = \cos x$.

Solution:

$$\sin\left(x - \frac{3\pi}{2}\right) = (\sin x)\cos\frac{3\pi}{2} - (\cos x)\sin\frac{3\pi}{2} = (\sin x)(0) - \cos x(-1) = \cos x$$

29. Verify $\dfrac{\cos 3x - \cos x}{\sin 3x - \sin x} = -\tan 2x$.

Solution:

$$\frac{\cos 3x - \cos x}{\sin 3x - \sin x} = \frac{-2\sin\left(\dfrac{3x+x}{2}\right)\sin\left(\dfrac{3x-x}{2}\right)}{2\cos\left(\dfrac{3x+x}{2}\right)\sin\left(\dfrac{3x-x}{2}\right)}$$

$$= \frac{-2(\sin 2x)\sin x}{2(\cos 2x)\sin x} = -\frac{\sin 2x}{\cos 2x} = -\tan 2x$$

31. Verify $2\sin y\cos y\sec 2y = \tan 2y$.

Solution:

$$2\sin y\cos y\sec 2y = \sin 2y\sec 2y = \sin 2y\left(\frac{1}{\cos 2y}\right) = \frac{\sin 2y}{\cos 2y} = \tan 2y$$

33. Verify $\tan^2 x = \dfrac{1-\cos 2x}{1+\cos 2x}$.

Solution:

$$\frac{1-\cos 2x}{1+\cos 2x} = \frac{1-(1-2\sin^2 x)}{1+(2\cos^2 x - 1)} = \frac{2\sin^2 x}{2\cos^2 x} = \tan^2 x$$

35. Verify $1 + \cos 2x + \cos 4x + \cos 6x = 4 \cos x \cos 2x \cos 3x$.

Solution:

$$\begin{aligned}4 \cos x \cos 2x \cos 3x &= 4 \cos x \left[\tfrac{1}{2}(\cos(-x) + \cos(5x))\right] = 2 \cos x(\cos x + \cos 5x)\\ &= 2\cos^2 x + 2 \cos x \cos 5x = 2\cos^2 x + 2\left[\tfrac{1}{2}(\cos(-4x) + \cos 6x)\right]\\ &= 2\cos^2 x + \cos(-4x) + \cos 6x = 1 + \cos 2x + \cos 4x + \cos 6x\end{aligned}$$

39. Using the sum, difference, or half-angle formulas, find the exact value of

$$\cos(157°30') = \cos \frac{315°}{2}.$$

Solution:

$$\begin{aligned}\cos(157°30') &= \cos \frac{315°}{2} = -\sqrt{\frac{1 + \cos 315°}{2}} = -\sqrt{\frac{1 + \cos 45°}{2}}\\ &= -\sqrt{\frac{1 + \frac{\sqrt{2}}{2}}{2}} = -\sqrt{\frac{2 + \sqrt{2}}{4}} = -\frac{\sqrt{2 + \sqrt{2}}}{2}\end{aligned}$$

43. Find the exact value of $\cos(u - v)$, given that $\sin u = 3/4$, $\cos v = -5/13$, and u and v are in Quadrant II.

Solution:

Since u and v are in Quadrant II,

$$\sin u = \frac{3}{4} \quad \Rightarrow \quad \cos u = -\frac{\sqrt{7}}{4}$$

$$\cos v = -\frac{5}{13} \quad \Rightarrow \quad \sin v = \frac{12}{13}.$$

$$\cos(u - v) = \cos u \cos v + \sin u \sin v = \left(-\frac{\sqrt{7}}{4}\right)\left(-\frac{5}{13}\right) + \left(\frac{3}{4}\right)\left(\frac{12}{13}\right) = \frac{5\sqrt{7} + 36}{52}$$

47. Determine if the statement is true or false. If it is false, make the necessary correction.

If $\frac{\pi}{2} < \theta < \pi$, then $\cos \frac{\theta}{2} < 0$.

Solution:

False; if $\pi/2 < \theta < \pi$, then $\pi/4 < \theta/2 < \pi/2$ and $\cos(\theta/2) > 0$ since $\theta/2$ is in Quadrant I.

51. Find all solutions of

$$\sin x - \tan x = 0$$

in the interval $[0,\ 2\pi)$.

Solution:

$$\sin x - \tan x = 0$$

$$\sin x - \frac{\sin x}{\cos x} = 0$$

$$(\sin x)\cos x - \sin x = 0$$

$$\sin x(\cos x - 1) = 0$$

$\sin x = 0$ or $\cos x = 1$

$x = 0,\ \pi$ $\qquad x = 0$

55. Find all solutions of

$$\sin 2x + \sqrt{2}\sin x = 0$$

in the interval $[0,\ 2\pi)$.

Solution:

$$\sin 2x + \sqrt{2}\sin x = 0$$

$$2(\sin x)\cos x + \sqrt{2}\sin x = 0$$

$$\sin x(2\cos x + \sqrt{2}) = 0$$

$\sin x = 0$ or $\cos x = -\dfrac{\sqrt{2}}{2}$

$x = 0,\ \pi$ $\qquad x = \dfrac{3\pi}{4},\ \dfrac{5\pi}{4}$

59. Find all solutions of $\tan^3 x - \tan^2 x + 3\tan x - 3 = 0$ in the interval $[0,\ 2\pi)$.

Solution:

$$\tan^3 x - \tan^2 x + 3\tan x - 3 = 0$$

$$(\tan x - 1)(\tan^2 x + 3) = 0$$

$\tan x = 1$ or $\tan^2 x = -3$

$x = \dfrac{\pi}{4},\ \dfrac{5\pi}{4}$ $\qquad$ No real solutions

63. Write $\sin 3\alpha \sin 2\alpha$ as a sum or difference.

Solution:

$$\sin 3\alpha \sin 2\alpha = \tfrac{1}{2}[\cos(3\alpha - 2\alpha) - \cos(3\alpha + 2\alpha)] = \tfrac{1}{2}[\cos\alpha - \cos 5\alpha]$$

67. The rate of change of the function $f(x) = 2\sqrt{\sin x}$ with respect to change in the variable x is given by the expression $\sin^{-1/2} x \cos x$. Show that the expression for the rate of change can also be given by $\cot x\sqrt{\sin x}$.

Solution:

$$\sin^{-1/2} x \cos x = \frac{\cos x}{\sqrt{\sin x}}$$

$$= \frac{\cos x}{\sqrt{\sin x}} \cdot \frac{\sqrt{\sin x}}{\sqrt{\sin x}} = \frac{\cos x\sqrt{\sin x}}{\sin x} = \frac{\cos x}{\sin x}\sqrt{\sin x} = \cot x\sqrt{\sin x}$$

Practice Test for Chapter 6

1. Find the value of the other five trigonometric functions, given $\tan x = \frac{4}{11}$, $\sec x < 0$.

2. Simplify $\dfrac{\sec^2 x + \csc^2 x}{\csc^2 x(1 + \tan^2 x)}$.

3. Rewrite as a single logarithm and simplify $\ln|\tan\theta| - \ln|\cot\theta|$.

4. True or false:

$$\cos\left(\frac{\pi}{2} - x\right) = \frac{1}{\csc x}$$

5. Factor and simplify :

$$\sin^4 x + (\sin^2 x)\cos^2 x$$

6. Multiply and simplify:

$$(\csc x + 1)(\csc x - 1)$$

7. Rationalize the denominator and simplify:

$$\frac{\cos^2 x}{1 - \sin x}$$

8. Verify:

$$\frac{1 + \cos\theta}{\sin\theta} + \frac{\sin\theta}{1 + \cos\theta} = 2\csc\theta$$

9. Verify:

$$\tan^4 x + 2\tan^2 x + 1 = \sec^4 x$$

10. Use the sum or difference formulas to determine:

(a) $\sin 105°$ (b) $\tan 15°$

11. Simplify:

$$(\sin 42°)\cos 38° - (\cos 42°)\sin 38°$$

12. Verify $\tan\left(\theta + \frac{\pi}{4}\right) = \dfrac{1 + \tan\theta}{1 - \tan\theta}$.

13. Write $\sin(\arcsin x - \arccos x)$ as an algebraic expression in x.

14. Use the double-angle formulas to determine:

(a) $\cos 120°$ (b) $\tan 300°$

15. Use the half-angle formulas to determine:

(a) $\sin 22.5°$ (b) $\tan\dfrac{\pi}{12}$

16. Given $\sin = 4/5$, θ lies in Quadrant II, find $\cos\theta/2$.

17. Use the power-reducing identities to write $(\sin^2 x)\cos^2 x$ in terms of the first power of cosine.

18. Rewrite as a sum:

$$6(\sin 5\theta)\cos 2\theta.$$

19. Rewrite as a product:

$$\sin(x + \pi) + \sin(x - \pi).$$

20. Verify $\dfrac{\sin 9x + \sin 5x}{\cos 9x - \cos 5x} = -\cot 2x.$

21. Verify:

$$(\cos u)\sin v = \tfrac{1}{2}[\sin(u+v) - \sin(u-v)].$$

22. Find all solutions in the interval $[0,\ 2\pi)$:

$$4\sin^2 x = 1$$

23. Find all solutions in the interval $[0,\ 2\pi)$:

$$\tan^2\theta + (\sqrt{3} - 1)\tan\theta - \sqrt{3} = 0$$

24. Find all solutions in the interval $[0,\ 2\pi)$:

$$\sin 2x = \cos x$$

25. Use the quadratic formula to find all solutions in the interval $[0,\ 2\pi)$:

$$\tan^2 x - 6\tan x + 4 = 0$$

CHAPTER 7

Additional Applications of Trigonometry

Section 7.1 Law of Sines 298

Section 7.2 Law of Cosines 302

Section 7.3 Vectors in the Plane 307

Section 7.4 Trigonometric Form of a Complex Number 313

Section 7.5 DeMoivre's Theorem and nth Roots 318

Review Exercises 321

Practice Test 326

SECTION 7.1

Law of Sines

- If ABC is any oblique triangle with sides a, b, and c, then

$$\frac{a}{\sin A} = \frac{b}{\sin B} = \frac{c}{\sin C}.$$

- You should be able to use the Law of Sines to solve an oblique triangle for the remaining three parts, given:
  (a) Two angles and any side (AAS or ASA)
  (b) Two sides and an angle opposite one of them (SSA)
    1. If A is acute and:
       (a) $a < h$, no triangle is possible.
       (b) $a = h$ or $a > b$, one triangle is possible.
       (c) $h < a < b$, two triangles are possible.
    2. If A is obtuse and:
       (a) $a \leq b$, no triangle is possible.
       (b) $a > b$, one triangle is possible.

- The area of any triangle equals one-half the product of the lengths of two sides times the sine of their included angle.

$$A = \tfrac{1}{2}ab\sin C = \tfrac{1}{2}ac\sin B = \tfrac{1}{2}bc\sin A$$

Solutions to Selected Exercises

3. Find the remaining sides and angles of the triangle.

Solution:

$A = 10°,\ B = 60°,\ a = 4.5$

$C = 180° - (10° + 60°) = 110°$

$$\frac{4.5}{\sin 10°} = \frac{b}{\sin 60°}$$

$$b = \sin 60° \left(\frac{4.5}{\sin 10°}\right) \approx 22.44$$

$$\frac{4.5}{\sin 10°} = \frac{c}{\sin 110°}$$

$$c = \sin 110° \left(\frac{4.5}{\sin 10°}\right) \approx 24.35$$

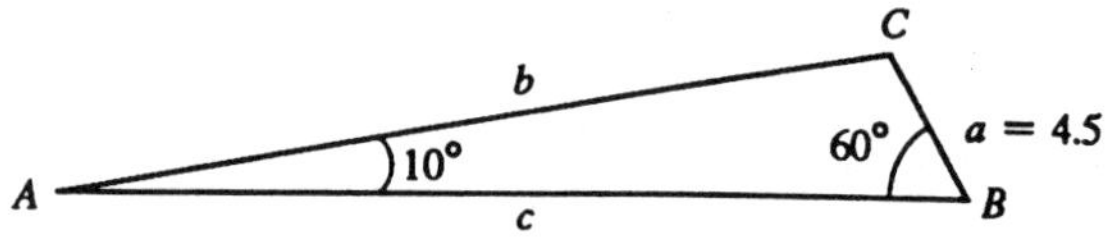

A, B, C, b, c, 10°, 60°, a = 4.5

7. Find the remaining sides and angles of the triangle, given $A = 150°$, $C = 20°$, $a = 200$.

Solution:

$A = 150°,\ C = 20°,\ a = 200$

$B = 180° - (150° + 20°) = 10°$

$$\frac{200}{\sin 150°} = \frac{b}{\sin 10°}$$

$$b = \sin 10° \left(\frac{200}{\sin 150°}\right) \approx 69.46$$

$$\frac{200}{\sin 150°} = \frac{c}{\sin 20°}$$

$$c = \sin 20° \left(\frac{200}{\sin 150°}\right) \approx 136.81$$

11. Find the remaining sides and angles of the triangle, given $B = 15°30'$, $a = 4.5$, $b = 6.8$.

Solution:

$B = 15°30',\ a = 4.5,\ b = 6.8$

$B = 15°30' = 15.5°$

$$\frac{6.8}{\sin 15.5°} = \frac{4.5}{\sin A}$$

$$\sin A = \frac{4.5 \sin 15.5}{6.8} \approx 0.17684$$

$$A \approx 10.19° \approx 10°11'$$

$$C = 180° - (10°11' + 15°30')$$

$$= 154°19'$$

$$\frac{6.8}{\sin 15.5°} = \frac{c}{\sin 154.32°}$$

$$c \approx 11.03$$

15. Find the remaining sides and angles of the triangle, given $A = 110°15'$, $a = 48$, $b = 16$.

Solution:

$A = 110°15',\ a = 48,\ b = 16$

$$\frac{48}{\sin 110.25°} = \frac{16}{\sin B}$$

$$\sin B \approx 0.3127$$

$$B \approx 18.22° \approx 18°13'$$

$$C = 180° - (110°15' + 18°13') = 51°32'$$

$$\frac{48}{\sin 110.25°} = \frac{c}{\sin 51.53°}$$

$$c \approx 40.06$$

19. Solve the triangle: $A = 58°$, $a = 4.5$, $b = 5$, (if possible). If two solutions exist, find both.

Solution:

$A = 58°$, $a = 4.5$, $b = 5$

$h = b \sin A = 5 \sin 58° = 4.24$

A is acute and $h < a < b$, so there are two possible solutions.

$$\frac{4.5}{\sin 58°} = \frac{5}{\sin B}$$

$$\sin B \approx 0.9423$$

$B \approx 70.4°$ or $B \approx 109.6°$

$C = 51.6°$ $\qquad$ $C = 12.4°$

$$\frac{4.5}{\sin 58°} = \frac{c}{\sin 51.6°} \qquad \frac{4.5}{\sin 58°} = \frac{c}{\sin 12.4°}$$

$c \approx 4.16$ $\qquad$ $c \approx 1.14$

25. Find the area of the triangle with $C = 120°$, $a = 4$, and $b = 6$.

Solution:

$$\text{Area} = \tfrac{1}{2}ab \sin C = \tfrac{1}{2}(4)(6)\sin 120° \approx 10.4 \text{ square units}$$

31. Find the length d of the brace required to support the streetlight shown in the figure.

Solution:

$$\frac{3}{\sin 30°} = \frac{5}{\sin \alpha}$$

$$\sin \alpha \approx 0.8333$$

$$\alpha \approx 56.44°$$

$$\beta = 93.56°$$

$$\frac{3}{\sin 30°} = \frac{d}{\sin 93.56°}$$

$$d \approx 6 \text{ feet}$$

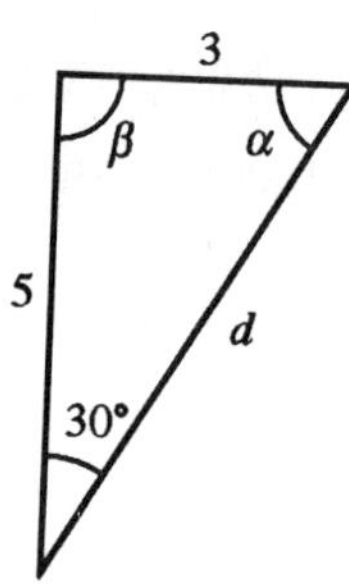

37. Two fire towers A and B are 18.5 miles apart. The bearing from A to B is N 65° E. A fire is spotted by the ranger in each tower, and its bearings from A and B are N 28° E and N 16.5° W, respectively (see figure). Find the distance of the fire from each tower.

Solution:

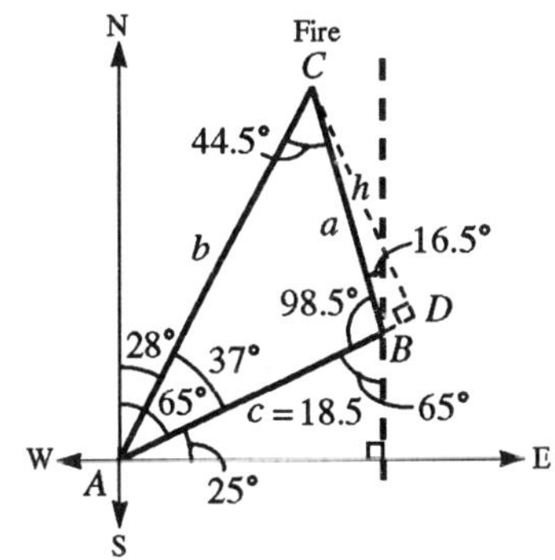

$A = 37°, \ B = 98.5°, \ C = 44.5°, \ c = 18.5$

$$\frac{b}{\sin 98.5°} = \frac{18.5}{\sin 44.5°}$$

$$b \approx 26.1 \text{ mi}$$

$$\frac{a}{\sin 37°} = \frac{18.5}{\sin 44.5°}$$

$$a \approx 15.9 \text{ mi}$$

41. The following information about a triangular parcel of land is given at a zoning board meeting: "One side is 450 feet long and another is 120 feet long. The angle opposite the shorter side is 30°." Could this information be correct?

Solution:

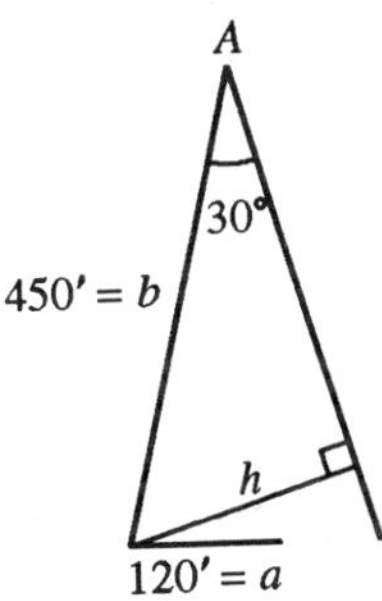

$$h = b \sin A = 450 \sin 30° = 225$$

$$a = 120$$

Since $a < h$, no such triangle is possible.

SECTION 7.2

Law of Cosines

- If ABC is any oblique triangle with sides a, b, and c, the following equations are valid.

  (a) $a^2 = b^2 + c^2 - 2bc\cos A$ or $\cos A = \dfrac{b^2 + c^2 - a^2}{2bc}$

  (b) $b^2 = a^2 + c^2 - 2ac\cos B$ or $\cos B = \dfrac{a^2 + c^2 - b^2}{2ac}$

  (c) $c^2 = a^2 + b^2 - 2ab\cos C$ or $\cos C = \dfrac{a^2 + b^2 - c^2}{2ab}$

- You should be able to use the Law of Cosines to solve an oblique triangle for the remaining three parts, given:

  (a) Three sides (SSS)

  (b) Two sides and their included angle (SAS)

- Given any triangle with sides of length a, b, and c, then the area of the triangle is

  $$\text{Area} = \sqrt{s(s-a)(s-b)(s-c)}, \text{ where } s = \frac{a+b+c}{2}. \qquad \text{(Heron's Formula)}$$

Solutions to Selected Exercises

5. Use the Law of Cosines to solve the triangle: $a = 9$, $b = 12$, $c = 15$.

Solution:

$$\cos A = \frac{144 + 225 - 81}{360} = 0.8$$

$$A \approx 36.9°$$

$$\cos B = \frac{81 + 225 - 144}{270} = 0.6$$

$$B \approx 53.1°$$

$$C \approx 180° - (36.9° + 53.1°) \approx 90°$$

7. Use the Law of Cosines to solve the triangle: $a = 75.4,\ b = 52,\ c = 52.$

Solution:

$$\cos A = \frac{(52)^2 + (52)^2 - (75.4)^2}{2(52)(52)} = -0.05125$$

$$A \approx 92.9°$$

Since $b = c$, the triangle is isosceles and $B = C$.

$$2B \approx 180° - 92.9°$$

$$B = C \approx 43.55°$$

13. Use the Law of Cosines to solve the triangle: $C = 125°40',\ a = 32,\ b = 32.$

Solution:

Since $a = b$, the triangle is isosceles and $A = B$.

$$2A = 180° - 125°40'$$

$$A = B = 27°10'$$

$$c^2 = (32)^2 + (32)^2 - 2(32)(32)\cos 125°40'$$

$$c \approx 56.9$$

17. Solve the parallelogram shown in the figure, given $a = 10,\ b = 14,$ and $c = 20.$

Solution:

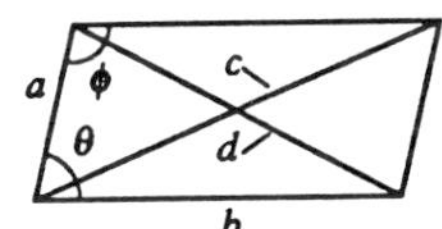

$a = 10,\ b = 14,\ c = 20$

$$\cos\phi = \frac{a^2 + b^2 - c^2}{2ab}$$

$$= \frac{100 + 196 - 400}{2(10)(14)} = -\frac{104}{280}$$

$$\phi \approx 111.8°$$

$$2\theta = 360° - 2\phi$$

$$\theta = \frac{360° - 2(11.8°)}{2} \approx 68.2°$$

$$d^2 = a^2 + b^2 - 2ab\cos\theta$$

$$= 100 + 196 - 280\cos 68.2°$$

$$d \approx 13.86$$

23. Use Heron's Formula to find the area of the triangle: $a = 12,\ b = 15,\ c = 9$.

Solution:

$$s = \tfrac{1}{2}(12 + 15 + 9) = 18$$

$$\text{Area} = \sqrt{18(18-12)(18-15)(18-9)} = \sqrt{2916} = 54 \text{ square units}$$

27. The lengths of the sides of a triangular parcel of land are approximately 400 feet, 500 feet, and 700 feet. Approximate the area of the parcel.

Solution:

$a = 400,\ b = 500,\ c = 700$

$$s = \tfrac{1}{2}(400 + 500 + 700) = 800$$

$$\text{Area} = \sqrt{800(800-400)(800-500)(800-700)}$$

$$= \sqrt{9{,}600{,}000{,}000} = 40{,}000\sqrt{6} \approx 97{,}979.6 \text{ square feet}$$

31. Two ships leave a port at 9 A. M. One travels at a bearing of N 53° W at 12 miles per hour and the other at a bearing of S 67° W at 16 miles per hour. Approximately how far apart are they at noon of that day?

Solution:

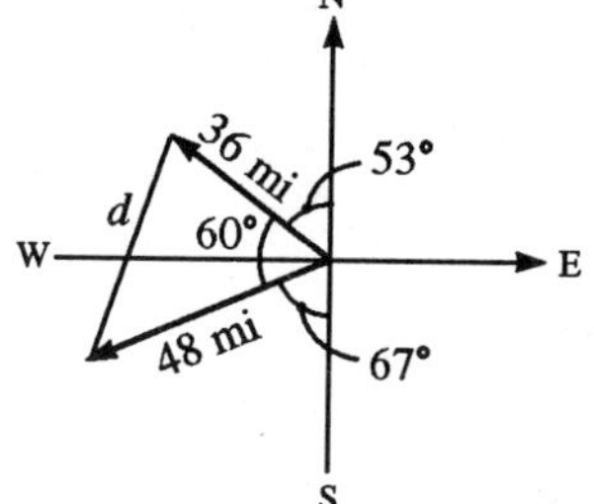

By noon, the first ship has traveled $3(12) = 36$ miles, and the second ship has traveled $3(16) = 48$ miles.

$$d^2 = 36^2 + 48^2 - 2(36)(48)\cos 60°$$

$$d^2 = 1872$$

$$d \approx 43.3 \text{ miles}$$

35. Determine the angle θ as shown on the streetlight in the figure.

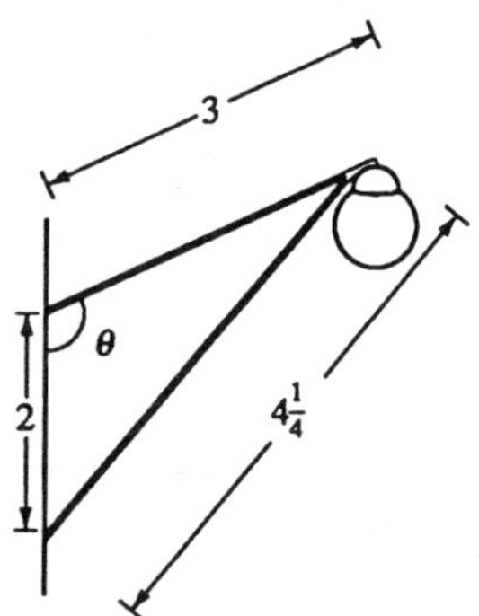

Solution:

$$\cos\theta = \frac{3^2 + 2^2 - 4.25^2}{2(3)(2)} = -0.421875$$

$$\theta \approx 114.95°$$

39. On a certain map, Orlando is 7 inches due south of Niagara Falls, Denver is 10.75 inches from Orlando, and Denver is 9.25 inches from Niagara Falls (see figure).

(a) Find the bearing of Denver from Orlando.

(b) Find the bearing of Denver from Niagara Falls.

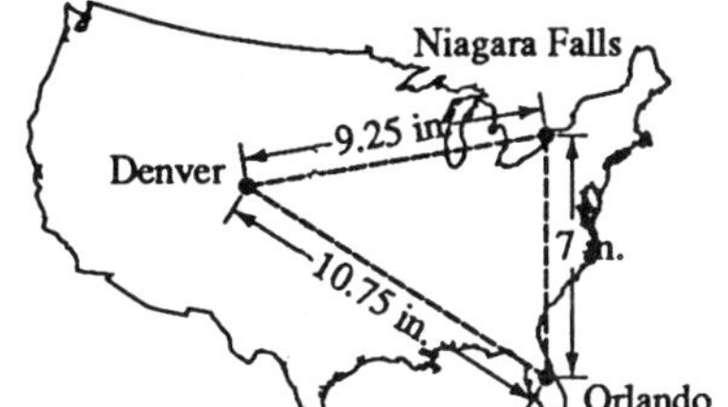

Solution:

(a) $\cos\theta = \dfrac{(10.75)^2 + 7^2 - (9.25)^2}{2(10.75)(7)} = \dfrac{79}{150.5}$

$\theta \approx 58.3°$

Bearing: N 58.3° W

(b) $\cos\phi = \dfrac{(9.25)^2 + 7^2 - (10.75)^2}{2(9.25)(7)} = \dfrac{19}{129.5}$

$\phi \approx 81.6°$

Bearing: S 81.6° W

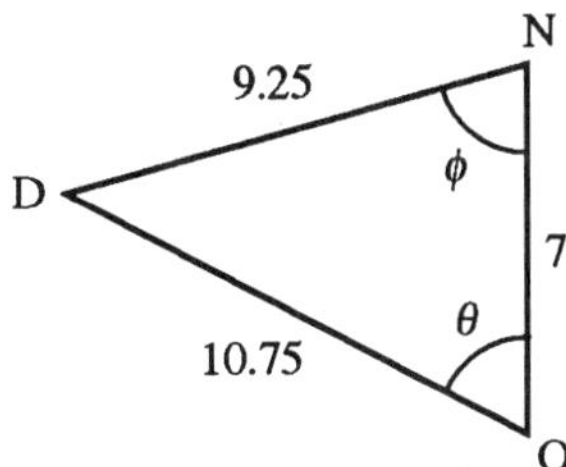

41. In a (square) baseball diamond with 90-foot sides the pitcher's mound is 60 feet from home plate.

(a) How far is it from the pitcher's mound to third base?

(b) When a runner is halfway from second to third, how far is the runner from the pitcher's mound?

Solution:

(a) $x^2 = 90^2 + 60^2 - 2(90)(60)\cos 45°$

$x \approx 63.7$ feet

(b) $\dfrac{60}{\sin\alpha} = \dfrac{63.7}{\sin 45°}$

$\sin\alpha = \dfrac{60\sin 45°}{63.7}$

$\alpha \approx 41.76°$

$\beta = 90° - \alpha = 48.24°$

$y^2 = 45^2 + 63.7^2 - 2(45)(63.7)\cos 48.24°$

$y \approx 47.6$ feet

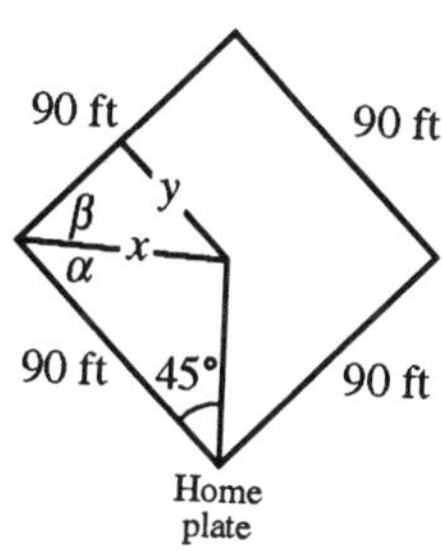

47. Use the Law of Cosines to prove that

$$\frac{1}{2}bc(1+\cos A) = \frac{a+b+c}{2} \cdot \frac{-a+b+c}{2}.$$

Solution:

$$\begin{aligned}
\frac{1}{2}bc(1+\cos A) &= \frac{1}{2}bc\left[1+\frac{b^2+c^2-a^2}{2bc}\right] \\
&= \frac{1}{2}bc\left[\frac{2bc+b^2+c^2-a^2}{2bc}\right] \\
&= \frac{1}{4}[(b+c)^2-a^2] \\
&= \frac{1}{4}[(b+c)+a][(b+c)-a] \\
&= \frac{b+c+a}{2} \cdot \frac{b+c-a}{2} \\
&= \frac{a+b+c}{2} \cdot \frac{-a+b+c}{2}
\end{aligned}$$

SECTION 7.3

Vectors in the Plane

- A vector $\mathbf{v}$ is the collection of all directed line segments that are equivalent to a given directed line segment $\overrightarrow{PQ}$.
- You should be able to *geometrically* perform the operations of vector addition and scalar multiplication.
- The component form of the vector with initial point $P = (p_1, p_2)$ and terminal point $Q = (q_1, q_2)$ is

  $$\overrightarrow{PQ} = \langle q_1 - p_1, q_2 - p_2 \rangle = \langle v_1, v_2 \rangle = \mathbf{v}.$$

- The magnitude of $\mathbf{v} = \langle v_1, v_2 \rangle$ is given by $\|\mathbf{v}\| = \sqrt{v_1{}^2 + v_2{}^2}$.
- You should be able to perform the operations of scalar multiplication and vector addition in component form.
- You should know the following properties of vector addition and scalar multiplication.

  (a) $\mathbf{u} + \mathbf{v} = \mathbf{v} + \mathbf{u}$
  (b) $(\mathbf{u} + \mathbf{v}) + \mathbf{w} = \mathbf{u} + (\mathbf{v} + \mathbf{w})$
  (c) $\mathbf{u} + \mathbf{0} = \mathbf{u}$
  (d) $\mathbf{u} + (-\mathbf{u}) = \mathbf{0}$
  (e) $c(d\mathbf{u}) = (cd)\mathbf{u}$
  (f) $(c + d)\mathbf{u} = c\mathbf{u} + d\mathbf{u}$
  (g) $c(\mathbf{u} + \mathbf{v}) = c\mathbf{u} + c\mathbf{v}$
  (h) $1(\mathbf{u}) = \mathbf{u}$, $0\mathbf{u} = \mathbf{0}$
  (i) $\|c\mathbf{v}\| = |c|\,\|\mathbf{v}\|$

- A unit vector in the direction of $\mathbf{v}$ is given by $\mathbf{u} = \dfrac{\mathbf{v}}{\|\mathbf{v}\|}$.
- The standard unit vectors are $\mathbf{i} = \langle 1, 0 \rangle$ and $\mathbf{j} = \langle 0, 1 \rangle$. $\mathbf{v} = \langle v_1, v_2 \rangle$ can be written as $\mathbf{v} = v_1\mathbf{i} + v_2\mathbf{j}$.
- A vector $\mathbf{v}$ with magnitude $\|\mathbf{v}\|$ and direction θ can be written as $\mathbf{v} = a\mathbf{i} + b\mathbf{j} = \|\mathbf{v}\|(\cos\theta)\mathbf{i} + \|\mathbf{v}\|(\sin\theta)\mathbf{j}$ where $\tan\theta = b/a$.

Solutions to Selected Exercises

3. Use the figure to sketch a graph of $\mathbf{u}+\mathbf{v}$.

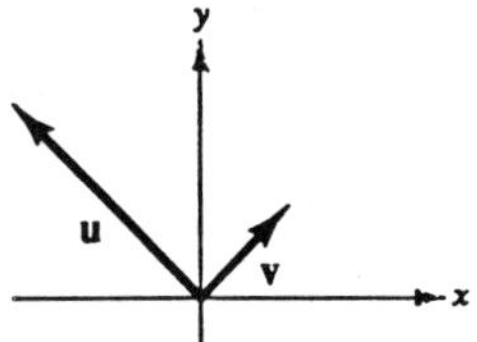

Solution:

Move the initial point of $\mathbf{v}$ to the terminal point of $\mathbf{u}$.

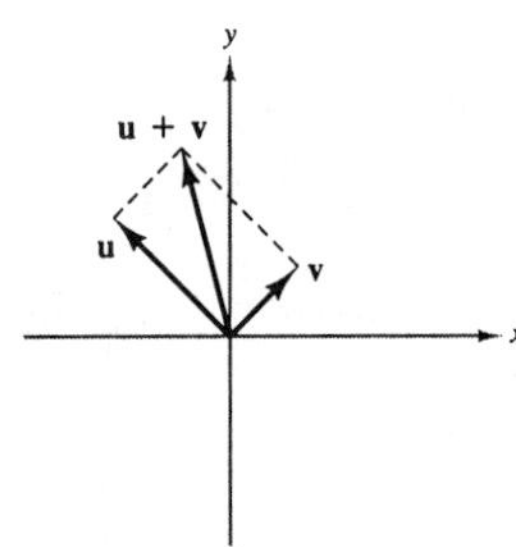

9. Find the component form and the magnitude of the vector $\mathbf{v}$.

Solution:

$$\mathbf{v} = \langle -1-2,\ 3-1\rangle = \langle -3,\ 2\rangle$$

$$\|\mathbf{v}\| = \sqrt{(-3)^2+(2)^2} = \sqrt{13}$$

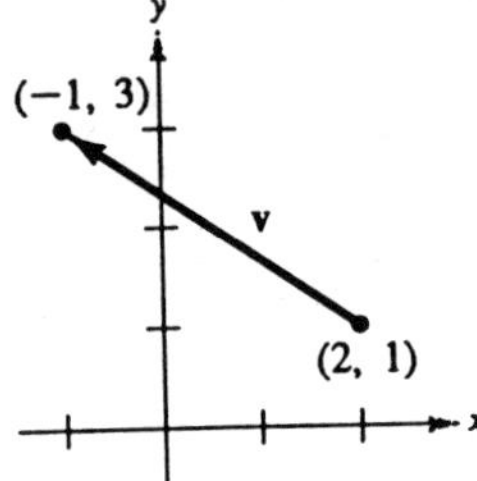

13. Find the component form and the magnitude of the vector $\mathbf{v}$ with initial point $(-1,\ 5)$ and terminal point $(15,\ 2)$.

Solution:

$$\mathbf{v} = \langle 15-(-1),\ 2-5\rangle = \langle 16,\ -3\rangle$$

$$\|\mathbf{v}\| = \sqrt{(16)^2+(-3)^2} = \sqrt{265}$$

19. Find (a) $\mathbf{u}+\mathbf{v}$, (b) $\mathbf{u}-\mathbf{v}$, and (c) $2\mathbf{u}-3\mathbf{v}$ for $\mathbf{u}=\langle -2,\ 3\rangle$, $\mathbf{v}=\langle -2,\ 1\rangle$.

Solution:

(a) $\mathbf{u}+\mathbf{v}=\langle -2+(-2),\ 3+1\rangle=\langle -4,\ 4\rangle$

(b) $\mathbf{u}-\mathbf{v}=\langle -2-(-2),\ 3-1\rangle=\langle 0,\ 2\rangle$

(c) $2\mathbf{u}-3\mathbf{v}=\langle 2(-2)-3(-2),\ 2(3)-3(1)\rangle=\langle 2,\ 3\rangle$

25. Find (a) $\mathbf{u}+\mathbf{v}$, (b) $\mathbf{u}-\mathbf{v}$, and (c) $2\mathbf{u}-3\mathbf{v}$ for $\mathbf{u}=2\mathbf{i}$, $\mathbf{v}=\mathbf{j}$.

Solution:

(a) $\mathbf{u}+\mathbf{v}=2\mathbf{i}+\mathbf{j}$

(b) $\mathbf{u}-\mathbf{v}=2\mathbf{i}-\mathbf{j}$

(c) $2\mathbf{u}-3\mathbf{v}=4\mathbf{i}-3\mathbf{j}$

29. Find the magnitude and direction angle for $\mathbf{v}=6\mathbf{i}-6\mathbf{j}$.

Solution:

$$\mathbf{v}=6\mathbf{i}-6\mathbf{j}$$

$$\|\mathbf{v}\|=\sqrt{6^2+(-6)^2}=\sqrt{72}=6\sqrt{2}$$

$$\tan\theta=-\tfrac{6}{6}=-1$$

Since $\mathbf{v}$ lies in Quadrant IV, $\theta=315°$.

33. Sketch $\mathbf{v}$ and find its component form given $\|\mathbf{v}\|=1$, $\theta=150°$. (Assume θ is measured counterclockwise from the x-axis to the vector.)

Solution:

$$\mathbf{v}=1\cos 150°\mathbf{i}+1\sin 150°\mathbf{j}$$

$$=-\frac{\sqrt{3}}{2}\mathbf{i}+\frac{1}{2}\mathbf{j}$$

$$=\left\langle -\frac{\sqrt{3}}{2},\ \frac{1}{2}\right\rangle$$

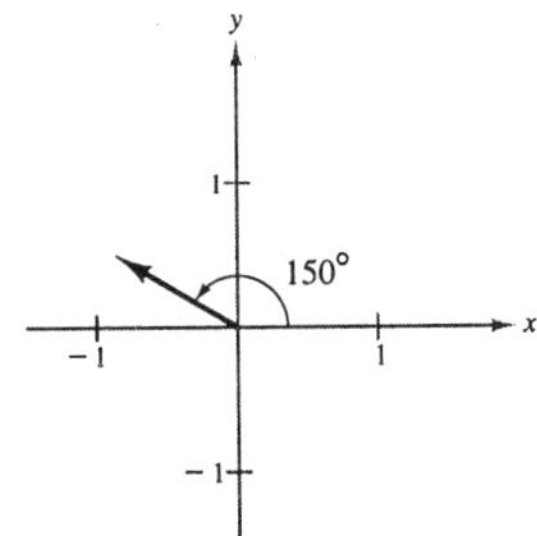

37. Sketch $\mathbf{v}$ and find its component form given $\|\mathbf{v}\| = 2$, and $\mathbf{v}$ is in the direction of $\mathbf{i} + 3\mathbf{j}$. (Assume θ is measured counterclockwise from the x-axis to the vector.)

Solution:

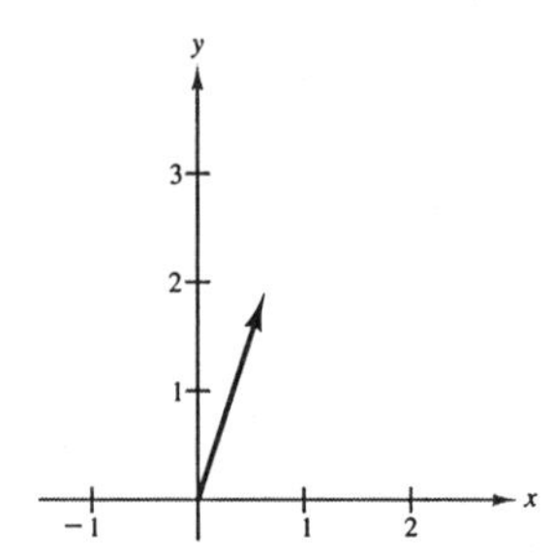

$$\tan\theta = \frac{3}{1} \Rightarrow \sin\theta = \frac{3\sqrt{10}}{10} \text{ and } \cos\theta = \frac{\sqrt{10}}{10}$$

$$\mathbf{v} = 2\left(\frac{\sqrt{10}}{10}\right)\mathbf{i} + 2\left(\frac{3\sqrt{10}}{10}\right)\mathbf{j}$$

$$= \left\langle \frac{\sqrt{10}}{5}, \frac{3\sqrt{10}}{5} \right\rangle$$

41. Find the component form of $\mathbf{v} = \mathbf{u} + 2\mathbf{w}$ and sketch the indicated vector operations geometrically, where $\mathbf{u} = 2\mathbf{i} - \mathbf{j}$ and $\mathbf{w} = \mathbf{i} + 2\mathbf{j}$.

Solution:

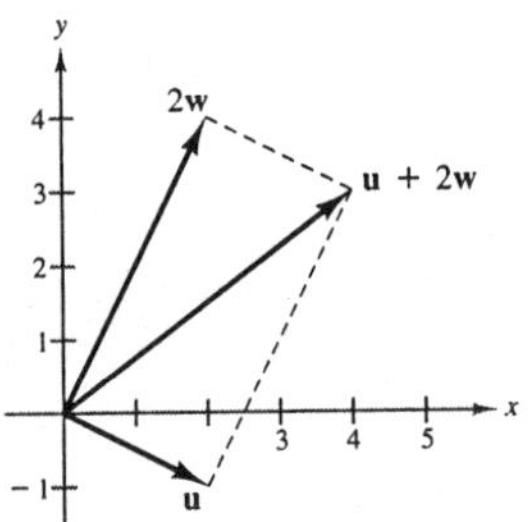

$$\mathbf{v} = \mathbf{u} + 2\mathbf{w}$$

$$= 4\mathbf{i} + 3\mathbf{j}$$

$$= \langle 4, 3 \rangle$$

45. Find the component form of the sum of the vectors $\mathbf{u}$ and $\mathbf{v}$ with direction angles $\theta_{\mathbf{u}}$ and $\theta_{\mathbf{v}}$, respectively, given $\|\mathbf{u}\| = 5$, $\theta_{\mathbf{u}} = 0°$, and $\|\mathbf{v}\| = 5$, $\theta_{\mathbf{v}} = 90°$.

Solution:

$$\|\mathbf{u}\| = 5,\ \theta_{\mathbf{u}} = 0° \Rightarrow \mathbf{u} = 5\mathbf{i}$$

$$\|\mathbf{v}\| = 5,\ \theta_{\mathbf{v}} = 90° \Rightarrow \mathbf{v} = 5\mathbf{j}$$

$$\mathbf{u} + \mathbf{v} = 5\mathbf{i} + 5\mathbf{j} = \langle 5, 5 \rangle$$

49. Find a unit vector in the direction of $\mathbf{v} = 4\mathbf{i} - 3\mathbf{j}$.

Solution:

$$\|\mathbf{v}\| = \sqrt{(4)^2 + (-3)^2} = 5$$

$$\frac{\mathbf{v}}{\|\mathbf{v}\|} = \frac{4\mathbf{i} - 3\mathbf{j}}{5} = \left\langle \frac{4}{5}, -\frac{3}{5} \right\rangle$$

53. Use the Law of Cosines to find the angle α between the vectors $\mathbf{v} = \mathbf{i} + \mathbf{j}$ and $\mathbf{w} = 2(\mathbf{i} - \mathbf{j})$. (Assume $0° \leq \alpha \leq 180°$.)

Solution:

$\mathbf{v} = \mathbf{i} + \mathbf{j}, \quad \mathbf{w} = 2(\mathbf{i} - \mathbf{j})$

$\mathbf{u} = \mathbf{w} - \mathbf{v} = \mathbf{i} - 3\mathbf{j}$

$$\|\mathbf{v}\| = \sqrt{2}$$

$$\|\mathbf{w}\| = 2\sqrt{2}$$

$$\|\mathbf{u}\| = \sqrt{10}$$

$$\cos\alpha = \frac{2 + 8 - 10}{2(\sqrt{2})(2\sqrt{2})} = 0$$

$$\alpha = 90°$$

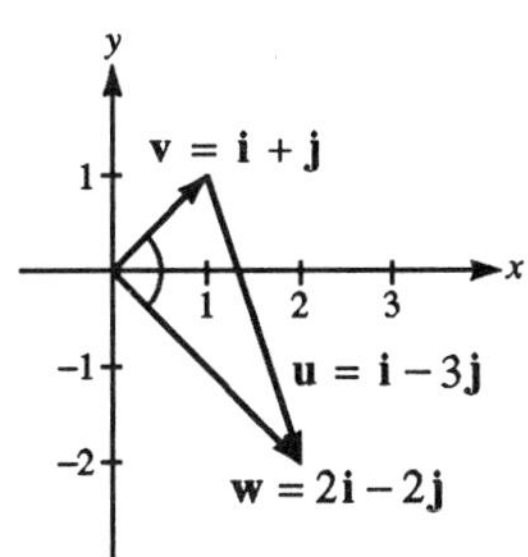

59. Forces with magnitudes of 35 pounds and 50 pounds act on a hook. The angle between the two forces is 30°. Find the direction and magnitude of the resultant (vector sum) of these two forces.

Solution:

$$\mathbf{u} = 50\mathbf{i}$$

$$\mathbf{v} = 35(\cos 30°\mathbf{i} + \sin 30°\mathbf{j}) = 35\left(\frac{\sqrt{3}}{2}\mathbf{i} + \frac{1}{2}\mathbf{j}\right)$$

$$\mathbf{r} = \mathbf{u} + \mathbf{v} = \left(50 + \frac{35\sqrt{3}}{2}\right)\mathbf{i} + \frac{35}{2}\mathbf{j}$$

$$\|\mathbf{r}\| \approx 82.2 \text{ lb}$$

$$\tan\theta = \frac{35/2}{50 + (35\sqrt{3}/2)}$$

$$\theta \approx 12.3°$$

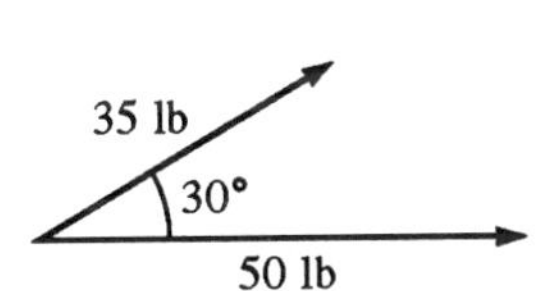

63. A ball is thrown with an initial velocity of 80 feet per second at an angle of 50° with the horizontal (see figure). Find the vertical and horizontal components of the velocity.

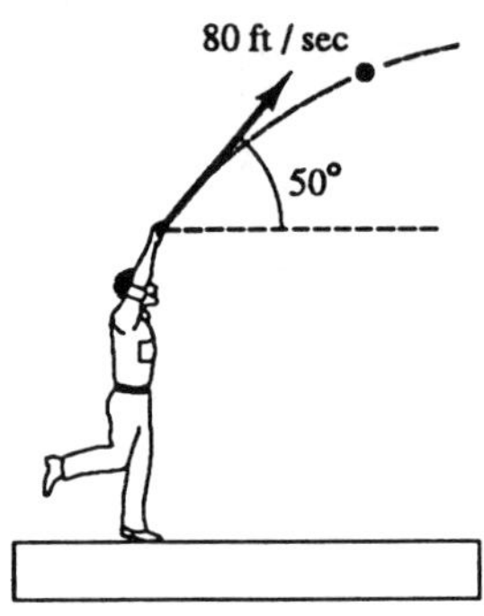

Solution:

Vertical component: $80 \sin 50° \approx 61.28$ ft/sec
Horizontal component: $80 \cos 50° \approx 51.42$ ft/sec

69. An airplane is flying in the direction S 32° E, with an airspeed of 540 miles per hour. Because of the wind, its groundspeed and direction are 500 miles per hour and S 40° E, respectively. Find the direction and speed of the wind.

Solution:

$$a = \|\overrightarrow{BC}\| = \text{speed of wind}$$

$$a^2 = 500^2 + 540^2 - 2(500)(540)\cos 8°$$

$$a \approx 82.8 \text{ mi/hr}$$

$$\cos C \approx \frac{82.8^2 + 500^2 - 540^2}{2(82.8)(500)}$$

$$C \approx 114.8°$$

$$C + \theta + 40 \approx 180°$$

$$\theta \approx \text{N } 25.2° \text{ E}$$

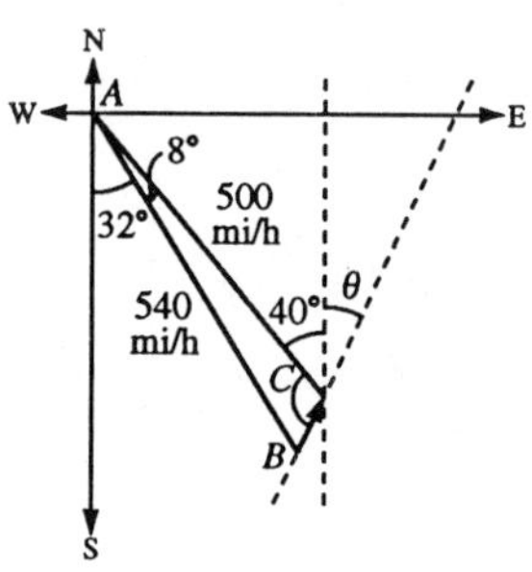

71. A heavy implement is pulled 10 feet across the floor, using a force of 85 pounds. Find the work done if the direction of the force is 60° above the horizontal (see figure). (Use the formula for work, $W = FD$, where F is the component of the force in the direction of motion and D is the distance.)

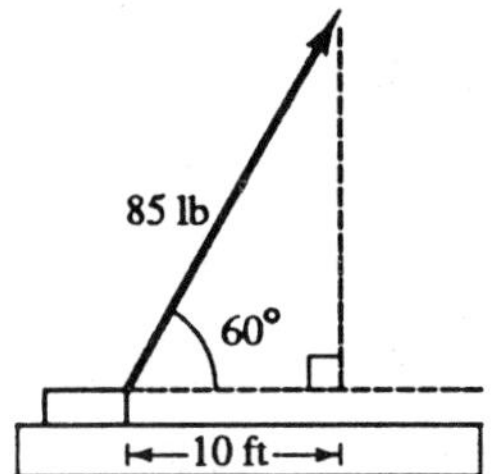

Solution:

The horizontal component of the force is $85 \cos 60° = \frac{85}{2}$.

$$W = FD = \tfrac{85}{2}(10) = 425 \text{ ft-lb}$$

SECTION 7.4

Trigonometric Form of a Complex Number

- You should be able to graphically represent complex numbers and know the following facts about them.
- The absolute value of the complex number $z = a + bi$ is $|z| = \sqrt{a^2 + b^2}$.
- The trigonometric form of the complex number $z = a + bi$ is $z = r(\cos\theta + i\sin\theta)$ where
  (a) $a = r\cos\theta$
  (b) $b = r\sin\theta$
  (c) $r = \sqrt{a^2 + b^2}$; r is called the modulus of z.
  (d) $\tan\theta = b/a$; θ is called the argument of z.
- Given $z_1 = r_1(\cos\theta_1 + i\sin\theta_1)$ and $z_2 = r_2(\cos\theta_2 + i\sin\theta_2)$:
  (a) $z_1 z_2 = r_1 r_2[\cos(\theta_1 + \theta_2) + i\sin(\theta_1 + \theta_2)]$
  (b) $\dfrac{z_1}{z_2} = \dfrac{r_1}{r_2}[\cos(\theta_1 - \theta_2) + i\sin(\theta_1 - \theta_2)]$, $z_2 \neq 0$

Solutions to Selected Exercises

5. Represent the complex number $6 - 7i$ graphically and find its absolute value.

Solution:

$$\begin{aligned}|6 - 7i| &= \sqrt{(6)^2 + (-7)^2}\\ &= \sqrt{36 + 49}\\ &= \sqrt{85}\end{aligned}$$

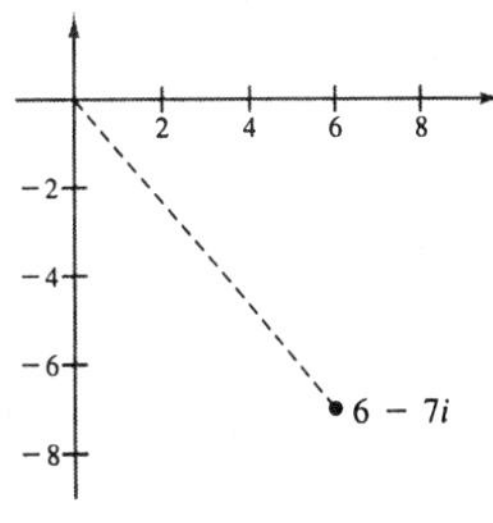

9. Express the complex number $-3-3i$ in trigonometric form.

Solution:

$$z=-3-3i$$

$$r=\sqrt{(-3)^2+(-3)^2}=\sqrt{18}=3\sqrt{2}$$

$$\tan\theta=\frac{-3}{-3}=1,\ \theta \text{ is in Quadrant III}$$

$$\theta=225^\circ \text{ or } \frac{5\pi}{4}$$

$$z=3\sqrt{2}\left(\cos\frac{5\pi}{4}+i\sin\frac{5\pi}{4}\right)$$

13. Represent $\sqrt{3}+i$ graphically, and find the trigonometric form of the number.

Solution:

$$z=\sqrt{3}+i$$

$$r=\sqrt{(\sqrt{3})^2+1^2}=2$$

$$\tan\theta=\frac{1}{\sqrt{3}}$$

$$\theta=30^\circ \text{ or } \frac{\pi}{6}$$

$$z=2\left(\cos\frac{\pi}{6}+i\sin\frac{\pi}{6}\right)$$

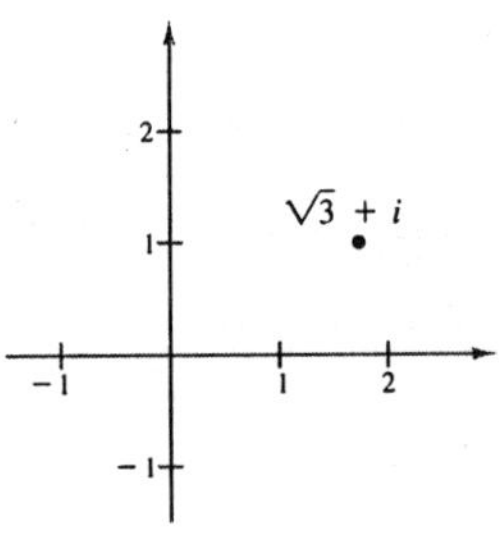

17. Represent $6i$ graphically, and find the trigonometric form of the number.

Solution:

$$z=6i$$

$$r=\sqrt{0^2+6^2}=6$$

$$\tan\theta=\frac{6}{0},\ \text{undefined}$$

$$\theta=\frac{\pi}{2}$$

$$z=6\left(\cos\frac{\pi}{2}+i\sin\frac{\pi}{2}\right)$$

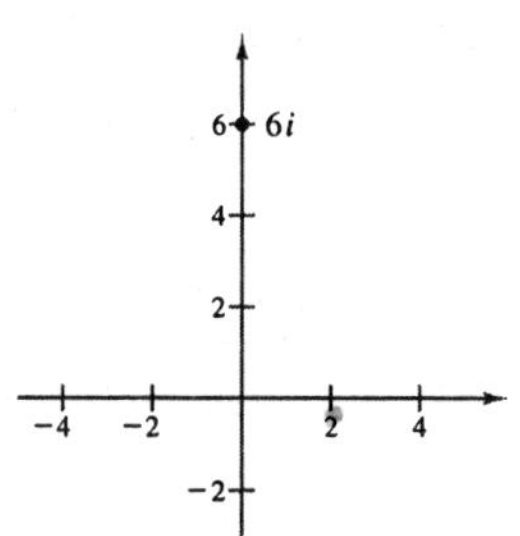

23. Represent $1+6i$ graphically, and find the trigonometric form of the number.

Solution:

$$z = 1+6i$$

$$r = \sqrt{37}$$

$$\tan\theta = 6$$

$$\theta \approx 1.41 \text{ rad}$$

$$z = \sqrt{37}[\cos(1.41) + i\sin(1.41)]$$

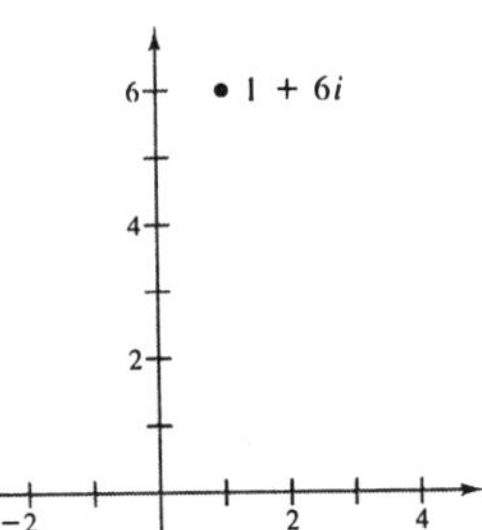

27. Represent $2(\cos 150° + i\sin 150°)$ graphically, and find the standard form of the number.

Solution:

$$z = 2(\cos 150° + i\sin 150°)$$

$$= 2\left(-\frac{\sqrt{3}}{2} + \frac{1}{2}i\right)$$

$$= -\sqrt{3} + i$$

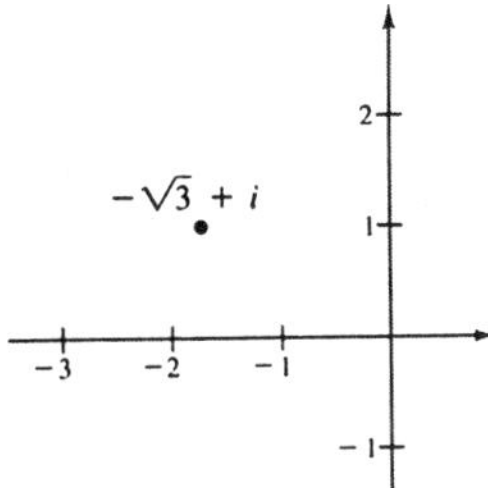

31. Represent $3.75\,(\cos(3\pi/4) + i\sin(3\pi/4))$ graphically, and find the standard form of the number.

Solution:

$$z = 3.75\left(\cos\frac{3\pi}{4} + i\sin\frac{3\pi}{4}\right)$$

$$= \frac{15}{4}\left(-\frac{\sqrt{2}}{2} + \frac{\sqrt{2}}{2}i\right)$$

$$= \frac{-15\sqrt{2}}{8} + \frac{15\sqrt{2}}{8}i$$

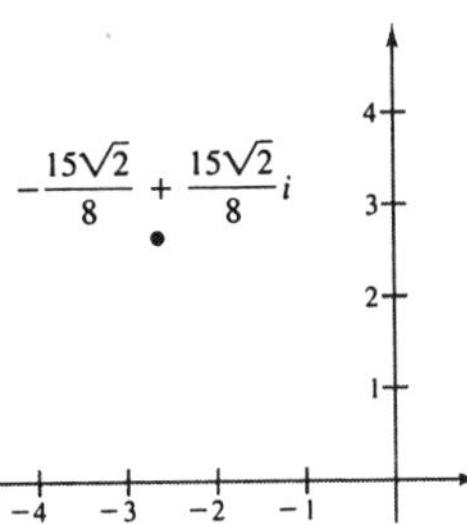

35. Represent $3[\cos(18°45') + i\sin(18°45')]$ graphically, and find the standard form of the number.

Solution:

$$\begin{aligned} z &= 3(\cos 18°45' + i\sin 18°45') \\ &= 3\cos 18°45' + 3i\sin 18°45' \\ &\approx 2.8408 + 0.9643i \end{aligned}$$

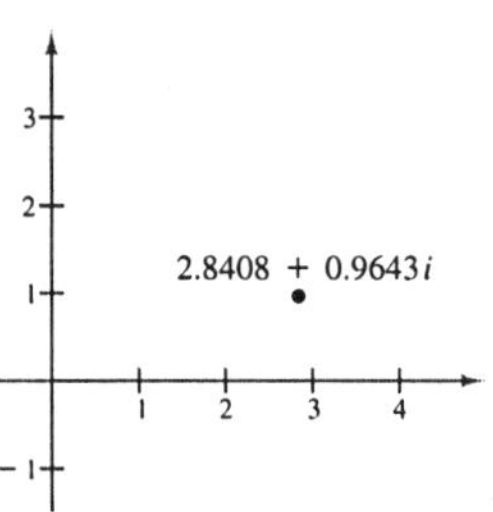

39. Perform the indicated operation and leave the result in trigonometric form.

$$\left[\tfrac{5}{3}(\cos 140° + i\sin 140°)\right]\left[\tfrac{2}{3}(\cos 60° + i\sin 60°)\right]$$

Solution:

$$\begin{aligned} &\left[\tfrac{5}{3}(\cos 140° + i\sin 140°)\right]\left[\tfrac{2}{3}(\cos 60° + i\sin 60°)\right] \\ &\quad = \left(\tfrac{5}{3}\right)\left(\tfrac{2}{3}\right)[\cos(140° + 60°) + i\sin(140° + 60°)] \\ &\quad = \tfrac{10}{9}[\cos 200° + i\sin 200°] \end{aligned}$$

45. Perform the indicated operation and leave the result in trigonometric form.

$$\frac{\cos(5\pi/3) + i\sin(5\pi/3)}{\cos\pi + i\sin\pi}$$

Solution:

$$\frac{\cos(5\pi/3) + i\sin(5\pi/3)}{\cos\pi + i\sin\pi} = \cos\left(\frac{5\pi}{3} - \pi\right) + i\sin\left(\frac{5\pi}{3} - \pi\right) = \cos\frac{2\pi}{3} + i\sin\frac{2\pi}{3}$$

49. For $(2 + 2i)(1 - i)$, (a) give the trigonometric form of the complex numbers, (b) perform the indicated operation using the trigonometric form, and (c) perform the indicated operation using the standard form and check your result with the answer in part (b).

Solution:

(a) Trigonometric form: $\left[2\sqrt{2}(\cos 45° + i\sin 45°)\right]\left[\sqrt{2}(\cos 315° + i\sin 315°)\right]$

(b) Operation in trigonometric form:

$$\begin{aligned} (2\sqrt{2})(\sqrt{2})\left[\cos(45° + 315°) + i\sin(45° + 315°)\right] &= \cos 360° + i\sin 360° \\ &= 4[\cos(360°) + i\sin(360°)] = 4 \end{aligned}$$

(c) $(2 + 2i)(1 - i) = 2 - 2i + 2i - 2i^2 = 4$

53. For $5/(2+3i)$, (a) give the trigonometric form of the complex numbers, (b) perform the indicated operation using the trigonometric form, and (c) perform the indicated operation using the standard form and check your result with the answer in part (b).

Solution:

(a) Trigonometric form: $\dfrac{5[\cos 0° + i\sin 0°]}{\sqrt{13}[\cos 56.31° + i\sin 56.31°]}$

(b) Operation in trigonometric form:

$$\frac{5}{\sqrt{13}}[\cos(-56.31°) + i\sin(-56.31°)] = \frac{5\sqrt{13}}{13}[\cos(-56.31°) + i\sin(-56.31°)] \approx \frac{5}{13}(2-3i)$$

(c) $\dfrac{5}{2+3i} \cdot \dfrac{2-3i}{2-3i} = \dfrac{5(2-3i)}{13} = \dfrac{5}{13}(2-3i)$

57. Use the trigonometric form $z = r(\cos\theta + i\sin\theta)$ and $\overline{z} = r[\cos(-\theta) + i\sin(-\theta)]$ to find (a) $z\overline{z}$ and (b) $z/\overline{z}$, $z \neq 0$.

Solution:

(a) $z\overline{z} = [r(\cos\theta + i\sin\theta)][r\cos(-\theta) + i\sin(-\theta)]$

$= r^2[\cos(\theta - \theta) + i\sin(\theta - \theta)]$

$= r^2$

(b) $\dfrac{z}{\overline{z}} = \dfrac{r}{r}[\cos(\theta - (-\theta)) + i\sin(\theta - (-\theta))]$

$= \cos 2\theta + i\sin 2\theta$

SECTION 7.5

DeMoivre's Theorem and nth Roots

- You should know DeMoivre's Theorem: If $z = r(\cos\theta + i\sin\theta)$, then for any positive integer n,

$$z^n = r^n(\cos n\theta + i\sin n\theta).$$

- You should know that for any positive integer n, $z = r(\cos\theta + i\sin\theta)$ has n distinct nth roots given by

$$\sqrt[n]{r}\left[\cos\left(\frac{\theta + 2\pi k}{n}\right) + i\sin\left(\frac{\theta + 2\pi k}{n}\right)\right]$$

where $k = 0,\ 1,\ 2,\ \ldots,\ n-1$.

Solutions to Selected Exercises

3. Use DeMoivre's Theorem to find $(-1+i)^{10}$. Express the result in standard form.

Solution:

$$\begin{aligned}
(-1+i)^{10} &= \left[\sqrt{2}\left(\cos\frac{3\pi}{4} + i\sin\frac{3\pi}{4}\right)\right]^{10} \\
&= (\sqrt{2})^{10}\left[\cos 10\left(\frac{3\pi}{4}\right) + i\sin 10\left(\frac{3\pi}{4}\right)\right] \\
&= 32\left[\cos\frac{15\pi}{2} + i\sin\frac{15\pi}{2}\right] \\
&= 32[0 - i] \\
&= -32i
\end{aligned}$$

7. Use DeMoivre's Theorem to find $[5(\cos 20^\circ + i\sin 20^\circ)]^3$. Express the result in standard form.

Solution:

$$\begin{aligned}
[5(\cos 20^\circ + i\sin 20^\circ)]^3 &= 5^3[\cos 60^\circ + i\sin 60^\circ] \\
&= 125\left(\frac{1}{2} + \frac{\sqrt{3}}{2}i\right) \\
&= \frac{125}{2} + \frac{125\sqrt{3}}{2}i
\end{aligned}$$

11. Use DeMoivre's Theorem to find $[5(\cos 3.2 + i \sin 3.2)]^4$. Express the result in standard form.

Solution:

$$\begin{aligned}[5(\cos 3.2 + i\sin 3.2)]^4 &= 5^4[\cos 12.8 + i \sin 12.8] \\ &\approx 625(0.97283 + 0.2315i) \\ &\approx 608.02 + 144.69i\end{aligned}$$

15. (a) Use DeMoivre's Theorem to find the fourth roots of

$$16\left(\cos\frac{4\pi}{3} + i\sin\frac{4\pi}{3}\right),$$

(b) represent each of the roots graphically, and (c) express each of the roots in standard form.

Solution:

(a) & (c) $n = 4$

$$k = 0: \ 2\left(\cos\frac{\pi}{3} + i\sin\frac{\pi}{3}\right) = 1 + \sqrt{3}i$$

$$k = 1: \ 2\left(\cos\frac{5\pi}{6} + i\sin\frac{5\pi}{6}\right) = -\sqrt{3} + i$$

$$k = 2: \ 2\left(\cos\frac{4\pi}{3} + i\sin\frac{4\pi}{3}\right) = -1 - \sqrt{3}i$$

$$k = 3: \ 2\left(\cos\frac{11\pi}{6} + i\sin\frac{11\pi}{6}\right) = \sqrt{3} - i$$

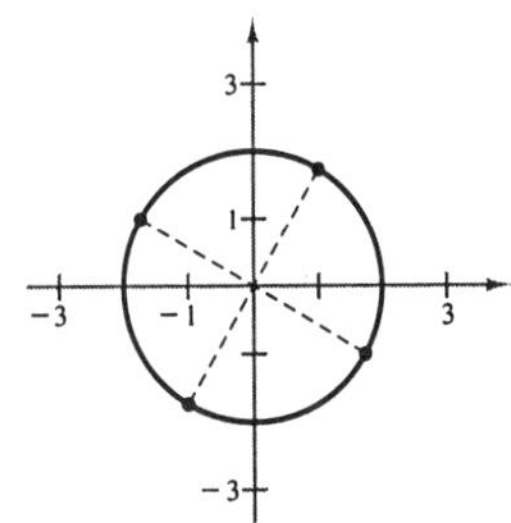

19. (a) Use DeMoivre's Theorem to find the cube roots of $-\frac{125}{2}(1 + \sqrt{3}i)$, (b) represent each of the roots graphically, and (c) express each of the roots in standard form.

Solution:

$$-\frac{125}{2}(1 + \sqrt{3}i) = 125\left(\cos\frac{4\pi}{3} + i\sin\frac{4\pi}{3}\right)$$

(a) & (c) $n = 3$

$$k = 0: \ 5\left(\cos\frac{4\pi}{9} + i\sin\frac{4\pi}{9}\right) \approx 0.868 + 4.924i$$

$$k = 1: \ 5\left(\cos\frac{10\pi}{9} + i\sin\frac{10\pi}{9}\right) \approx -4.698 - 1.710i$$

$$k = 2: \ 5\left(\cos\frac{16\pi}{9} + i\sin\frac{16\pi}{9}\right) \approx 3.830 - 3.214i$$

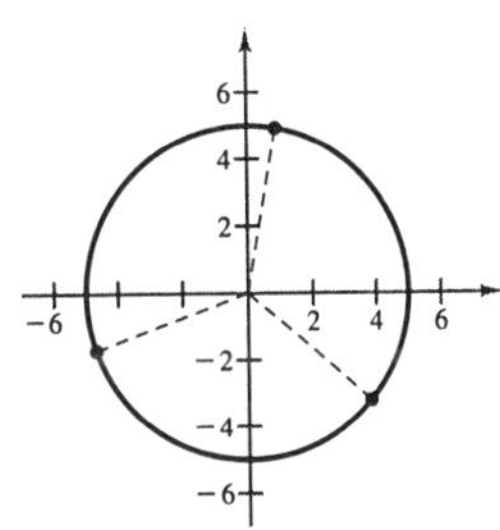

25. Find all the solutions of $x^4 - i = 0$ and represent your solutions graphically.

Solution:

$$x^4 - i = 0$$

$$x^4 = i$$

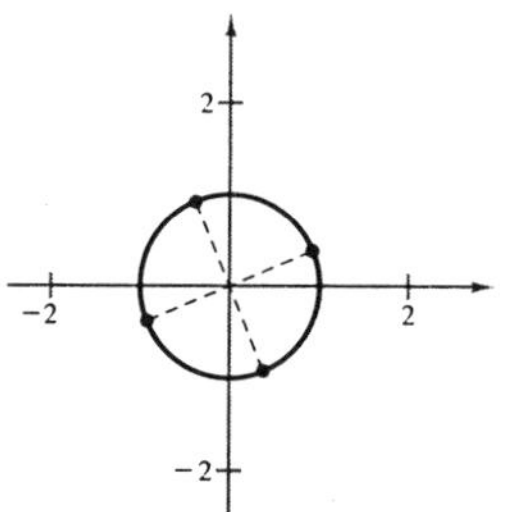

Find the fourth roots of $i = \cos\frac{\pi}{2} + i\sin\frac{\pi}{2}$.

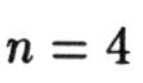

$n = 4$

$k = 0: \ \cos\frac{\pi}{8} + i\sin\frac{\pi}{8}$

$k = 1: \ \cos\frac{5\pi}{8} + i\sin\frac{5\pi}{8}$

$k = 2: \ \cos\frac{9\pi}{8} + i\sin\frac{9\pi}{8}$

$k = 3: \ \cos\frac{13\pi}{8} + i\sin\frac{13\pi}{8}$

29. Find all the solutions of $x^3 + 64i = 0$ and represent your solutions graphically.

Solution:

$$x^3 + 64i = 0$$

$$x^3 = -64i$$

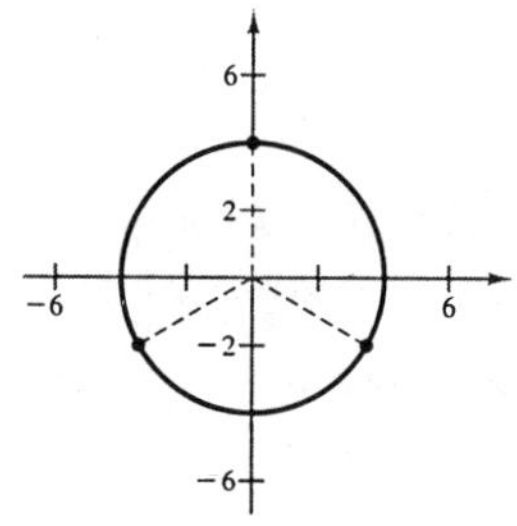

Find the cube roots of $-64i = 64\left(\cos\frac{3\pi}{2} + i\sin\frac{3\pi}{2}\right)$.

$n = 3$

$k = 0: \ 4\left(\cos\frac{\pi}{2} + i\sin\frac{\pi}{2}\right)$

$k = 1: \ 4\left(\cos\frac{7\pi}{6} + i\sin\frac{7\pi}{6}\right)$

$k = 2: \ 4\left(\cos\frac{11\pi}{6} + i\sin\frac{11\pi}{6}\right)$

REVIEW EXERCISES FOR CHAPTER 7

Solutions to Selected Exercises

5. Solve the triangle, given
$B = 110°,\ a = 4,\ c = 4.$

Solution:

Since the triangle is isosceles,

$$A = C = \frac{1}{2}(180 - 110) = 35°.$$

By the Law of Sines:

$$\frac{4}{\sin 35°} = \frac{b}{\sin 110°}$$
$$b \approx 6.6$$

9. Solve the triangle, given
$B = 115°,\ a = 7,\ b = 14.5.$

Solution:

By the Law of Sines:

$$\frac{\sin A}{7} = \frac{\sin 115°}{14.5}$$
$$\sin A \approx 0.4375$$
$$A \approx 25.9°$$
$$C \approx 180 - (115 + 25.9) \approx 39.1°$$
$$\frac{c}{\sin 39.1°} = \frac{14.5}{\sin 115°}$$
$$c \approx 10.1$$

13. Solve the triangle, given
$B = 150°,\ a = 10,\ c = 20.$

Solution:

By the Law of Cosines:

$$b^2 = 10^2 + 20^2 - 2(10)(20)\cos 150°$$
$$b \approx 29.1$$

By the Law of Sines:

$$\frac{\sin A}{10} = \frac{\sin 150°}{29.1}$$
$$\sin A \approx 0.1718$$
$$A \approx 9.9°$$
$$C \approx 180 - (150 + 9.9) \approx 20.1°$$

17. Find the area of the triangle with
$a = 4,\ b = 5,$ and $c = 7.$

Solution:

$$s = \frac{4 + 5 + 7}{2} = 8$$
$$A = \sqrt{8(8-4)(8-5)(8-7)}$$
$$= \sqrt{96} = 4\sqrt{6} \approx 9.798 \text{ square units}$$

21. Find the height of a tree that stands on a hillside of slope 32° (from the horizontal) if from a point 75 feet downhill the angle of elevation to the top of the tree is 48° (see figure).

Solution:

Let $x =$ the height of the hillside.

$$\sin 32^\circ = \frac{x}{75}$$

$$x = 75 \sin 32 \approx 39.7439 \text{ feet}$$

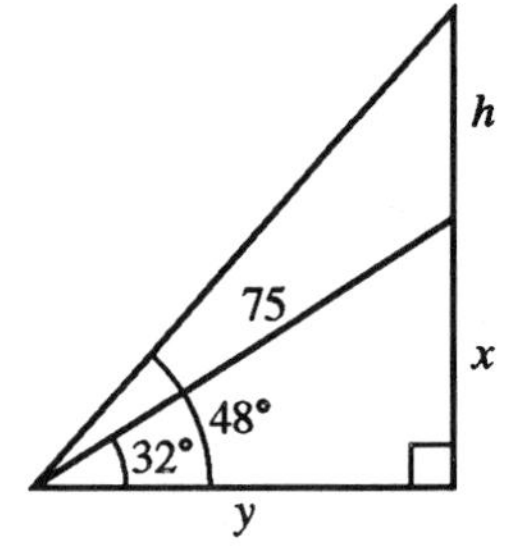

Let $y =$ the horizontal distance.

$$y = \sqrt{75^2 - x^2} = \sqrt{75^2 - 39.7439^2} \approx 63.6036 \text{ feet}$$

Let $h =$ the height of the tree.

$$\tan 48^\circ = \frac{x + h}{y} = \frac{39.7439 + h}{63.6036}$$

$$h = 63.6036 \tan 48^\circ - 39.7439 \approx 31 \text{ feet}$$

23. From a certain distance, the angle of elevation of the top of a building is 17°. At a point 50 meters closer to the building, the angle of elevation is 31°. Approximate the height of the building.

Solution:

$$\tan 17^\circ = \frac{h}{50 + y} \Rightarrow h = (50 + y) \tan 17^\circ$$

$$\tan 31^\circ = \frac{h}{y} \Rightarrow h = y \tan 31^\circ$$

$$(50 + y) \tan 17^\circ = y \tan 31^\circ$$

$$50 \tan 17^\circ + y \tan 17^\circ = y \tan 31^\circ$$

$$y(\tan 17^\circ - \tan 31^\circ) = -50 \tan 17^\circ$$

$$y = \frac{-50 \tan 17^\circ}{\tan 17^\circ - \tan 31^\circ} \approx 51.7959 \text{ m}$$

$$h = y \tan 31^\circ \approx 51.7959 \tan 31^\circ \approx 31.1 \text{ m}$$

27. Find the component form of the vector **v** with initial point (0, 10), and terminal point (7, 3).

Solution:

$$\mathbf{v} = \langle 7 - 0,\ 3 - 10 \rangle = \langle 7,\ -7 \rangle$$

33. Find the component form of $4\mathbf{u} - 5\mathbf{v}$ and sketch its graph given that $\mathbf{u} = 6\mathbf{i} - 5\mathbf{j}$ and $\mathbf{v} = 10\mathbf{i} + 3\mathbf{j}$.

Solution:

$$\begin{aligned} 4\mathbf{u} - 5\mathbf{v} &= (24\mathbf{i} - 20\mathbf{j}) - (50\mathbf{i} + 15\mathbf{j}) \\ &= -26\mathbf{i} - 35\mathbf{j} \\ &= \langle -26, -35 \rangle \end{aligned}$$

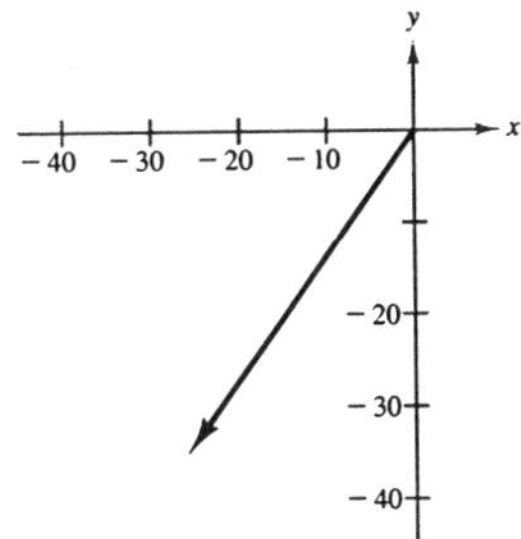

35. Find the direction and magnitude of the resultant of the three forces shown in the figure.

Solution:

$\tan\alpha = \frac{12}{5} \Rightarrow \sin\alpha = \frac{12}{13}$ and $\cos\alpha = \frac{5}{13}$

$\tan\beta = \frac{3}{4} \Rightarrow \sin(180° - \beta) = \frac{3}{5}$ and $\cos(180° - \beta) = -\frac{4}{5}$

$$\begin{aligned} \mathbf{u} &= 300\left(\tfrac{5}{13}\mathbf{i} + \tfrac{12}{13}\mathbf{j}\right) \\ \mathbf{v} &= 150\left(-\tfrac{4}{5}\mathbf{i} + \tfrac{3}{5}\mathbf{j}\right) \\ \mathbf{w} &= 250(0\mathbf{i} - \mathbf{j}) \\ \mathbf{r} &= \mathbf{u} + \mathbf{v} + \mathbf{w} \\ &= \left(\tfrac{1500}{13} - 120 + 0\right)\mathbf{i} + \left(\tfrac{3600}{13} + 90 - 250\right)\mathbf{j} \\ &= \tfrac{-60}{13}\mathbf{i} + \tfrac{1520}{13}\mathbf{j} \end{aligned}$$

$$\|\mathbf{r}\| = \sqrt{\left(-\tfrac{60}{13}\right)^2 + \left(\tfrac{1520}{13}\right)^2} \approx 117.0 \text{ lb}$$

$$\theta = 180° - \arctan\frac{1520}{60} \approx 92.3°$$

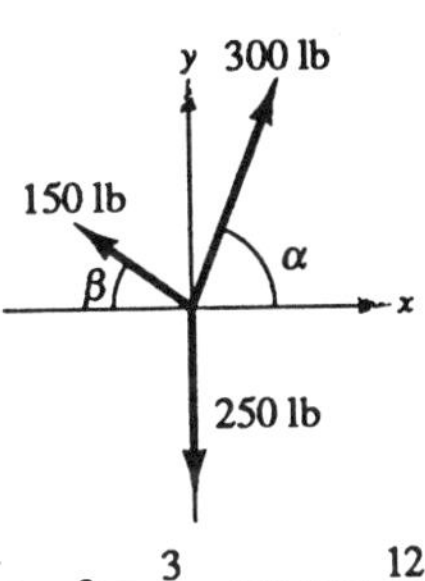

41. Find the trigonometric form of $5 - 5i$.

Solution:

$$\begin{aligned} z &= 5 - 5i \\ |z| &= \sqrt{5^2 + (-5)^2} = 5\sqrt{2} \\ \tan\theta &= -\tfrac{5}{5} = -1, \ \theta \text{ is in Quadrant IV} \\ \theta &= 315° \\ z &= 5\sqrt{2}(\cos 315° + i\sin 315°) \end{aligned}$$

45. Write $100(\cos 240° + i\sin 240°)$ in standard form.

Solution:

$$\begin{aligned} z &= 100(\cos 240° + i\sin 240°) \\ &= 100\left(-\frac{1}{2} - \frac{\sqrt{3}}{2}i\right) \\ &= -50 - 50\sqrt{3}i \end{aligned}$$

49. (a) Express the two complex numbers in trigonometric form, and (b) use the trigonometric form to find z_1z_2 and z_1/z_2. ($z_1 = -5,\ z_2 = 5i$)

Solution:

(a) $z_1 = -5 = -5 + 0i$

$r = 5, \quad \theta = \pi$

$z_1 = 5(\cos \pi + i \sin \pi)$

$z_2 = 5i = 0 + 5i$

$r = 5, \quad \theta = \dfrac{\pi}{2}$

$z_2 = 5\left(\cos \dfrac{\pi}{2} + i \sin \dfrac{\pi}{2}\right)$

(b) $$z_1z_2 = 5(\cos \pi + i \sin \pi) \bullet 5\left(\cos \frac{\pi}{2} + i \sin \frac{\pi}{2}\right)$$

$$= 25\left[\cos\left(\pi + \frac{\pi}{2}\right) + i \sin\left(\pi + \frac{\pi}{2}\right)\right] = 25\left(\cos \frac{3\pi}{2} + i \sin \frac{3\pi}{2}\right)$$

$$\frac{z_1}{z_2} = \frac{5(\cos \pi + i \sin \pi)}{5(\cos(\pi/2) + i \sin(\pi/2))} = \cos\left(\pi - \frac{\pi}{2}\right) + i \sin\left(\pi - \frac{\pi}{2}\right) = \cos \frac{\pi}{2} + i \sin \frac{\pi}{2}$$

53. Use DeMoivre's Theorem to find the indicated power of the following complex number. Express the result in standard form.

$$\left[5\left(\cos \frac{\pi}{12} + i \sin \frac{\pi}{12}\right)\right]^4$$

Solution:

$$\left[5\left(\cos \frac{\pi}{12} + i \sin \frac{\pi}{12}\right)\right]^4 = 625\left(\cos \frac{\pi}{3} + i \sin \frac{\pi}{3}\right) = \frac{625}{2} + \frac{625\sqrt{3}}{2}i$$

57. Use DeMoivre's Theorem to find the sixth roots of $-729i$.

Solution:

Find the sixth roots of $-729i = 729\left(\cos\frac{3\pi}{2} + i\sin\frac{3\pi}{2}\right)$.

$n = 6$

$k = 0: \ 3\left(\cos\frac{\pi}{4} + i\sin\frac{\pi}{4}\right)$

$k = 1: \ 3\left(\cos\frac{7\pi}{12} + i\sin\frac{7\pi}{12}\right)$

$k = 2: \ 3\left(\cos\frac{11\pi}{12} + i\sin\frac{11\pi}{12}\right)$

$k = 3: \ 3\left(\cos\frac{5\pi}{4} + i\sin\frac{5\pi}{4}\right)$

$k = 4: \ 3\left(\cos\frac{19\pi}{12} + i\sin\frac{19\pi}{12}\right)$

$k = 5: \ 3\left(\cos\frac{23\pi}{12} + i\sin\frac{23\pi}{12}\right)$

61. Find all solutions to $x^4 + 81 = 0$ and represent the solutions graphically.

Solution:

$$x^4 + 81 = 0$$

$$x^4 = -81$$ Find the fourth roots of -81.

$$-81 = 81(\cos\pi + i\sin\pi)$$

$$\sqrt[4]{-81} = \sqrt[4]{81}\left[\cos\left(\frac{\pi + 2\pi k}{4}\right) + i\sin\left(\frac{\pi + 2\pi k}{4}\right)\right], \quad k = 0,\ 1,\ 2,\ 3$$

$k = 0: \ 3\left(\cos\frac{\pi}{4} + i\sin\frac{\pi}{4}\right) = \frac{3\sqrt{2}}{2} + \frac{3\sqrt{2}}{2}i$

$k = 1: \ 3\left(\cos\frac{3\pi}{4} + i\sin\frac{3\pi}{4}\right) = -\frac{3\sqrt{2}}{2} + \frac{3\sqrt{2}}{2}i$

$k = 2: \ 3\left(\cos\frac{5\pi}{4} + i\sin\frac{5\pi}{4}\right) = -\frac{3\sqrt{2}}{2} - \frac{3\sqrt{2}}{2}i$

$k = 3: \ 3\left(\cos\frac{7\pi}{4} + i\sin\frac{7\pi}{4}\right) = \frac{3\sqrt{2}}{2} - \frac{3\sqrt{2}}{2}i$

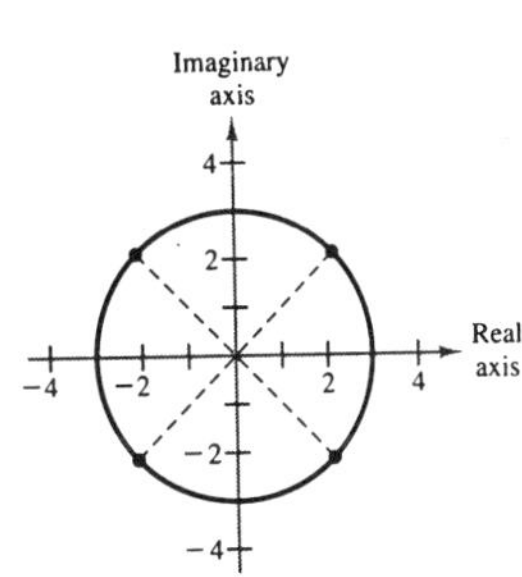

Practice Test for Chapter 7

For Exercises 1 and 2, use the Law of Sines to find the remaining sides and angles of the triangle.

1. $A = 40°,\ B = 12°,\ b = 100$

2. $C = 150°,\ a = 5,\ c = 20$

3. Find the area of the triangle: $a = 3,\ b = 5,\ C = 130°$.

4. Determine the number of solutions to the triangle: $a = 10,\ b = 35,\ A = 22.5°$.

For Exercises 5 and 6, use the Law of Cosines to find the remaining sides and angles of the triangle.

5. $a = 49,\ b = 53,\ c = 38$

6. $C = 29°,\ a = 100,\ b = 300$

7. Use Heron's Formula to find the area of the triangle: $a = 4.1,\ b = 6.8,\ c = 5.5$.

8. A ship travels 40 miles due east, then adjusts its course 12° southward. After traveling 70 miles in that direction, how far is the ship from its point of departure?

9. $\mathbf{w}$ is $4\mathbf{u} - 7\mathbf{v}$ where $\mathbf{u} = 3\mathbf{i} + \mathbf{j}$ and $\mathbf{v} = -\mathbf{i} + 2\mathbf{j}$. Find $\mathbf{w}$.

10. Find a unit vector in the direction of $\mathbf{v} = 5\mathbf{i} - 3\mathbf{j}$.

11. Find the angle between $\mathbf{u} = 6\mathbf{i} + 5\mathbf{j}$ and $\mathbf{v} = 2\mathbf{i} - 3\mathbf{j}$.

12. $\mathbf{v}$ is a vector of magnitude 4 making an angle of 30° with the positive x-axis. Find $\mathbf{v}$.

13. Give the trigonometric form of $z = 5 - 5i$.

14. Give the standard form of $z = 6(\cos 225° + i \sin 225°)$.

15. Multiply $[7(\cos 23° + i \sin 23°)][4(\cos 7° + i \sin 7°)]$.

16. Divide $\dfrac{9\left(\cos \frac{5\pi}{4} + i \sin \frac{5\pi}{4}\right)}{3(\cos \pi + i \sin \pi)}$.

17. Find $(2 + 2i)^8$.

18. Find the cube roots of $8\left(\cos \frac{\pi}{3} + i \sin \frac{\pi}{3}\right)$.

19. Find all the solutions to $x^3 + 125 = 0$.

20. Find all the solutions to $x^4 + i = 0$.

CHAPTER 8

Systems of Equations and Inequalities

Section 8.1 Systems of Equations 328

Section 8.2 Systems of Linear Equations in Two Variables 334

Section 8.3 Linear Systems in More Than Two Variables 339

Section 8.4 Systems of Inequalities 346

Section 8.5 Linear Programming 352

Review Exercises 358

Practice Test 365

SECTION 8.1

Systems of Equations

- You should be able to solve systems of equations by the method of substitution.
  1. Solve one of the equations for one of the variables.
  2. Substitute this expression into the other equation and solve.
  3. Back substitute into the first equation to find the value of the other variable.
- You should be able to find solutions graphically. (See Example 5 in the textbook.)

Solutions to Selected Exercises

5. Solve the following system by the method of substitution.

$$x + 3y = 15$$
$$x^2 + y^2 = 25$$

Solution:

$$x + 3y = 15 \Rightarrow x = 15 - 3y$$
$$x^2 + y^2 = 25$$
$$(15 - 3y)^2 + y^2 = 25$$
$$225 - 90y + 10y^2 = 25$$
$$10y^2 - 90y + 200 = 0$$
$$10(y^2 - 9y + 20) = 0$$
$$10(y - 4)(y - 5) = 0$$

$y = 4$ or $y = 5$

$x = 3$ $\quad x = 0$

Solutions: $(3, 4)$, $(0, 5)$

9. Solve the following system by the method of substitution.

$$x - 3y = -4$$
$$x^2 - y^3 = 0$$

Solution:

$$x - 3y = -4 \Rightarrow x = 3y - 4$$
$$x^2 - y^3 = 0$$
$$(3y - 4)^2 - y^3 = 0$$
$$9y^2 - 24y + 16 - y^3 = 0$$
$$y^3 - 9y^2 + 24y - 16 = 0$$ Multiply both sides by -1.
$$(y - 1)(y - 4)^2 = 0$$ See Section 4.5.

$y = 1$ or $y = 4$

$x = -1$ $\quad x = 8$

Solutions: $(-1, 1)$, $(8, 4)$

13. Solve the following system by the method of substitution.

$$2x - y + 2 = 0$$
$$4x + y - 5 = 0$$

Solution:

$$2x - y + 2 = 0 \Rightarrow y = 2x + 2$$
$$4x + y - 5 = 0$$
$$4x + (2x + 2) - 5 = 0$$
$$6x = 3$$
$$x = \tfrac{1}{2}$$
$$y = 3$$

Solution: $(\frac{1}{2},\ 3)$

17. Solve the following system by the method of substitution.

$$\tfrac{1}{5}x + \tfrac{1}{2}y = 8$$
$$x + y = 20$$

Solution:

$$\tfrac{1}{5}x + \tfrac{1}{2}y = 8 \Rightarrow 2x + 5y = 80$$
$$x + y = 20 \Rightarrow y = 20 - x$$
$$2x + 5(20 - x) = 80$$
$$-3x = -20$$
$$x = \tfrac{20}{3}$$
$$y = \tfrac{40}{3}$$

Solution: $(\frac{20}{3},\ \frac{40}{3})$

23. Solve the following system by the method of substitution.

$$3x - 7y + 6 = 0$$
$$x^2 - y^2 = 4$$

Solution:

$$3x - 7y + 6 = 0 \Rightarrow x = \frac{7y - 6}{3}$$
$$x^2 - y^2 = 4$$
$$\left(\frac{7y - 6}{3}\right)^2 - y^2 = 4$$
$$\frac{49y^2 - 84y + 36}{9} - y^2 = 4$$
$$49y^2 - 84y + 36 - 9y^2 = 36$$
$$40y^2 - 84y = 0$$
$$4y(10y - 21) = 0$$

$$y = 0 \qquad \text{or} \qquad y = \frac{21}{10}$$
$$x = -2 \qquad\qquad x = \frac{29}{10}$$

Solutions: $(-2,\ 0)$, $(\frac{29}{10},\ \frac{21}{10})$

27. Solve the following system by the method of substitution.

$$y = x^4 - 2x^2 + 1$$
$$y = 1 - x^2$$

Solution:

$$x^4 - 2x^2 + 1 = 1 - x^2$$
$$x^4 - x^2 = 0$$
$$x^2(x^2 - 1) = 0$$
$$x^2(x + 1)(x - 1) = 0$$

$x = 0$ or $x = -1$ or $x = 1$

$y = 1$ $\quad y = 0$ $\quad y = 0$

Solutions: $(0, 1)$, $(\pm 1, 0)$

29. Solve the following system by the method of substitution.

$$xy - 1 = 0$$
$$2x - 4y + 7 = 0$$

Solution:

$$xy - 1 = 0$$
$$2x - 4y + 7 = 0 \Rightarrow x = \frac{4y - 7}{2}$$
$$\left(\frac{4y - 7}{2}\right) y - 1 = 0$$
$$4y^2 - 7y - 2 = 0$$
$$(4y + 1)(y - 2) = 0$$

$y = -\frac{1}{4}$ or $y = 2$

$x = -4$ $\quad x = \frac{1}{2}$

Solutions: $\left(-4, -\frac{1}{4}\right)$, $\left(\frac{1}{2}, 2\right)$

33. Find all points of intersection of the graphs of the given pair of equations. [*Hint:* A graphical approach, as demonstrated in Example 5, may be helpful.]

$$2x - y + 3 = 0$$
$$x^2 + y^2 - 4x = 0$$

Solution:

$2x - y + 3 = 0 \Rightarrow y = 2x + 3$ (Line)

$x^2 + y^2 - 4x = 0 \Rightarrow (x - 2)^2 + y^2 = 4$ (Circle)

No points of intersection

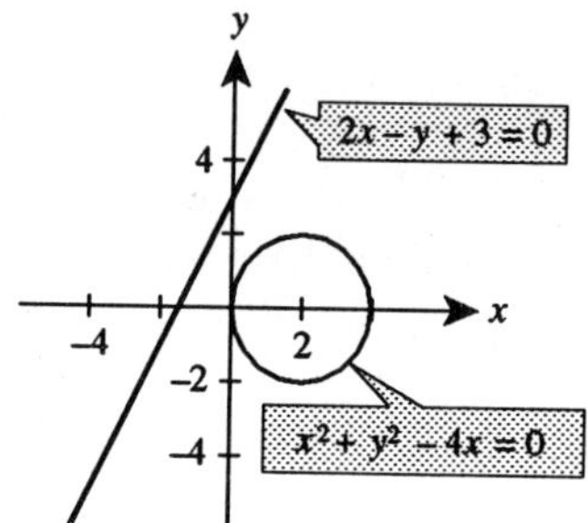

37. Find all points of intersection of the graphs of the given pair of equations. [*Hint:* A graphical approach, as demonstrated in Example 5, may be helpful.]

$$y = e^x$$
$$x - y + 1 = 0$$

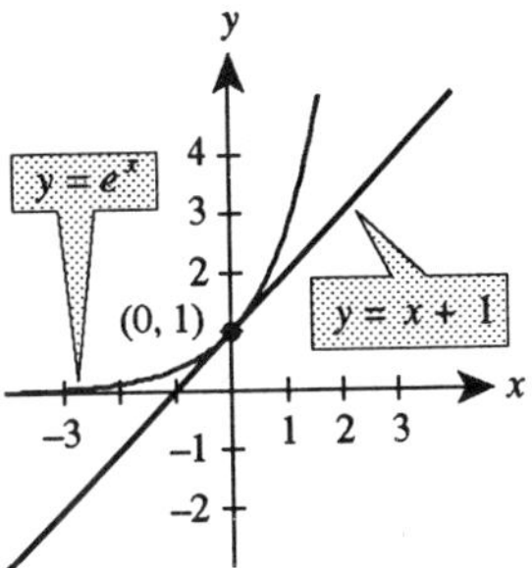

Solution:

$$y = e^x \qquad \text{(Exponential curve)}$$
$$x - y + 1 = 0 \Rightarrow y = x + 1 \qquad \text{(Line)}$$

The graphs intersect at the point $(0,\ 1)$.

41. Find all points of intersection of the graphs of the given pair of equations. [*Hint:* A graphical approach, as demonstrated in Example 5, may be helpful.]

$$x^2 + y^2 = 169$$
$$x^2 - 8y = 104$$

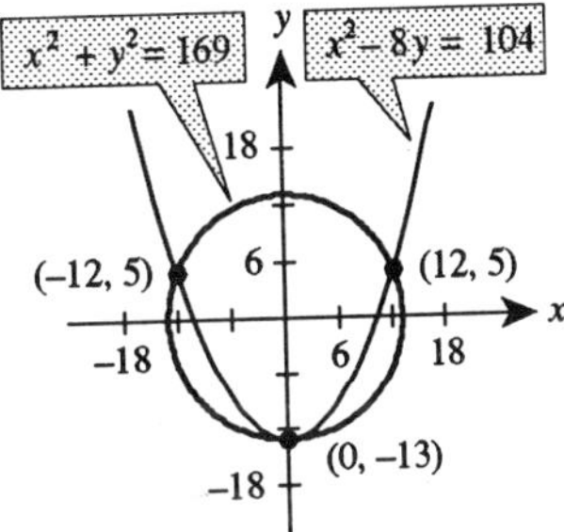

Solution:

$$x^2 + y^2 = 169 \Rightarrow x^2 = 8(y + 13) \qquad \text{(Circle)}$$
$$x^2 - 8y = 104 \qquad \text{(Parabola)}$$

The points of intersection are $(0,\ -13)$ and $(\pm 12,\ 5)$.

43. Find the sales necessary to break even ($R = C$) for $C = 8650x + 250{,}000$ and $R = 9950x$. (Round your answer to the nearest whole unit.)

Solution:

$$R = C$$
$$9950x = 8650x + 250{,}000$$
$$1300x = 250{,}000$$
$$x \approx 192 \text{ units}$$

47. Suppose you are setting up a small business and have invested \$16,000 to produce an item that will sell for \$5.95. If each unit can be produced for \$3.45, how many units must be sold to break even?

Solution:

Let $x =$ the number of units.

$$C = 3.45x + 16{,}000$$

$$R = 5.95x$$

To break even: $R = C$

$$5.95x = 3.45x + 16,000$$

$$2.5x = 16{,}000$$

$$x = 6400 \text{ units}$$

51. Suppose you are offered two different jobs selling dental supplies. One company offers a straight commission of 6% of sales. The other company offers a salary of \$250 per week *plus* 3% of the sales. How much would you have to sell in a week in order to make the straight commission offer better?

Solution:

Let $x =$ total sales.

Job #1: $y = 0.06x$

Job #2: $y = 250 + 0.03x$

You break even when

$$0.06x = 250 + 0.03x$$

$$0.03x = 250$$

$$x \approx \$8333.33.$$

If you sell more than \$8333.33 in a week, the straight commission offer is better.

55. What are the dimensions of a rectangular tract of land if its perimeter is 40 miles and its area is 96 square miles?

Solution:

Let l = the length of the rectangle and w = the width of the rectangle.

Perimeter: $2l + 2w = 40 \Rightarrow w = 20 - l$

Area: $lw = 96 \Rightarrow l(20 - l) = 96$

$$20l - l^2 = 96$$

$$0 = l^2 - 20l + 96$$

$$0 = (l - 8)(l - 12)$$

$l = 8$ or $l = 12$

$w = 12$ $\quad$ $w = 8$

The dimensions are 12 miles by 8 miles.

SECTION 8.2

Systems of Linear Equations in Two Variables

- You should be able to solve a linear system by the method of elimination.
- You should know that for a system of two linear equations, one of the following is true.
  (a) There are infinitely many solutions; the lines are identical.
  (b) There is no solution; the lines are parallel.
  (c) There is one solution; the lines intersect at one point.

Solutions to Selected Exercises

5. Solve the linear system by elimination. Identify and label each line with the appropriate equation.

$$x - y = 1$$
$$-2x + 2y = 5$$

Solution:

$$\begin{aligned} x - y = 1 &\Rightarrow \quad 2x - 2y = 2 \\ -2x + 2y = 5 &\Rightarrow \underline{-2x + 2y = 5} \\ & \qquad\qquad 0 = 7 \end{aligned}$$

Inconsistent; no solution

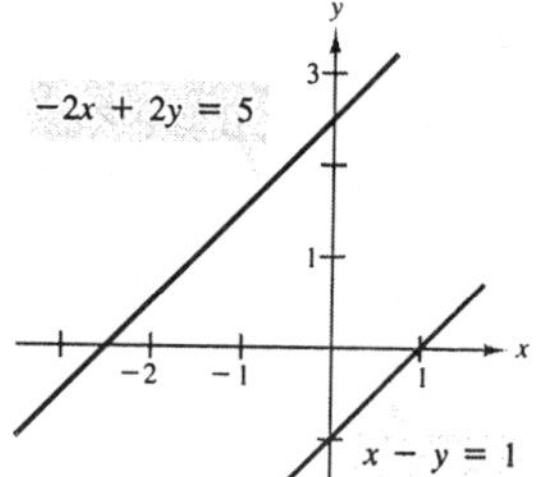

9. Solve the linear system by elimination. Identify and label each line with the appropriate equation.

$$9x - 3y = -1$$
$$3x + 6y = -5$$

Solution:

$$\begin{aligned} 9x - 3y = -1 &\Rightarrow 18x - 6y = -2 \\ 3x + 6y = -5 &\Rightarrow \underline{\ \ 3x + 6y = -5} \\ & \quad 21x \qquad = -7 \\ & \qquad\qquad x = -\tfrac{1}{3} \\ & \qquad\qquad y = -\tfrac{2}{3} \end{aligned}$$

Consistent; one solution
Solution: $\left(-\frac{1}{3}, -\frac{2}{3}\right)$

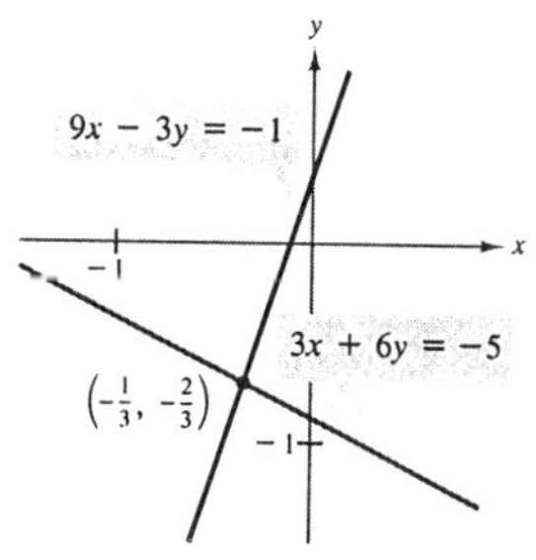

13. Solve the system by elimination.

$$\begin{aligned} 2x + 3y &= 18 \\ 5x - y &= 11 \end{aligned}$$

Solution:

$$\begin{aligned} 2x + 3y = 18 &\Rightarrow 2x + 3y = 18 \\ 5x - y = 11 &\Rightarrow \underline{15x - 3y = 33} \\ & 17x = 51 \\ & x = 3 \\ & y = 4 \end{aligned}$$

Consistent; one solution
Solution: (3, 4)

17. Solve the system by elimination.

$$\begin{aligned} 2u + v &= 120 \\ u + 2v &= 120 \end{aligned}$$

Solution:

$$\begin{aligned} 2u + v = 120 &\Rightarrow -4u - 2v = -240 \\ u + 2v = 120 &\Rightarrow \underline{u + 2v = 120} \\ & -3u = -120 \\ & u = 40 \\ & v = 40 \end{aligned}$$

Consistent; one solution
Solution: (40, 40)

23. Solve the system by elimination.

$$\frac{x+3}{4} + \frac{y-1}{3} = 1$$
$$2x - y = 12$$

Solution:

$$\begin{aligned} \frac{x+3}{4} + \frac{y-1}{3} = 1 \Rightarrow 3(x+3) + 4(y-1) = 12 &\Rightarrow 3x + 4y = 7 \\ 2x - y = 12 &\Rightarrow \underline{8x - 4y = 48} \\ & 11x = 55 \\ & x = 5 \\ & y = -2 \end{aligned}$$

Consistent; one solution
Solution: (5, −2)

25. Solve the system by elimination.

$$\begin{aligned} 2.5x - 3y &= 1.5 \\ 10x - 12y &= 6 \end{aligned}$$

Solution:

$$\begin{aligned} 2.5x - 3y = 1.5 &\Rightarrow 25x - 30y = 15 \Rightarrow 5x - 6y = 3 \\ 10x - 12y = 6 &\Rightarrow 5x - 6y = 3 \Rightarrow \underline{-5x + 6y = -3} \\ & 0 = 0 \end{aligned}$$

Consistent; infinite solutions of the form $\left(a,\ \frac{5}{6}a - \frac{1}{2}\right)$

29. Solve the system by elimination.

$$\begin{aligned} 4b + 3m &= 3 \\ 3b + 11m &= 13 \end{aligned}$$

Solution:

$$\begin{aligned} 4b + 3m = 3 &\Rightarrow 44b + 33m = 33 \\ 3b + 11m = 13 &\Rightarrow -9b - 33m = -39 \\ & \quad 35b = -6 \\ & \quad b = -\tfrac{6}{35} \\ & \quad m = \tfrac{43}{35} \end{aligned}$$

Consistent; one solution
Solution: $\left(-\frac{6}{35}, \frac{43}{35}\right)$

33. An airplane flying into a headwind travels the 1800-mile flying distance between two cities in 3 hours and 36 minutes. On the return flight, the distance is traveled in 3 hours. Find the ground speed of the plane and the speed of the wind, assuming that both remain constant.

Solution:

Use (rate)(time) = distance. Let g = ground speed and w = wind speed.

$$\begin{aligned} (g - w)\left(3 + \tfrac{36}{60}\right) = 1800 &\Rightarrow g - w = 500 \\ (g + w)(3) = 1800 &\Rightarrow g + w = 600 \\ & \quad 2g = 1100 \\ & \quad g = 550 \\ & \quad w = 50 \end{aligned}$$

The ground speed of the plane is 550 mph and the speed of the wind is 50 mph.

35. Ten gallons of a 30% acid solution are obtained by mixing a 20% solution with a 50% solution. How much of each must be used?

Solution:

Let x = amount of 20% solution and y = amount of 50% solution.

$$\begin{aligned} x + y = 10 &\Rightarrow -2x - 2y = -20 \\ 0.2x + 0.5y = 0.3(10) &\Rightarrow 2x + 5y = 30 \\ & \quad 3y = 10 \\ & \quad y = \tfrac{10}{3} \\ & \quad x = \tfrac{20}{3} \end{aligned}$$

Solution: $\frac{20}{3}$ gallons of 20% solution, $\frac{10}{3}$ gallons of 50% solution

39. Five hundred tickets were sold for a certain performance of a play. The tickets for adults and children sold for \$7.50 and \$4.00, respectively, and the receipts for the performance were \$3312.50. How many of each kind of ticket were sold?

Solution:

Let x = number of adult tickets and y = number of child tickets. Then

$$\begin{aligned} x + y &= 500 &&\Rightarrow -4x - 4y = -2000 \\ \text{and } 7.50x + 4.00y &= 3312.50 &&\Rightarrow \underline{7.5x + 4y = 3312.50} \\ &&& 3.5x = 1312.50 \\ &&& x = 375 \\ &&& y = 125 \end{aligned}$$

Thus, 375 adult tickets and 125 child tickets were sold.

43. Find the point of equilibrium.

Demand: $p = 300 - x$

Supply: $p = 100 + x$

Solution:

$$\begin{aligned} \text{Supply} &= \text{Demand} \\ 100 + x &= 300 - x \\ 2x &= 200 \\ x &= 100 \\ p &= 200 \end{aligned}$$

The point of equilibrium occurs when $x = 100$ units and the price per unit is \$200 (assuming that the price is in dollars).

47. On a 300 mile trip two people do the driving. One person drives three times as far as the other. Find the distance that each person drives.

Solution:

Let x = distance of the first driver and y = distance of the second driver. Then

$$\begin{aligned} \text{Then } x + y = 300 &\Rightarrow 3x + 3y = 900 \\ \text{and } x = 3y &\Rightarrow \underline{x - 3y = 0} \\ & 4x = 900 \\ & x = 225 \\ & y = 75 \end{aligned}$$

One person drives 225 miles and the other person drives 75 miles.

51. Find the least squares regression line $y = ax + b$. To find the line, solve the following system for a and b.

$$\begin{aligned} 7b + 21a &= 35.1 \\ 21b + 91a &= 114.2 \end{aligned}$$

–CONTINUED ON NEXT PAGE–

51. –CONTINUED–

Solution:

$$\begin{aligned} 7b + 21a &= 35.1 \Rightarrow -21b - 63a = -105.3 \\ 21b + 91a &= 114.2 \Rightarrow \underline{\;\;21b + 91a = \;\;114.2} \end{aligned}$$

$$\begin{aligned} 28a &= 8.9 \\ a &= \tfrac{8.9}{28} = \tfrac{89}{280} \\ b &= \tfrac{1137}{280} \end{aligned}$$

Thus, $y = \frac{89}{280}x + \frac{1137}{280} \approx 0.318x + 4.061$.

55. Find the least squares regression line, $y = ax + b$ for the points $(0,\ 4)$, $(1,\ 3)$, $(1,\ 1)$, $(2,\ 0)$. The *least squares regression line*, $y = ax + b$, for the points $(x_1,\ y_1)$, $(x_2,\ y_2)$, $\ldots$, $(x_n,\ y_n)$ is obtained by solving the following system of linear equations for a and b.

$$nb + \left(\sum_{i=1}^{n} x_i\right) a = \sum_{i=1}^{n} y_i$$

$$\left(\sum_{i=1}^{n} x_i\right) b + \left(\sum_{i=1}^{n} {x_i}^2\right) a = \sum_{i=1}^{n} x_i y_i$$

Solution:

$$n = 4$$

$$\sum_{i=1}^{4} x_i = 4$$

$$\sum_{i=1}^{4} {x_i}^2 = 6$$

$$\sum_{i=1}^{4} y_i = 8$$

$$\sum_{i=1}^{4} x_i y_i = 4$$

$$\begin{aligned} 4b + 4a &= 8 \\ \underline{-(4b + 6a} &\underline{= 4)} \\ -2a &= 4 \\ a &= -2 \\ b &= 4 \end{aligned}$$

$$y = -2x + 4$$

59. Find a system of linear equations having the solution $\left(3,\ \frac{5}{2}\right)$. (The answer is not unique.)

Solution:

Since $x = 3$ and $y = \frac{5}{2}$, one possible system of linear equations is

$$\begin{aligned} x + 2y &= 8 \\ x - 2y &= -2. \end{aligned}$$

SECTION 8.3

Linear Systems in More Than Two Variables

- You should know the operations that lead to equivalent systems of equations:
  (a) Interchange any two equations.
  (b) Multiply all terms of an equation by a nonzero constant.
  (c) Replace an equation by the sum of itself and a constant multiple of any other equation in the system.
- You should be able to use the method of elimination.

Solutions to Selected Exercises

3. Solve the following system of linear equations.

$$\begin{aligned} 4x + y - 3z &= 11 \\ 2x - 3y + 2z &= 9 \\ x + y + z &= -3 \end{aligned}$$

Solution:

$$\begin{array}{r} 4x + y - 3z = 11 \\ -4x + 6y - 4z = -18 \\ \hline 7y - 7z = -7 \\ y - z = -1 \end{array} \qquad \begin{array}{r} 2x - 3y + 2z = 9 \\ -2x - 2y - 2z = 6 \\ \hline -5y = 15 \end{array}$$

$y = -3,\ z = -2,\ x = 2$

Solution: $(2, -3, -2)$

7. Solve the following system of linear equations.

$$\begin{aligned} 3x - 2y + 4z &= 1 \\ x + y - 2z &= 3 \\ 2x - 3y + 6z &= 8 \end{aligned}$$

Solution:

$$\begin{array}{r} 3x - 2y + 4z = 1 \\ -3x - 3y + 6z = -9 \\ \hline -5y + 10z = -8 \end{array} \qquad \begin{array}{r} -2x - 2y + 4z = -6 \\ 2x - 3y + 6z = 8 \\ \hline -5y + 10z = 2 \end{array} \qquad \begin{array}{r} -5y + 10z = -8 \\ 5y - 10z = -2 \\ \hline 0 = -10 \end{array}$$

Inconsistent; no solution

11. Solve the following system of linear equations.

$$\begin{aligned} x + 2y - 7z &= -4 \\ 2x + y + z &= 13 \\ 3x + 9y - 36z &= -33 \end{aligned}$$

Solution:

$$\begin{array}{r} 2x + 4y - 14z = -8 \\ \underline{-2x - y - z = -13} \\ 3y - 15z = -21 \\ y - 5z = -7 \end{array} \qquad \begin{array}{r} 3x + 6y - 21z = -12 \\ \underline{-3x - 9y + 36z = 33} \\ -3y + 15z = 21 \\ y - 5z = -7 \end{array} \qquad \begin{array}{r} y - 5z = -7 \\ \underline{-y + 5z = 7} \\ 0 = 0 \end{array}$$

Consistent; infinite solutions

Let $z = a$. Then

$$y = 5a - 7$$

$$x = -4 - 2(5a - 7) + 7a = -3a + 10.$$

Solution: $(-3a + 10,\ 5a - 7,\ a)$, a is any real number.

15. Solve the following system of linear equations.

$$\begin{aligned} x - 2y + 5z &= 2 \\ 3x + 2y - z &= -2 \end{aligned}$$

Solution:

$$\begin{array}{r} x - 2y + 5z = 2 \\ \underline{3x + 2y - z = -2} \\ 4x \qquad + 4z = 0 \end{array} \Rightarrow x = -z$$

Let $z = a$. Then

$$x = -a$$

$$y = \tfrac{1}{2}[-a + 5a - 2] = 2a - 1.$$

Solution: $(-a,\ 2a - 1,\ a)$, a is any real number.

21. Solve the following system of linear equations.

$$\begin{aligned} x \quad\;\; + 4z &= 1 \\ x + y + 10z &= 10 \\ 2x - y + \;2z &= -5 \end{aligned}$$

Solution:

$$\begin{array}{rrr} -x \quad - 4z = -1 & -2x \quad - 8z = -2 & y + 6z = 9 \\ \underline{x + y + 10z = 10} & \underline{2x - y + 2z = -5} & \underline{-y - 6z = -7} \\ y + 6z = 9 & -y - 6z = -7 & 0 = 2 \end{array}$$

Inconsistent; no solution

23. Solve the following system of linear equations.

$$\begin{aligned} 4x + 3y + 17z &= 0 \\ 5x + 4y + 22z &= 0 \\ 4x + 2y + 19z &= 0 \end{aligned}$$

Solution:

$$\begin{array}{rr} 20x + 15y + 85z = 0 & 4x + 3y + 17z = 0 \\ \underline{-20x - 16y - 88z = 0} & \underline{-4x - 2y - 19z = 0} \\ -y - 3z = 0 & y - 2z = 0 \\ & \underline{-y - 3z = 0} \\ & -5z = 0 \\ & z = 0,\; y = 0,\; x = 0 \end{array}$$

Solution: (0, 0, 0)

27. Find the equation of the parabola $y = ax^2 + bx + c$ that passes through the points $(0, -4)$, $(1, 1)$, and $(2, 10)$.

Solution:

$$\begin{aligned} -4 &= a(0)^2 + b(0) + c \Rightarrow -4 = c \\ 1 &= a(1)^2 + b(1) + c \Rightarrow 1 = a + b + c \\ 10 &= a(2)^2 + b(2) + c \Rightarrow 10 = 4a + 2b + c \end{aligned}$$

$$\begin{aligned} 1 &= a + b - 4 \Rightarrow a + b = 5 \Rightarrow -a - b = -5 \\ 10 &= 4a + 2b - 4 \Rightarrow 4a + 2b = 14 \Rightarrow \underline{2a + b = 7} \\ & \qquad a = 2 \\ & \qquad b = 3 \end{aligned}$$

Thus, $y = 2x^2 + 3x - 4$.

31. Find the equation of the circle $x^2 + y^2 + Dx + Ey + F = 0$ that passes through the points $(0, 0)$, $(2, -2)$, and $(4, 0)$.

Solution:

$$\begin{aligned}
(0)^2 + (0)^2 + D(0) + E(0) + F = 0 &\Rightarrow \quad F = 0\\
(2)^2 + (-2)^2 + D(2) + E(-2) + F = 0 &\Rightarrow 2D - 2E + F = -8\\
(4)^2 + (0)^2 + D(4) + E(0) + F = 0 &\Rightarrow 4D \quad + F = -16
\end{aligned}$$

$$\begin{aligned}
2D - 2E + 0 = -8 &\Rightarrow D - E = -4\\
4D \quad + 0 = -16 &\Rightarrow D \quad = -4 \Rightarrow E = 0
\end{aligned}$$

Thus, $x^2 + y^2 - 4x + 0y + 0 = 0$

$$x^2 + y^2 - 4x = 0.$$

35. Find a, v_0, and s_0 in the position equation $s = \frac{1}{2}at^2 + v_0t + s_0$.

At $t = 1$ second, $s = 128$ feet.
At $t = 2$ seconds, $s = 80$ feet.
At $t = 3$ seconds, $s = 0$ feet.

Solution:

$s = \frac{1}{2}at^2 + v_0t + s_0$

$(1, 128)$, $(2, 80)$, $(3, 0)$

$$\begin{aligned}
128 = \tfrac{1}{2}a + v_0 + s_0 &\Rightarrow a + 2v_0 + 2s_0 = 256\\
80 = 2a + 2v_0 + s_0 &\Rightarrow 2a + 2v_0 + s_0 = 80\\
0 = \tfrac{9}{2}a + 3v_0 + s_0 &\Rightarrow 9a + 6v_0 + 2s_0 = 0
\end{aligned}$$

$$\begin{aligned}
2a + 4v_0 + 4s_0 &= 512\\
-2a - 2v_0 - s_0 &= -80\\
\hline
2v_0 + 3s_0 &= 432
\end{aligned}$$

$$\begin{aligned}
18a + 18v_0 + 9s_0 &= 720\\
-18a - 12v_0 - 4s_0 &= 0\\
\hline
6v_0 + 5s_0 &= 720\\
-6v_0 - 9s_0 &= -1296\\
\hline
-4s_0 &= -576\\
s_0 &= 144\\
v_0 &= 0\\
a &= -32
\end{aligned}$$

Thus, $s = \frac{1}{2}(-32)t^2 + (0)t + 144$

$$= -16t^2 + 144.$$

39. An inheritance of \$16,000 was divided among three investments yielding a total of \$900 in interest per year. The interest rates for the three investments were 5%, 6%, and 7%. Find the amount placed in each investment if the 5% and 6% investments were \$3000 and \$2000 less than the 7% investment, respectively.

Solution:

Let x = amount at 5%, y = amount at 6%, and z = amount at 7%.

$$x + y + z = 16{,}000$$

$$0.05x + 0.06y + 0.07z = 990$$

$$x = z - 3000$$

$$y = z - 2000$$

$$(z - 3000) + (z - 2000) + z = 16{,}000$$

$$3z = 21{,}000$$

$$z = \$7000 \text{ at } 7\%$$

$$y = \$5000 \text{ at } 6\%$$

$$x = \$4000 \text{ at } 5\%$$

Check: $0.05(4000) + 0.06(5000) + 0.07(7000) = 990$

43. Consider an investor with a portfolio totaling \$500,000 that is to be allocated among the following types of investments: (1) certificates of deposit, (2) municipal bonds, (3) blue-chip stocks, and (4) growth or speculative stocks. How much should be allocated to each type of investment? The certificates of deposit pay 10% annually, and the municipal bonds pay 8% annually. Over a five-year period, the investor expects the blue-chip stocks to return to 12% annually, and expects the growth stocks to return to 13% annually. The investor wants a combined annual return of 10% and also wants to have only one-fourth of the portfolio invested in stocks.

Solution:

Let x = amount in certificates of deposit, y = amount in municipal bonds, z = amount in blue-chip stocks, and w = amount in growth or speculative stocks.

$$x + y + z + w = 500{,}000$$

$$0.10x + 0.08y + 0.12z + 0.13w = 0.10(500{,}000)$$

$$z + w = 0.25(500{,}000)$$

–CONTINUED ON NEXT PAGE–

43. –CONTINUED–

Using elimination, we find **infinitely** many solutions:

$$w = a$$

$$z = 125{,}000 - a$$

$$y = 125{,}000 + \tfrac{1}{2}a$$

$$x = 250{,}000 - \tfrac{1}{2}a$$

One **possible** solution is to let $a = 50{,}000$.

Certificates of deposit:	\$225,000
Municipal bonds:	\$150,000
Blue-chip stocks:	\$75,000
Growth or speculative stocks:	\$50,000

47. A small company that manufactures products A and B has an order for 15 units of product A and 16 units of product B. The company has trucks of three different sizes that can haul the products, as shown in the following table. How many trucks of each size are needed to deliver the order? (Give *two* possible solutions.)

	Product	
Truck	A	B
Large	6	3
Medium	4	4
Small	0	3

Solution:

Possible solutions:

(1) 4 medium trucks

(2) 2 large trucks, 1 medium truck, 2 small trucks

(3) 3 large trucks, 1 medium truck, 1 small truck

(4) 3 large trucks, 3 small trucks

51. Use a system of linear equations to decompose the following rational fraction into partial fractions. (See Example 9 in this section.)

$$\frac{1}{x^3 - x} = \frac{A}{x} + \frac{B}{x-1} + \frac{C}{x+1}$$

Solution:

$$\frac{1}{x^3 - x} = \frac{A}{x} + \frac{B}{x-1} + \frac{C}{x+1}$$

$$1 = A(x+1)(x-1) + Bx(x+1) + Cx(x-1)$$

$$1 = Ax^2 - A + Bx^2 + Bx + Cx^2 - Cx$$

$$1 = (A+B+C)x^2 + (B-C)x - A$$

By equating coefficients, we have

$$\begin{aligned} 0 &= A + B + C \\ 0 &= B - C \\ 1 &= -A \end{aligned} \quad \Rightarrow \quad \begin{aligned} A &= -1 \\ B + C &= 1 \\ B - C &= 0 \\ \hline 2B &= 1 \Rightarrow B = \tfrac{1}{2},\ C = \tfrac{1}{2}. \end{aligned}$$

$$\frac{A}{x} + \frac{B}{x-1} + \frac{C}{x+1} = \frac{-1}{x} + \frac{1/2}{x-1} + \frac{1/2}{x+1} = \frac{1}{2}\left(-\frac{2}{x} + \frac{1}{x-1} + \frac{1}{x+1}\right)$$

57. Find the least squares regression parabola using the points (0, 0), (2, 2), (3, 6), (4, 12). *The least squares regression parabola,* $y = ax^2 + bx + c$, for the points $(x_1, y_1), (x_2, y_2), \ldots, (x_n, y_n)$ is obtained by solving the following system of linear equations for a, b, and c.

$$\begin{aligned} nc + \left(\sum_{i=1}^{n} x_i\right) b + \left(\sum_{i=1}^{n} x_i^2\right) a &= \sum_{i=1}^{n} y_i \\ \left(\sum_{i=1}^{n} x_i\right) c + \left(\sum_{i=1}^{n} x_i^2\right) b + \left(\sum_{i=1}^{n} x_i^3\right) a &= \sum_{i=1}^{n} x_i y_i \\ \left(\sum_{i=1}^{n} x_i^2\right) c + \left(\sum_{i=1}^{n} x_i^3\right) b + \left(\sum_{i=1}^{n} x_i^4\right) a &= \sum_{i=1}^{n} x_i^2 y_i \end{aligned}$$

Solution:

$n = 4$

$\sum x_i = 9$ $\quad \sum y_i = 20$

$\sum x_i^2 = 29$ $\quad \sum x_i^3 = 99$

$\sum x_i^4 = 353$ $\quad \sum x_i y_i = 70$

$\sum x_i^2 y_i = 254$

$$\begin{aligned} 353a + 99b + 29c &= 254 \\ 99a + 29b + 9c &= 70 \\ 29a + 9b + 4c &= 20 \end{aligned}$$

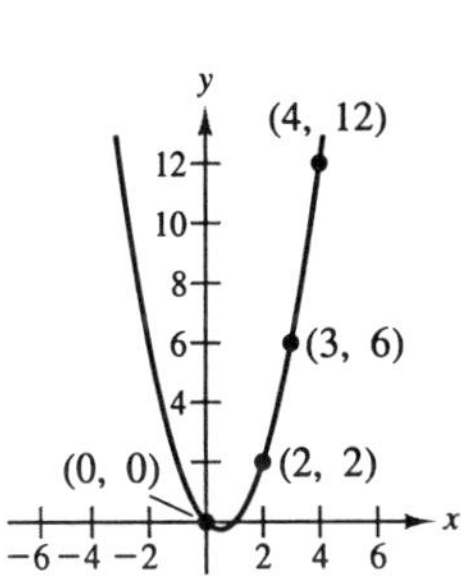

By solving, we get $a = 1,\ b = -1,\ c = 0$. Thus, $y = x^2 - x$.

SECTION 8.4
Systems of Inequalities

- You should be able to sketch the graph of an inequality in two variables:
  (a) Replace the inequality with an equal sign and graph the equation. Use a dashed line for $<$ or $>$, a solid line for $\leq$ or $\geq$.
  (b) Test a point in each region formed by the graph. If the point satisfies the inequality, shade the whole region.

Solutions to Selected Exercises

5. Match $x^2 + y^2 < 4$ with its graph.

Solution:

Since $x^2 + y^2 = 4$ is a circle with center (0, 0) and radius $r = 2$, it matches graph a.

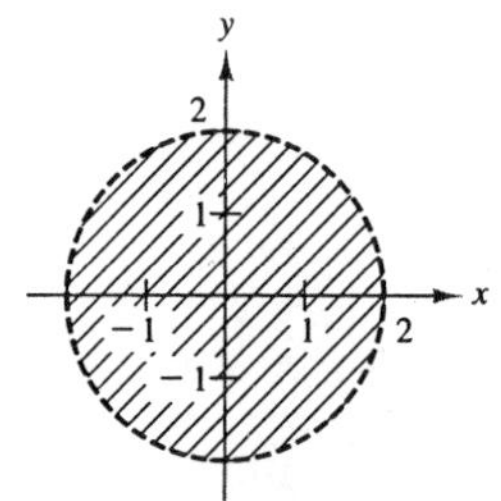

9. Sketch the graph of $x \geq 2$.

Solution:

Using a solid line, sketch the graph of the vertical line $x = 2$. Test point (3, 0). Shade the half-plane to the right of $x = 2$.

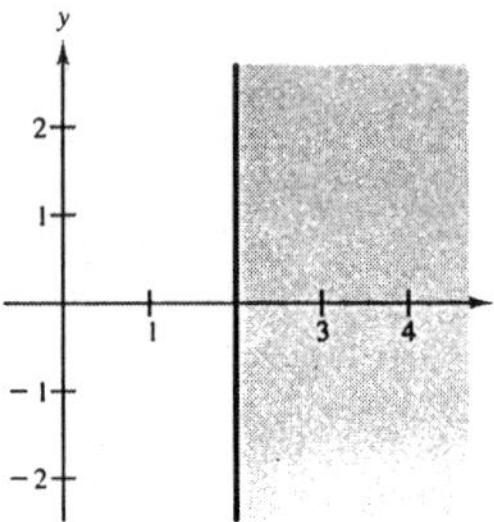

13. Sketch the graph of $y < 2 - x$.

Solution:

Using a dashed line, graph $x + y = 2$, and then shade the half-plane below the line. [Use (0, 0) as a test point.]

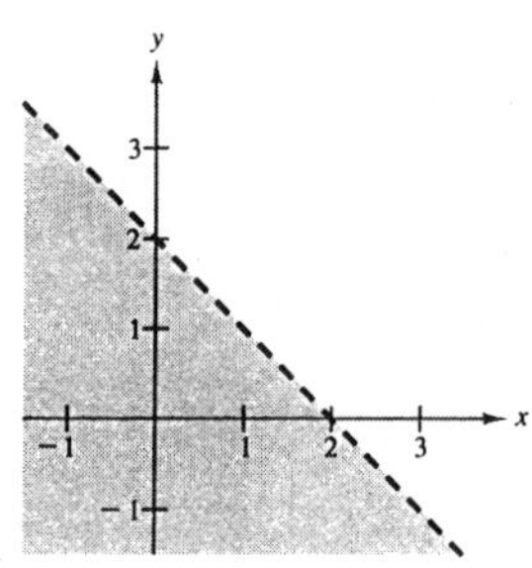

17. Sketch the graph of $(x+1)^2 + (y-2)^2 < 9$.

Solution:

Using a dashed line, sketch the circle $(x+1)^2 + (y-2)^2 = 9$.

Center: $(-1,\ 2)$
Radius: 3

Test Point: $(0,\ 0)$. Shade the inside of the circle.

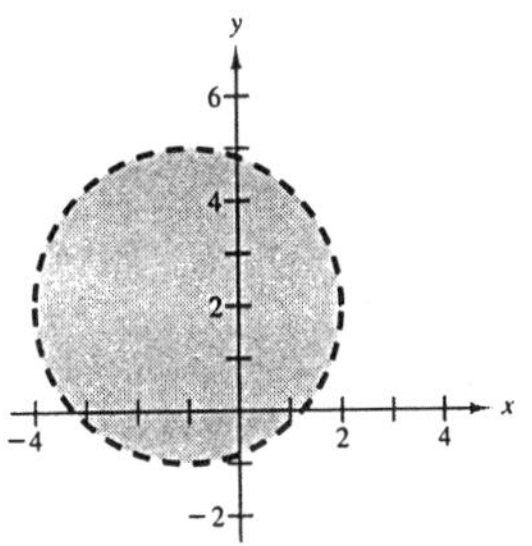

21. Sketch the graph of the solution of the system of inequalities.

$$x + y \le 1$$
$$-x + y \le 1$$
$$y \ge 0$$

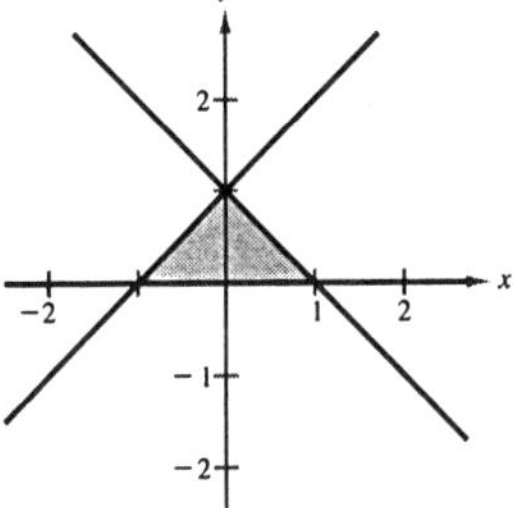

Solution:

First, find the points of intersection of each pair of equations.

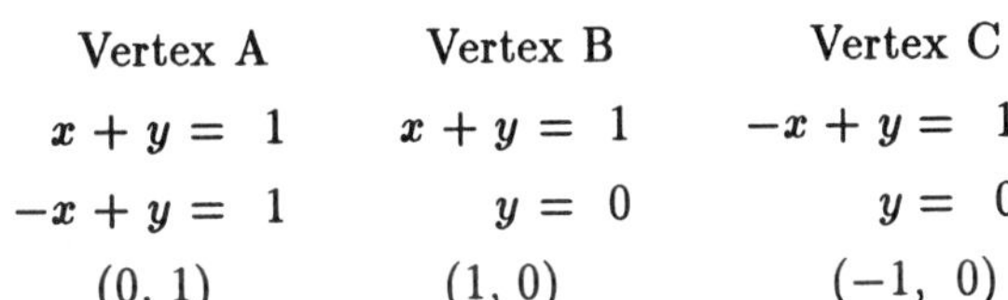

Vertex A	Vertex B	Vertex C
$x + y = 1$	$x + y = 1$	$-x + y = 1$
$-x + y = 1$	$y = 0$	$y = 0$
$(0,\ 1)$	$(1,\ 0)$	$(-1,\ 0)$

25. Sketch the graph of the solution of the system of inequalities.

$$-3x + 2y < 6$$
$$x + 4y > -2$$
$$2x + y < 3$$

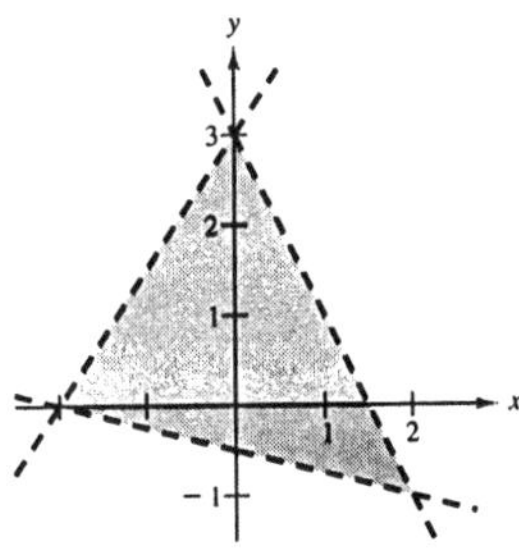

Solution:

First, find the points of intersection of each pair of equations.

Vertex A	Vertex B	Vertex C
$-3x + 2y = 6$	$-3x + 2y = 6$	$x + 4y = -2$
$x + 4y = -2$	$2x + y = 3$	$2x + y = 3$
$(-2,\ 0)$	$(0,\ 3)$	$(2,\ -1)$

29. Sketch the graph of the solution of the system of inequalities.

$$x \geq 1$$

$$x - 2y \leq 3$$

$$3x + 2y \geq 9$$

$$x + y \leq 6$$

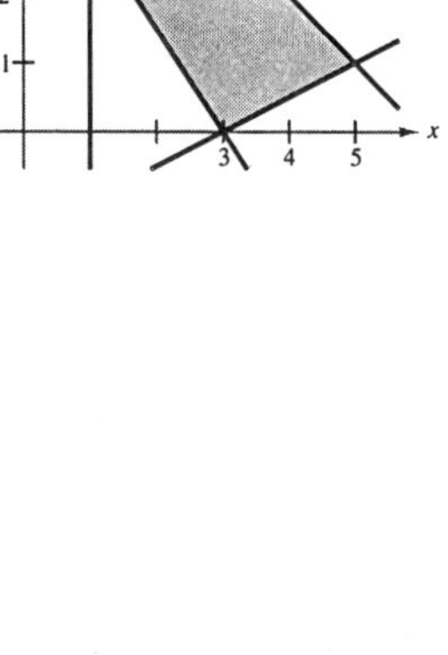

Solution:

First, find the points of intersection of each pair of equations.

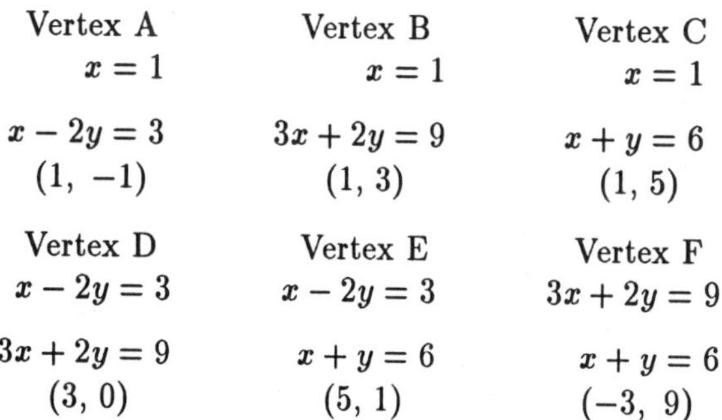

Vertex A	Vertex B	Vertex C
$x = 1$	$x = 1$	$x = 1$
$x - 2y = 3$	$3x + 2y = 9$	$x + y = 6$
$(1, -1)$	$(1, 3)$	$(1, 5)$

Vertex D	Vertex E	Vertex F
$x - 2y = 3$	$x - 2y = 3$	$3x + 2y = 9$
$3x + 2y = 9$	$x + y = 6$	$x + y = 6$
$(3, 0)$	$(5, 1)$	$(-3, 9)$

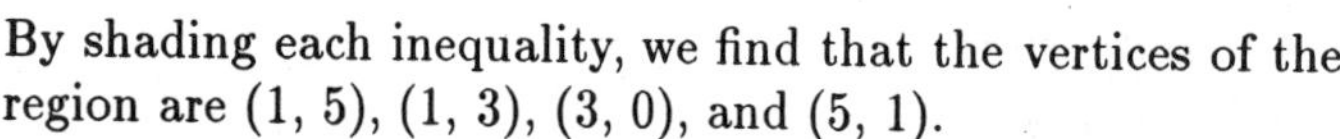
By shading each inequality, we find that the vertices of the region are $(1, 5)$, $(1, 3)$, $(3, 0)$, and $(5, 1)$.

33. Sketch the graph of the solution of the system of inequalities.

$$x > y^2$$

$$x < y \ + 2$$

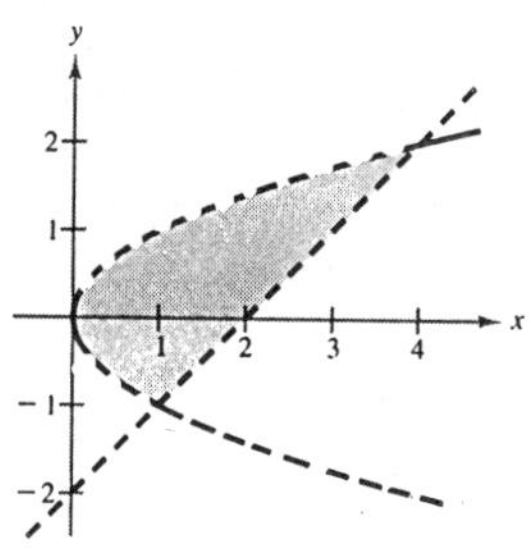

Solution:

Points of intersection:

$$y^2 = y + 2$$

$$y^2 - y - 2 = 0$$

$$(y + 1)(y - 2) = 0$$

$$y = -1, \ 2$$

$(1, -1)$, $(4, 2)$

37. Sketch the graph of the solution of the system of inequalities.

$y < x^3 - 2x + 1$

$y > -2x$

$x \leq 1$

Solution:

$y = x^3 - 2x + 1$

x	-2	-1	0	1	2
y	-3	2	1	0	5

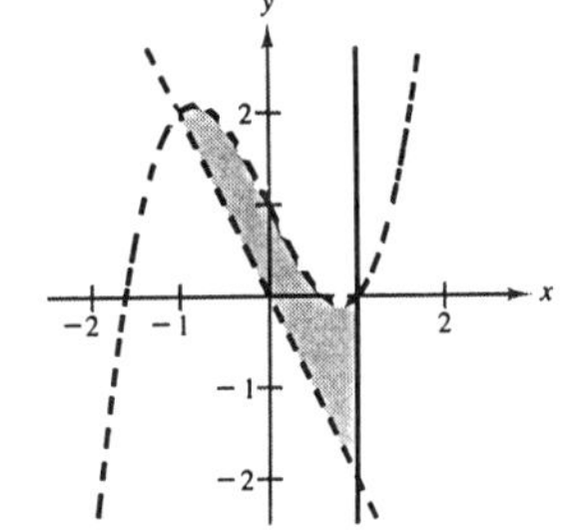

Points of intersection:

$x^3 - 2x + 1 = -2x$ $\quad$ $x = 1$ $\quad$ $x = 1$

$x^3 + 1 = 0$ $\quad$ $y = x^3 - 2x + 1$ $\quad$ $y = -2x$

$x = -1$ $\quad$ $(1,\ 0)$ $\quad$ $(1,\ -2)$

$(-1,\ 2)$

41. Derive a set of inequalities to describe the rectangular region with vertices at (2, 1), (5, 1), (5, 7), and (2, 7).

Solution:

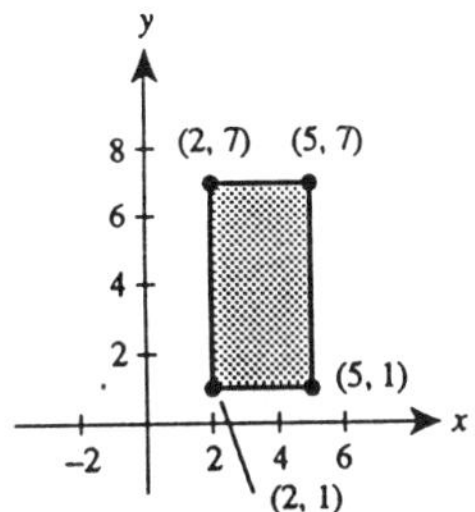

$x \geq 2$

$x \leq 5$

$y \geq 1$

$y \leq 7$

Thus, $2 \leq x \leq 5,\ 1 \leq y \leq 7$.

47. A furniture company can sell all the tables and chairs it produces. Each table requires 1 hour in the assembly center and $1\frac{1}{3}$ hours in the finishing center. Each chair requires $1\frac{1}{2}$ hours in the assembly center and $1\frac{1}{2}$ hours in the finishing center. The company's assembly center is available 12 hours per day, and its finishing center is available 15 hours per day. If x is the number of tables produced per day and y is the number of chairs, find a system of inequalities describing all possible production levels. Sketch the graph of the system.

Solution:

Assembly center constraint: $x + \frac{3}{2}y \leq 12$

Finishing center constraint: $\frac{4}{3}x + \frac{3}{2}y \leq 15$

Point of intersection: $(9,\ 2)$

Physical constraints: $x \geq 0$ and $y \geq 0$

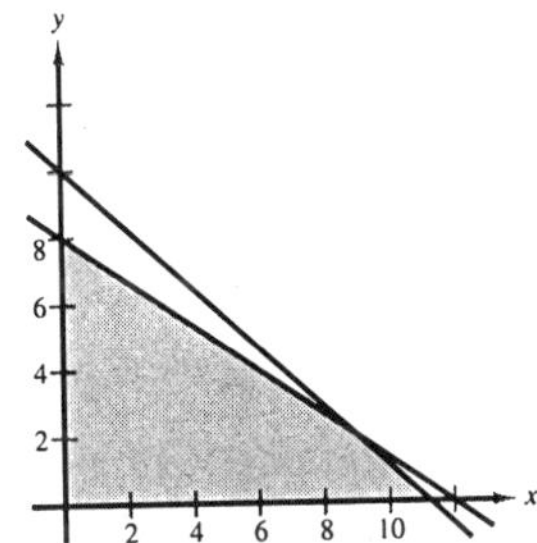

49. A person plans to invest \$20,000 in two different interest bearing accounts. Each account is to contain at least \$5000. Moreover, one account should have at least twice the amount that is in the other account. Find a system of inequalities to describe the various amounts that can be deposited in each account, and sketch the graph of the system.

Solution:

Account constraints:

$$x \geq 5{,}000$$
$$y \geq 5{,}000$$
$$2x \leq y$$
$$x + y \leq 20{,}000$$

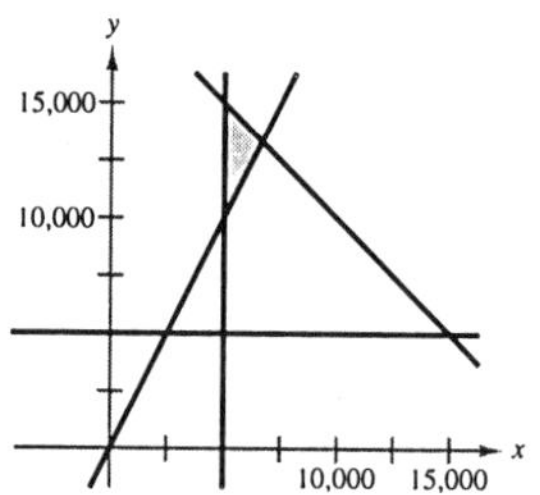

53. Find the consumer surplus and producer surplus for the given pair of equations.

Demand: $p = 50 - 0.5x$
Supply: $p = 0.125x$

Solution:

$$\text{Demand} = \text{Supply}$$
$$50 - 0.5x = 0.125x$$
$$50 = 0.625x$$
$$80 = x$$
$$10 = p$$

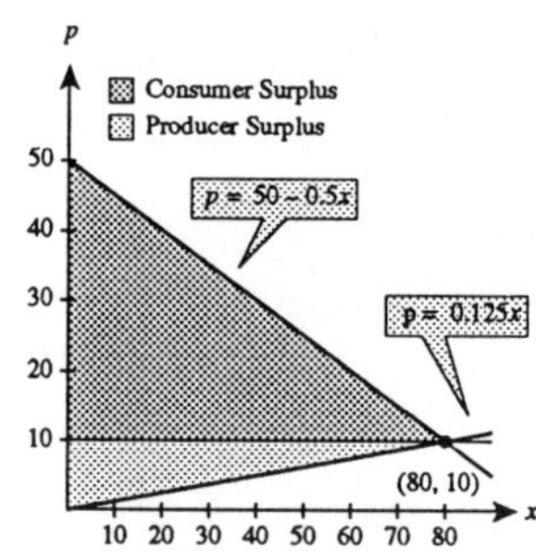

Point of equilibrium: $(80,\ 10)$
The consumer surplus is the area of the triangle bounded by

$$p \leq 50 - 0.5x$$
$$p \geq 10$$
$$x \geq 0.$$

Consumer surplus $= \frac{1}{2}(\text{base})(\text{height}) = \frac{1}{2}(80)(40) = \$1600$
The producer surplus is the area of the triangle bounded by

$$p \geq 0.125x$$
$$p \leq 10$$
$$x \geq 0.$$

Producer surplus $= \frac{1}{2}(\text{base})(\text{height}) = \frac{1}{2}(80)(10) = \$400$

57. Find the consumer surplus and producer surplus for the given pair of equations.

Demand: $p = 140 - 0.00002x$
Supply: $p = 80 + 0.00001x$

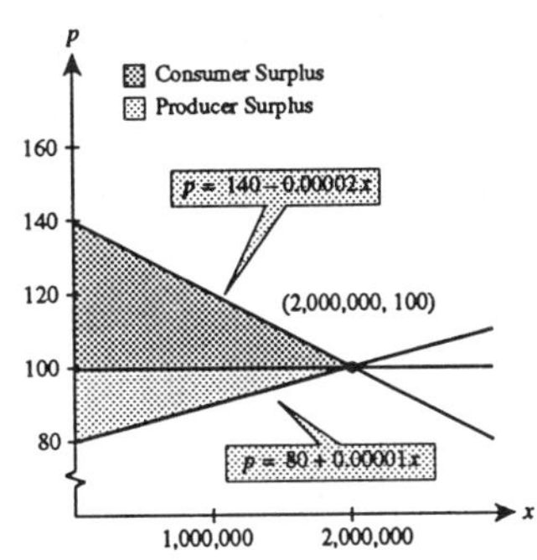

Solution:

$$\text{Demand} = \text{Supply}$$
$$140 - 0.00002x = 80 + 0.00001x$$
$$60 = 0.00003x$$
$$2{,}000{,}000 = x$$
$$100 = p$$

Point of equilibrium: (2,000,000, 100)
The consumer surplus is the area of the triangle bounded by

$p \le 140 - 0.00002x$

$p \ge 100$

$x \ge 0.$

Consumer surplus $= \frac{1}{2}(\text{base})(\text{height}) = \frac{1}{2}(2{,}000{,}000)(40) = \$40{,}000{,}000$ or \$40 million

The producer surplus is the area of the triangle bounded by

$p \ge 80 + 0.00001x$

$p \le 100$

$x \ge 0.$

Producer surplus $= \frac{1}{2}(\text{base})(\text{height}) = \frac{1}{2}(2{,}000{,}000)(20) = \$20{,}000{,}000$ or \$20 million

SECTION 8.5
Linear Programming

- To solve a linear programming problem:
  1. Sketch the solution set for the system of constraints.
  2. Find the vertices of the region.
  3. Test the objective function at each of the vertices.

Solutions to Selected Exercises

5. Find the minimum and maximum values of the objective function, $z = 3x + 2y$, subject to the following constraints.

$$x \geq 0$$
$$y \geq 0$$
$$x + 3y \leq 15$$
$$4x + y \leq 16$$

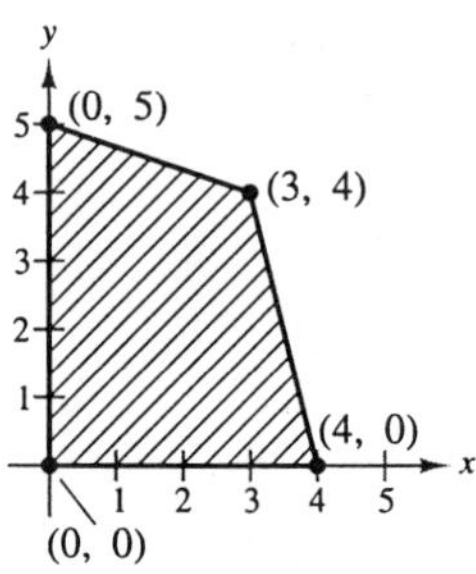

Solution:

Minimum and maximum values occur at the vertices of the constrained region.

Vertex	Value of $z = 3x + 2y$
(0, 0)	$z = 0$, minimum value
(4, 0)	$z = 12$
(0, 5)	$z = 10$
(3, 4)	$z = 17$, maximum value

11. Find the minimum and maximum values of the objective function, $z = 25x + 30y$, subject to the following constraints.

$$0 \leq x \leq 60$$
$$0 \leq y \leq 45$$
$$5x + 6y \leq 420$$

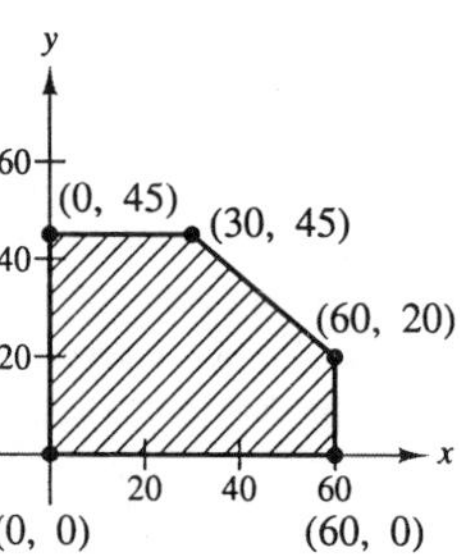

–CONTINUED ON NEXT PAGE–

11. –CONTINUED–

Solution:

Vertex	Value of $z = 25x + 30y$
$(0, 0)$	$z = 0$, minimum value
$(60, 0)$	$z = 1500$
$(60, 20)$	$z = 2100$, maximum value
$(30, 45)$	$z = 2100$, maximum value
$(0, 45)$	$z = 1350$

Maximum of 2100 occurs at any point on the line segment connecting $(60, 20)$ and $(30, 45)$.

15. Sketch the region determined by the constraints. Then find the minimum and maximum values of the objective function, $z = 9x + 24y$, subject to the following constraints.

$$x \geq 0$$
$$y \geq 0$$
$$2x + 5y \leq 10$$

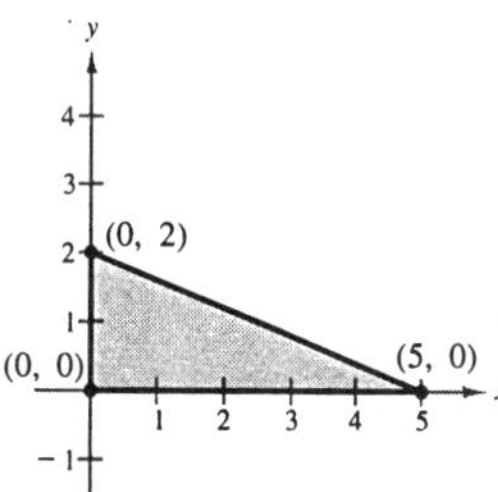

Solution:

Vertex	Value of $z = 9x + 24y$
$(0, 0)$	$z = 0$, minimum value
$(0, 2)$	$z = 48$, maximum value
$(5, 0)$	$z = 45$

21. Sketch the region determined by the constraints. Then find the minimum and maximum values of the objective function, $z = 4x + y$, subject to the following constraints.

$$x \geq 0$$
$$y \geq 0$$
$$x + 2y \leq 40$$
$$x + y \geq 30$$
$$2x + 3y \geq 72$$

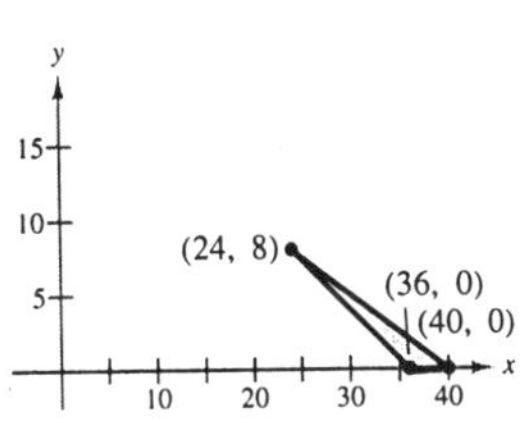

Solution:

Vertex	Value of $z = 4x + y$
$(36, 0)$	$z = 144$
$(40, 0)$	$z = 160$, maximum value
$(24, 8)$	$z = 104$, minimum value

25. Maximize the objective function, $z = 2x + y$, subject to the following constraints.

$$3x + y \leq 15$$
$$4x + 3y \leq 30$$
$$x \geq 0$$
$$y \geq 0$$

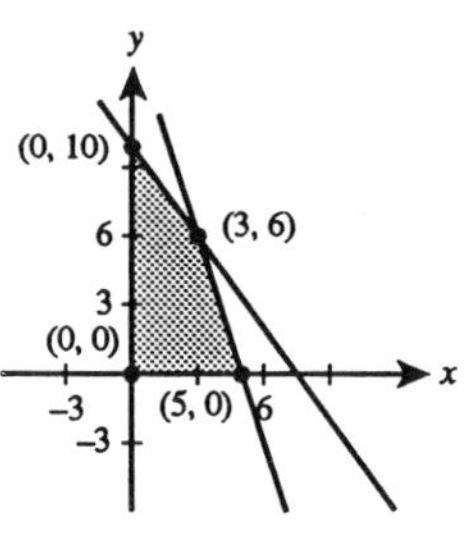

Solution:

Vertex	Value of $z = 2x + y$
(0, 0)	$z = 0$
(0, 10)	$z = 10$
(3, 6)	$z = 12$, maximum value
(5, 0)	$z = 10$

29. Maximize the objective function, $z = x + 5y$, subject to the following constraints.

$$x + 4y \leq 20$$
$$x + y \leq 8$$
$$3x + 2y \leq 21$$
$$x \geq 0$$
$$y \geq 0.$$

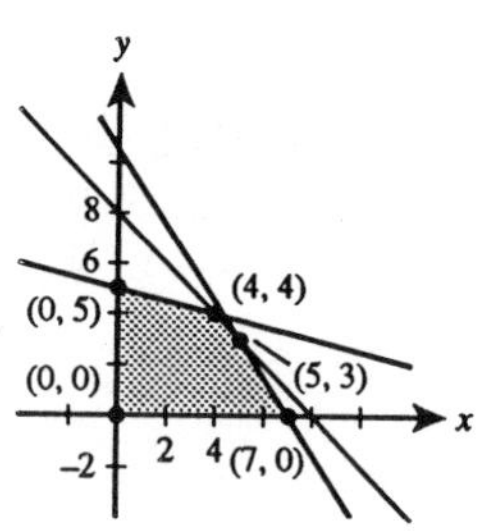

Solution:

Vertex	Value of $z = x + 5y$
(0, 0)	$z = 0$
(0, 5)	$z = 25$, maximum value
(4, 4)	$z = 24$
(5, 3)	$z = 20$
(7, 0)	$z = 7$

33. A merchant plans to sell two models of home computers at costs of \$250 and \$400, respectively. The \$250 model yields a profit of \$45 and the \$400 model yields a profit of \$50. The merchant estimates that the total monthly demand will not exceed 250 units. Find the number of units of each model that should be stocked in order to maximize profit. Assume that the merchant does not want to invest more than \$70,000 in computer inventory.

Solution:

Objective function: Maximize $P = 45x + 50y$.

Constraints: $x \geq 0$

$y \geq 0$

$x + y \leq 250$

$250x + 400y \leq 70{,}000$

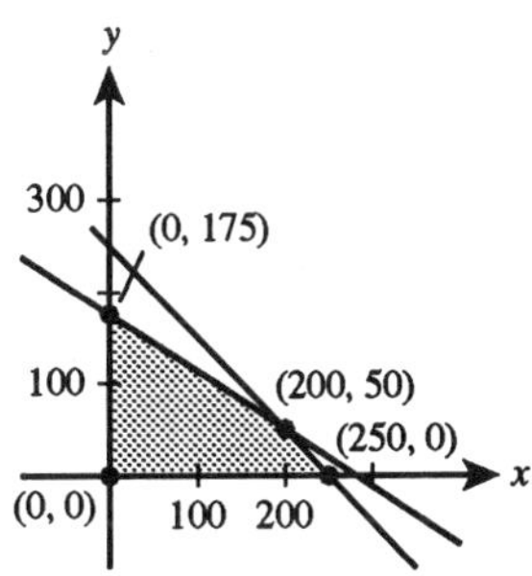

Testing the vertices shows that the profit is maximized when $x = 200$ units and $y = 50$ units.

35. A farmer mixes two brands of cattle feed. Brand X costs \$25 per bag and contains 2 units of nutritional element A, 2 units of element B, and 2 units of element C. Brand Y costs \$20 per bag and contains 1 unit of nutritional element A, 9 units of element B, and 3 units of element C. Find the number of bags of each brand that should be mixed to produce a mixture having a minimum cost per bag. The minimum requirements of nutrients A, B, and C are 12 units, 36 units, and 24 units, respectively.

Solution:

Objective function: Minimize $z = 25x + 20y$.

Constraints: $2x + y \geq 12$

$2x + 9y \geq 36$

$2x + 3y \geq 24$

$x \geq 0, \; y \geq 0$

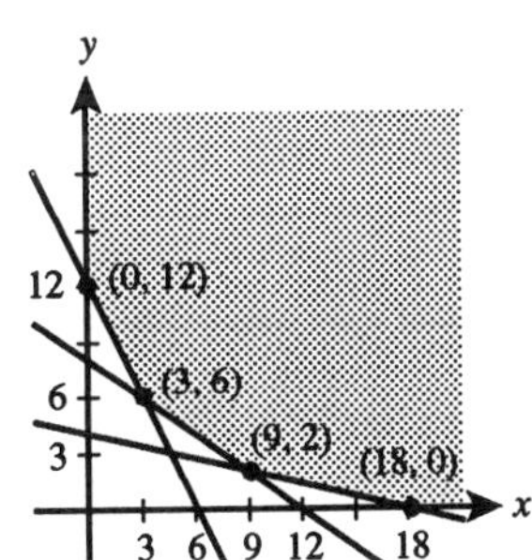

Vertex	Value of $z = 25x + 20y$
(0, 12)	$z = 240$
(3, 6)	$z = 195$, minimum value
(9, 2)	$z = 265$
(18, 0)	$z = 450$

To minimize cost, use three bags of Brand X and six bags of Brand Y.

39. An accounting firm has 900 hours of staff time and 100 hours of reviewing time available each week. The firm charges \$2000 for an audit and \$300 for a tax return. Each audit requires 100 hours of staff time and 10 hours of review time. Each tax return requires 12.5 hours of staff time and 2.5 hours of review time. What number of audits and tax returns will yield the maximum revenue?

Solution:

$x =$ the number of audits, and $y =$ the number of tax returns.

Objective function: Maximize $R = 2000x + 300y$.

Constraints: $100x + 12.5y \leq 900$

$$10x + 2.5y \leq 100$$

$$x \geq 0, \ y \geq 0$$

Vertex	Value of $R = 2000x + 300y$
(0, 0)	$R = 2000(0) + 300(0) = 0$
(0, 40)	$R = 2000(0) + 300(40) = 12{,}000$
(8, 8)	$R = 2000(8) + 300(8) = 18{,}400$, maximum value
(9, 0)	$R = 2000(9) + 300(0) = 18{,}000$

The revenue will be maximum if the firm does 8 audits and 8 tax returns each week.

43. Sketch a graph of the solution region for the given linear programming problem and describe its unusual characteristic. (The objective function, $z = -x + 2y$, is to be maximized.)

Constraints: $x \geq 0$

$$y \geq 0$$

$$x \leq 10$$

$$x + y \leq 7$$

Solution:

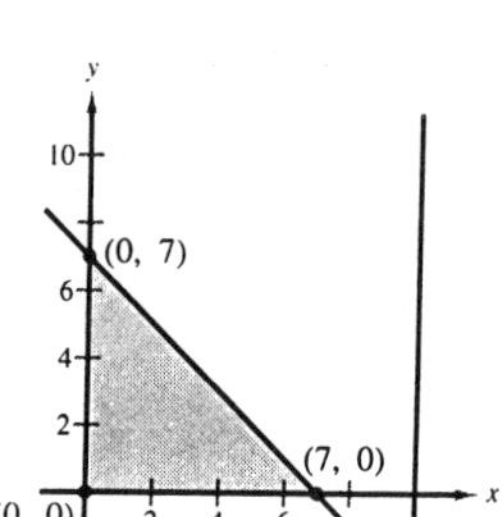

The constraint $x \leq 10$ is extraneous.

Vertex	Value of $z = -x + 2y$
(0, 0)	$z = 0$
(7, 0)	$z = -7$
(0, 7)	$z = 14$, maximum value

47. Determine the values of t such that the objective function, $z = 3x + ty$, has a maximum value at the indicated vertex.

Constraints: $x \geq 0$

$y \geq 0$

$x + 3y \leq 15$

$4x + y \leq 16$

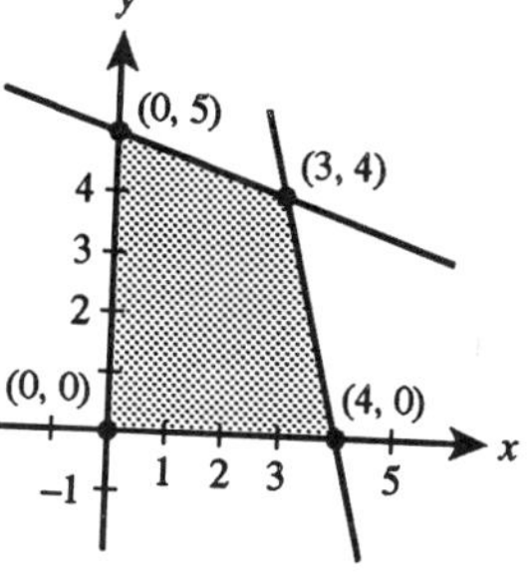

(a) (0, 5) (b) (3, 4)

Solution:

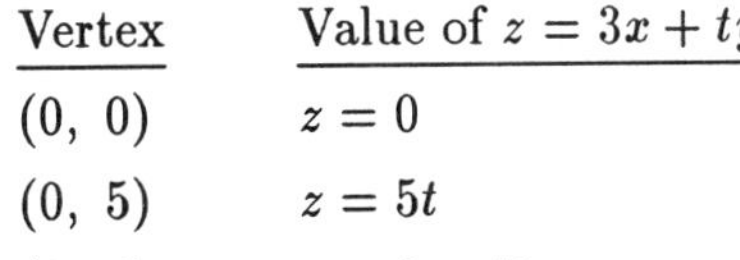

Vertex	Value of $z = 3x + ty$
(0, 0)	$z = 0$
(0, 5)	$z = 5t$
(3, 4)	$z = 9 + 4t$
(4, 0)	$z = 12$

(a) For the maximum value to be at (0, 5), $z = 5t$ must be greater than $z = 9 + 4t$ and $z = 12$.

$$5t > 9 + 4t \quad \text{and} \quad 5t > 12$$

$$t > 9 \qquad\qquad t > \tfrac{12}{5}$$

Thus, $t > 9$.

(b) For the maximum value to be at (3, 4), $z = 9 + 4t$ must be greater than $z = 5t$ and $z = 12$.

$$9 + 4t > 5t \quad \text{and} \quad 9 + 4t > 12$$

$$9 > t \qquad\qquad t > 3$$

$$t > \tfrac{3}{4}$$

Thus, $\frac{3}{4} < t < 9$.

REVIEW EXERCISES FOR CHAPTER 8

Solutions to Selected Exercises

3. Solve the system of equations by the method of substitution.

$$x^2 - y^2 = 9$$
$$x - y = 1$$

Solution:

$$x^2 - y^2 = 9$$
$$x - y = 1 \Rightarrow x = y + 1$$
$$(y+1)^2 - y^2 = 9$$
$$2y + 1 = 9$$
$$y = 4$$
$$x = 5$$

Solution: $(5, 4)$

7. Solve the system of equations by the method of substitution.

$$y^2 - 2y + x = 0$$
$$x + y = 0$$

Solution:

$$y^2 - 2y + x = 0$$
$$x + y = 0 \Rightarrow x = -y$$
$$y^2 - 2y - y = 0$$
$$y(y - 3) = 0$$
$$y = 0 \quad \text{or} \quad y = 3$$
$$x = 0 \qquad\quad x = -3$$

Solutions: $(0, 0)$, $(-3, 3)$

13. Solve the system of equations by elimination.

$$0.2x + 0.3y = 0.14$$
$$0.4x + 0.5y = 0.20$$

Solution:

$$\begin{aligned} 0.2x + 0.3y = 0.14 \Rightarrow 20x + 30y = 14 \Rightarrow&\quad 20x + 30y = 14 \\ 0.4x + 0.5y = 0.20 \Rightarrow 4x + 5y = 2 \Rightarrow&\quad \underline{-20x - 25y = -10} \\ &\quad 5y = 4 \\ &\quad y = \tfrac{4}{5} \\ &\quad x = -\tfrac{1}{2} \end{aligned}$$

Solution: $\left(-\frac{1}{2}, \frac{4}{5}\right)$ or $(-0.5, 0.8)$

19. One hundred gallons of a 60% acid solution are obtained by mixing a 75% solution with a 50% solution. How many gallons of each must be used to obtain the desired mixture?

Solution:

Let $x =$ the amount of 75% solution, and $y =$ amount of 50% solution.

$$x + y = 100 \Rightarrow y = 100 - x$$

$$0.75x + 0.50y = 0.60(100)$$

$$0.75x + 0.50(100 - x) = 60$$

$$0.75x + 50 - 0.50x = 60$$

$$0.25x = 10$$

$$x = 40$$

$$y = 100 - x = 60$$

Answer: 40 gallons of 75% solution, 60 gallons of 60% solution

23. Find the point of equilibrium.

Demand: $p = 37 - 0.0002x$
Supply: $p = 22 + 0.00001x$

Solution:

$$\text{Demand} = \text{Supply}$$

$$37 - 0.0002x = 22 + 0.00001x$$

$$15 = 0.00021x$$

$$x \approx 71{,}429 \text{ units}$$

$$p = \$22.71$$

Point of equilibrium: (71,429, \$22.71)

25. Solve the system of equations.

$$\begin{aligned} x + 2y + 6z &= 4 \\ -3x + 2y - z &= -4 \\ 4x \qquad + 2z &= 16 \end{aligned}$$

Solution:

$$\begin{array}{r} 3x + 6y + 18z = 12 \\ -3x + 2y - z = -4 \\ \hline 8y + 17z = 8 \end{array} \qquad \begin{array}{r} 2x + 4y + 12z = 8 \\ -2x \qquad - z = -8 \\ \hline 4y + 11z = 0 \end{array} \qquad \begin{array}{r} 8y + 17z = 8 \\ -8y - 22z = 0 \\ \hline -5z = 8 \end{array}$$

$$\begin{aligned} z &= -\tfrac{8}{5} = -1.6 \\ y &= \tfrac{1}{8}[8 - 17(-1.6)] = 4.4 \\ x &= \tfrac{1}{2}[8 - (-1.6)] = 4.8 \end{aligned}$$

Solution: $(4.8, 4.4, -1.6)$

29. Solve the system of equations.

$$\begin{aligned} 2x + 5y - 19z &= 34 \\ 3x + 8y - 31z &= 54 \end{aligned}$$

Solution:

$$\begin{array}{rr} 2x + 5y - 19z = 34 \Rightarrow & 6x + 15y - 57z = 102 \\ 3x + 8y - 31z = 54 \Rightarrow & -6x - 16y + 62z = -108 \\ \hline & -y + 5z = -6 \end{array}$$

Let $z = a$. Then,

$$y = 5a + 6$$

$$x = \tfrac{1}{2}[34 - 5(5a + 6) + 19a] = -3a + 2.$$

Solution: $(-3a + 2, 5a + 6, a)$ where a is any real number.

31. Find the equation of the parabola $y = ax^2 + bx + c$ that passes through the points $(0, -6)$, $(1, -3)$, and $(2, 4)$.

Solution:

$$y = ax^2 + bx + c$$

At $(0, -6)$: $-6 = c$

At $(1, -3)$: $-3 = a + b + c \Rightarrow a + b = 3 \Rightarrow -a - b = -3$

At $(2, 4)$: $4 = 4a + 2b + c \Rightarrow 4a + 2b = 10 \Rightarrow 2a + b = 5$

$$a = 2$$

$$b = 1$$

Thus, $y = 2x^2 + x - 6$.

35. A mixture of 6 gallons of chemical A, 8 gallons of chemical B, and 13 gallons of chemical C is required to kill a certain destructive crop insect. Commercial spray X contains 1, 2, and 2 parts, respectively, of these chemicals. Commercial spray Y contains only chemical C. Commercial spray Z contains chemicals A, B, and C in equal amounts. How much of each type of commerical spray is needed to get the desired mixture?

Solution:

From the following chart we obtain our system of equations.

	A	B	C
Mixture X	$\frac{1}{5}$	$\frac{2}{5}$	$\frac{2}{5}$
Mixture Y	0	0	1
Mixture Z	$\frac{1}{3}$	$\frac{1}{3}$	$\frac{1}{3}$
Desired Mixture	$\frac{6}{27}$	$\frac{8}{27}$	$\frac{13}{27}$

$$\left.\begin{aligned}\tfrac{1}{5}x + \tfrac{1}{3}z &= \tfrac{6}{27}\\ \tfrac{2}{5}x + \tfrac{1}{3}z &= \tfrac{8}{27}\end{aligned}\right\} x = \tfrac{10}{27},\ z = \tfrac{12}{27}$$

$$\tfrac{2}{5}x + y + \tfrac{1}{3}z = \tfrac{13}{27} \Rightarrow y = \tfrac{5}{27}$$

To obtain the desired mixture, use 10 gallons of X, 5 gallons of Y, and 12 gallons of Z.

37. Find the least squares regression line $y = ax + b$ for the points $(x_1,\ y_1),\ (x_2,\ y_2), \ldots, (x_n,\ y_n)$. To find the line, solve the following system of linear equations for a and b.

$$5b + 10a = 17.8$$
$$10b + 30a = 45.7$$

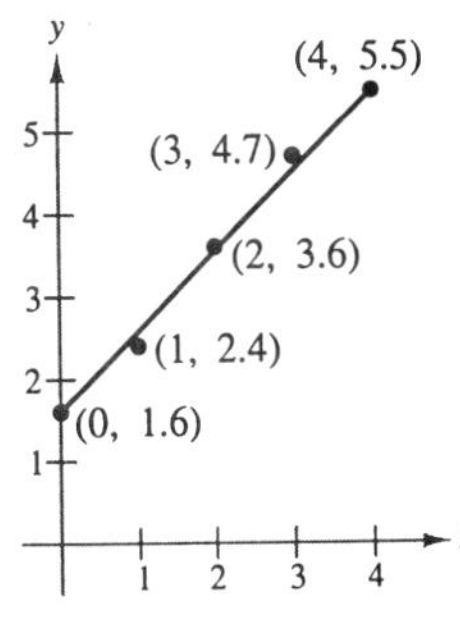

Solution:

$$\begin{aligned} 5b + 10a = 17.8 &\Rightarrow -10b - 20a = -35.6\\ 10b + 30a = 45.7 &\Rightarrow \underline{\quad 10b + 30a = \quad 45.7}\\ &\qquad\qquad 10a = \quad 10.1\\ &\qquad\qquad\quad a = \quad 1.01\\ &\qquad\qquad\quad b = \quad 1.54\end{aligned}$$

Least squares regression line: $y = 1.01x + 1.54$

41. Sketch the graph of the solution set of the system of inequalities.

$$3x + 2y \geq 24$$
$$x + 2y \geq 12$$
$$2 \leq x \leq 15$$
$$y \leq 15$$

Solution:

Vertex A	Vertex B	Vertex C
$3x + 2y = 24$	$3x + 2y = 24$	$3x + 2y = 24$
$x + 2y = 12$	$x = 2$	$x = 15$
$(6, 3)$	$(2, 9)$	$(15, -\frac{21}{2})$ Outside the region

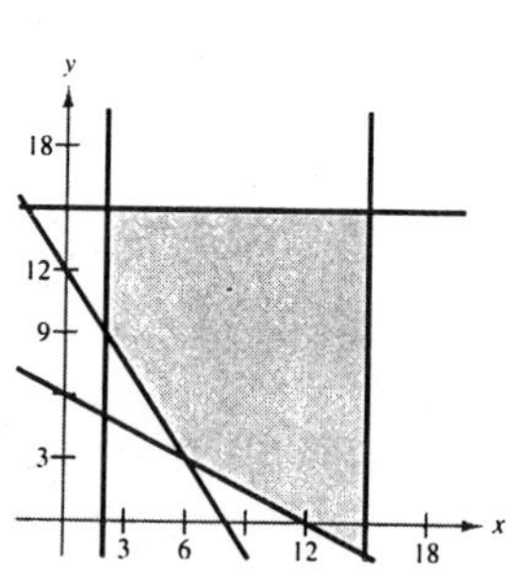

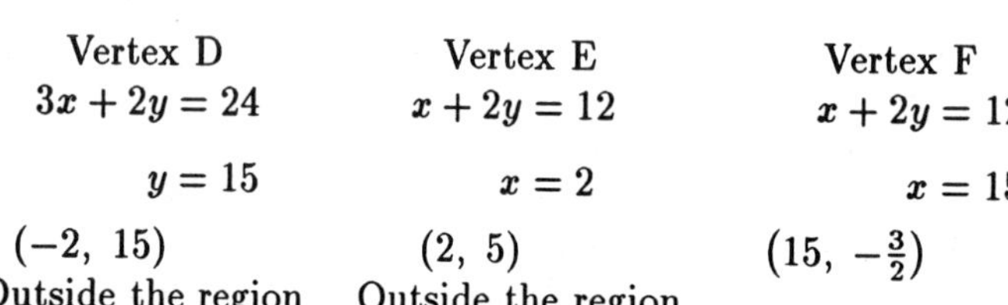

Vertex D	Vertex E	Vertex F
$3x + 2y = 24$	$x + 2y = 12$	$x + 2y = 12$
$y = 15$	$x = 2$	$x = 15$
$(-2, 15)$ Outside the region	$(2, 5)$ Outside the region	$(15, -\frac{3}{2})$

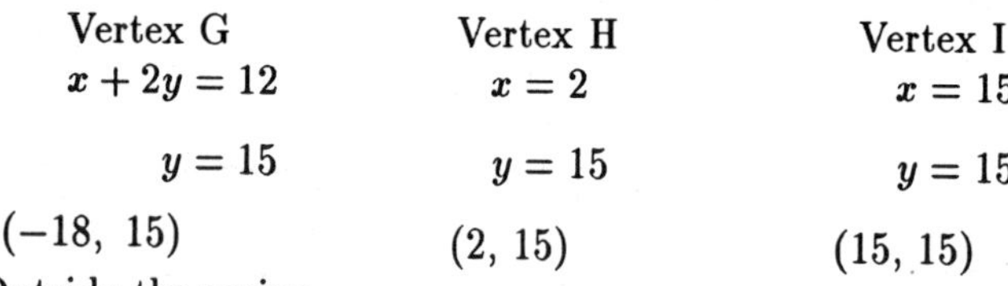

Vertex G	Vertex H	Vertex I
$x + 2y = 12$	$x = 2$	$x = 15$
$y = 15$	$y = 15$	$y = 15$
$(-18, 15)$ Outside the region	$(2, 15)$	$(15, 15)$

45. Sketch the graph of the solution set of the system of inequalities.

$$2x - 3y \geq 0$$
$$2x - y \leq 8$$
$$y \geq 0$$

Solution:

Vertex A	Vertex B	Vertex C
$2x - 3y = 0$	$2x - 3y = 0$	$2x - y = 8$
$2x - y = 8$	$y = 0$	$y = 0$
$(6, 4)$	$(0, 0)$	$(4, 0)$

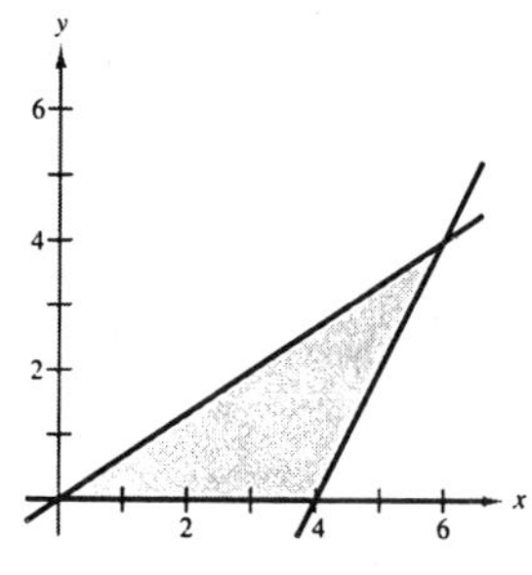

49. A Pennsylvania fruit grower has 1500 bushels of apples that are to be divided between markets in Harrisburg and Philadelphia. These two markets need at least 400 bushels and 600 bushels, respectively. Determine a system of inequalities and sketch a graph of the solution of the system.

Solution:

Let $x =$ the number of bushels for Harrisburg, and $y =$ the number of bushels for Philadelphia.

$$x \geq 400$$

$$y \geq 600$$

$$x + y \leq 1500$$

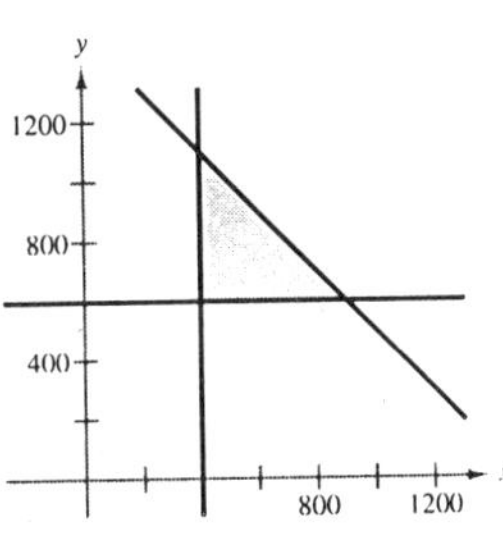

51. Find the consumer surplus and producer surplus.

Demand: $p = 160 - 0.0001x$
Supply: $p = 70 + 0.0002x$

Solution:

$$\text{Demand} = \text{Supply}$$

$$160 - 0.0001x = 70 + 0.0002x$$

$$90 = 0.0003x$$

$$x = 300{,}000 \text{ units}$$

$$p = \$130$$

Point of equilibrium: (300,000, 130)
Consumer surplus: $\frac{1}{2}(300{,}000)(30) = \$4{,}500{,}000$
Producer surplus: $\frac{1}{2}(300{,}000)(60) = \$9{,}000{,}000$

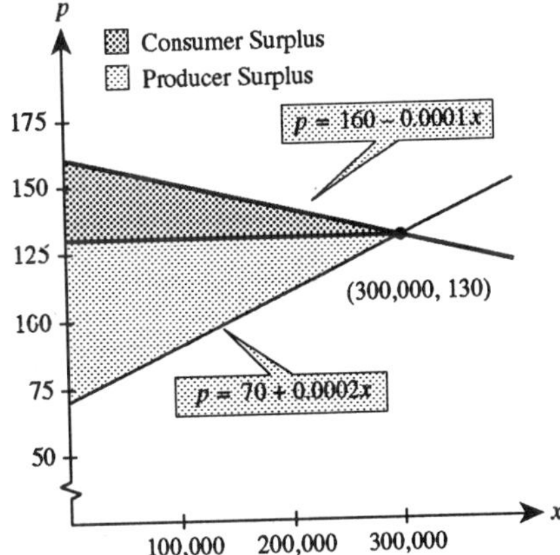

55. Minimize $z = 1.75x + 2.25y$ subject to the following constraints.

$$2x + y \geq 25$$
$$3x + 2y \geq 45$$
$$x \geq 0$$
$$y \geq 0$$

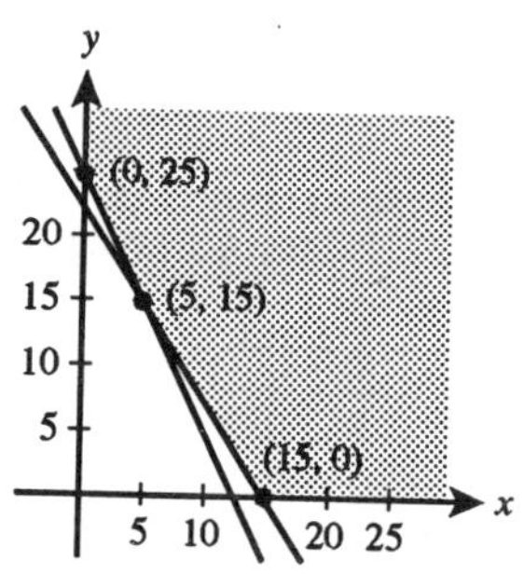

Solution:

Vertex	Value of $z = 1.75x + 2.25y$
(0, 25)	$z = 56.25$
(5, 15)	$z = 42.5$
(15, 0)	$z = 26.25$, minimum value

59. A pet supply company mixes two brands of dry dog food. Brand X costs \$15 per bag and contains 8 units of nutritional element A, 1 unit of nutritional element B, and 2 units of nutritional element C. Brand Y costs \$30 per bag and contains 2 units of nutritional element A, 1 unit of nutritional element B, and 7 units of nutritional element C. Each bag of dog food must contain at least 16 units, 5 units, and 20 units of nutritional elements A, B, and C, respectively. Find the number of bags of brands X and Y that should be mixed to produce a mixture meeting the minimum nutritional requirements and having a minimum cost per bag.

Solution:

Let $x =$ the number of bags of Brand X, and $y =$ the number of bags of Brand Y.

Objective function: Minimize $C = 15x + 30y$

Constraints: $8x + 2y \geq 16$

$$x + y \geq 5$$
$$2x + 7y \geq 20$$
$$x \geq 0, \; y \geq 0$$

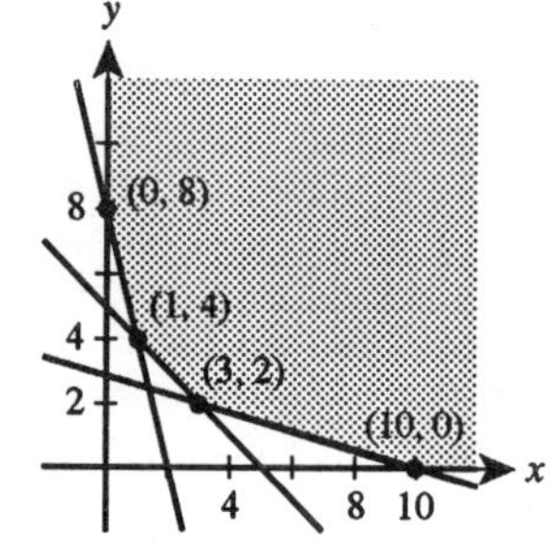

Vertex	Value of $C = 15x + 30y$
(0, 8)	$C = 15(0) + 30(8) = 240$
(1, 4)	$C = 15(1) + 30(4) = 135$
(3, 2)	$C = 15(3) + 30(2) = 105$, minimum value
(10, 0)	$C = 15(10) + 30(0) = 150$

To minimize cost, use three bags of Brand X and two bags of Brand Y.

Practice Test for Chapter 8

For Exercises 1–3, solve the given system by the method of substitution.

1. $\begin{aligned} x + y &= 1 \\ 3x - y &= 15 \end{aligned}$

2. $\begin{aligned} x - 3y &= -3 \\ x^2 + 6y &= 5 \end{aligned}$

3. $\begin{aligned} x + y + z &= 6 \\ 2x - y + 3z &= 0 \\ 5x + 2y - z &= -3 \end{aligned}$

4. Find two numbers whose sum is 110 and product is 2800.

5. Find the dimensions of a rectangle if its perimeter is 170 feet and its area is 2800 square feet.

For Exercises 6–8, solve the linear system by elimination.

6. $\begin{aligned} 2x + 15y &= 4 \\ x - 3y &= 23 \end{aligned}$

7. $\begin{aligned} x + y &= 2 \\ 38x - 19y &= 7 \end{aligned}$

8. $\begin{aligned} 0.4x + 0.5y &= 0.112 \\ 0.3x - 0.7y &= -0.131 \end{aligned}$

9. Herbert invests $17,000 in two funds that pay 11% and 13% simple interest, respectively. If he receives $2080 in yearly interest, how much is invested in each fund?

10. Find the least squares regression line for the points (4, 3), (1, 1), (−1, −2), and (−2, −1).

For Exercises 11–13, solve the system of equations.

11. $\begin{aligned} x + y &= -2 \\ 2x - y + z &= 11 \\ 4y - 3z &= -20 \end{aligned}$

12. $\begin{aligned} 4x - y + 5z &= 4 \\ 2x + y - z &= 0 \\ 2x + 4y + 8z &= 0 \end{aligned}$

13. $\begin{aligned} 3x + 2y - z &= 5 \\ 6x - y + 5z &= 2 \end{aligned}$

14. Find the equation of the parabola $y = ax^2 + bx + c$ passing through the points (0, −1), (1, 4) and (2, 13).

15. Find the position equation $s = \frac{1}{2}at^2 + v_0t + s_0$ given that $s = 12$ feet after 1 second, $s = 5$ feet after 2 seconds, and $s = 4$ feet after 3 seconds.

16. Graph $x^2 + y^2 \geq 9$.

17. Graph the solution of the system.

$$\begin{aligned} x + y &\leq 6 \\ x &\geq 2 \\ y &\geq 0 \end{aligned}$$

18. Derive a set of inequalities to describe the triangle with vertices (0, 0), (0, 7), and (2, 3).

19. Find the maximum value of the objective function, $z = 30x + 26y$, subject to the following constraints.

$$x \geq 0$$
$$y \geq 0$$
$$2x + 3y \leq 21$$
$$5x + 3y \leq 30$$

20. Graph the system of inequalities.

$$x^2 + y^2 \leq 4$$
$$(x-2)^2 + y^2 \geq 4$$

CHAPTER 9

Matrices and Determinants

Section 9.1 Matrices and Systems of Linear Equations 368
Section 9.2 Operations with Matrices 375
Section 9.3 The Inverse of a Square Matrix 380
Section 9.4 The Determinant of a Square Matrix 386
Section 9.5 Properties of Determinants 390
Section 9.6 Applications of Determinants and Matrices 395
Review Exercises 400
Practice Test 406

SECTION 9.1

Matrices and Systems of Linear Equations

■ You should be able to transform an augmented matrix into reduced row-echelon form. This is called Gauss-Jordan elimination. Then use back-substitution to solve the system.

Solutions to Selected Exercises

3. Determine the order of the matrix.

$$\begin{bmatrix} -9 \\ 2 \\ 36 \\ 11 \\ 3 \end{bmatrix}$$

Solution:

Since the matrix has five rows and one column, its order is 5×1.

9. Determine whether the following matrix is in row-echelon form.

$$\begin{bmatrix} 2 & 0 & 4 & 0 \\ 0 & -1 & 3 & 6 \\ 0 & 0 & 1 & 5 \end{bmatrix}$$

Solution:

The first nonzero entries in rows one and two are not 1; the matrix is *not* in row-echelon form.

13. Fill in the blanks using elementary row operations to form a row equivalent matrix.

$$\begin{bmatrix} 1 & 1 & 4 & -1 \\ 3 & 8 & 10 & 3 \\ -2 & 1 & 12 & 6 \end{bmatrix} \rightarrow \begin{bmatrix} 1 & 1 & 4 & -1 \\ 0 & 5 & \_ & \_ \\ 0 & 3 & \_ & \_ \end{bmatrix} \rightarrow \begin{bmatrix} 1 & 1 & 4 & -1 \\ 0 & 1 & \_ & \_ \\ 0 & 3 & 20 & 4 \end{bmatrix}$$

Solution:

$$\begin{bmatrix} 1 & 1 & 4 & -1 \\ 3 & 8 & 10 & 3 \\ -2 & 1 & 12 & 6 \end{bmatrix} \qquad \begin{matrix} \\ -3R_1 + R_2 \rightarrow \\ 2R_1 + R_3 \rightarrow \end{matrix} \begin{bmatrix} 1 & 1 & 4 & -1 \\ 0 & 5 & -2 & 6 \\ 0 & 3 & 20 & 4 \end{bmatrix}$$

$$\begin{matrix} \\ \frac{1}{5}R_2 \rightarrow \\ \\ \end{matrix} \begin{bmatrix} 1 & 1 & 4 & -1 \\ 0 & 1 & -\frac{2}{5} & \frac{6}{5} \\ 0 & 3 & 20 & 4 \end{bmatrix}$$

19. Write the following matrix in row-echelon form. Remember that the row-echelon form of a matrix is not unique.

$$\begin{bmatrix} 1 & -1 & -1 & 1 \\ 5 & -4 & 1 & 8 \\ -6 & 8 & 18 & 0 \end{bmatrix}$$

Solution:

$$\begin{bmatrix} 1 & -1 & -1 & 1 \\ 5 & -4 & 1 & 8 \\ -6 & 8 & 18 & 0 \end{bmatrix} \quad \begin{matrix} \\ -5R_1 + R_2 \rightarrow \\ 6R_1 + R_3 \rightarrow \end{matrix} \begin{bmatrix} 1 & -1 & -1 & 1 \\ 0 & 1 & 6 & 3 \\ 0 & 2 & 12 & 6 \end{bmatrix}$$

$$\begin{matrix} \\ \\ -2R_2 + R_3 \rightarrow \end{matrix} \begin{bmatrix} 1 & -1 & -1 & 1 \\ 0 & 1 & 6 & 3 \\ 0 & 0 & 0 & 0 \end{bmatrix}$$

23. Write the following matrix in *reduced* row-echelon form.

$$\begin{bmatrix} 1 & 2 & 3 & -5 \\ 1 & 2 & 4 & -9 \\ -2 & -4 & -4 & 3 \\ 4 & 8 & 11 & -14 \end{bmatrix}$$

Solution:

$$\begin{bmatrix} 1 & 2 & 3 & -5 \\ 1 & 2 & 4 & -9 \\ -2 & -4 & -4 & 3 \\ 4 & 8 & 11 & -14 \end{bmatrix} \quad \begin{matrix} \\ -R_1 + R_2 \rightarrow \\ 2R_1 + R_3 \rightarrow \\ -4R_1 + R_4 \rightarrow \end{matrix} \begin{bmatrix} 1 & 2 & 3 & -5 \\ 0 & 0 & 1 & -4 \\ 0 & 0 & 2 & -7 \\ 0 & 0 & -1 & 6 \end{bmatrix}$$

$$\begin{matrix} -3R_2 + R_1 \rightarrow \\ \\ -2R_2 + R_3 \rightarrow \\ R_2 + R_4 \rightarrow \end{matrix} \begin{bmatrix} 1 & 2 & 0 & 7 \\ 0 & 0 & 1 & -4 \\ 0 & 0 & 0 & 1 \\ 0 & 0 & 0 & 2 \end{bmatrix}$$

$$\begin{matrix} -7R_3 + R_1 \rightarrow \\ 4R_3 + R_2 \rightarrow \\ \\ -2R_3 + R_4 \rightarrow \end{matrix} \begin{bmatrix} 1 & 2 & 0 & 0 \\ 0 & 0 & 1 & 0 \\ 0 & 0 & 0 & 1 \\ 0 & 0 & 0 & 0 \end{bmatrix}$$

27. Write the system of linear equations represented by the augmented matrix.

$$\left[\begin{array}{ccc:c} 1 & 0 & 2 & -10 \\ 0 & 3 & -1 & 5 \\ 4 & 2 & 0 & 3 \end{array}\right]$$

Solution:

Row 1: $1x + 0y + 2z = -10 \Rightarrow x + 2z = -10$

Row 2: $0x + 3y - 1z = 5 \Rightarrow 3y - z = 5$

Row 3: $4x + 2y + 0z = 3 \Rightarrow 4x + 2y = 3$

31. Write the system of linear equations represented by the augmented matrix. Then use back-substitution to find the solution.

$$\begin{bmatrix} 1 & -1 & 2 & \vdots & 4 \\ 0 & 1 & -1 & \vdots & 2 \\ 0 & 0 & 1 & \vdots & -2 \end{bmatrix}$$

Solution:

$$\begin{aligned} \text{Row 1: } x - y + 2z &= 4 \\ \text{Row 2: } y - z &= 2 \\ \text{Row 3: } z &= -2 \\ y &= 2 + z = 2 + (-2) = 0 \\ x &= 4 + y - 2z = 4 + 0 - 2(-2) = 8 \end{aligned}$$

Answer: $(8, 0, -2)$

35. Write the solution represented by the augmented matrix.

$$\begin{bmatrix} 1 & 0 & 0 & \vdots & -4 \\ 0 & 1 & 0 & \vdots & -8 \\ 0 & 0 & 1 & \vdots & 2 \end{bmatrix}$$

Solution:

Row 1: $x = -4$

Row 2: $y = -8$

Row 3: $z = 2$

Answer: $(-4, -8, 2)$

39. Solve the system of equations. Use Gaussian elimination with back-substitution or Gauss-Jordan elimination.

$$\begin{aligned} -3x + 5y &= -22 \\ 3x + 4y &= 4 \\ 4x - 8y &= 32 \end{aligned}$$

–CONTINUED ON NEXT PAGE–

39. -CONTINUED-

Solution:

$$\begin{bmatrix} -3 & 5 & \vdots & -22 \\ 3 & 4 & \vdots & 4 \\ 4 & -8 & \vdots & 32 \end{bmatrix} \quad \begin{matrix} -\frac{1}{3}R_1 \rightarrow \\ -3R_1 + R_2 \rightarrow \\ -4R_1 + R_3 \rightarrow \end{matrix} \begin{bmatrix} 1 & -\frac{5}{3} & \vdots & \frac{22}{3} \\ 0 & 9 & \vdots & -18 \\ 0 & -\frac{4}{3} & \vdots & \frac{8}{3} \end{bmatrix}$$

$$\begin{matrix} \frac{5}{3}R_2 + R_1 \rightarrow \\ \frac{1}{9}R_2 \rightarrow \\ \frac{4}{3}R_2 + R_3 \rightarrow \end{matrix} \begin{bmatrix} 1 & 0 & \vdots & 4 \\ 0 & 1 & \vdots & -2 \\ 0 & 0 & \vdots & 0 \end{bmatrix}$$

Answer: $(4, \ -2)$

43. Solve the system of equations. Use Gaussian elimination with back-substitution or Gauss-Jordan elimination.

$$-x + 2y = 1.5$$
$$2x - 4y = 3$$

Solution:

$$\begin{bmatrix} -1 & 2 & \vdots & 1.5 \\ 2 & -4 & \vdots & 3 \end{bmatrix} \quad \begin{matrix} -R_1 \rightarrow \\ -2R_1 + R_2 \rightarrow \end{matrix} \begin{bmatrix} 1 & -2 & \vdots & -1.5 \\ 0 & 0 & \vdots & 6 \end{bmatrix}$$

The second line says $0 = 6$. This is inconsistent.

47. Solve the system of equations. Use Gaussian elimination with back-substitution or Gauss-Jordan elimination.

$$x + y - 5z = 3$$
$$x \qquad - 2z = 1$$
$$2x - y - \ \ z = 0$$

Solution:

$$\begin{bmatrix} 1 & 1 & -5 & \vdots & 3 \\ 1 & 0 & -2 & \vdots & 1 \\ 2 & -1 & -1 & \vdots & 0 \end{bmatrix} \quad \begin{matrix} \\ -R_1 + R_2 \rightarrow \\ -2R_1 + R_3 \rightarrow \end{matrix} \begin{bmatrix} 1 & 1 & -5 & \vdots & 3 \\ 0 & -1 & 3 & \vdots & -2 \\ 0 & -3 & 9 & \vdots & -6 \end{bmatrix}$$

$$\begin{matrix} R_2 + R_1 \rightarrow \\ -R_2 \rightarrow \\ 3R_2 + R_3 \rightarrow \end{matrix} \begin{bmatrix} 1 & 0 & -2 & \vdots & 1 \\ 0 & 1 & -3 & \vdots & 2 \\ 0 & 0 & 0 & \vdots & 0 \end{bmatrix}$$

Thus, $x - 2z = 1$ and $y - 3z = 2$. By letting $z = a$, we have $y = 3a + 2$ and $x = 2a + 1$.

Answer: $(2a + 1, \ 3a + 2, \ a)$

53. Solve the system of equations. Use Gaussian elimination with back-substitution or Gauss-Jordan elimination.

$$\begin{aligned} 2x + y - z + 2w &= -6 \\ 3x + 4y + w &= 1 \\ x + 5y + 2z + 6w &= -3 \\ 5x + 2y - z - w &= 3 \end{aligned}$$

Solution:

$$\begin{bmatrix} 2 & 1 & -1 & 2 & \vdots & -6 \\ 3 & 4 & 0 & 1 & \vdots & 1 \\ 1 & 5 & 2 & 6 & \vdots & -3 \\ 5 & 2 & -1 & -1 & \vdots & 3 \end{bmatrix} \quad \begin{matrix} R_3 \rightarrow \\ \\ R_1 \rightarrow \\ \\ \end{matrix} \begin{bmatrix} 1 & 5 & 2 & 6 & \vdots & -3 \\ 3 & 4 & 0 & 1 & \vdots & 1 \\ 2 & 1 & -1 & 2 & \vdots & -6 \\ 5 & 2 & -1 & -1 & \vdots & 3 \end{bmatrix}$$

$$\begin{matrix} \\ -3R_1 + R_2 \rightarrow \\ -2R_1 + R_3 \rightarrow \\ -5R_1 + R_4 \rightarrow \end{matrix} \begin{bmatrix} 1 & 5 & 2 & 6 & \vdots & -3 \\ 0 & -11 & -6 & -17 & \vdots & 10 \\ 0 & -9 & -5 & -10 & \vdots & 0 \\ 0 & -23 & -11 & -31 & \vdots & 18 \end{bmatrix}$$

$$\begin{matrix} 5R_4 + R_1 \rightarrow \\ -11R_4 + R_2 \rightarrow \\ -9R_4 + R_3 \rightarrow \\ -2R_2 + R_4 \rightarrow \end{matrix} \begin{bmatrix} 1 & 0 & 7 & 21 & \vdots & -13 \\ 0 & 0 & -17 & -50 & \vdots & 32 \\ 0 & 0 & -14 & -37 & \vdots & 18 \\ 0 & -1 & 1 & 3 & \vdots & -2 \end{bmatrix}$$

$$\begin{matrix} -7R_3 + R_1 \rightarrow \\ 17R_3 + R_2 \rightarrow \\ -\frac{1}{14}R_3 \rightarrow \\ -R_4 \rightarrow \end{matrix} \begin{bmatrix} 1 & 0 & 0 & \frac{5}{2} & \vdots & -4 \\ 0 & 0 & 0 & -\frac{71}{14} & \vdots & \frac{71}{7} \\ 0 & 0 & 1 & \frac{37}{14} & \vdots & -\frac{9}{7} \\ 0 & 1 & -1 & -3 & \vdots & 2 \end{bmatrix}$$

$$\begin{matrix} \\ -\frac{14}{71}R_2 \rightarrow \\ \\ \\ \end{matrix} \begin{bmatrix} 1 & 0 & 0 & \frac{5}{2} & \vdots & -4 \\ 0 & 0 & 0 & 1 & \vdots & -2 \\ 0 & 0 & 1 & \frac{37}{14} & \vdots & -\frac{9}{7} \\ 0 & 1 & -1 & -3 & \vdots & 2 \end{bmatrix}$$

$$\begin{aligned} x \qquad + \tfrac{5}{2}w &= -4 \\ y - z - 3w &= 2 \\ z + \tfrac{37}{14}w &= -\tfrac{9}{7} \\ w &= -2 \end{aligned} \qquad \begin{matrix} \\ R_4 \rightarrow \\ \\ R_2 \rightarrow \end{matrix} \begin{bmatrix} 1 & 0 & 0 & \frac{5}{2} & \vdots & -4 \\ 0 & 1 & -1 & -3 & \vdots & 2 \\ 0 & 0 & 1 & \frac{37}{14} & \vdots & -\frac{9}{7} \\ 0 & 0 & 0 & 1 & \vdots & -2 \end{bmatrix}$$

Thus,

$w = -2$

$x = -4 - \frac{5}{2}(-2) = 1$

$z = -\frac{9}{7} - \frac{37}{14}(-2) = 4$

$y = 2 + 4 + 3(-2) = 0.$

Answer: $(1, 0, 4, -2)$

57. Solve the system of equations. Use Gaussian elimination with back-substitution or Gauss-Jordan elimination.

$$\begin{aligned} x + y + z &= 0 \\ 2x + 3y + z &= 0 \\ 3x + 5y + z &= 0 \end{aligned}$$

Solution:

$$\left[\begin{array}{ccc:c} 1 & 1 & 1 & 0 \\ 2 & 3 & 1 & 0 \\ 3 & 5 & 1 & 0 \end{array}\right] \quad \begin{array}{r} \\ -2R_1 + R_2 \rightarrow \\ -3R_1 + R_3 \rightarrow \end{array} \left[\begin{array}{ccc:c} 1 & 1 & 1 & 0 \\ 0 & 1 & -1 & 0 \\ 0 & 2 & -2 & 0 \end{array}\right]$$

$$\begin{array}{r} -R_2 + R_1 \rightarrow \\ \\ -2R_2 + R_3 \rightarrow \end{array} \left[\begin{array}{ccc:c} 1 & 0 & 2 & 0 \\ 0 & 1 & -1 & 0 \\ 0 & 0 & 0 & 0 \end{array}\right]$$

Thus, $x + 2z = 0$ and $y - z = 0$. By letting $z = a$, we have $x = -2a$ and $y = a$.

Answer: $(-2a,\ a,\ a)$ where a is any real number.

59. A small corporation borrowed \$1,500,000 to expand its product line. Some of the money was borrowed at 8%, some at 9%, and some at 12%. How much was borrowed at each rate if the annual interest was \$133,000 and the amount borrowed at 8% was 4 times the amount borrowed at 12%?

Solution:

Let x = 8% amount, y = 9% amount, and z = 12% amount.

$$\begin{aligned} x + y + z &= 1{,}500{,}000 \\ 0.08x + 0.09y + 0.12z &= 133{,}000 \\ x - 4z &= 0 \end{aligned}$$

$$\left[\begin{array}{ccc:r} 1 & 1 & 1 & 1{,}500{,}000 \\ 8 & 9 & 12 & 13{,}300{,}000 \\ 1 & 0 & -4 & 0 \end{array}\right]$$

$$\begin{array}{r} \\ -8R_1 + R_2 \rightarrow \\ -R_1 + R_3 \rightarrow \end{array} \left[\begin{array}{ccc:r} 1 & 1 & 1 & 1{,}500{,}000 \\ 0 & 1 & 4 & 1{,}300{,}000 \\ 0 & -1 & -5 & -1{,}500{,}000 \end{array}\right]$$

$$\begin{array}{r} -R_2 + R_1 \rightarrow \\ \\ R_2 + R_3 \rightarrow \end{array} \left[\begin{array}{ccc:r} 1 & 0 & -3 & 200{,}000 \\ 0 & 1 & 4 & 1{,}300{,}000 \\ 0 & 0 & -1 & -200{,}000 \end{array}\right]$$

$$\begin{array}{r} 3R_3 + R_1 \rightarrow \\ -4R_3 + R_2 \rightarrow \\ -R_3 \rightarrow \end{array} \left[\begin{array}{ccc:r} 1 & 0 & 0 & 800{,}000 \\ 0 & 1 & 0 & 500{,}000 \\ 0 & 0 & 1 & 200{,}000 \end{array}\right]$$

Thus, $x = \$800{,}000$, $y = \$500{,}000$, and $z = \$200{,}000$.

63. Find D, E, and F such that (1, 1), (3, 3), and (4, 2) are solution points of the equation $x^2 + y^2 + Dx + Ey + F = 0$.

Solution:

$$\begin{aligned}
&\text{At } (1, 1)\text{:} \quad (1)^2 + (1)^2 + D(1) + E(1) + F = 0 \Rightarrow D + E + F = -2 \\
&\text{At } (3, 3)\text{:} \quad (3)^2 + (3)^2 + D(3) + E(3) + F = 0 \Rightarrow 3D + 3E + F = -18 \\
&\text{At } (4, 2)\text{:} \quad (4)^2 + (2)^2 + D(4) + E(2) + F = 0 \Rightarrow 4D + 2E + F = -20
\end{aligned}$$

$$\left[\begin{array}{ccc:c} 1 & 1 & 1 & -2 \\ 3 & 3 & 1 & -18 \\ 4 & 2 & 1 & -20 \end{array}\right] \quad \begin{array}{r} \\ -3R_1 + R_2 \rightarrow \\ -4R_1 + R_3 \rightarrow \end{array} \left[\begin{array}{ccc:c} 1 & 1 & 1 & -2 \\ 0 & 0 & -2 & -12 \\ 0 & -2 & -3 & -12 \end{array}\right]$$

$$\begin{array}{r} \\ R_3 \rightarrow \\ R_2 \rightarrow \end{array} \left[\begin{array}{ccc:c} 1 & 1 & 1 & -2 \\ 0 & -2 & -3 & -12 \\ 0 & 0 & -2 & -12 \end{array}\right]$$

$$\begin{array}{r} \\ -\frac{1}{2}R_2 \rightarrow \\ -\frac{1}{2}R_3 \rightarrow \end{array} \left[\begin{array}{ccc:c} 1 & 1 & 1 & -2 \\ 0 & 1 & \frac{3}{2} & 6 \\ 0 & 0 & 1 & 6 \end{array}\right]$$

$$\begin{aligned}
D + E + \phantom{\tfrac{3}{4}}F &= -2 \\
E + \tfrac{3}{4}F &= 6 \\
F &= 6
\end{aligned}$$

Thus,

$$F = 6$$

$$E = 6 - \tfrac{3}{2}(6) = -3$$

$$D = -2 - (-3) - 6 = -5$$

The equation of the circle is $x^2 + y^2 - 5x - 3y + 6 = 0$.

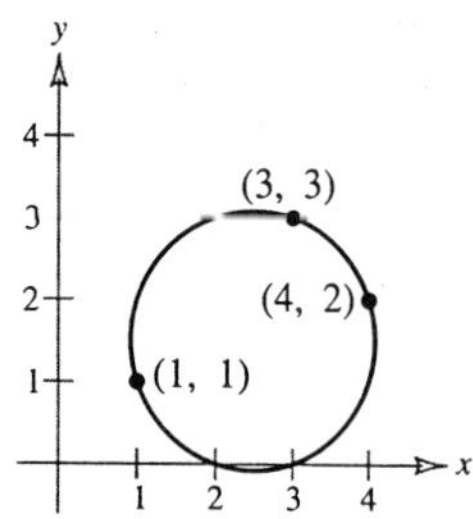

SECTION 9.2

Operations with Matrices

- $A = B$ if and only if they have the same order and $a_{ij} = b_{ij}$.
- You should be able to perform the operations of matrix addition, scalar multiplication, and matrix multiplication.
- Some properties of matrix addition, scalar multiplication, and matrix multiplication are:

  (a) $A + B = B + A$
  (b) $A + (B + C) = (A + B) + C$
  (c) $(cd)A = c(dA)$
  (d) $1A = A$
  (e) $c(A + B) = cA + cB$
  (f) $(c + d)A = cA + dA$
  (g) $A(BC) = (AB)C$
  (h) $A(B + C) = AB + AC$
  (i) $(A + B)C = AC + BC$
  (j) $c(AB) = (cA)B = A(cB)$

- You should remember that $AB \neq BA$ in general.

Solutions to Selected Exercises

3. Find x and y.

$$\begin{bmatrix} 16 & 4 & 5 & 4 \\ -3 & 13 & 15 & 6 \\ 0 & 2 & 4 & 0 \end{bmatrix} = \begin{bmatrix} 16 & 4 & 2x+1 & 4 \\ -3 & 13 & 15 & 3x \\ 0 & 2 & 3y-5 & 0 \end{bmatrix}$$

Solution:

$$\left.\begin{matrix} 5 = 2x + 1 \\ 6 = 3x \end{matrix}\right\} \Rightarrow x = 2$$

$$4 = 3y - 5 \quad \Rightarrow y = 3$$

Answer: $x = 2, \ y = 3$

9. Find (a) $A+B$, (b) $A-B$, (c) $3A$, and (d) $3A-2B$.

$$A = \begin{bmatrix} 2 & 2 & -1 & 0 & 1 \\ 1 & 1 & -2 & 0 & -1 \end{bmatrix}, \quad B = \begin{bmatrix} 1 & 1 & -1 & 1 & 0 \\ -3 & 4 & 9 & -6 & -7 \end{bmatrix}$$

Solution:

(a) $$A+B = \begin{bmatrix} 2+1 & 2+1 & -1+(-1) & 0+1 & 1+0 \\ 1+(-3) & 1+4 & -2+9 & 0+(-6) & -1+(-7) \end{bmatrix}$$

$$= \begin{bmatrix} 3 & 3 & -2 & 1 & 1 \\ -2 & 5 & 7 & -6 & -8 \end{bmatrix}$$

(b) $$A-B = \begin{bmatrix} 2-1 & 2-1 & -1-(-1) & 0-1 & 1-0 \\ 1-(-3) & 1-4 & -2-9 & 0-(-6) & -1-(-7) \end{bmatrix}$$

$$= \begin{bmatrix} 1 & 1 & 0 & -1 & 1 \\ 4 & -3 & -11 & 6 & 6 \end{bmatrix}$$

(c) $$3A = \begin{bmatrix} 3(2) & 3(2) & 3(-1) & 3(0) & 3(1) \\ 3(1) & 3(1) & 3(-2) & 3(0) & 3(-1) \end{bmatrix} = \begin{bmatrix} 6 & 6 & -3 & 0 & 3 \\ 3 & 3 & -6 & 0 & -3 \end{bmatrix}$$

(d) $$3A-2B = \begin{bmatrix} 6 & 6 & -3 & 0 & 3 \\ 3 & 3 & -6 & 0 & -3 \end{bmatrix} - \begin{bmatrix} 2 & 2 & -2 & 2 & 0 \\ -6 & 8 & 18 & -12 & -14 \end{bmatrix}$$

$$= \begin{bmatrix} 4 & 4 & -1 & -2 & 3 \\ 9 & -5 & -24 & 12 & 11 \end{bmatrix}$$

13. Find (a) AB, (b) BA, and if possible (c) A^2.

$$A = \begin{bmatrix} 3 & -1 \\ 1 & 3 \end{bmatrix}, \quad B = \begin{bmatrix} 1 & -3 \\ 3 & 1 \end{bmatrix}$$

Solution:

(a) $$AB = \begin{bmatrix} 3 & -1 \\ 1 & 3 \end{bmatrix}\begin{bmatrix} 1 & -3 \\ 3 & 1 \end{bmatrix} = \begin{bmatrix} 3(1)+(-1)(3) & 3(-3)+(-1)(1) \\ 1(1)+3(3) & 1(-3)+3(1) \end{bmatrix} = \begin{bmatrix} 0 & -10 \\ 10 & 0 \end{bmatrix}$$

(b) $$BA = \begin{bmatrix} 1 & -3 \\ 3 & 1 \end{bmatrix}\begin{bmatrix} 3 & -1 \\ 1 & 3 \end{bmatrix} = \begin{bmatrix} 1(3)+(-3)(1) & 1(-1)+(-3)(3) \\ 3(3)+1(1) & 3(-1)+1(3) \end{bmatrix} = \begin{bmatrix} 0 & -10 \\ 10 & 0 \end{bmatrix}$$

(c) $$A^2 = \begin{bmatrix} 3 & -1 \\ 1 & 3 \end{bmatrix}\begin{bmatrix} 3 & -1 \\ 1 & 3 \end{bmatrix} = \begin{bmatrix} 3(3)+(-1)(1) & 3(-1)+(-1)(3) \\ 1(3)+3(1) & 1(-1)+3(3) \end{bmatrix} = \begin{bmatrix} 8 & -6 \\ 6 & 8 \end{bmatrix}$$

17. Find AB, if possible.

$$A = \begin{bmatrix} 2 & 1 \\ -3 & 4 \\ 1 & 6 \end{bmatrix}, \quad B = \begin{bmatrix} 0 & -1 & 0 \\ 4 & 0 & 2 \\ 8 & -1 & 7 \end{bmatrix}$$

Solution:

A is 3×2 and B is 3×3. Since the number of columns of A does not equal the number of rows of B, the multiplication is not possible. **Note:** BA is possible.

21. Find AB, if possible.

$$A = \begin{bmatrix} 5 & 0 & 0 \\ 0 & -8 & 0 \\ 0 & 0 & 7 \end{bmatrix}, \quad B = \begin{bmatrix} \frac{1}{5} & 0 & 0 \\ 0 & -\frac{1}{8} & 0 \\ 0 & 0 & \frac{1}{2} \end{bmatrix}$$

Solution:

$$AB = \begin{bmatrix} 5 & 0 & 0 \\ 0 & -8 & 0 \\ 0 & 0 & 7 \end{bmatrix} \begin{bmatrix} \frac{1}{5} & 0 & 0 \\ 0 & -\frac{1}{8} & 0 \\ 0 & 0 & \frac{1}{2} \end{bmatrix} = \begin{bmatrix} 1 & 0 & 0 \\ 0 & 1 & 0 \\ 0 & 0 & \frac{7}{2} \end{bmatrix}$$

25. Solve for X in $X = 3A - 2B$, given

$$A = \begin{bmatrix} -2 & -1 \\ 1 & 0 \\ 3 & -4 \end{bmatrix} \quad \text{and} \quad B = \begin{bmatrix} 0 & 3 \\ 2 & 0 \\ -4 & -1 \end{bmatrix}.$$

Solution:

$$X = 3A - 2B = \begin{bmatrix} -6 & -3 \\ 3 & 0 \\ 9 & -12 \end{bmatrix} - \begin{bmatrix} 0 & 6 \\ 4 & 0 \\ -8 & -2 \end{bmatrix} = \begin{bmatrix} -6 & -9 \\ -1 & 0 \\ 17 & 10 \end{bmatrix}$$

29. Find matrices A, X, and B such that the given system of linear equations can be written as the matrix equation $AX = B$. Solve the system of equations.

$$-x + y = 4$$
$$-2x + y = 0$$

Solution:

$$A = \begin{bmatrix} -1 & 1 \\ -2 & 1 \end{bmatrix}, \quad X = \begin{bmatrix} x \\ y \end{bmatrix}, \quad B = \begin{bmatrix} 4 \\ 0 \end{bmatrix}$$

By Gauss-Jordan elimination on

$$\left[\begin{array}{cc:c} -1 & 1 & 4 \\ -2 & 1 & 0 \end{array}\right] \quad \begin{array}{r} -R_1 \rightarrow \\ 2R_1 + R_2 \rightarrow \end{array} \left[\begin{array}{cc:c} 1 & -1 & -4 \\ 0 & -1 & -8 \end{array}\right]$$

$$\begin{array}{r} R_2 + R_1 \rightarrow \\ -R_2 \rightarrow \end{array} \left[\begin{array}{cc:c} 1 & 0 & 4 \\ 0 & 1 & 8 \end{array}\right], \quad \text{we have } x = 4 \text{ and } y = 8.$$

33. Find $f(A)$, given

$$f(x) = x^2 - 5x + 2 \quad \text{and} \quad A = \begin{bmatrix} 2 & 0 \\ 4 & 5 \end{bmatrix}.$$

Solution:

$$\begin{aligned} f(A) &= \begin{bmatrix} 2 & 0 \\ 4 & 5 \end{bmatrix}\begin{bmatrix} 2 & 0 \\ 4 & 5 \end{bmatrix} - 5\begin{bmatrix} 2 & 0 \\ 4 & 5 \end{bmatrix} + 2\begin{bmatrix} 1 & 0 \\ 0 & 1 \end{bmatrix} \\ &= \begin{bmatrix} 4 & 0 \\ 28 & 25 \end{bmatrix} - \begin{bmatrix} 10 & 0 \\ 20 & 25 \end{bmatrix} + \begin{bmatrix} 2 & 0 \\ 0 & 2 \end{bmatrix} \\ &= \begin{bmatrix} -4 & 0 \\ 8 & 2 \end{bmatrix} \end{aligned}$$

37. If a, b, and c are real numbers such that $c \neq 0$ and $ac = bc$, then $a = b$. However, if A, B, and C are matrices such that $AC = BC$, then A is *not* necessarily equal to B. Illustrate this, using the following matrices.

$$A = \begin{bmatrix} 1 & 2 & 3 \\ 0 & 5 & 4 \\ 3 & -2 & 1 \end{bmatrix}, \quad B = \begin{bmatrix} 4 & -6 & 3 \\ 5 & 4 & 4 \\ -1 & 0 & 1 \end{bmatrix}, \quad \text{and } C = \begin{bmatrix} 0 & 0 & 0 \\ 0 & 0 & 0 \\ 4 & -2 & 3 \end{bmatrix}$$

Solution:

$$AC = \begin{bmatrix} 1 & 2 & 3 \\ 0 & 5 & 4 \\ 3 & -2 & 1 \end{bmatrix}\begin{bmatrix} 0 & 0 & 0 \\ 0 & 0 & 0 \\ 4 & -2 & 3 \end{bmatrix} = \begin{bmatrix} 12 & -6 & 9 \\ 16 & -8 & 12 \\ 4 & -2 & 3 \end{bmatrix}$$

$$BC = \begin{bmatrix} 4 & -6 & 3 \\ 5 & 4 & 4 \\ -1 & 0 & 1 \end{bmatrix}\begin{bmatrix} 0 & 0 & 0 \\ 0 & 0 & 0 \\ 4 & -2 & 3 \end{bmatrix} = \begin{bmatrix} 12 & -6 & 9 \\ 16 & -8 & 12 \\ 4 & -2 & 3 \end{bmatrix}$$

Thus, $AC = BC$ even though $A \neq B$.

41. A fruit grower raises two crops that are shipped to three outlets. The number of units of product i that are shipped to outlet j is represented by a_{ij} in the matrix

$$A = \begin{bmatrix} 100 & 75 & 75 \\ 125 & 150 & 100 \end{bmatrix}.$$

The profit per unit is represented by $B = [\$3.75 \quad \$7.00]$. Find the product BA, and state what each entry of the product represents.

Solution:

$$BA = [3.75 \quad 7.00]\begin{bmatrix} 100 & 75 & 75 \\ 125 & 150 & 100 \end{bmatrix} = [\$1250.00 \quad \$1331.25 \quad \$981.25]$$

The entries in the last matrix represent the profit for both crops at each of the three outlets.

45. The matrix

$$P = \overbrace{\begin{matrix} R & D & I \end{matrix}}^{\text{From}}$$

$$P = \begin{bmatrix} 0.6 & 0.1 & 0.1 \\ 0.2 & 0.7 & 0.1 \\ 0.2 & 0.2 & 0.8 \end{bmatrix} \begin{matrix} R \\ D \\ I \end{matrix} \left.\vphantom{\begin{matrix} R \\ D \\ I \end{matrix}}\right\} \text{To}$$

is called a stochastic matrix. Each entry p_{ij} $(i \neq j)$ represents the proportion of the voting population that changes from party i to party j, and p_{ii} represents the proportion that remains loyal to the party from one election to the next. Find P^2. (This matrix gives the transition probabilities from the first election to the third.)

Solution:

$$P^2 = \begin{bmatrix} 0.6 & 0.1 & 0.1 \\ 0.2 & 0.7 & 0.1 \\ 0.2 & 0.2 & 0.8 \end{bmatrix} \begin{bmatrix} 0.6 & 0.1 & 0.1 \\ 0.2 & 0.7 & 0.1 \\ 0.2 & 0.2 & 0.8 \end{bmatrix} = \begin{bmatrix} 0.40 & 0.15 & 0.15 \\ 0.28 & 0.53 & 0.17 \\ 0.32 & 0.32 & 0.68 \end{bmatrix}$$

SECTION 9.3

The Inverse of a Square Matrix

- You should be able to find the inverse, if it exists, of a square matrix.
- You should be able to use inverse matrices to solve systems of equations.

Solutions to Selected Exercises

3. Show that B is the inverse of A where

$$A = \begin{bmatrix} 1 & 2 \\ 3 & 4 \end{bmatrix} \quad \text{and} \quad B = \begin{bmatrix} -2 & 1 \\ \frac{3}{2} & -\frac{1}{2} \end{bmatrix}.$$

Solution:

$$AB = \begin{bmatrix} 1 & 2 \\ 3 & 4 \end{bmatrix}\begin{bmatrix} -2 & 1 \\ \frac{3}{2} & -\frac{1}{2} \end{bmatrix} = \begin{bmatrix} -2+3 & 1-1 \\ -6+6 & 3-2 \end{bmatrix} = \begin{bmatrix} 1 & 0 \\ 0 & 1 \end{bmatrix}$$

$$BA = \begin{bmatrix} -2 & 1 \\ \frac{3}{2} & -\frac{1}{2} \end{bmatrix}\begin{bmatrix} 1 & 2 \\ 3 & 4 \end{bmatrix} = \begin{bmatrix} -2+3 & -4+4 \\ \frac{3}{2}-\frac{3}{2} & 3-2 \end{bmatrix} = \begin{bmatrix} 1 & 0 \\ 0 & 1 \end{bmatrix}$$

7. Show that B is the inverse of A where

$$A = \begin{bmatrix} 2 & 0 & 1 & 1 \\ 3 & 0 & 0 & 1 \\ -1 & 1 & -2 & 1 \\ 4 & -1 & 1 & 0 \end{bmatrix} \quad \text{and} \quad B = \begin{bmatrix} -1 & 2 & -1 & -1 \\ -4 & 9 & -5 & -6 \\ 0 & 1 & -1 & -1 \\ 3 & -5 & 3 & 3 \end{bmatrix}.$$

Solution:

$$AB = \begin{bmatrix} 2 & 0 & 1 & 1 \\ 3 & 0 & 0 & 1 \\ -1 & 1 & -2 & 1 \\ 4 & -1 & 1 & 0 \end{bmatrix}\begin{bmatrix} -1 & 2 & -1 & -1 \\ -4 & 9 & -5 & -6 \\ 0 & 1 & -1 & -1 \\ 3 & -5 & 3 & 3 \end{bmatrix}$$

$$= \begin{bmatrix} -2+0+0+3 & 4+0+1+(-5) & -2+0+(-1)+3 & -2+0+(-1)+3 \\ -3+0+0+3 & 6+0+0+(-5) & -3+0+0+3 & -3+0+0+3 \\ 1+(-4)+0+3 & -2+9+(-2)+(-5) & 1+(-5)+2+3 & 1+(-6)+2+3 \\ -4+4+0+0 & 8+(-9)+1+0 & -4+5+(-1)+0 & -4+6+(-1)+0 \end{bmatrix}$$

$$= \begin{bmatrix} 1 & 0 & 0 & 0 \\ 0 & 1 & 0 & 0 \\ 0 & 0 & 1 & 0 \\ 0 & 0 & 0 & 1 \end{bmatrix}$$

–CONTINUED ON NEXT PAGE–

7. –CONTINUED–

$$BA = \begin{bmatrix} -1 & 2 & -1 & -1 \\ -4 & 9 & -5 & -6 \\ 0 & 1 & -1 & -1 \\ 3 & -5 & 3 & 3 \end{bmatrix} \begin{bmatrix} 2 & 0 & 1 & 1 \\ 3 & 0 & 0 & 1 \\ -1 & 1 & -2 & 1 \\ 4 & -1 & 1 & 0 \end{bmatrix}$$

$$= \begin{bmatrix} -2+6+1+(-4) & 0+0+(-1)+1 & -1+0+2+(-1) & -1+2+(-1)+0 \\ -8+27+5+(-24) & 0+0+(-5)+6 & -4+0+10+(-6) & -4+9+(-5)+0 \\ 0+3+1+(-4) & 0+0+(-1)+1 & 0+0+2+(-1) & 0+1+(-1)+0 \\ 6+(-15)+(-3)+12 & 0+0+3+(-3) & 3+0+(-6)+3 & 3+(-5)+3+0 \end{bmatrix}$$

$$= \begin{bmatrix} 1 & 0 & 0 & 0 \\ 0 & 1 & 0 & 0 \\ 0 & 0 & 1 & 0 \\ 0 & 0 & 0 & 1 \end{bmatrix}$$

11. Find the inverse of the following matrix (if it exists).

$$\begin{bmatrix} 1 & -2 \\ 2 & -3 \end{bmatrix}$$

Solution:

$$\left[\begin{array}{cc:cc} 1 & -2 & 1 & 0 \\ 2 & -3 & 0 & 1 \end{array}\right] \quad \begin{array}{r} \\ -2R_1 + R_2 \rightarrow \end{array} \left[\begin{array}{cc:cc} 1 & -2 & 1 & 0 \\ 0 & 1 & -2 & 1 \end{array}\right]$$

$$2R_2 + R_1 \rightarrow \left[\begin{array}{cc:cc} 1 & 0 & -3 & 2 \\ 0 & 1 & -2 & 1 \end{array}\right]$$

$$A^{-1} = \begin{bmatrix} -3 & 2 \\ -2 & 1 \end{bmatrix}$$

15. Find the inverse of the following matrix (if it exists).

$$\begin{bmatrix} 2 & 4 \\ 4 & 8 \end{bmatrix}$$

Solution:

$$\left[\begin{array}{cc:cc} 2 & 4 & 1 & 0 \\ 4 & 8 & 0 & 1 \end{array}\right] \quad \begin{array}{r} \frac{1}{2}R_1 \rightarrow \\ -4R_1 + R_2 \rightarrow \end{array} \left[\begin{array}{cc:cc} 1 & 2 & \frac{1}{2} & 0 \\ 0 & 0 & -2 & 1 \end{array}\right]$$

Since the left side does not reduce to I_2, the inverse does not exist.

19. Find the inverse of the following matrix (if it exists).

$$\begin{bmatrix} 1 & 1 & 1 \\ 3 & 5 & 4 \\ 3 & 6 & 5 \end{bmatrix}$$

–CONTINUED ON NEXT PAGE–

19. –CONTINUED–

Solution:

$$\left[\begin{array}{ccc:ccc} 1 & 1 & 1 & 1 & 0 & 0 \\ 3 & 5 & 4 & 0 & 1 & 0 \\ 3 & 6 & 5 & 0 & 0 & 1 \end{array}\right] \quad \begin{array}{r} \\ -3R_1 + R_2 \rightarrow \\ -3R_1 + R_3 \rightarrow \end{array} \left[\begin{array}{ccc:ccc} 1 & 1 & 1 & 1 & 0 & 0 \\ 0 & 2 & 1 & -3 & 1 & 0 \\ 0 & 3 & 2 & -3 & 0 & 1 \end{array}\right]$$

$$\begin{array}{r} -R_2 + R_1 \rightarrow \\ \frac{1}{2}R_2 \rightarrow \\ -3R_2 + R_3 \rightarrow \end{array} \left[\begin{array}{ccc:ccc} 1 & 0 & \frac{1}{2} & \frac{5}{2} & -\frac{1}{2} & 0 \\ 0 & 1 & \frac{1}{2} & -\frac{3}{2} & \frac{1}{2} & 0 \\ 0 & 0 & \frac{1}{2} & \frac{3}{2} & -\frac{3}{2} & 1 \end{array}\right]$$

$$\begin{array}{r} -R_3 + R_1 \rightarrow \\ -R_3 + R_2 \rightarrow \\ 2R_3 \rightarrow \end{array} \left[\begin{array}{ccc:ccc} 1 & 0 & 0 & 1 & 1 & -1 \\ 0 & 1 & 0 & -3 & 2 & -1 \\ 0 & 0 & 1 & 3 & -3 & 2 \end{array}\right]$$

$$A^{-1} = \begin{bmatrix} 1 & 1 & -1 \\ -3 & 2 & -1 \\ 3 & -3 & 2 \end{bmatrix}$$

23. Find the inverse of the following matrix (if it exists).

$$\begin{bmatrix} 0.1 & 0.2 & 0.3 \\ -0.3 & 0.2 & 0.2 \\ 0.5 & 0.4 & 0.4 \end{bmatrix}$$

Solution:

$$\left[\begin{array}{ccc:ccc} 0.1 & 0.2 & 0.3 & 1 & 0 & 0 \\ -0.3 & 0.2 & 0.2 & 0 & 1 & 0 \\ 0.5 & 0.4 & 0.4 & 0 & 0 & 1 \end{array}\right] \quad \begin{array}{r} 10R_1 \rightarrow \\ 10R_2 \rightarrow \\ 10R_3 \rightarrow \end{array} \left[\begin{array}{ccc:ccc} 1 & 2 & 3 & 10 & 0 & 0 \\ -3 & 2 & 2 & 0 & 10 & 0 \\ 5 & 4 & 4 & 0 & 0 & 10 \end{array}\right]$$

$$\begin{array}{r} \\ 3R_1 + R_2 \rightarrow \\ -5R_1 + R_3 \rightarrow \end{array} \left[\begin{array}{ccc:ccc} 1 & 2 & 3 & 10 & 0 & 0 \\ 0 & 8 & 11 & 30 & 10 & 0 \\ 0 & -6 & -11 & -50 & 0 & 10 \end{array}\right]$$

$$\begin{array}{r} \\ R_3 + R_2 \rightarrow \\ 3R_2 + R_3 \rightarrow \end{array} \left[\begin{array}{ccc:ccc} 1 & 2 & 3 & 10 & 0 & 0 \\ 0 & 2 & 0 & -20 & 10 & 10 \\ 0 & 0 & -11 & -110 & 30 & 40 \end{array}\right]$$

$$\begin{array}{r} -R_2 + R_1 \rightarrow \\ \frac{1}{2}R_2 \rightarrow \\ -\frac{1}{11}R_3 \rightarrow \end{array} \left[\begin{array}{ccc:ccc} 1 & 0 & 3 & 30 & -10 & -10 \\ 0 & 1 & 0 & -10 & 5 & 5 \\ 0 & 0 & 1 & 10 & -\frac{30}{11} & -\frac{40}{11} \end{array}\right]$$

$$\begin{array}{r} -3R_3 + R_1 \rightarrow \\ \\ \\ \end{array} \left[\begin{array}{ccc:ccc} 1 & 0 & 0 & 0 & -\frac{20}{11} & \frac{10}{11} \\ 0 & 1 & 0 & -10 & 5 & 5 \\ 0 & 0 & 1 & 10 & -\frac{30}{11} & -\frac{40}{11} \end{array}\right]$$

$$A^{-1} = \begin{bmatrix} 0 & -\frac{20}{11} & \frac{10}{11} \\ -10 & 5 & 5 \\ 10 & -\frac{30}{11} & -\frac{40}{11} \end{bmatrix} = \frac{5}{11}\begin{bmatrix} 0 & -4 & 2 \\ -22 & 11 & 11 \\ 22 & -6 & -8 \end{bmatrix}$$

25. Find the inverse of $\begin{bmatrix} 1 & 0 & 0 \\ 3 & 4 & 0 \\ 2 & 5 & 5 \end{bmatrix}$ (if it exists).

Solution:

$$\left[\begin{array}{ccc:ccc} 1 & 0 & 0 & 1 & 0 & 0 \\ 3 & 4 & 0 & 0 & 1 & 0 \\ 2 & 5 & 5 & 0 & 0 & 1 \end{array}\right] \begin{array}{r} \\ -3R_1 + R_2 \rightarrow \\ -2R_1 + R_3 \rightarrow \end{array} \left[\begin{array}{ccc:ccc} 1 & 0 & 0 & 1 & 0 & 0 \\ 0 & 4 & 0 & -3 & 1 & 0 \\ 0 & 5 & 5 & -2 & 0 & 1 \end{array}\right]$$

$$\begin{array}{r} \\ \frac{1}{4}R_2 \rightarrow \\ -5R_2 + R_3 \rightarrow \end{array} \left[\begin{array}{ccc:ccc} 1 & 0 & 0 & 1 & 0 & 0 \\ 0 & 1 & 0 & -0.75 & 0.25 & 0 \\ 0 & 0 & 5 & 1.75 & -1.25 & 1 \end{array}\right]$$

$$\begin{array}{r} \\ \\ \frac{1}{5}R_3 \rightarrow \end{array} \left[\begin{array}{ccc:ccc} 1 & 0 & 0 & 1 & 0 & 0 \\ 0 & 1 & 0 & -0.75 & 0.25 & 0 \\ 0 & 0 & 1 & 0.35 & -0.25 & 0.2 \end{array}\right]$$

$$A^{-1} = \begin{bmatrix} 1 & 0 & 0 \\ -0.75 & 0.25 & 0 \\ 0.35 & -0.25 & 0.2 \end{bmatrix}$$

29. Find the inverse of the following matrix.

$$\begin{bmatrix} 1 & -2 & -1 & -2 \\ 3 & -5 & -2 & -3 \\ 2 & -5 & -2 & -5 \\ -1 & 4 & 4 & 11 \end{bmatrix}$$

Solution:

$$\left[\begin{array}{cccc:cccc} 1 & -2 & -1 & -2 & 1 & 0 & 0 & 0 \\ 3 & -5 & -2 & -3 & 0 & 1 & 0 & 0 \\ 2 & -5 & -2 & -5 & 0 & 0 & 1 & 0 \\ -1 & 4 & 4 & 11 & 0 & 0 & 0 & 1 \end{array}\right] \begin{array}{r} \\ -3R_1 + R_2 \rightarrow \\ -2R_1 + R_3 \rightarrow \\ R_1 + R_4 \rightarrow \end{array} \left[\begin{array}{cccc:cccc} 1 & -2 & -1 & -2 & 1 & 0 & 0 & 0 \\ 0 & 1 & 1 & 3 & -3 & 1 & 0 & 0 \\ 0 & -1 & 0 & -1 & -2 & 0 & 1 & 0 \\ 0 & 2 & 3 & 9 & 1 & 0 & 0 & 1 \end{array}\right]$$

$$\begin{array}{r} 2R_2 + R_1 \rightarrow \\ \\ R_2 + R_3 \rightarrow \\ -2R_2 + R_4 \rightarrow \end{array} \left[\begin{array}{cccc:cccc} 1 & 0 & 1 & 4 & -5 & 2 & 0 & 0 \\ 0 & 1 & 1 & 3 & -3 & 1 & 0 & 0 \\ 0 & 0 & 1 & 2 & -5 & 1 & 1 & 0 \\ 0 & 0 & 1 & 3 & 7 & -2 & 0 & 1 \end{array}\right]$$

$$\begin{array}{r} -R_3 + R_1 \rightarrow \\ -R_3 + R_2 \rightarrow \\ \\ -R_3 + R_4 \rightarrow \end{array} \left[\begin{array}{cccc:cccc} 1 & 0 & 0 & 2 & 0 & 1 & -1 & 0 \\ 0 & 1 & 0 & 1 & 2 & 0 & -1 & 0 \\ 0 & 0 & 1 & 2 & -5 & 1 & 1 & 0 \\ 0 & 0 & 0 & 1 & 12 & -3 & -1 & 1 \end{array}\right]$$

$$\begin{array}{r} -2R_4 + R_1 \rightarrow \\ -R_4 + R_2 \rightarrow \\ -2R_4 + R_3 \rightarrow \\ \\ \end{array} \left[\begin{array}{cccc:cccc} 1 & 0 & 0 & 0 & -24 & 7 & 1 & -2 \\ 0 & 1 & 0 & 0 & -10 & 3 & 0 & -1 \\ 0 & 0 & 1 & 0 & -29 & 7 & 3 & -2 \\ 0 & 0 & 0 & 1 & 12 & -3 & -1 & 1 \end{array}\right]$$

$$A^{-1} = \begin{bmatrix} -24 & 7 & 1 & -2 \\ -10 & 3 & 0 & -1 \\ -29 & 7 & 3 & -2 \\ 12 & -3 & -1 & 1 \end{bmatrix}$$

33. Use an inverse matrix to solve the given system of linear equations.

$$x - 2y = 4$$
$$2x - 3y = 2$$

Solution:

$$A = \begin{bmatrix} 1 & -2 \\ 2 & -3 \end{bmatrix}$$

From Exercise 11 we have

$$A^{-1} = \begin{bmatrix} -3 & 2 \\ -2 & 1 \end{bmatrix}.$$

$$X = A^{-1}B = \begin{bmatrix} -3 & 2 \\ -2 & 1 \end{bmatrix} \begin{bmatrix} 4 \\ 2 \end{bmatrix}$$
$$= \begin{bmatrix} -8 \\ -6 \end{bmatrix}$$

Answer: $(-8, -6)$

37. Use an inverse matrix to solve the given system of linear equations.

$$-x + y = 20$$
$$-2x + y = 10$$

Solution:

$$A = \begin{bmatrix} -1 & 1 \\ -2 & 1 \end{bmatrix}$$

From Exercise 13 we have

$$A^{-1} = \begin{bmatrix} 1 & -1 \\ 2 & -1 \end{bmatrix}.$$

$$X = A^{-1}B = \begin{bmatrix} 1 & -1 \\ 2 & -1 \end{bmatrix} \begin{bmatrix} 20 \\ 10 \end{bmatrix}$$
$$= \begin{bmatrix} 10 \\ 30 \end{bmatrix}$$

Answer: $(10, 30)$.

41. Use an inverse matrix to solve the following system.

$$\begin{aligned} x_1 - 2x_2 - x_3 - 2x_4 &= 0 \\ 3x_1 - 5x_2 - 2x_3 - 3x_4 &= 1 \\ 2x_1 - 5x_2 - 2x_3 - 5x_4 &= -1 \\ -x_1 + 4x_2 + 4x_3 + 11x_4 &= 2 \end{aligned}$$

Solution:

$$A = \begin{bmatrix} 1 & -2 & -1 & -2 \\ 3 & -5 & -2 & -3 \\ 2 & -5 & -2 & -5 \\ -1 & 4 & 4 & 11 \end{bmatrix}$$

From Exercise 29 we have

$$A^{-1} = \begin{bmatrix} -24 & 7 & 1 & -2 \\ -10 & 3 & 0 & -1 \\ -29 & 7 & 3 & -2 \\ 12 & -3 & -1 & 1 \end{bmatrix}.$$

$$X = A^{-1}B = \begin{bmatrix} -24 & 7 & 1 & -2 \\ -10 & 3 & 0 & -1 \\ -29 & 7 & 3 & -2 \\ 12 & -3 & -1 & 1 \end{bmatrix} \begin{bmatrix} 0 \\ 1 \\ -1 \\ 2 \end{bmatrix} = \begin{bmatrix} 2 \\ 1 \\ 0 \\ 0 \end{bmatrix}$$

Answer: $(2, 1, 0, 0)$

45. Consider a person who invests in AAA-rated bonds, A-rated bonds, and B-rated bonds. The average yield is 6.5% on AAA-bonds, 7% on A-bonds, and 9% on B-bonds. Suppose the person invests twice as much in B-bonds as in A-bonds. A system of linear equations (where x, y, and z represent the amounts invested in AAA-, A-, and B-bonds, respectively) is as follows.

$$\begin{aligned} x + \quad y + \quad z &= \text{(total investment)} \\ 0.065x + 0.07y + 0.09z &= \text{(annual return)} \\ 2y - \quad z &= 0 \end{aligned}$$

Use the inverse of the coefficient matrix of this system to find the amount invested in each type of bond if the total investment is \$12,000, and the annual return is \$835.

Solution:

$$A = \begin{bmatrix} 1 & 1 & 1 \\ 0.065 & 0.07 & 0.09 \\ 0 & 2 & -1 \end{bmatrix}$$

Using the methods of this section, we have

$$A^{-1} = \frac{1}{11}\begin{bmatrix} 50 & -600 & -4 \\ -13 & 200 & 5 \\ -26 & 400 & -1 \end{bmatrix}.$$

$$X = A^{-1}B = \frac{1}{11}\begin{bmatrix} 50 & -600 & -4 \\ -13 & 200 & 5 \\ -26 & 400 & -1 \end{bmatrix}\begin{bmatrix} 12{,}000 \\ 835 \\ 0 \end{bmatrix} = \begin{bmatrix} 9000 \\ 1000 \\ 2000 \end{bmatrix}$$

Answer: \$9000 in AAA-bonds, \$1000 in A-bonds, \$2000 in B-bonds

SECTION 9.4

The Determinant of a Square Matrix

- You should be able to determine the determinant of a matrix of order 2 or of order 3 by using the products of the diagonals.
- You should be able to use expansion by cofactors to find the determinant of a matrix of order 3 or greater.
- The determinant of a triangular matrix equals the product of the entries on the main diagonal.

Solutions to Selected Exercises

7. Find the determinant of $\begin{bmatrix} -7 & 6 \\ \frac{1}{2} & 3 \end{bmatrix}$.

Solution:

$$\begin{vmatrix} -7 & 6 \\ \frac{1}{2} & 3 \end{vmatrix} = -7(3) - 6(\tfrac{1}{2}) = -24$$

11. Find the determinant of $\begin{bmatrix} 2 & -1 & 0 \\ 4 & 2 & 1 \\ 4 & 2 & 1 \end{bmatrix}$.

Solution:

$$\left|\begin{matrix} 2 & -1 & 0 \\ 4 & 2 & 1 \\ 4 & 2 & 1 \end{matrix}\right|\begin{matrix} 2 & -1 \\ 4 & 2 \\ 4 & 2 \end{matrix} = 4 + (-4) + 0 - 0 - 4 - (-4) = 0$$

15. Find the determinant of $\begin{bmatrix} 1 & 4 & -2 \\ 3 & 6 & -6 \\ -2 & 1 & 4 \end{bmatrix}$.

Solution:

$$\left|\begin{matrix} 1 & 4 & -2 \\ 3 & 6 & -6 \\ -2 & 1 & 4 \end{matrix}\right|\begin{matrix} 1 & 4 \\ 3 & 6 \\ -2 & 1 \end{matrix} = 24 + 48 + (-6) - 24 - (-6) - 48 = 0$$

19. Find the determinant of

$$\begin{bmatrix} -1 & 2 & -5 \\ 0 & 3 & 4 \\ 0 & 0 & 3 \end{bmatrix}$$

Solution:

$$\begin{vmatrix} -1 & 2 & -5 \\ 0 & 3 & 4 \\ 0 & 0 & 3 \end{vmatrix} = (-1)(3)(3) = -9 \quad \text{(Upper Triangular)}$$

23. Find the determinant of

$$\begin{bmatrix} x & y & 1 \\ -2 & -2 & 1 \\ 1 & 5 & 1 \end{bmatrix}.$$

Solution:

$$\begin{vmatrix} x & y & 1 \\ -2 & -2 & 1 \\ 1 & 5 & 1 \end{vmatrix} \begin{matrix} x & y \\ -2 & -2 \\ 1 & 5 \end{matrix} = -2x + y + (-10) - (-2) - 5x - (-2y) = -7x + 3y - 8$$

25. Find (a) all minors, and (b) cofactors for

$$\begin{bmatrix} 3 & 4 \\ 2 & -5 \end{bmatrix}.$$

Solution:

$$\begin{bmatrix} 3 & 4 \\ 2 & -5 \end{bmatrix}$$

(a) $M_{11} = -5$

$M_{12} = 2$

$M_{21} = 4$

$M_{22} = 3$

(b) $C_{11} = M_{11} = -5$

$C_{12} = -M_{12} = -2$

$C_{21} = -M_{21} = -4$

$C_{22} = M_{22} = 3$

29. Find the determinant of the following matrix by the method of expansion by cofactors. Expand using (a) Row 1 and (b) Column 2.

$$\begin{bmatrix} -3 & 2 & 1 \\ 4 & 5 & 6 \\ 2 & -3 & 1 \end{bmatrix}$$

Solution:

(a) Expansion along the first row:

$$\begin{vmatrix} -3 & 2 & 1 \\ 4 & 5 & 6 \\ 2 & -3 & 1 \end{vmatrix} = -3C_{11} + 2C_{12} + 1C_{13}$$
$$= -3(23) + 2(8) + 1(-22) = -69 + 16 - 22 = -75$$

(b) Expansion along the second column:

$$\begin{vmatrix} -3 & 2 & 1 \\ 4 & 5 & 6 \\ 2 & -3 & 1 \end{vmatrix} = 2C_{12} + 5C_{22} - 3C_{32}$$
$$= 2(8) + 5(-5) - 3(22) = 16 - 25 - 66 = -75$$

35. Find the determinant of

$$\begin{bmatrix} 1 & 4 & -2 \\ 3 & 2 & 0 \\ -1 & 4 & 3 \end{bmatrix}.$$

Solution:

Expansion along the third column:

$$\begin{vmatrix} 1 & 4 & -2 \\ 3 & 2 & 0 \\ -1 & 4 & 3 \end{vmatrix} = -2\begin{vmatrix} 3 & 2 \\ -1 & 4 \end{vmatrix} + 0 + 3\begin{vmatrix} 1 & 4 \\ 3 & 2 \end{vmatrix}$$
$$= -2(14) + 3(-10) = -28 - 30 = -58$$

39. Find the determinant of

$$\begin{bmatrix} 3 & 6 & -5 & 4 \\ -2 & 0 & 6 & 0 \\ 1 & 1 & 2 & 2 \\ 0 & 3 & -1 & -1 \end{bmatrix}.$$

Solution:

Expansion along the second row:

$$\begin{vmatrix} 3 & 6 & -5 & 4 \\ -2 & 0 & 6 & 0 \\ 1 & 1 & 2 & 2 \\ 0 & 3 & -1 & -1 \end{vmatrix} = -(-2)\begin{vmatrix} 6 & -5 & 4 \\ 1 & 2 & 2 \\ 3 & -1 & -1 \end{vmatrix} - 6\begin{vmatrix} 3 & 6 & 4 \\ 1 & 1 & 2 \\ 0 & 3 & -1 \end{vmatrix}$$
$$= 2[6(0) - 1(9) + 3(-18)] - 6[3(-7) - (-18)]$$
$$= 2[-9 - 54] - 6[-21 + 18] = -126 + 18 = -108$$

43. Find the determinant of

$$\begin{bmatrix} 3 & 2 & 4 & -1 & 5 \\ -2 & 0 & 1 & 3 & 2 \\ 1 & 0 & 0 & 4 & 0 \\ 6 & 0 & 2 & -1 & 0 \\ 3 & 0 & 5 & 1 & 0 \end{bmatrix}.$$

Solution:

Expansion along the second column:

$$\begin{vmatrix} 3 & 2 & 4 & -1 & 5 \\ -2 & 0 & 1 & 3 & 2 \\ 1 & 0 & 0 & 4 & 0 \\ 6 & 0 & 2 & -1 & 0 \\ 3 & 0 & 5 & 1 & 0 \end{vmatrix} = -2\begin{vmatrix} -2 & 1 & 3 & 2 \\ 1 & 0 & 4 & 0 \\ 6 & 2 & -1 & 0 \\ 3 & 5 & 1 & 0 \end{vmatrix}$$

$$= -2(-2)\begin{vmatrix} 1 & 0 & 4 \\ 6 & 2 & -1 \\ 3 & 5 & 1 \end{vmatrix} \quad \text{Expansion along fourth column}$$

$$= 4[1(7) - 0 + 4(24)] = 4[7 + 96] = 412$$

45. Solve for x.

$$\begin{vmatrix} x-1 & 2 \\ 3 & x-2 \end{vmatrix} = 0$$

Solution:

$$\begin{vmatrix} x-1 & 2 \\ 3 & x-2 \end{vmatrix} = 0$$

$$(x-1)(x-2) - 6 = 0$$

$$x^2 - 3x - 4 = 0$$

$$(x+1)(x-4) = 0$$

$$x = -1 \quad \text{or} \quad x = 4$$

47. Evaluate the determinant of $\begin{bmatrix} 4u & -1 \\ -1 & 2v \end{bmatrix}$. Determinants of this type occur in calculus.

Solution:

$$\begin{vmatrix} 4u & -1 \\ -1 & 2v \end{vmatrix} = 8uv - 1$$

49. Evaluate the determinant of $\begin{bmatrix} e^{2x} & e^{3x} \\ 2e^{2x} & 3e^{3x} \end{bmatrix}$. Determinants of this type occur in calculus.

Solution:

$$\begin{vmatrix} e^{2x} & e^{3x} \\ 2e^{2x} & 3e^{3x} \end{vmatrix} = 3e^{5x} - 2e^{5x} = e^{5x}$$

SECTION 9.5

Properties of Determinants

- You should know what effect each elementary row (column) operation has on the determinant of a matrix.
- You should know what conditions yield a determinant of zero.
- You should be able to use determinants to determine if a matrix has an inverse.

Solutions to Selected Exercises

5. State the property of determinants that verifies the equation.

$$\begin{vmatrix} 1 & 3 & 4 \\ -7 & 2 & -5 \\ 6 & 1 & 2 \end{vmatrix} = - \begin{vmatrix} 1 & 4 & 3 \\ -7 & -5 & 2 \\ 6 & 2 & 1 \end{vmatrix}$$

Solution:

Interchanging Columns 2 and 3 results in a change of sign of the determinant.

9. State the property of determinants that verifies the equation.

$$\begin{vmatrix} 5 & 0 & 10 \\ 25 & -30 & 40 \\ -15 & 5 & 20 \end{vmatrix} = 5^3 \begin{vmatrix} 1 & 0 & 2 \\ 5 & -6 & 8 \\ -3 & 1 & 4 \end{vmatrix}$$

Solution:

Multiplying the entries of all three rows by 5 produces a determinant that is 5^3 the determinant of the second matrix.

13. State the property of determinants that verifies the equation.

$$\begin{vmatrix} 3 & 2 & 4 \\ -2 & 1 & 5 \\ 5 & -7 & -20 \end{vmatrix} = \begin{vmatrix} 7 & 2 & -6 \\ 0 & 1 & 0 \\ -9 & -7 & 15 \end{vmatrix}$$

Solution:

Adding multiples of Column 2 to Columns 1 and 3 leaves the determinant unchanged.

19. Use elementary row (or column) operations as aids for evaluating

$$\begin{vmatrix} 3 & 8 & -7 \\ 0 & -5 & 4 \\ 8 & 1 & 6 \end{vmatrix}.$$

Solution:

$$\begin{vmatrix} 3 & 8 & -7 \\ 0 & -5 & 4 \\ 8 & 1 & 6 \end{vmatrix} = 3(-34) + 8(-3) = -126$$

Expansion along Column 1

23. Use elementary row (or column) operations as aids for evaluating

$$\begin{vmatrix} 7 & 0 & -14 \\ -2 & 5 & 4 \\ -6 & 2 & 12 \end{vmatrix}.$$

Solution:

$$\begin{vmatrix} 7 & 0 & -14 \\ -2 & 5 & 4 \\ -6 & 2 & 12 \end{vmatrix} = \begin{vmatrix} 7 & 0 & 0 \\ -2 & 5 & 0 \\ 6 & 2 & 0 \end{vmatrix} = 0$$

29. Use elementary row (or column) operations as aids for evaluating

$$\begin{vmatrix} 3 & -2 & 4 & 3 & 1 \\ -1 & 0 & 2 & 1 & 0 \\ 5 & -1 & 0 & 3 & 2 \\ 4 & 7 & -8 & 0 & 0 \\ 1 & 2 & 3 & 0 & 2 \end{vmatrix}.$$

–CONTINUED ON NEXT PAGE–

29. –CONTINUED–

Solution:

$$\begin{vmatrix} 3 & -2 & 4 & 3 & 1 \\ -1 & 0 & 2 & 1 & 0 \\ 5 & -1 & 0 & 3 & 2 \\ 4 & 7 & -8 & 0 & 0 \\ 1 & 2 & 3 & 0 & 2 \end{vmatrix} = \begin{vmatrix} 3 & -2 & 4 & 3 & 1 \\ -1 & 0 & 2 & 1 & 0 \\ -1 & 3 & -8 & -3 & 0 \\ 4 & 7 & -8 & 0 & 0 \\ -4 & 3 & 3 & -3 & 0 \end{vmatrix} \begin{matrix} \\ \\ \leftarrow -2R_1 + R_3 \\ \\ \leftarrow -R_3 + R_5 \end{matrix}$$

$$= \begin{vmatrix} -1 & 0 & 2 & 1 \\ -1 & 3 & -8 & -3 \\ 4 & 7 & -8 & 0 \\ -4 & 3 & 3 & -3 \end{vmatrix} \quad \text{(Expansion along Column 5)}$$

$$= \begin{vmatrix} -1 & 0 & 2 & 1 \\ -4 & 3 & -2 & 0 \\ 4 & 7 & -8 & 0 \\ -3 & 0 & 11 & 0 \end{vmatrix} \begin{matrix} \\ \leftarrow 3R_1 + R_2 \\ \\ \leftarrow -R_2 + R_4 \end{matrix}$$

$$= -\begin{vmatrix} -4 & 3 & -2 \\ 4 & 7 & -8 \\ -3 & 0 & 11 \end{vmatrix} \quad \text{(Expansion along Column 4)}$$

$$= -\begin{vmatrix} -4 & 3 & -2 \\ 0 & 10 & -10 \\ -3 & 0 & 11 \end{vmatrix} \begin{matrix} \\ \leftarrow R_1 + R_2 \\ \\ \end{matrix}$$

$$= -[-4(110) - 3(-10)] \quad \text{(Expansion along Column 1)}$$

$$= -[-440 + 30]$$

$$= 410$$

33. Use a determinant to determine whether the following matrix is invertible.

$$\begin{bmatrix} 14 & 7 & 0 \\ 2 & 3 & 0 \\ 1 & -5 & 2 \end{bmatrix}$$

Solution:

$$\begin{vmatrix} 14 & 7 & 0 \\ 2 & 3 & 0 \\ 1 & -5 & 2 \end{vmatrix} = 2\begin{vmatrix} 14 & 7 \\ 2 & 3 \end{vmatrix} = 2(42 - 14) = 56 \neq 0$$

The matrix *is* invertible.

37. Find the value(s) of k so that $\begin{bmatrix} k-1 & 3 \\ 2 & k-2 \end{bmatrix}$ is singular.

Solution:

$$\begin{vmatrix} k-1 & 3 \\ 2 & k-2 \end{vmatrix} = 0$$

$$(k-1)(k-2) - 6 = 0$$

$$k^2 - 3k - 4 = 0$$

$$(k+1)(k-4) = 0$$

$$k = -1 \text{ or } k = 4$$

39. Verify $\begin{vmatrix} w & x \\ y & z \end{vmatrix} = -\begin{vmatrix} y & z \\ w & x \end{vmatrix}$.

Solution:

$$\begin{vmatrix} w & x \\ y & z \end{vmatrix} = wz - xy$$

$$-\begin{vmatrix} y & z \\ w & x \end{vmatrix} = -[xy - wz] = wz - xy$$

Therefore, $\begin{vmatrix} w & x \\ y & z \end{vmatrix} = -\begin{vmatrix} y & z \\ w & x \end{vmatrix}$.

43. Show that $\begin{vmatrix} 1 & x & x^2 \\ 1 & y & y^2 \\ 1 & z & z^2 \end{vmatrix} = (y-x)(z-x)(z-y)$.

Solution:

$$\begin{aligned}
\begin{vmatrix} 1 & x & x^2 \\ 1 & y & y^2 \\ 1 & z & z^2 \end{vmatrix} &= \begin{vmatrix} y & y^2 \\ z & z^2 \end{vmatrix} - \begin{vmatrix} x & x^2 \\ z & z^2 \end{vmatrix} + \begin{vmatrix} x & x^2 \\ y & y^2 \end{vmatrix} \\
&= (yz^2 - y^2z) - (xz^2 - x^2z) + (xy^2 - x^2y) \\
&= yz^2 - xz^2 - y^2z + x^2z + xy(y-x) \\
&= z^2(y-x) - z(y^2 - x^2) + xy(y-x) \\
&= z^2(y-x) - z(y-x)(y+x) + xy(y-x) \\
&= (y-x)[z^2 - z(y+x) + xy] \\
&= (y-x)[z^2 - zy - zx + xy] \\
&= (y-x)[z^2 - zx - zy + xy] \\
&= (y-x)[z(z-x) - y(z-x)] \\
&= (y-x)(z-x)(z-y)
\end{aligned}$$

47. Find (a) $|A|$, (b) $|B|$, (c) AB, and (d) $|AB|$.

$$A = \begin{bmatrix} -1 & 2 & 1 \\ 1 & 0 & 1 \\ 0 & 1 & 0 \end{bmatrix}, \qquad B = \begin{bmatrix} -1 & 0 & 0 \\ 0 & 2 & 0 \\ 0 & 0 & 3 \end{bmatrix}$$

Solution:

(a) $$|A| = \begin{vmatrix} -1 & 2 & 1 \\ 1 & 0 & 1 \\ 0 & 1 & 0 \end{vmatrix} = \begin{vmatrix} -1 & 2 & 1 \\ 0 & 2 & 2 \\ 0 & 1 & 0 \end{vmatrix} \leftarrow R_1 + R_2$$

$$= -1\begin{vmatrix} 2 & 2 \\ 1 & 0 \end{vmatrix} = 2$$

(b) $$|B| = \begin{vmatrix} -1 & 0 & 0 \\ 0 & 2 & 0 \\ 0 & 0 & 3 \end{vmatrix} = (-1)(2)(3) = -6 \quad \text{(Triangular)}$$

(c) $$AB = \begin{bmatrix} -1 & 2 & 1 \\ 1 & 0 & 1 \\ 0 & 1 & 0 \end{bmatrix}\begin{bmatrix} -1 & 0 & 0 \\ 0 & 2 & 0 \\ 0 & 0 & 3 \end{bmatrix} = \begin{bmatrix} 1 & 4 & 3 \\ -1 & 0 & 3 \\ 0 & 2 & 0 \end{bmatrix}$$

(d) $$|AB| = \begin{vmatrix} 1 & 4 & 3 \\ -1 & 0 & 3 \\ 0 & 2 & 0 \end{vmatrix} = -2\begin{vmatrix} 1 & 3 \\ -1 & 3 \end{vmatrix} = -12 \quad \text{(Expansion along Row 3)}$$

49. Find square matrices A and B to demonstrate that $|A+B| \neq |A| + |B|$.

Solution:

Let $A = \begin{bmatrix} 1 & 0 \\ 0 & 1 \end{bmatrix}$ and $B = \begin{bmatrix} -1 & 0 \\ 0 & -1 \end{bmatrix}$. Then $A + B = \begin{bmatrix} 0 & 0 \\ 0 & 0 \end{bmatrix}$.

$$|A+B| = 0 \quad \text{and} \quad |A| + |B| = 1 + 1 = 2$$

Thus, $|A+B| \neq |A| + |B|$.

SECTION 9.6

Applications of Determinants and Matrices

- You should be able to use Cramer's Rule to solve a system of linear equations.
- Now you should be able to solve a system of linear equations by substitution, elimination, elementary row operations on an augmented matrix, using the inverse matrix, or Cramer's Rule.
- You should be able to find the area of a triangle in the xy-plane given the vertices.
- You should be able to use determinants to determine if three points are collinear.
- You should be able to use determinants to find the equation of a line through two distinct points.

Solutions to Selected Exercises

5. Use Cramer's Rule to solve the system of equations.

$$20x + 8y = 11$$
$$12x - 24y = 21$$

Solution:

$$x = \frac{\begin{vmatrix} 11 & 8 \\ 21 & -24 \end{vmatrix}}{\begin{vmatrix} 20 & 8 \\ 12 & -24 \end{vmatrix}} = \frac{-432}{-576} = \frac{3}{4}$$

$$y = \frac{\begin{vmatrix} 20 & 11 \\ 12 & 21 \end{vmatrix}}{\begin{vmatrix} 20 & 8 \\ 12 & -24 \end{vmatrix}} = \frac{288}{-576} = -\frac{1}{2}$$

Answer: $\left(\frac{3}{4}, -\frac{1}{2}\right)$

9. Use Cramer's Rule to solve the system of equations.

$$3x + 6y = 5$$
$$6x + 14y = 11$$

Solution:

$$x = \frac{\begin{vmatrix} 5 & 6 \\ 11 & 14 \end{vmatrix}}{\begin{vmatrix} 3 & 6 \\ 6 & 14 \end{vmatrix}} = \frac{4}{6} = \frac{2}{3}, \quad y = \frac{\begin{vmatrix} 3 & 5 \\ 6 & 11 \end{vmatrix}}{\begin{vmatrix} 3 & 6 \\ 6 & 14 \end{vmatrix}} = \frac{3}{6} = \frac{1}{2}$$

Answer: $\left(\frac{2}{3}, \frac{1}{2}\right)$

13. Use Cramer's Rule to solve the system of equations for x.

$$3x + 4y + 4z = 11$$
$$4x - 4y + 6z = 11$$
$$6x - 6y \qquad = 3$$

Solution:

$$x = \frac{\begin{vmatrix} 11 & 4 & 4 \\ 11 & -4 & 6 \\ 3 & -6 & 0 \end{vmatrix}}{\begin{vmatrix} 3 & 4 & 4 \\ 4 & -4 & 6 \\ 6 & -6 & 0 \end{vmatrix}} = \frac{252}{252} = 1$$

19. Use Cramer's Rule to solve the system of equations for x.

$$7x - 3y \qquad + 2w = 41$$
$$-2x + y \qquad - w = -13$$
$$4x \qquad + z - 2w = 12$$
$$-x + y \qquad - w = -8$$

Solution:

$$x = \frac{\begin{vmatrix} 41 & -3 & 0 & 2 \\ -13 & 1 & 0 & -1 \\ 12 & 0 & 1 & -2 \\ -8 & 1 & 0 & -1 \end{vmatrix}}{\begin{vmatrix} 7 & -3 & 0 & 2 \\ -2 & 1 & 0 & -1 \\ 4 & 0 & 1 & -2 \\ -1 & 1 & 0 & -1 \end{vmatrix}} = \frac{\begin{vmatrix} 41 & -3 & 2 \\ -13 & 1 & -1 \\ -8 & 1 & -1 \end{vmatrix}}{\begin{vmatrix} 7 & -3 & 2 \\ -2 & 1 & -1 \\ -1 & 1 & -1 \end{vmatrix}} = \frac{5}{1} = 5 \quad \text{(Expansion along Column 3)}$$

23. The maximum Social Security contributions for an employee between 1981 and 1989 are shown in the accompanying figure. (The figure shows the amount contributed by the *employee.* This amount is matched by the employer.) The least squares regression line $y = a + bt$ for this data is found by solving the system

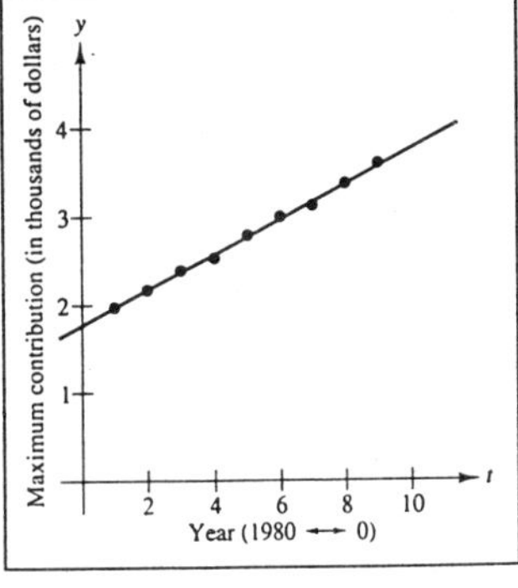

$$9a + 45b = 24.983$$
$$45a + 285b = 137.012$$

where y is the contribution in 1000s of dollars and t is the calendar year with $t = 1$ corresponding to 1981. Use Cramer's Rule to solve this system, and use the result to approximate the maximum Social Security contribution in 1992. (*Source:* U.S. Social Security Administration)

Solution:

$$a = \frac{\begin{vmatrix} 24.983 & 45 \\ 137.012 & 285 \end{vmatrix}}{\begin{vmatrix} 9 & 45 \\ 45 & 285 \end{vmatrix}} = \frac{954.615}{540} \approx 1.768$$

Using back-substitution in the first equation we have $b \approx 0.202$. Thus, $y \approx 1.768 + 0.202t$. When $t = 12$, $y \approx 4.2$ thousand dollars.

27. Use a determinant to find the area of the triangle with vertices $(-2, -3)$, $(2, -3)$, and $(0, 4)$.

Solution:

$$A = \frac{1}{2}\begin{vmatrix} -2 & -3 & 1 \\ 2 & -3 & 1 \\ 0 & 4 & 1 \end{vmatrix} = 14 \text{ square units}$$

31. Use a determinant to find the area of the triangle with vertices $(-2, 4)$, $(2, 3)$, $(-1, 5)$.

Solution:

$$A = \frac{1}{2}\begin{vmatrix} -2 & 4 & 1 \\ 2 & 3 & 1 \\ -1 & 5 & 1 \end{vmatrix} = \frac{5}{2} \text{ square units}$$

35. A large region of forest has been infested with gypsy moths. The region is roughly triangular, as shown in the figure. From the northernmost vertex A of the region, the distance to Vertex B is 25 miles south and 10 miles east, and the distance to vertex C is 20 miles south and 28 miles east. Approximate the number of square miles in this region.

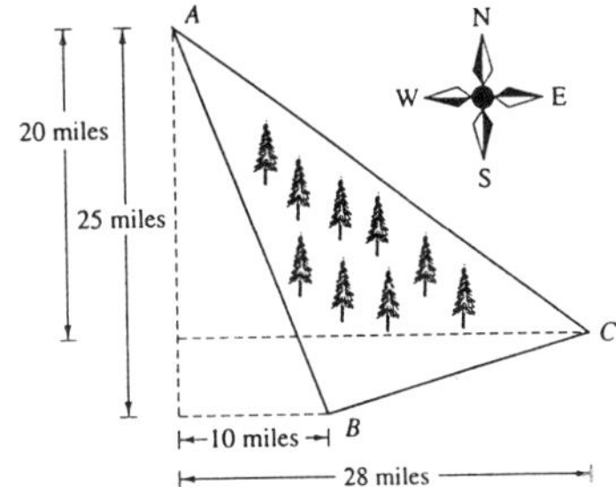

Solution: Vertices: $(0, 25)$, $(10, 0)$, $(28, 5)$

$$A = \frac{1}{2}\begin{vmatrix} 0 & 25 & 1 \\ 10 & 0 & 1 \\ 28 & 5 & 1 \end{vmatrix} = \frac{500}{2} = 250 \text{ square miles}$$

39. Use a determinant to determine if the points $(2, -\frac{1}{2})$, $(-4, 4)$, and $(6, -3)$ are collinear.

Solution:

The points are *not* collinear since $\begin{vmatrix} 2 & -\frac{1}{2} & 1 \\ -4 & 4 & 1 \\ 6 & -3 & 1 \end{vmatrix} = -3 \neq 0.$

45. Use a determinant to find an equation of the line through the points $(-4, 3)$ and $(2, 1)$.

Solution:

$$\begin{vmatrix} x & y & 1 \\ -4 & 3 & 1 \\ 2 & 1 & 1 \end{vmatrix} = 0$$

$$2x + 6y - 10 = 0$$

$$x + 3y - 5 = 0$$

49. Write a cryptogram for LANDING SUCCESSFUL using the following matrix.

$$A = \begin{bmatrix} 1 & 2 & 2 \\ 3 & 7 & 9 \\ -1 & -4 & -7 \end{bmatrix}$$

Solution:

L A N D I N G _ S U C C E S S F U L

[12 1 14] [4 9 14] [7 0 19] [21 3 3] [5 19 19] [6 21 12]

$$[12 \quad 1 \quad 14]A = [1 \quad -25 \quad -65]$$

$$[4 \quad 9 \quad 14]A = [17 \quad 15 \quad -9]$$

$$[7 \quad 0 \quad 19]A = [-12 \quad -62 \quad -119]$$

$$[21 \quad 3 \quad 3]A = [27 \quad 51 \quad 48]$$

$$[5 \quad 19 \quad 19]A = [43 \quad 67 \quad 48]$$

$$[6 \quad 21 \quad 12]A = [57 \quad 111 \quad 117]$$

Cryptogram: 1 −25 −65 17 15 −9 −12 −62 −119 27 51 48 43 67 48 57 111 117

53. Decode the cryptogram

$$20 \quad 17 \quad -15 \quad -12 \quad -56 \quad -104 \quad 1 \quad -25 \quad -65 \quad 62 \quad 143 \quad 181$$

using the inverse of the following matrix.

$$A = \begin{bmatrix} 1 & 2 & 2 \\ 3 & 7 & 9 \\ -1 & -4 & -7 \end{bmatrix}$$

Solution:

To find A^{-1}, use Gauss-Jordan elimination.

$$\left[\begin{array}{ccc:ccc} 1 & 2 & 2 & 1 & 0 & 0 \\ 3 & 7 & 9 & 0 & 1 & 0 \\ -1 & -4 & -7 & 0 & 0 & 1 \end{array}\right] \rightarrow \left[\begin{array}{ccc:ccc} 1 & 0 & 0 & -13 & 6 & 4 \\ 0 & 1 & 0 & 12 & -5 & -3 \\ 0 & 0 & 1 & -5 & 2 & 1 \end{array}\right]$$

$$A^{-1} = \begin{bmatrix} -13 & 6 & 4 \\ 12 & -5 & -3 \\ -5 & 2 & 1 \end{bmatrix}$$

$$[20 \quad 17 \quad -15]\,A^{-1} = [19 \quad 5 \quad 14]$$

$$[-12 \quad -56 \quad -104]\,A^{-1} = [4 \quad 0 \quad 16]$$

$$[1 \quad -25 \quad -65]\,A^{-1} = [12 \quad 1 \quad 14]$$

$$[62 \quad 143 \quad 181]\,A^{-1} = [5 \quad 19 \quad 0]$$

19	5	14	4	0	16	12	1	14	5	19	0
S	E	N	D	_	P	L	A	N	E	S	_

REVIEW EXERCISES FOR CHAPTER 9

Solutions to Selected Exercises

3. Use matrices and elementary row operations to solve the system of equations.

$$0.2x - 0.1y = 0.07$$

$$0.4x - 0.5y = -0.01$$

Solution:

$$\left[\begin{array}{cc:c} 0.2 & -0.1 & 0.07 \\ 0.4 & -0.5 & -0.01 \end{array}\right] \quad \begin{array}{r} 5R_1 \rightarrow \\ -2R_1 + R_2 \rightarrow \end{array} \left[\begin{array}{cc:c} 1 & -0.5 & 0.35 \\ 0 & -0.3 & -0.15 \end{array}\right]$$

$$\begin{array}{r} 0.5R_2 + R_1 \rightarrow \\ -\frac{1}{0.3}R_2 \rightarrow \end{array} \left[\begin{array}{cc:c} 1 & 0 & 0.6 \\ 0 & 1 & 0.5 \end{array}\right]$$

$$x = 0.6$$

$$y = 0.5$$

Answer: $(0.6, 0.5)$

7. Use matrices and elementary row operations to solve the system of equations.

$$2x + 3y + 3z = 3$$

$$6x + 6y + 12z = 13$$

$$12x + 9y - z = 2$$

Solution:

$$\left[\begin{array}{ccc:c} 2 & 3 & 3 & 3 \\ 6 & 6 & 12 & 13 \\ 12 & 9 & -1 & 2 \end{array}\right] \quad \begin{array}{r} \\ -3R_1 + R_2 \rightarrow \\ -2R_2 + R_3 \rightarrow \end{array} \left[\begin{array}{ccc:c} 2 & 3 & 3 & 3 \\ 0 & -3 & 3 & 4 \\ 0 & -3 & -25 & -24 \end{array}\right]$$

$$\begin{array}{r} R_2 + R_1 \rightarrow \\ \\ -R_2 + R_3 \rightarrow \end{array} \left[\begin{array}{ccc:c} 2 & 0 & 6 & 7 \\ 0 & -3 & 3 & 4 \\ 0 & 0 & -28 & -28 \end{array}\right]$$

$$\begin{array}{r} \frac{1}{2}R_1 \rightarrow \\ -\frac{1}{3}R_2 \rightarrow \\ -\frac{1}{28}R_3 \rightarrow \end{array} \left[\begin{array}{ccc:c} 1 & 0 & 3 & \frac{7}{2} \\ 0 & 1 & -1 & -\frac{4}{3} \\ 0 & 0 & 1 & 1 \end{array}\right]$$

$$z = 1$$

$$x - 3z = \frac{7}{2} \Rightarrow x = \frac{1}{2}$$

$$y - z = -\frac{4}{3} \Rightarrow y = -\frac{1}{3}$$

Answer: $\left(\frac{1}{2}, -\frac{1}{3}, 1\right)$

11. Use matrices and elementary row operations to solve the system of equations.

$$\begin{aligned} x + 2y + \ 6z &= \ \ 1 \\ 2x + 5y + 15z &= \ \ 4 \\ 3x + \ \ y + \ 3z &= -6 \end{aligned}$$

Solution:

$$\left[\begin{array}{ccc|c} 1 & 2 & 6 & 1 \\ 2 & 5 & 15 & 4 \\ 3 & 1 & 3 & -6 \end{array}\right] \begin{array}{r} \\ -2R_1 + R_2 \rightarrow \\ -3R_1 + R_3 \rightarrow \end{array} \left[\begin{array}{ccc|c} 1 & 2 & 6 & 1 \\ 0 & 1 & 3 & 2 \\ 0 & -5 & -15 & -9 \end{array}\right]$$

$$\begin{array}{r} -2R_2 + R_1 \rightarrow \\ \\ 5R_2 + R_3 \rightarrow \end{array} \left[\begin{array}{ccc|c} 1 & 0 & 0 & -3 \\ 0 & 1 & 3 & 2 \\ 0 & 0 & 0 & 1 \end{array}\right]$$

$$x = -3$$

$$y + 3z = 2$$

$$0 = 1, \quad \text{Inconsistent, no solution}$$

15. Perform the indicated matrix operation.

$$\begin{bmatrix} 1 & 2 \\ 5 & -4 \\ 6 & 0 \end{bmatrix} \begin{bmatrix} 6 & -2 & 8 \\ 4 & 0 & 0 \end{bmatrix}$$

Solution:

$$\begin{bmatrix} 1 & 2 \\ 5 & -4 \\ 6 & 0 \end{bmatrix} \begin{bmatrix} 6 & -2 & 8 \\ 4 & 0 & 0 \end{bmatrix} = \begin{bmatrix} 1(6) + 2(4) & 1(-2) + 2(0) & 1(8) + 2(0) \\ 5(6) + (-4)(4) & 5(-2) + (-4)(0) & 5(8) + (-4)(0) \\ 6(6) + (0)(4) & 6(-2) + (0)(0) & 6(8) + (0)(0) \end{bmatrix}$$

$$= \begin{bmatrix} 14 & -2 & 8 \\ 14 & -10 & 40 \\ 36 & -12 & 48 \end{bmatrix}$$

19. Perform the indicated matrix operation.

$$\begin{bmatrix} 1 & 3 & 2 \\ 0 & 2 & -4 \\ 0 & 0 & 3 \end{bmatrix} \begin{bmatrix} 4 & -3 & 2 \\ 0 & 3 & -1 \\ 0 & 0 & 2 \end{bmatrix}$$

Solution:

$$\begin{bmatrix} 1 & 3 & 2 \\ 0 & 2 & -4 \\ 0 & 0 & 3 \end{bmatrix} \begin{bmatrix} 4 & -3 & 2 \\ 0 & 3 & -1 \\ 0 & 0 & 2 \end{bmatrix} = \begin{bmatrix} 1(4) & 1(-3) + 3(3) & 1(2) + 3(-1) + 2(2) \\ 0 & 2(3) & 2(-1) + (-4)(2) \\ 0 & 0 & 3(2) \end{bmatrix}$$

$$= \begin{bmatrix} 4 & 6 & 3 \\ 0 & 6 & -10 \\ 0 & 0 & 6 \end{bmatrix}$$

23. Solve for X in $3X + 2A = B$, given

$$A = \begin{bmatrix} -4 & 0 \\ 1 & -5 \\ -3 & 2 \end{bmatrix} \quad \text{and} \quad B = \begin{bmatrix} 1 & 2 \\ -2 & 1 \\ 4 & 4 \end{bmatrix}.$$

Solution:

$$X = \frac{1}{3}[B - 2A] = \frac{1}{3}\left(\begin{bmatrix} 1 & 2 \\ -2 & 1 \\ 4 & 4 \end{bmatrix} - 2\begin{bmatrix} -4 & 0 \\ 1 & -5 \\ -3 & 2 \end{bmatrix}\right) = \frac{1}{3}\begin{bmatrix} 9 & 2 \\ -4 & 11 \\ 10 & 0 \end{bmatrix}$$

25. Write the system of linear equations represented by the matrix equation.

$$\begin{bmatrix} 5 & 4 \\ -1 & 1 \end{bmatrix}\begin{bmatrix} x \\ y \end{bmatrix} = \begin{bmatrix} 2 \\ -22 \end{bmatrix}$$

Solution:

$$\begin{bmatrix} 5 & 4 \\ -1 & 1 \end{bmatrix}\begin{bmatrix} x \\ y \end{bmatrix} = \begin{bmatrix} 2 \\ -22 \end{bmatrix}$$

$$\begin{bmatrix} 5x + 4y \\ -x + y \end{bmatrix} = \begin{bmatrix} 2 \\ -22 \end{bmatrix}$$

$$5x + 4y = 2$$

$$-x + y = -22$$

29. Find the inverse of the following matrix.

$$\begin{bmatrix} 2 & 0 & 3 \\ -1 & 1 & 1 \\ 2 & -2 & 1 \end{bmatrix}$$

Solution:

$$\left[\begin{array}{ccc:ccc} 2 & 0 & 3 & 1 & 0 & 0 \\ -1 & 1 & 1 & 0 & 1 & 0 \\ 2 & -2 & 1 & 0 & 0 & 1 \end{array}\right] \begin{array}{r} R_2 + R_1 \to \\ R_1 + R_2 \to \\ -2R_1 + R_3 \to \end{array} \left[\begin{array}{ccc:ccc} 1 & 1 & 4 & 1 & 1 & 0 \\ 0 & 2 & 5 & 1 & 2 & 0 \\ 0 & -4 & -7 & -2 & -2 & 1 \end{array}\right]$$

$$\begin{array}{r} -R_2 + R_1 \to \\ \frac{1}{2}R_2 \to \\ 4R_2 + R_3 \to \end{array} \left[\begin{array}{ccc:ccc} 1 & 0 & \frac{3}{2} & \frac{1}{2} & 0 & 0 \\ 0 & 1 & \frac{5}{2} & \frac{1}{2} & 1 & 0 \\ 0 & 0 & 3 & 0 & 2 & 1 \end{array}\right]$$

$$\begin{array}{r} -\frac{3}{2}R_3 + R_1 \to \\ -\frac{5}{2}R_3 + R_2 \to \\ \frac{1}{3}R_3 \to \end{array} \left[\begin{array}{ccc:ccc} 1 & 0 & 0 & \frac{1}{2} & -1 & -\frac{1}{2} \\ 0 & 1 & 0 & \frac{1}{2} & -\frac{2}{3} & -\frac{5}{6} \\ 0 & 0 & 1 & 0 & \frac{2}{3} & \frac{1}{3} \end{array}\right]$$

Inverse: $\begin{bmatrix} \frac{1}{2} & -1 & -\frac{1}{2} \\ \frac{1}{2} & -\frac{2}{3} & -\frac{5}{6} \\ 0 & \frac{2}{3} & \frac{1}{3} \end{bmatrix}$

33. Evaluate the determinant.

$$\begin{vmatrix} 3 & 0 & -4 & 0 \\ 0 & 8 & 1 & 2 \\ 6 & 1 & 8 & 2 \\ 0 & 3 & -4 & 1 \end{vmatrix}$$

Solution:

$$\begin{vmatrix} 3 & 0 & -4 & 0 \\ 0 & 8 & 1 & 2 \\ 6 & 1 & 8 & 2 \\ 0 & 3 & -4 & 1 \end{vmatrix} = 3\begin{vmatrix} 8 & 1 & 2 \\ 1 & 8 & 2 \\ 3 & -4 & 1 \end{vmatrix} + (-4)\begin{vmatrix} 0 & 8 & 2 \\ 6 & 1 & 2 \\ 0 & 3 & 1 \end{vmatrix} \quad \text{(Expansion along Row 1)}$$

$$= 3[8(8-(-8)) - 1(1-6) + 2(-4-24)] - 4[0 - 6(8-6) + 0]$$

$$= 3[128 + 5 - 56] - 4[-12]$$

$$= 279$$

37. Solve the system of linear equations using (a) the inverse of the coefficient matrix and (b) Cramer's Rule.

$$\begin{aligned} -3x - 3y - 4z &= 2 \\ y + z &= -1 \\ 4x + 3y + 4z &= -1 \end{aligned}$$

Solution:

(a) $\left[\begin{array}{ccc|ccc} -3 & -3 & -4 & 1 & 0 & 0 \\ 0 & 1 & 1 & 0 & 1 & 0 \\ 4 & 3 & 4 & 0 & 0 & 1 \end{array}\right]$ $\begin{array}{r} R_3 + R_1 \rightarrow \\ \\ -4R_1 + R_3 \rightarrow \end{array} \left[\begin{array}{ccc|ccc} 1 & 0 & 0 & 1 & 0 & 1 \\ 0 & 1 & 1 & 0 & 1 & 0 \\ 0 & 3 & 4 & -4 & 0 & -3 \end{array}\right]$

$$\begin{array}{r} \\ -R_3 + R_2 \rightarrow \\ -3R_2 + R_3 \rightarrow \end{array} \left[\begin{array}{ccc|ccc} 1 & 0 & 0 & 1 & 0 & 1 \\ 0 & 1 & 0 & 4 & 4 & 3 \\ 0 & 0 & 1 & -4 & -3 & -3 \end{array}\right]$$

$$\begin{bmatrix} x \\ y \\ z \end{bmatrix} = \begin{bmatrix} 1 & 0 & 1 \\ 4 & 4 & 3 \\ -4 & -3 & -3 \end{bmatrix} \begin{bmatrix} 2 \\ -1 \\ -1 \end{bmatrix} = \begin{bmatrix} 1 \\ 1 \\ -2 \end{bmatrix}$$

Answer: $(1, 1, -2)$

–CONTINUED ON NEXT PAGE–

37. –CONTINUED–

$$\text{(b) } x = \frac{\begin{vmatrix} 2 & -3 & -4 \\ -1 & 1 & 1 \\ -1 & 3 & 4 \end{vmatrix}}{\begin{vmatrix} -3 & -3 & -4 \\ 0 & 1 & 1 \\ 4 & 3 & 4 \end{vmatrix}} = \frac{1}{1} = 1$$

$$y = \frac{\begin{vmatrix} -3 & 2 & -4 \\ 0 & -1 & 1 \\ 4 & -1 & 4 \end{vmatrix}}{\begin{vmatrix} -3 & -3 & -4 \\ 0 & 1 & 1 \\ 4 & 3 & 4 \end{vmatrix}} = \frac{1}{1} = 1$$

$$z = \frac{\begin{vmatrix} -3 & -3 & 2 \\ 0 & 1 & -1 \\ 4 & 3 & -1 \end{vmatrix}}{\begin{vmatrix} -3 & -3 & -4 \\ 0 & 1 & 1 \\ 4 & 3 & 4 \end{vmatrix}} = \frac{-2}{1} = -2$$

Answer: $(1, 1, -2)$

43. Use a determinant to find the area of the triangle with vertices (1, 0), (5, 0), and (5, 8).

Solution:

$$\text{Area} = \tfrac{1}{2}\begin{vmatrix} 1 & 0 & 1 \\ 5 & 0 & 1 \\ 5 & 8 & 1 \end{vmatrix} = \tfrac{1}{2}(32) = 16 \text{ square units}$$

47. Use a determinant to find the equation of the line through the points (−4, 0) and (4, 4).

Solution:

$$\begin{vmatrix} x & y & 1 \\ -4 & 0 & 1 \\ 4 & 4 & 1 \end{vmatrix} = 0$$

$$-4x + 8y - 16 = 0$$

$$x - 2y + 4 = 0$$

51. A florist wants to arrange a dozen flowers consisting of two varieties—carnations and roses. Carnations cost \$0.75 each and roses cost \$1.50 each. How many of each should the florist use in order for the arrangement to cost \$12.00?

Solution:

Let $x =$ the number of carnations, and $y =$ the number of roses. Then,

$$x + y = 12$$
$$0.75x + 1.50y = \$12.00.$$

By Cramer's Rule we have

$$x = \frac{\begin{vmatrix} 12 & 1 \\ 12 & 1.50 \end{vmatrix}}{\begin{vmatrix} 1 & 1 \\ 0.75 & 1.50 \end{vmatrix}} = \frac{6}{0.75} = 8$$

Using back-substitution in the first equation yields $y = 4$. The florist should use 8 carnations and 4 roses.

55. If A is a 3×3 matrix and $|A| = 2$, what is the value of $|4A|$? Give a reason for your answer.

Solution:

$|4A| = 4^3(2) = 128$, since $4A$ means that each of the three rows of A was multiplied by 4.

Practice Test for Chapter 9

1. Put the matrix in reduced echelon form.

$$\begin{bmatrix} 1 & -2 & 4 \\ 3 & -5 & 9 \end{bmatrix}$$

For Exercises 2–4, use matrices to solve the system of equations.

2. $\begin{aligned} 3x + 5y &= 3 \\ 2x - y &= -11 \end{aligned}$

3. $\begin{aligned} 2x + 3y &= -3 \\ 3x + 2y &= 8 \\ x + y &= 1 \end{aligned}$

4. $\begin{aligned} x + 3z &= -5 \\ 2x + y &= 0 \\ 3x + y - z &= 3 \end{aligned}$

5. Multiply $\begin{bmatrix} 1 & 4 & 5 \\ 2 & 0 & -3 \end{bmatrix} \begin{bmatrix} 1 & 6 \\ 0 & -7 \\ -1 & 2 \end{bmatrix}$.

6. Given $A = \begin{bmatrix} 9 & 1 \\ -4 & 8 \end{bmatrix}$ and $B = \begin{bmatrix} 6 & -2 \\ 3 & 5 \end{bmatrix}$, find $3A - 5B$.

7. Find $f(A)$:

$$f(x) = x^2 - 7x + 8, \quad A = \begin{bmatrix} 3 & 0 \\ 7 & 1 \end{bmatrix}.$$

8. True or false:

$(A + B)(A + 3B) = A^2 + 4AB + 3B^2$ where A and B are matrices.

(Assume that A^2, AB, and B^2 exist.)

For Exercises 9–10, find the inverse of the matrix, if it exists.

9. $\begin{bmatrix} 1 & 2 \\ 3 & 5 \end{bmatrix}$

10. $\begin{bmatrix} 1 & 1 & 1 \\ 3 & 6 & 5 \\ 6 & 10 & 8 \end{bmatrix}$

11. Use an inverse matrix to solve the systems.

(a) $\begin{aligned} x + 2y &= 4 \\ 3x + 5y &= 1 \end{aligned}$

(b) $\begin{aligned} x + 2y &= 3 \\ 3x + 5y &= -2 \end{aligned}$

For Exercises 12–14, find the determinant of the matrix.

12. $\begin{bmatrix} 6 & -1 \\ 3 & 4 \end{bmatrix}$

13. $\begin{bmatrix} 1 & 3 & -1 \\ 5 & 9 & 0 \\ 6 & 2 & -5 \end{bmatrix}$

14. $\begin{bmatrix} 1 & 4 & 2 & 3 \\ 0 & 1 & -2 & 0 \\ 3 & 5 & -1 & 1 \\ 2 & 0 & 6 & 1 \end{bmatrix}$

15. True or false:

$$\begin{vmatrix} 3 & 0 & 0 \\ 0 & 3 & 0 \\ 0 & 0 & 3 \end{vmatrix} = -3^3 \begin{vmatrix} 1 & 0 & 0 \\ 0 & 0 & 1 \\ 0 & 1 & 0 \end{vmatrix}$$

16. Evaluate:

$$\begin{bmatrix} 6 & 4 & 3 & 0 & 6 \\ 0 & 5 & 1 & 4 & 8 \\ 0 & 0 & 2 & 7 & 3 \\ 0 & 0 & 0 & 9 & 2 \\ 0 & 0 & 0 & 0 & 1 \end{bmatrix}$$

17. Use a determinant to find the area of the triangle with vertices (0, 7), (5, 0), and (3, 9).

For Exercises 18–20, use Cramer's Rule to find the indicated value.

18. Find x.

$$\begin{aligned} 6x - 7y &= 4 \\ 2x + 5y &= 11 \end{aligned}$$

19. Find z.

$$\begin{aligned} 3x \quad\quad + z &= 1 \\ y + 4z &= 3 \\ x - y \quad\quad &= 2 \end{aligned}$$

20. Find y.

$$\begin{aligned} 721.4x - 29.1y &= 33.77 \\ 45.9x + 105.6y &= 19.85 \end{aligned}$$

CHAPTER 10

Sequences, Counting Principles, and Probability

Section 10.1 Sequences and Summation Notation **409**

Section 10.2 Arithmetic Sequences **414**

Section 10.3 Geometric Sequences **418**

Section 10.4 Mathematical Induction **423**

Section 10.5 The Binomial Theorem **428**

Section 10.6 Counting Principles, Permutations, and Combinations **433**

Section 10.7 Probability **437**

Review Exercises **441**

Practice Test **445**

SECTION 10.1

Sequences and Summation Notation

- Given the general nth term in a sequence, you should be able to find, or list, some of the terms.
- You should be able to find an expression for the nth term of a sequence.
- You should be able to use and evaluate factorials.
- You should be able to use sigma notation for a sum.

Solutions to Selected Exercises

3. Write the first five terms of the following sequence. (Assume n begins with 1.)

$$a_n = 2^n$$

Solution:

$$a_n = 2^n$$

$$a_1 = 2^1 = 2$$

$$a_2 = 2^2 = 4$$

$$a_3 = 2^3 = 8$$

$$a_4 = 2^4 = 16$$

$$a_5 = 2^5 = 32$$

Terms: 2, 4, 8, 16, 32

7. Write the first five terms of the following sequence. (Assume n begins with 1.)

$$a_n = \frac{1+(-1)^n}{n}$$

Solution:

$$a_n = \frac{1+(-1)^n}{n}$$

$$a_1 = \frac{1+(-1)}{1} = \frac{0}{1} = 0$$

$$a_2 = \frac{1+(-1)^2}{2} = \frac{2}{2} = 1$$

$$a_3 = \frac{1+(-1)^3}{3} = \frac{0}{3} = 0$$

$$a_4 = \frac{1+(-1)^4}{4} = \frac{2}{4} = \frac{1}{2}$$

$$a_5 = \frac{1+(-1)^5}{5} = \frac{0}{5} = 0$$

Terms: 0, 1, 0, $\frac{1}{2}$, 0

13. Write the first five terms of the following sequence. (Assume n begins with 1.)

$$a_n = \frac{3^n}{n!}$$

Solution:

$$a_n = \frac{3^n}{n!}$$

$$a_1 = \frac{3^1}{1!} = 3$$

$$a_2 = \frac{3^2}{2!} = \frac{9}{2}$$

$$a_3 = \frac{3^3}{3!} = \frac{27}{6} = \frac{9}{2}$$

$$a_4 = \frac{3^4}{4!} = \frac{81}{24} = \frac{27}{8}$$

$$a_5 = \frac{3^5}{5!} = \frac{243}{120} = \frac{81}{40}$$

Terms: $3, \frac{9}{2}, \frac{9}{2}, \frac{27}{8}, \frac{81}{40}$

17. Write the first five terms of the sequence $a_1 = 3$ and $a_{k+1} = 2(a_k - 1)$. (Assume n begins with 1.)

Solution:

$$a_1 = 3 \text{ and } a_{k+1} = 2(a_k - 1)$$

$$a_1 = 3$$

$$a_2 = 2(3 - 1) = 4$$

$$a_3 = 2(4 - 1) = 6$$

$$a_4 = 2(6 - 1) = 10$$

$$a_5 = 2(10 - 1) = 18$$

Terms: 3, 4, 6, 10, 18

19. Simplify the ratio $\frac{4!}{6!}$.

Solution:

$$\frac{4!}{6!} = \frac{4!}{6 \cdot 5 \cdot 4!}$$

$$= \frac{1}{6 \cdot 5}$$

$$= \frac{1}{30}$$

23. Simplify the ratio

$$\frac{(2n-1)!}{(2n+1)!}.$$

Solution:

$$\frac{(2n-1)!}{(2n+1)!} = \frac{(2n-1)!}{(2n+1)(2n)(2n-1)!}$$

$$= \frac{1}{(2n+1)(2n)}$$

$$= \frac{1}{2n(2n+1)}$$

27. Write an expression for the nth term of the sequence 0, 3, 8, 15, 24, ... (Assume n begins with 1.)

Solution:

$$a_1 = 0 = 1^2 - 1$$
$$a_2 = 3 = 2^2 - 1$$
$$a_3 = 8 = 3^2 - 1$$
$$a_4 = 15 = 4^2 - 1$$
$$a_5 = 24 = 5^2 - 1$$

Therefore, $a_n = n^2 - 1$.

31. Write an expression for the nth term of the sequence $1+\frac{1}{1}, 1+\frac{1}{2}, 1+\frac{1}{3}, 1+\frac{1}{4}, 1+\frac{1}{5}, \ldots$. (Assume n begins with 1.)

Solution:

$$a_1 = 1 + \tfrac{1}{1}$$
$$a_2 = 1 + \tfrac{1}{2}$$
$$a_3 = 1 + \tfrac{1}{3}$$
$$a_4 = 1 + \tfrac{1}{4}$$
$$a_5 = 1 + \tfrac{1}{5}$$

Therefore, $a_n = 1 + \dfrac{1}{n}$.

35. Write an expression for the nth term of the sequence 1, −1, 1, −1, 1, ... (Assume n begins with 1.)

Solution:

$$a_1 = 1 = (-1)^{1-1} \quad \text{or} \quad (-1)^{1+1}$$
$$a_2 = -1 = (-1)^{2-1} \quad \text{or} \quad (-1)^{2+1}$$
$$a_3 = 1 = (-1)^{3-1} \quad \text{or} \quad (-1)^{3+1}$$
$$a_4 = -1 = (-1)^{4-1} \quad \text{or} \quad (-1)^{4+1}$$
$$a_5 = 1 = (-1)^{5-1} \quad \text{or} \quad (-1)^{5+1}$$

Therefore, $a_n = (-1)^{n-1}$ or $a_n = (-1)^{n+1}$.

39. Find the sum.

$$\sum_{k=1}^{4} 10$$

Solution:

$$\sum_{k=1}^{4} 10 = 10 + 10 + 10 + 10$$
$$= 40$$

43. Find the sum.

$$\sum_{k=0}^{3} \frac{1}{k^2+1}$$

Solution:

$$\sum_{k=0}^{3} \frac{1}{k^2+1} = \frac{1}{(0)^2+1} + \frac{1}{(1)^2+1} + \frac{1}{(2)^2+1} + \frac{1}{(3)^2+1}$$
$$= 1 + \frac{1}{2} + \frac{1}{5} + \frac{1}{10}$$
$$= \frac{10+5+2+1}{10}$$
$$= \frac{18}{10} = \frac{9}{5}$$

47. Find the sum.

$$\sum_{i=1}^{4}(9+2i)$$

Solution:

$$\sum_{i=1}^{4}(9+2i) = (9+2)+(9+4)+(9+6)+(9+8) = 56$$

51. Use sigma notation to write the sum

$$\frac{1}{3(1)}+\frac{1}{3(2)}+\frac{1}{3(3)}+\cdots+\frac{1}{3(9)}.$$

Solution:

$$\frac{1}{3(1)}+\frac{1}{3(2)}+\frac{1}{3(3)}+\cdots+\frac{1}{3(9)} = \sum_{i=1}^{9}\frac{1}{3i}$$

55. Use sigma notation to write the sum $3-9+27-81+243-729$.

Solution:

$$3-9+27-81+243-729 = 3^1-3^2+3^3-3^4+3^5-3^6 = \sum_{i=1}^{6}(-1)^{i+1}3^i$$

59. Use sigma notation to write the sum

$$\frac{1}{4}+\frac{3}{8}+\frac{7}{16}+\frac{15}{32}+\frac{31}{64}.$$

Solution:

$$\frac{1}{4}+\frac{3}{8}+\frac{7}{16}+\frac{15}{32}+\frac{31}{64} = \frac{2^1-1}{2^2}+\frac{2^2-1}{2^3}+\frac{2^3-1}{2^4}+\frac{2^4-1}{2^5}+\frac{2^5-1}{2^6}$$

$$= \sum_{i=1}^{5}\frac{2^i-1}{2^{i+1}}$$

63. The average cost of a day in a hospital from 1980 to 1987 is given by the model

$$a_n = 242.67 + 42.67n, \quad n = 0,\ 1,\ 2,\ \ldots,\ 7$$

where a_n is the average cost in dollars and n is the year with $n = 0$ corresponding to 1980. (*Source:* American Hospital Association) Find the terms of this finite sequence and construct a bar graph that represents the sequence.

Solution:

$$a_n = 242.67 + 42.67n,\ n = 0,\ 1,\ 2,\ \ldots,\ 7$$

$$a_0 = 242.67$$

$$a_1 = 285.34$$

$$a_2 = 328.01$$

$$a_3 = 370.68$$

$$a_4 = 413.35$$

$$a_5 = 456.02$$

$$a_6 = 498.69$$

$$a_7 = 541.36$$

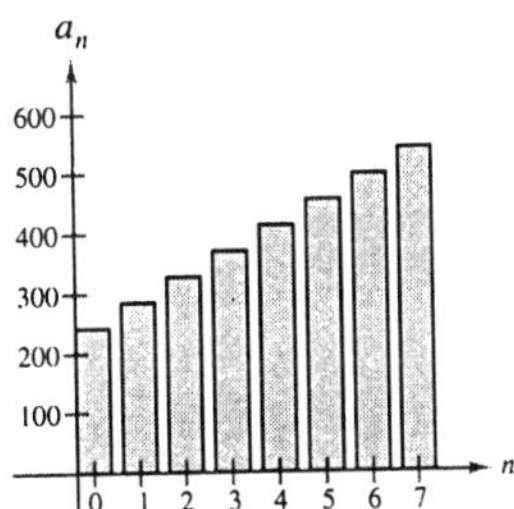

67. Prove that

$$\sum_{i=1}^{n}(x_i - \overline{x}) = 0, \quad \text{where } \overline{x} = \frac{1}{n}\sum_{i=1}^{n} x_i.$$

Solution:

$$\begin{aligned}\sum_{i=1}^{n}(x_i - \overline{x}) &= \sum_{i=1}^{n} x_i - \sum_{i=1}^{n} \overline{x} \\ &= \sum_{i=1}^{n} x_i - n\overline{x} \\ &= \sum_{i=1}^{n} x_i - n\left(\frac{1}{n}\sum_{i=1}^{n} x_i\right) \\ &= 0\end{aligned}$$

SECTION 10.2

Arithmetic Sequences

- You should be able to recognize an arithmetic sequence, find its common difference, and find its nth term.
- You should be able to find the nth partial sum of an arithmetic sequence with common difference d using the formula

  $$S_n = \frac{n}{2}(a_1 + a_n).$$

- You should know that the arithmetic mean of a and b is $\dfrac{a+b}{2}$.

Solutions to Selected Exercises

5. Determine whether the sequence $\frac{9}{4}, 2, \frac{7}{4}, \frac{3}{2}, \frac{5}{4}, \ldots$ is arithmetic. If it is, find the common difference.

Solution:

$$\frac{9}{4}, 2, \frac{7}{4}, \frac{3}{2}, \frac{5}{4}, \ldots = \frac{9}{4}, \frac{8}{4}, \frac{7}{4}, \frac{6}{4}, \frac{5}{4}, \ldots$$

$$a_n = \frac{10}{4} - \frac{1}{4}n$$

Therefore, the sequence *is* arithmetic with $d = -\frac{1}{4}$.

9. Determine whether the sequence 5.3, 5.7, 6.1, 6.5, 6.9, ... is arithmetic. If it is, find the common difference.

Solution:

$$5.3, 5.7, 6.1, 6.5, 6.9, \ldots = 4.9 + 0.4,\ 4.9 + 2(0.4),\ 4.9 + 3(0.4),$$

$$4.9 + 4(0.4),\ 4.9 + 5(0.4),\ \ldots$$

$$a_n = 4.9 + 0.4n$$

Therefore, the sequence *is* arithmetic with $d = 0.4$.

13. Write the first five terms of the sequence $a_n = 1/(n+1)$. Determine whether the sequence is arithmetic. If it is, find the common difference.

Solution:

$$a_n = \frac{1}{n+1}$$
$$a_1 = \frac{1}{1+1} = \frac{1}{2}$$
$$a_2 = \frac{1}{2+1} = \frac{1}{3}$$
$$a_3 = \frac{1}{3+1} = \frac{1}{4}$$
$$a_4 = \frac{1}{4+1} = \frac{1}{5}$$
$$a_5 = \frac{1}{5+1} = \frac{1}{6}$$

The sequence is *not* arithmetic.

17. Write the first five terms of the sequence $a_1 = 1,\ a_2 = 1,\ a_n = a_{n-1} + a_{n-2}$, $n \geq 3$. Determine whether the sequence is arithmetic. If it is, find the common difference.

Solution:

$$a_1 = 1,\ a_2 = 1,\ a_n = a_{n-1} + a_{n-2},\ n \geq 3$$
$$a_1 = 1$$
$$a_2 = 1$$
$$a_3 = a_2 + a_1 = 1 + 1 = 2$$
$$a_4 = a_3 + a_2 = 2 + 1 = 3$$
$$a_5 = a_4 + a_3 = 3 + 2 = 5$$

The sequence is *not* arithmetic.

21. Find a formula for a_n for the arithmetic sequence $a_1 = 100,\ d = -8$.

Solution:

$$a_1 = 100,\ d = -8$$
$$a_n = dn + c = -8n + c$$
$$a_1 = 100 = -8(1) + c \Rightarrow c = 108$$

Thus, $a_n = -8n + 108$.

25. Find a formula for a_n for the arithmetic sequence $4,\ \frac{3}{2},\ -1,\ -\frac{7}{2},\ \ldots$.

Solution:

$$4,\ \tfrac{3}{2},\ -1,\ -\tfrac{7}{2},\ \ldots,\ d = -\tfrac{5}{2}$$
$$a_n = dn + c = -\tfrac{5}{2}n + c$$
$$a_1 = 4 = -\tfrac{5}{2}(1) + c \quad \Rightarrow \quad c = \tfrac{13}{2}$$

Thus, $a_n = -\frac{5}{2}n + \frac{13}{2}$.

29. Find a formula for a_n for the arithmetic sequence $a_3 = 94,\ a_6 = 85$.

Solution:

$$a_n = dn + c$$
$$a_3 = 94 = d(3) + c \qquad a_6 = 85 = d(6) + c$$
$$3d + c = 94$$
$$6d + c = 85$$

Solving this system of equations yields $d = -3$ and $c = 103$. Thus, $a_n = -3n + 103$.

33. Write the first five terms of the arithmetic sequence given $a_1 = -2.6$ and $d = -0.4$.

Solution:

$$a_1 = -2.6,\ d = -0.4$$
$$a_2 = -2.6 - 0.4, = -3$$
$$a_3 = -3 - 0.4 = -3.4$$
$$a_4 = -3.4 - 0.4 = -3.8$$
$$a_5 = -3.8 - 0.4 = -4.2$$

Terms: $-2.6,\ -3,\ -3.4,\ -3.8,\ -4.2$

39. Write the first five terms of the arithmetic sequence given $a_8 = 26$ and $a_{12} = 42$.

Solution:

$$a_8 = 26, \ a_{12} = 42$$

$$a_8 = 26 = d(8) + c, \qquad a_{12} = 42 = d(12) + c$$

$$8d + c = 26$$
$$12d + c = 42$$

Solving this system yields $d = 4$ and $c = -6$. Thus, $a_n = 4n - 6$.

$$a_1 = -2$$
$$a_2 = 2$$
$$a_3 = 6$$
$$a_4 = 10$$
$$a_5 = 14$$

Terms: $-2, \ 2, \ 6, \ 10, \ 14$

43. Find the nth partial sum of the arithmetic sequence $-6, \ -2, \ 2, \ 6, \ \ldots, \ n = 50$.

Solution:

$-6, \ -2, \ 2, \ 6, \ \ldots, \ n = 50$

$$a_n = 4n - 10$$
$$a_1 = -6 \text{ and } a_{50} = 190$$
$$S_{50} = \tfrac{50}{2}(-6 + 190) = 4600$$

47. Find the nth partial sum of the arithmetic sequence $a_1 = 100, \ a_{25} = 220, \ n = 25$.

Solution:

$a_1 = 100, \ a_{25} = 220, \ n = 25$

$$S_{25} = \tfrac{25}{2}(100 + 220) = \tfrac{25}{2}(320) = 4000$$

51. Find the sum.

$$\sum_{n=1}^{100} 5n$$

Solution:

$$a_n = 5n$$
$$a_1 = 5, \ a_{100} = 500$$
$$S_{100} = \tfrac{100}{2}(5 + 500) = 25{,}250$$

55. Find the sum.

$$\sum_{n=1}^{500} (n + 3)$$

Solution:

$$a_n = n + 3$$
$$a_1 = 4, \ a_{500} = 503$$
$$S_{500} = \tfrac{500}{2}(4 + 503)$$
$$= 126{,}750$$

59. Find the sum.

$$\sum_{n=0}^{50}(1000-5n)$$

Solution:

$$\begin{aligned}\sum_{n=0}^{50}(1000-5n) &= 1000+\sum_{n=1}^{50}(1000-5n)\\ &= 1000+\tfrac{50}{2}(995+750)\\ &= 1000+43{,}625\\ &= 44{,}625\end{aligned}$$

63. Insert three arithmetic means between the pair of numbers 3 and 6.

Solution:

$3,\ 6;\ k=3$

$$3,\ m_1,\ m_2,\ m_3,\ 6$$

$$a_5 = 6 = 3+4d$$

$$d=\tfrac{3}{4}$$

$$m_1 = 3+\tfrac{3}{4}=\tfrac{15}{4}$$

$$m_2=\tfrac{15}{4}+\tfrac{3}{4}=\tfrac{18}{4}=\tfrac{9}{2}$$

$$m_3=\tfrac{18}{4}+\tfrac{3}{4}=\tfrac{21}{4}$$

67. A person accepts a position with a company and will receive a salary of $27,500 for the first year. The person is guaranteed a raise of $1,500 per year for the first five years.

(a) Determine the person's salary during the sixth year of employment.

(b) Determine the person's total compensation from the company through six full years of employment.

Solution:

(a)
$$\begin{aligned}a_n &= a_1+(n-1)d\\ &= 27{,}500+(n-1)(1500)\\ &= 1500n+26{,}000\\ a_6 &= 1500(6)+26{,}000\\ &= \$35{,}000\end{aligned}$$

(b)
$$\begin{aligned}S_6 &= \tfrac{6}{2}(27{,}500+35{,}000)\\ &= \$187{,}500\end{aligned}$$

69. Determine the seating capacity of an auditorium with 30 rows of seats if there are 20 seats in the first row, 24 seats in the second row, 28 seats in the third row, and so on.

Solution:

$$a_n = 16+4n$$

$$a_1=20,\ a_{30}=136$$

$$S_{30}=\tfrac{30}{2}(20+136)=2340 \text{ seats}$$

SECTION 10.3

Geometric Sequences

- You should be able to identify a geometric sequence, find its common ratio, and find the nth term.
- You should be able to find the nth partial sum of a geometric sequence with common ratio r using the formula

$$S_n = \frac{a_1(1-r^n)}{1-r}, \quad r \neq 1.$$

- You should know that if $|r| < 1$, then

$$\sum_{n=0}^{\infty} a_1 r^n = \sum_{n=1}^{\infty} a_1 r^{n-1} = \frac{a_1}{1-r}.$$

Solutions to Selected Exercises

3. Determine whether the sequence 3, 12, 21, 30, ... is geometric. If it is, find its common ratio.

Solution:

$$3, 12, 21, 30, \ldots$$

$$a_n = -6 + 9n$$

This is an arithmetic sequence, *not* a geometric sequence.

7. Determine whether the sequence $\frac{1}{2}, \frac{2}{3}, \frac{3}{4}, \frac{4}{5}, \ldots$ is geometric. If it is, find its common ratio.

Solution:

$$\frac{1}{2}, \frac{2}{3}, \frac{3}{4}, \frac{4}{5}, \ldots$$

$$a_n = \frac{n}{n+1}$$

This is *not* a geometric sequence.

11. Write the first five terms of the geometric sequence $a_1 = 2,\ r = 3$.

Solution:

$$a_1 = 2,\ r = 3$$
$$a_2 = 2(3) = 6$$
$$a_3 = 2(3)^2 = 18$$
$$a_4 = 2(3)^3 = 54$$
$$a_5 = 2(3)^4 = 162$$

Terms: 2, 6, 18, 54, 162

15. Write the first five terms of the geometric sequence $a_1 = 5,\ r = -\frac{1}{10}$.

Solution:

$$a_1 = 5,\ r = -\frac{1}{10}$$
$$a_2 = 5\left(-\frac{1}{10}\right) = -\frac{1}{2}$$
$$a_3 = 5\left(-\frac{1}{10}\right)^2 = \frac{1}{20}$$
$$a_4 = 5\left(-\frac{1}{10}\right)^3 = -\frac{1}{200}$$
$$a_5 = 5\left(-\frac{1}{10}\right)^4 = \frac{1}{2000}$$

Terms: $5,\ -\frac{1}{2},\ \frac{1}{20},\ -\frac{1}{200},\ \frac{1}{2000}$

21. Find the nth term of the geometric sequence $a_1 = 4,\ r = \frac{1}{2},\ n = 10$.

Solution:

$$a_1 = 4,\ r = \frac{1}{2},\ n = 10$$
$$a_{10} = 4\left(\frac{1}{2}\right)^9 = \frac{1}{128} = \left(\frac{1}{2}\right)^7$$

25. Find the nth term of the geometric sequence $a_1 = 100,\ r = e^x,\ n = 9$.

Solution:

$$a_1 = 100,\ r = e^x,\ n = 9$$
$$a_9 = 100(e^x)^8 = 100e^{8x}$$

29. Find the nth term of the geometric sequence $a_1 = 16,\ a_4 = \frac{27}{4},\ n = 3$.

Solution:

$$a_1 = 16,\ a_4 = \frac{27}{4},\ n = 3$$
$$a_4 = 16r^3 = \frac{27}{4}$$
$$r^3 = \frac{27}{64}$$
$$r = \frac{3}{4}$$
$$a_3 = 16\left(\frac{3}{4}\right)^2 = 16\left(\frac{9}{16}\right) = 9$$

33. A sum of \$1000 is invested at 10% interest. Find the amount after 10 years if the interest is compounded (a) annually, (b) semiannually, (c) quarterly, (d) monthly, and (e) daily.

Solution:

$$A = P\left(1+\frac{r}{n}\right)^{nt} = 1000\left(1+\frac{0.10}{n}\right)^{n(10)}$$

(a) $n = 1,\quad A = 1000(1+0.10)^{10} \approx \$2593.74$

(b) $n = 2,\quad A = 1000\left(1+\frac{0.10}{2}\right)^{2(10)} \approx \$2653.30$

(c) $n = 4,\quad A = 1000\left(1+\frac{0.10}{4}\right)^{4(10)} \approx \$2685.06$

(d) $n = 12,\quad A = 1000\left(1+\frac{0.10}{12}\right)^{12(10)} \approx \$2707.04$

(e) $n = 365,\quad A = 1000\left(1+\frac{0.10}{365}\right)^{365(10)} \approx \$2717.91$

37. Find the sum.

$$\sum_{n=1}^{9} 2^{n-1}$$

Solution:

$$\sum_{n=1}^{9} 2^{n-1} = \frac{1(1-2^9)}{1-2}$$

$$= 511$$

43. Find the sum.

$$\sum_{n=0}^{20} 3\left(\frac{3}{2}\right)^n$$

Solution:

$$\sum_{n=0}^{20} 3\left(\frac{3}{2}\right)^n = \frac{3(1-(3/2)^{21})}{1-(3/2)}$$

$$\approx 29{,}921.31$$

47. A deposit of \$100 is made at the beginning of each month for five years in an account that pays 10%, compounded monthly. What is the balance A in the account at the end of five years?

$$A = 100\left(1+\frac{0.01}{12}\right)^1 + \cdots + 100\left(1+\frac{0.10}{12}\right)^{60}$$

Solution:

$$A = \sum_{n=1}^{60} 100\left(1+\frac{0.10}{12}\right)^n = 100\left(1+\frac{0.10}{12}\right) \cdot \frac{\left[1-\left(1+\frac{0.10}{12}\right)^{60}\right]}{\left[1-\left(1+\frac{0.10}{12}\right)\right]} \approx \$7808.24$$

49. A deposit of P dollars is made at the beginning of each month in an account at an annual interest rate r compounded monthly. The balance A after t years is

$$A = P\left(1+\frac{r}{12}\right) + P\left(1+\frac{r}{12}\right)^2 + \cdots + P\left(1+\frac{r}{12}\right)^{12t}.$$

Show that the balance is given by

$$A = P\left[\left(1+\frac{r}{12}\right)^{12t} - 1\right]\left(1+\frac{12}{r}\right).$$

Solution:

Let $N = 12t$ be the total number of deposits.

$$\begin{aligned}
A &= P\left(1+\frac{r}{12}\right) + P\left(1+\frac{r}{12}\right)^2 + \cdots + P\left(1+\frac{r}{12}\right)^N \\
&= \left(1+\frac{r}{12}\right)\left[P + P\left(1+\frac{r}{12}\right) + \cdots + P\left(1+\frac{r}{12}\right)^{N-1}\right] \\
&= P\left(1+\frac{r}{12}\right)\sum_{n=1}^{N}\left(1+\frac{r}{12}\right)^{n-1} \\
&= P\left(1+\frac{r}{12}\right)\frac{1-\left(1+\frac{r}{12}\right)^N}{1-\left(1+\frac{r}{12}\right)} \\
&= P\left(1+\frac{r}{12}\right)\left(-\frac{12}{r}\right)\left[1-\left(1+\frac{r}{12}\right)^N\right] \\
&= P\left(\frac{12}{r}+1\right)\left[-1+\left(1+\frac{r}{12}\right)^N\right] \\
&= P\left[\left(1+\frac{r}{12}\right)^N - 1\right]\left(1+\frac{12}{r}\right) \\
&= P\left[\left(1+\frac{r}{12}\right)^{12t} - 1\right]\left(1+\frac{12}{r}\right)
\end{aligned}$$

53. Consider making monthly deposits of P dollars into a savings account at an annual interest rate r. Use the results of Exercises 49 and 50 to find the balance A after t years if the interest is compounded (a) monthly, and (b) continuously.

$$P = \$100, \qquad r = 10\%, \qquad t = 40 \text{ years}$$

Solution:

(a) $A = 100\left[\left(1+\dfrac{0.10}{12}\right)^{12(40)} - 1\right]\left(1+\dfrac{12}{0.10}\right) \approx \$637{,}678.02$

(b) $A = \dfrac{100e^{0.10/12}\left(e^{(0.10)(40)}-1\right)}{e^{0.10/12}-1} \approx \$645{,}861.43$

57. You accept a job with a salary of $30,000 for the first year. Suppose that during the next 39 years you receive a 5% raise each year. What would your total compensation be over the 40-year period?

Solution:

$$T = \sum_{n=0}^{39} 30{,}000(1.05)^n = \frac{30{,}000(1 - 1.05^{40})}{1 - 1.05} \approx \$3{,}623{,}993.23$$

59. Find the sum of the infinite geometric series.

$$\sum_{n=0}^{\infty} \left(\frac{1}{2}\right)^n = 1 + \frac{1}{2} + \frac{1}{4} + \frac{1}{8} + \cdots$$

Solution:

$$\sum_{n=0}^{\infty} \left(\frac{1}{2}\right)^n = 1 + \frac{1}{2} + \frac{1}{4} + \frac{1}{8} + \cdots = \frac{1}{1 - (1/2)} = 2$$

63. Find the sum of the infinite geometric series.

$$\sum_{n=0}^{\infty} 4\left(\frac{1}{4}\right)^n = 4 + 1 + \frac{1}{4} + \frac{1}{16} + \cdots$$

Solution:

$$\sum_{n=0}^{\infty} 4\left(\frac{1}{4}\right)^n = 4 + 1 + \frac{1}{4} + \frac{1}{16} + \cdots = \frac{4}{1 - (1/4)} = \frac{16}{3}$$

67. Find the sum of the infinite geometric series $4 - 2 + 1 - \frac{1}{2} + \cdots$.

Solution:

$$4 - 2 + 1 - \frac{1}{2} + \cdots = \sum_{n=0}^{\infty} 4\left(-\frac{1}{2}\right)^n = \frac{4}{1 - (-1/2)} = \frac{8}{3}$$

69. A ball is dropped from a height of 16 feet. Each time it drops h feet, it rebounds $0.81h$ feet. Find the total distance traveled by the ball.

Solution:

$$\text{Total distance} = \left[\sum_{n=0}^{\infty} 32(0.81)^n\right] - 16 = \frac{32}{1 - 0.81} - 16 \approx 152.42 \text{ feet}$$

SECTION 10.4

Mathematical Induction

- You should be sure that you understand the principle of mathematical induction. If P_n is a statement involving the positive integer n, where P_1 is true and the truth of P_k implies the truth of P_{k+1}, then P_n is true for all positive integers n.
- You should be able to verify (by induction) the formulas for the sums of powers of integers and be able to use these formulas.

Solutions to Selected Exercises

1. Find the following sum using the formulas for the sums of powers of integers.

$$\sum_{n=1}^{20} n$$

Solution:

$$\sum_{n=1}^{N} n = \frac{N(N+1)}{2}$$

$$\sum_{n=1}^{20} n = \frac{20(21)}{2} = 210$$

7. Find the following sum using the formulas for the sums of powers of integers.

$$\sum_{n=1}^{6} n^4$$

Solution:

$$\sum_{n=1}^{N} n^4 = \frac{N(N+1)(2N+1)(3N^2+3N-1)}{30}$$

$$\sum_{n=1}^{6} n^4 = \frac{6(7)(13)(125)}{30} = 2275$$

11. Find S_{k+1} for $S_k = \dfrac{5}{k(k+1)}$.

Solution:

$$S_k = \frac{5}{k(k+1)}$$

$$S_{k+1} = \frac{5}{(k+1)((k+1)+1)}$$

$$= \frac{5}{(k+1)(k+2)}$$

13. Find S_{k+1} for $S_k = \dfrac{k^2(k+1)^2}{4}$.

Solution:

$$S_k = \frac{k^2(k+1)^2}{4}$$

$$S_{k+1} = \frac{(k+1)^2((k+1)+1)^2}{4}$$

$$= \frac{(k+1)^2(k+2)^2}{4}$$

17. Use mathematical induction to prove the formula for every positive integer n.

$$2 + 7 + 12 + 17 + \cdots + (5n - 3) = \frac{n}{2}(5n - 1)$$

Solution:

1.) When $n = 1$,

$$S_1 = 2 = \frac{1}{2}(5(1) - 1).$$

2.) Assume that

$$S_k = 2 + 7 + 12 + 17 + \cdots + (5k - 3) = \frac{k}{2}(5k - 1).$$

Then,

$$\begin{aligned}
S_{k+1} &= 2 + 7 + 12 + 17 + \cdots + (5k - 3) + [5(k+1) - 3] \\
&= S_k + (5k + 5 - 3) \\
&= \frac{k}{2}(5k - 1) + 5k + 2 \\
&= \frac{5k^2 - k + 10k + 4}{2} \\
&= \frac{5k^2 + 9k + 4}{2} \\
&= \frac{(k+1)(5k+4)}{2} \\
&= \frac{(k+1)}{2}[5(k+1) - 1].
\end{aligned}$$

We conclude by mathematical induction that the formula is valid for all positive integer values of n.

21. Use mathematical induction to prove the formula for every positive integer n.

$$1+2+3+4+\cdots+n=\frac{n(n+1)}{2}$$

Solution:

1.) When $n=1$, $S_1=1=\frac{1(1+1)}{2}$.

2.) Assume that $S_k=1+2+3+4+\cdots+k=\frac{k(k+1)}{2}$.

Then,

$$\begin{aligned}S_{k+1}&=1+2+3+4+\cdots+k+(k+1)\\&=S_k+(k+1)=\frac{k(k+1)}{2}+\frac{2(k+1)}{2}=\frac{(k+1)(k+2)}{2}.\end{aligned}$$

Therefore, we conclude that this formula holds for all positive integer values of n.

23. Use mathematical induction to prove the formula for every positive integer n.

$$1^3+2^3+3^3+4^3+\cdots+n^3=\frac{n^2(n+1)^2}{4}$$

Solution:

1.) When $n=1$, $S_1=1^3=1=\frac{1(1+1)^2}{4}$.

2.) Assume that $S_k=1^3+2^3+3^3+4^3+\cdots+k^3=\frac{k^2(k+1)^2}{4}$.

Then,

$$\begin{aligned}S_{k+1}&=1^3+2^3+3^3+4^3+\cdots+k^3+(k+1)^3\\&=S_k+(k+1)^3\\&=\frac{k^2(k+1)^2}{4}+(k+1)^3\\&=\frac{k^2(k+1)^2+4(k+1)^3}{4}\\&=\frac{(k+1)^2[k^2+4(k+1)]}{4}\\&=\frac{(k+1)^2(k^2+4k+4)}{4}\\&=\frac{(k+1)^2(k+2)^2}{4}.\end{aligned}$$

Therefore, we conclude that this formula holds for all positive integer values of n.

27. Use mathematical induction to prove the formula for every positive integer n.

$$\sum_{i=1}^{n} i(i+1) = \frac{n(n+1)(n+2)}{3}$$

Solution:

1.) When $n = 1$, $S_1 = 2 = \dfrac{1(2)(3)}{3}$.

2.) Assume that

$$S_k = 1(2) + 2(3) + 3(4) + \cdots + k(k+1) = \frac{k(k+1)(k+2)}{3}.$$

Then,

$$\begin{aligned} S_{k+1} &= 1(2) + 2(3) + 3(4) + \cdots + k(k+1) + (k+1)(k+2) \\ &= S_k + (k+1)(k+2) \\ &= \frac{k(k+1)(k+2)}{3} + \frac{3(k+1)(k+2)}{3} \\ &= \frac{(k+1)(k+2)(k+3)}{3}. \end{aligned}$$

Thus, this formula is valid for all positive integer values of n.

29. Use mathematical induction to prove the inequality, $\left(\frac{4}{3}\right)^n > n$, $n \geq 7$.

Solution:

1.) When $n = 7$, $\left(\dfrac{4}{3}\right)^7 \approx 7.4915 > 7$.

2.) Assume that $\left(\dfrac{4}{3}\right)^k > k$, $k > 7$.

Then, $\left(\dfrac{4}{3}\right)^{k+1} = \left(\dfrac{4}{3}\right)^k \left(\dfrac{4}{3}\right) > k\left(\dfrac{4}{3}\right) = k + \dfrac{k}{3} > k + 1$ for $k > 7$.

Thus, $\left(\dfrac{4}{3}\right)^{k+1} > k + 1$. Therefore, $\left(\dfrac{4}{3}\right)^n > n$.

33. Use mathematical induction to prove the property $(ab)^n = a^n b^n$ for all positive integers n.

Solution:

1.) When $n = 1$, $(ab)^1 = a^1 b^1 = ab$.

2.) Assume that $(ab)^k = a^k b^k$. Then,

$$\begin{aligned}(ab)^{k+1} &= (ab)^k(ab)\\ &= a^k b^k ab\\ &= a^{k+1}b^{k+1}.\end{aligned}$$

Thus, $(ab)^n = a^n b^n$.

37. Use mathematical induction to prove the Generalized Distributive Law:

$$x(y_1 + y_2 + \cdots + y_n) = xy_1 + xy_2 + \cdots + xy_n.$$

Solution:

1.) When $n = 1$, $x(y_1) = xy_1$.

2.) Assume that $x(y_1 + y_2 + \cdots + y_k) = xy_1 + xy_2 + \cdots + xy_k$. Then,

$$\begin{aligned}xy_1 + xy_2 + \cdots + xy_k + xy_{k+1} &= x(y_1 + y_2 + \cdots + y_k) + xy_{k+1}\\ &= x[(y_1 + y_2 + \cdots + y_k) + y_{k+1}]\\ &= x(y_1 + y_2 + \cdots + y_k + y_{k+1}).\end{aligned}$$

Hence, the formula holds.

41. Use mathematical induction to prove that a factor of $(2^{2n-1} + 3^{2n-1})$ is 5.

Solution:

1.) When $n = 1$, $(2^{2(1)-1} + 3^{2(1)-1}) = 2 + 3 = 5$ and 5 is a factor.

2.) Assume that 5 is a factor of $(2^{2k-1} + 3^{2k-1})$. Then,

$$\begin{aligned}\left(2^{2(k+1)-1} + 3^{2(k+1)-1}\right) &= (2^{2k+2-1} + 3^{2k+2-1})\\ &= (2^{2k-1}2^2 + 3^{2k-1}3^2)\\ &= (4 \cdot 2^{2k-1} + 9 \cdot 3^{2k-1})\\ &= (2^{2k-1} + 3^{2k-1}) + (2^{2k-1} + 3^{2k-1})\\ &\quad + (2^{2k-1} + 3^{2k-1}) + (2^{2k-1} + 3^{2k-1}) + 5 \cdot 3^{2k-1}.\end{aligned}$$

Since 5 is a factor of each set of parenthesis and 5 is a factor of $5 \cdot 3^{2k-1}$, then 5 is a factor of the whole sum. Thus, 5 is a factor of $(2^{2n-1} + 3^{2n-1})$ for every positive integer n.

SECTION 10.5

The Binomial Theorem

- You should be able to use the formula

$$(x+y)^n = x^n + nx^{n-1}y + \frac{n(n-1)}{2!}x^{n-2}y^2 + \cdots + {}_nC_r{}^n x^{n-r}y^r + \cdots + y^n$$

where ${}_nC_r = \dfrac{n!}{(n-r)!r!}$, to expand $(x+y)^n$.

- You should be able to use Pascal's Triangle in binomial expansion.

Solutions to Selected Exercises

3. Evaluate ${}_{12}C_0$.

Solution:

$${}_{12}C_0 = \frac{12!}{(12-0)!0!} = \frac{12!}{(12!)(1)} = 1$$

7. Evaluate ${}_{100}C_{98}$.

Solution:

$${}_{100}C_{98} = \frac{100!}{(100-98)!98!} = \frac{100 \cdot 99 \cdot 98!}{2! \cdot 98!} = \frac{100 \cdot 99}{2} = 4950$$

11. Use the Binomial Theorem to expand $(x+1)^4$. Simplify your answer.

Solution:

$$\begin{aligned}(x+1)^4 &= {}_4C_0x^4 + {}_4C_1x^3(1) + {}_4C_2x^2(1)^2 + {}_4C_3x(1)^3 + {}_4C_4(1)^4 \\ &= x^4 + 4x^3 + 6x^2 + 4x + 1\end{aligned}$$

17. Use the Binomial Theorem to expand $(x+y)^5$. Simplify your answer.

Solution:

$$\begin{aligned}(x+y)^5 &= {}_5C_0x^5 + {}_5C_1x^4y + {}_5C_2x^3y^2 + {}_5C_3x^2y^3 + {}_5C_4xy^4 + {}_5C_5y^5 \\ &= x^5 + 5x^4y + 10x^3y^2 + 10x^2y^3 + 5xy^4 + y^5\end{aligned}$$

19. Use the Binomial Theorem to expand $(r+3s)^6$. Simplify your answer.

Solution:

$$\begin{aligned}(r+3s)^6 &= {}_6C_0r^6 + {}_6C_1r^5(3s) + {}_6C_2r^4(3s)^2 + {}_6C_3r^3(3s)^3 + {}_6C_4r^2(3s)^4 \\ &\quad + {}_6C_5r(3s)^5 + {}_6C_6(3s)^6 \\ &= r^6 + 18r^5s + 135r^4s^2 + 540r^3s^3 + 1215r^2s^4 + 1458rs^5 + 729s^6\end{aligned}$$

23. Use the Binomial Theorem to expand $(1-2x)^3$. Simplify your answer.

Solution:

$$\begin{aligned}(1-2x)^3 &= [1+(-2x)]^3 \\ &= {}_3C_01^3 + {}_3C_1(1)^2(-2x) + {}_3C_2(1)(-2x)^2 + {}_3C_3(-2x)^3 \\ &= 1 - 6x + 12x^2 - 8x^3\end{aligned}$$

27. Use the Binomial Theorem to expand the following. Simplify your answer.

$$\left(\frac{1}{x}+y\right)^5$$

Solution:

$$\begin{aligned}\left(\frac{1}{x}+y\right)^5 &= {}_5C_0\left(\frac{1}{x}\right)^5 + {}_5C_1\left(\frac{1}{x}\right)^4 y + {}_5C_2\left(\frac{1}{x}\right)^3 y^2 + {}_5C_3\left(\frac{1}{x}\right)^2 y^3 \\ &\quad + {}_5C_4\left(\frac{1}{x}\right) y^4 + {}_5C_5y^5 \\ &= \frac{1}{x^5} + \frac{5y}{x^4} + \frac{10y^2}{x^3} + \frac{10y^3}{x^2} + \frac{5y^4}{x} + y^5\end{aligned}$$

33. Use the Binomial Theorem to expand $(2-3i)^6$. Simplify your answer by recalling that $i^2 = -1$.

Solution:

$$\begin{aligned}(2-3i)^6 &= {}_6C_02^6 - {}_6C_1(2)^5(3i) + {}_6C_2(2)^4(3i)^2 - {}_6C_3(2)^3(3i)^3 + {}_6C_4(2)^2(3i)^4 \\ &\quad - {}_6C_5(2)(3i)^5 + {}_6C_6(3i)^6 \\ &= 64 - 576i - 2160 + 4320i + 4860 - 2916i - 729 \\ &= 2035 + 828i\end{aligned}$$

37. Expand $(2t - s)^5$, using Pascal's Triangle to determine the coefficients.

Solution:

$$\begin{array}{ccccccccccc} & & & & & 1 & & & & & \\ & & & & 1 & & 1 & & & & \\ & & & 1 & & 2 & & 1 & & & \\ & & 1 & & 3 & & 3 & & 1 & & \\ & 1 & & 4 & & 6 & & 4 & & 1 & \\ 1 & & 5 & & 10 & & 10 & & 5 & & 1 \end{array}$$

$$\begin{aligned}(2t - s)^5 &= (2t)^5 - 5(2t)^4 s + 10(2t)^3 s^2 - 10(2t)^2 s^3 + 5(2t)s^4 - s^5 \\ &= 32t^5 - 80t^4 s + 80t^3 s^2 - 40t^2 s^3 + 10ts^4 - s^5\end{aligned}$$

41. Find the coefficient a of the term ax^5 in the expansion of $(x+3)^{12}$.

Solution:

$$_{12}C_7 x^5 (3)^7 = \frac{12!3^7 x^5}{(12-7)!7!} = 1{,}732{,}104x^5$$

$$a = 1{,}732{,}104$$

45. Find the coefficient a of the term ax^4y^5 in the expansion of $(3x - 2y)^9$.

Solution:

$$_9C_5(3x)^4(-2y)^5 = \frac{9!}{5!4!}(81x^4)(-32y^5) = -326{,}592x^4y^5$$

$$a = -326{,}592$$

51. Use the Binomial Theorem to expand $\left(\frac{1}{3} + \frac{2}{3}\right)^8$. In the study of probability, it is sometimes necessary to use the expansion of $(p+q)^n$, where $p + q = 1$.

Solution:

$$\begin{aligned}\left(\frac{1}{3} + \frac{2}{3}\right)^8 &= \left(\frac{1}{3}\right)^8 + 8\left(\frac{1}{3}\right)^7\left(\frac{2}{3}\right) + 28\left(\frac{1}{3}\right)^6\left(\frac{2}{3}\right)^2 + 56\left(\frac{1}{3}\right)^5\left(\frac{2}{3}\right)^3 + 70\left(\frac{1}{3}\right)^4\left(\frac{2}{3}\right)^4 \\ &\quad + 56\left(\frac{1}{3}\right)^3\left(\frac{2}{3}\right)^5 + 28\left(\frac{1}{3}\right)^2\left(\frac{2}{3}\right)^6 + 8\left(\frac{1}{3}\right)\left(\frac{2}{3}\right)^7 + \left(\frac{2}{3}\right)^8 \\ &= \frac{1}{6561} + \frac{16}{6561} + \frac{112}{6561} + \frac{448}{6561} + \frac{1120}{6561} + \frac{1792}{6561} + \frac{1792}{6561} + \frac{1024}{6561} + \frac{256}{6561}\end{aligned}$$

55. Use the Binomial Theorem to approximate $(1.02)^8$ accurate to three decimal places. For example, $(1.02)^8 = (1 + 0.02)^8 = 1 + 8(0.02) + 28(0.02)^2 + \cdots$.

Solution:

$$\begin{aligned}(1.02)^8 = (1+0.02)^8 &= 1 + 8(0.02) + 28(0.02)^2 + 56(0.02)^3 + 70(0.02)^4 + 56(0.02)^5 \\ &\quad + 28(0.02)^6 + 8(0.02)^7 + (0.02)^8 \\ &= 1 + 0.16 + 0.0112 + 0.000448 + \cdots \\ &\approx 1.172\end{aligned}$$

59. Shift the graph of $f(x) = -x^2 + 3x + 2$ four units to the left to form the graph of g. Then write the polynomial function g in standard form.

Solution:

$$\begin{aligned}f(x) &= -x^2 + 3x + 2 \\ g(x) &= f(x+4) \\ &= -(x+4)^2 + 3(x+4) + 2 \\ &= -(x^2 + 8x + 16) + 3x + 12 + 2 \\ &= -x^2 - 5x - 2\end{aligned}$$

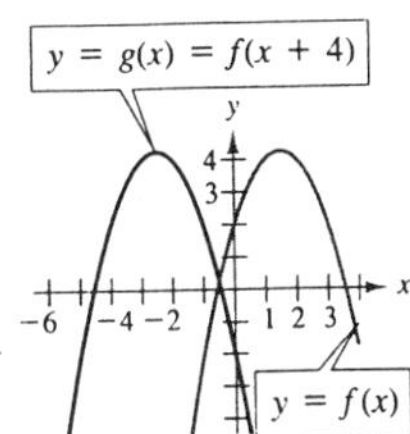

63. The average amount of life insurance per household (in households that carry life insurance) from 1970 through 1988 can be approximated by the model

$$f(t) = 0.2187t^2 + 0.6715t + 26.67, \quad 0 \le t \le 18.$$

In this model, $f(t)$ represents the amount of life insurance (in 1000s of dollars) and t represents the calendar year with $t = 0$ corresponding to 1970 (see figure). You want to adjust this model so that $t = 0$ corresponds to 1980 rather than 1970. To do this, you shift the graph of f ten units to the *left* and obtain

$$g(t) = f(t+10) = 0.2187(t+10)^2 + 0.6715(t+10) + 26.67.$$

Write this new polynomial function in standard form. (*Source:* American Council of Life Insurance)

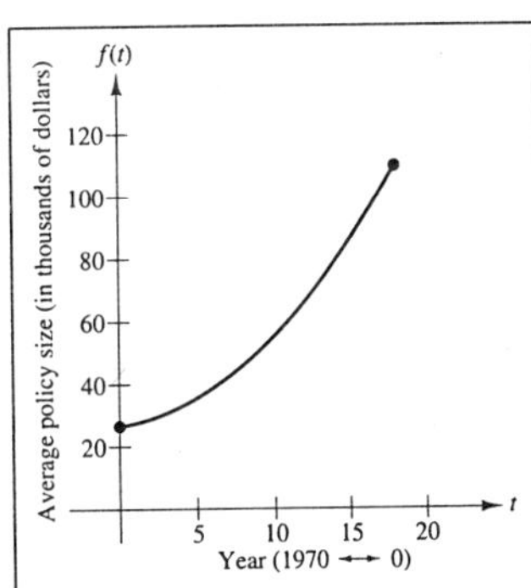

–CONTINUED ON NEXT PAGE–

63. –CONTINUED–

Solution:

$$\begin{aligned} f(t) &= 0.2187t^2 + 0.6715t + 26.67 \\ g(t) &= f(t+10) \\ &= 0.2187(t+10)^2 + 0.6715(t+10) + 26.67 \\ &= 0.2187(t^2 + 20t + 100) + 0.6715t + 6.715 + 26.67 \\ &= 0.2187t^2 + 5.0455t + 55.255 \end{aligned}$$

67. Prove ${}_{n+1}C_m = {}_nC_m + {}_nC_{m-1}$ for all integers m and n, $0 \le m \le n$.

Solution:

$$\begin{aligned} {}_nC_m + {}_nC_{m-1} &= \frac{n!}{(n-m)!m!} + \frac{n!}{(n-m+1)!(m-1)!} \\ &= \frac{n!(n-m+1)!(m-1)! + n!(n-m)!m!}{(n-m)!m!(n-m+1)!(m-1)!} \\ &= \frac{n![(n-m+1)!(m-1)! + m!(n-m)!]}{(n-m)!m!(n-m+1)!(m-1)!} \\ &= \frac{n!(m-1)![(n-m+1)! + m(n-m)!]}{(n-m)!m!(n-m+1)!(m-1)!} \\ &= \frac{n!(n-m)![(n-m+1)+m]}{(n-m)!m!(n-m+1)!} \\ &= \frac{n![n+1]}{m!(n-m+1)!} \\ &= \frac{(n+1)!}{[(n+1)-m]!m!} \\ &= {}_{n+1}C_m \end{aligned}$$

SECTION 10.6

Counting Principles, Permutations, Combinations

- You should know The Fundamental Principle of Counting.
- ${}_nP_r = \dfrac{n!}{(n-r)!}$ is the number of permutations of n elements taken r at a time.
- Given a set of n objects that has n_1 of one kind, n_2 of a second kind, and so on, the number of distinguishable permutations is
$$\frac{n!}{n_1!n_2!\ldots n_k!}.$$
- ${}_nC_r = \dfrac{n!}{(n-r)!r!}$ is the number of combinations of n elements taken r at a time.

Solutions to Selected Exercises

3. A small college needs two additional faculty members, a chemist and a statistician. In how many ways can these positions be filled if there are three applicants for the chemistry position and four applicants for the position in statistics?

Solution:

$3 \cdot 4 = 12$ ways to fill the positions

7. In a certain state the automobile license plates consist of two letters followed by a four-digit number. How many distinct license plate numbers can be formed?

Solution:

$26 \cdot 26 \cdot 10 \cdot 10 \cdot 10 \cdot 10 = 6,760,000$ distinct license plate numbers

11. How many three-digit numbers can be formed under the following conditions?

(a) The leading digit cannot be zero.
(b) The leading digit cannot be zero and no repetition of digits is allowed.
(c) The leading digit cannot be zero and the number must be a multiple of 5.
(d) The number is at least 400.

Solution:

(a) $9 \bullet 10 \bullet 10 = 900$ ways
(b) $9 \bullet 9 \bullet 8 = 648$ ways
(c) $9 \bullet 10 \bullet 2 = 180$ ways (The last digit must be 0 or 5.)
(d) $6 \bullet 10 \bullet 10 = 600$ ways (The first digit must be greater than or equal to 4.)

15. Three couples have reserved seats in a given row for a concert. In how many different ways can they be seated, given the following conditions?

(a) There are no seating restrictions.
(b) The two members of each couple wish to sit together.

Solution:

(a) $6! = 720$ different ways
(b) $6 \bullet 4 \bullet 2 = 48$ different ways

19. Evaluate ${}_8P_3$.

Solution:

$$\begin{aligned} {}_8P_3 &= \frac{8!}{(8-3)!} \\ &= \frac{8!}{5!} \\ &= 8 \bullet 7 \bullet 6 \\ &= 336 \end{aligned}$$

23. Evaluate ${}_{100}P_2$.

Solution:

$$\begin{aligned} {}_{100}P_2 &= \frac{100!}{(100-2)!} \\ &= \frac{100!}{98!} \\ &= 100 \bullet 99 \\ &= 9900 \end{aligned}$$

29. In how many ways can five children line up in one row to have their picture taken?

Solution:

$5! = 120$ ways

31. From a pool of 12 candidates, the offices of president, vice-president, secretary, and treasurer will be filled. In how many different ways can the offices be filled, if each of the 12 candidates can hold any office?

Solution:

${}_{12}P_4 = 12 \bullet 11 \bullet 10 \bullet 9 = 11{,}880$ ways

35. Find the number of distinguishable permutations of the letters A, A, Y, Y, Y, Y, X, X, X.

Solution:

$$\frac{9!}{2!4!3!} = \frac{9 \cdot 8 \cdot 7 \cdot 6 \cdot 5}{2 \cdot 3 \cdot 2} = 1260 \text{ distinguishable permutations}$$

41. In order to conduct a certain experiment, four students are randomly selected from a class of 20. How many different groups of four students are possible?

Solution:

$$_{20}C_4 = \frac{20!}{(20-4)!4!} = \frac{20!}{16!4!} = \frac{20 \cdot 19 \cdot 18 \cdot 17}{4 \cdot 3 \cdot 2} = 4845 \text{ different groups}$$

45. How many subsets of four elements can be formed from a set of 100 elements?

Solution:

$$_{100}C_4 = \frac{100!}{(100-4)!4!} = \frac{100!}{96!4!} = \frac{100 \cdot 99 \cdot 98 \cdot 97}{4 \cdot 3 \cdot 2} = 3{,}921{,}225 \text{ subsets}$$

49. An employer interviews eight people for four openings in the company. Three of the eight people are women. If all eight are qualified, in how many ways could the employer fill the four positions if (a) the selection is random and (b) exactly two are women?

Solution:

(a) $_8C_4 = \dfrac{8!}{(8-4)!4!} = \dfrac{8!}{4!4!} = \dfrac{8 \cdot 7 \cdot 6 \cdot 5}{4 \cdot 3 \cdot 2} = 70$ ways

(b) $_3C_2 \cdot {_5C_2} = \dfrac{3!}{(3-2)!2!} \cdot \dfrac{5!}{(5-2)!2!} = 3 \cdot 10 = 30$ ways

51. Four people are to be selected at random from a group of four couples. In how many ways can this be done, given the following conditions?

(a) There are no restrictions.

(b) There is to be at least one couple in the group of four.

(c) The selection must include one member from each couple.

Solution:

(a) $_8C_4 = \dfrac{8!}{4!4!} = 70$ ways

(b) There are 16 ways that a group of four can be formed without any couples in the group. Therefore, if at least one couple is to be in the group, there are $70 - 16 = 54$ ways that could occur.

(c) $2 \cdot 2 \cdot 2 \cdot 2 = 16$ ways

55. Find the number of diagonals of an octagon.

Solution:

$$_8C_2 - 8 = \frac{8!}{6!2!} - 8 = 28 - 8 = 20 \text{ diagonals}$$

61. Prove $_nC_{n-1} = {}_nC_1$.

Solution:

$$_nC_{n-1} = \frac{n!}{(n-(n-1))!(n-1)!} = \frac{n!}{(1)!(n-1)!} = \frac{n!}{(n-1)!1!} = {}_nC_1$$

SECTION 10.7

Probability

You should know the following basic principles of probability.

- If an event A has $n(A)$ equally likely outcomes and its sample space has $n(S)$ equally likely outcomes, then the probability of event A is
$$P(A) = \frac{n(A)}{n(S)}, \text{ where } 0 \leq P(A) \leq 1.$$
- If A and B are mutually exclusive events, then $P(A \cup B) = P(A) + P(B)$.
If A and B are not mutually exclusive events, then
$P(A \cup B) = P(A) + P(B) - P(A \cap B)$.
- If A and B are independent events, then the probability that both A and B will occur is $P(A)P(B)$.
- The complement of an event A is $P(A') = 1 - P(A)$.

Solutions to Selected Exercises

5. Two county supervisors are selected from five supervisors, A, B, C, D, and E, to study a recycling plan. Determine the sample space.

Solution:

S = {AB, AC, AD, AE, BC, BD, BE, CD, CE, DE}

9. A coin is tossed three times. Find the probability of getting at least one head.

Solution:

$$S = \{HHH,\ HHT,\ HTH,\ HTT,\ THH,\ THT,\ TTH,\ TTT\}$$
$$P(TTT) = \tfrac{1}{8}$$
$$P(\text{at least one head}) = 1 - \tfrac{1}{8} = \tfrac{7}{8}$$

13. One card is selected from a standard deck of 52 playing cards. Find the probability of getting a black card that is not a face card.

Solution:

Twenty–six of the cards are black. Six of these are face cards (J, Q, K of clubs and spades). Therefore, there are 20 black cards that are not face cards.

$$P(E) = \tfrac{20}{52} = \tfrac{5}{13}$$

15. A six-sided die is tossed twice. Find the probability that the sum is 4.

Solution:

$$n(S) = 6(6) = 36$$

$$E = \{(1,\ 3),\ (2,\ 2),\ (3,\ 1)\}$$

$$P(E) = \tfrac{3}{36} = \tfrac{1}{12}$$

19. A six-sided die is tossed twice. Find the probability that the sum is odd and no more than 7.

Solution:

$$n(S) = 6 \cdot 6 = 36$$

$$E = \{(1,\ 2),\ (1,\ 4),\ (1,\ 6),\ (2,\ 1),\ (2,\ 3),\ (2,\ 5),\ (3,\ 2),\ (3,\ 4),$$
$$(4,\ 1),\ (4,\ 3),\ (5,\ 2),\ (6,\ 1)\}$$

$$P(E) = \tfrac{12}{36} = \tfrac{1}{3}$$

23. Two marbles are drawn (the first is *not* replaced before the second is drawn) from a bag containing one green, two yellow, and three red marbles. Find the probability of drawing neither yellow marble.

Solution:

$$P(E) = \frac{{}_4C_2}{{}_6C_2} = \frac{\dfrac{4!}{2!2!}}{\dfrac{6!}{4!2!}} = \frac{6}{15} = \frac{2}{5}$$

27. The probability that an event *will not* happen is $p = 0.15$. Find the probability that the event *will* happen.

Solution:

$$1 - p = 1 - 0.15 = 0.85$$

31. Taylor, Moore, and Jenkins are candidates for public office. It is estimated that Moore and Jenkins have about the same probability of winning, and Taylor is believed to be twice as likely to win as either of the others. Find the probability of each candidate winning the election.

Solution:

$$p + p + 2p = 1$$

$$p = 0.25$$

Taylor: 0.50
Moore: 0.25
Jenkins: 0.25

35. Four letters and envelopes are addressed to four different people. If the letters are randomly inserted into the envelopes, what is the probability that (a) exactly one will be inserted in the correct envelope and (b) at least one will be inserted in the correct envelope?

Solution:

Total ways to insert letters: $4! = 24$ ways

4 correct: 1 way
3 correct: not possible
2 correct: 6 ways
1 correct: 8 ways
0 correct: 9 ways

(a) $\dfrac{8}{24} = \dfrac{1}{3}$

(b) $\dfrac{8+6+1}{24} = \dfrac{15}{24} = \dfrac{5}{8}$

39. Two cards are selected at random from an ordinary deck of 52 playing cards. Find the probability that two aces are selected, given the following conditions.

(a) The cards are drawn in sequence, with the first card being replaced and the deck reshuffled prior to the second drawing.

(b) The two cards are drawn consecutively, without replacement.

Solution:

(a) $\left(\frac{4}{52}\right)\left(\frac{4}{52}\right) = \frac{1}{169}$

(b) $\left(\frac{4}{52}\right)\left(\frac{3}{51}\right) = \frac{1}{221}$

43. Two integers (between 1 and 30 inclusive) are chosen by a random number generator on a computer. What is the probability that (a) the numbers are both even, (b) one number is even and one is odd, (c) both numbers are less than 10, and (d) the same number is chosen twice?

Solution:

(a) $P(EE) = \frac{15}{30} \cdot \frac{15}{30} = \frac{1}{4}$

(b) $P(EO \text{ or } OE) = 2\left(\frac{15}{30}\right)\left(\frac{15}{30}\right) = \frac{1}{2}$

(c) $P(N_1 < 10,\ N_2 < 10) = \frac{9}{30} \cdot \frac{9}{30} = \frac{9}{100}$

(d) $P(N_1 N_1) = \frac{30}{30} \cdot \frac{1}{30} = \frac{1}{30}$

45. A space vehicle has an independent back-up system for one of its communication networks. The probability that either system will function satisfactorily for the duration of a flight is 0.985. What is the probability that during a given flight (a) both systems function satisfactorily, (b) at least one system functions satisfactorily, and (c) both systems fail?

Solution:

(a) $P(SS) = (0.985)^2 \approx 0.9702$

(b) $P(S) = 1 - P(FF) = 1 - (0.015)^2 \approx 0.9998$

(c) $P(FF) = (0.015)^2 \approx 0.0002$

49. Assume that the probability of the birth of a child of a particular sex is 50%. In a family with four children, what is the probability that (a) all the children are boys, (b) all the children are the same sex, and (c) there is at least one boy?

Solution:

(a) $P(BBBB) = (\frac{1}{2})(\frac{1}{2})(\frac{1}{2})(\frac{1}{2}) = \frac{1}{16}$

(b) $P(BBBB) + P(GGGG) = \frac{1}{16} + \frac{1}{16} = \frac{1}{8}$

(c) $1 - P(GGGG) = 1 - \frac{1}{16} = \frac{15}{16}$

REVIEW EXERCISES FOR CHAPTER 10

Solutions to Selected Exercises

3. Use sigma notation to write the sum $\frac{1}{2} + \frac{2}{3} + \frac{3}{4} + \cdots + \frac{9}{10}$.

Solution:

$$\frac{1}{2} + \frac{2}{3} + \frac{3}{4} + \cdots + \frac{9}{10} = \sum_{k=1}^{9} \frac{k}{k+1}$$

7. Find the sum.

$$\sum_{i=0}^{6} 2^i$$

Solution:

$$\sum_{i=0}^{6} 2^i = 2^0 + 2^1 + 2^2 + 2^3 + 2^4 + 2^5 + 2^6 = 1 + 2 + 4 + 8 + 16 + 32 + 64 = 127$$

13. Find the sum.

$$\sum_{n=0}^{10} (n^2 + 3)$$

Solution:

$$\sum_{n=0}^{10} (n^2 + 3) = \sum_{n=0}^{10} n^2 + \sum_{n=0}^{10} 3 = \frac{10(11)(21)}{6} + 3(11) = 418$$

17. Write out the first five terms of the arithmetic sequence given $a_4 = 10$ and $a_{10} = 28$.

Solution:

$$a_4 = 10 = d(4) + c, \qquad a_{10} = 28 = d(10) + c$$

$$4d + c = 10$$

$$10d + c = 28$$

Solving this system yields $d = 3$ and $c = -2$. Thus,

$$a_n = 3n - 2$$

$$a_1 = 1$$

$$a_2 = 4$$

$$a_3 = 7$$

$$a_4 = 10$$

$$a_5 = 13$$

21. Find the sum for the first 100 multiples of 5.

Solution:

$$\sum_{i=1}^{100} 5i = 5\sum_{i=1}^{100} i = 5\left(\frac{100(101)}{2}\right) = 25{,}250$$

23. Write the first five terms of the geometric sequence, $a_1 = 4$, $r = -\frac{1}{4}$.

Solution:

$$a_1 = 4$$

$$a_2 = 4\left(-\frac{1}{4}\right) = -1$$

$$a_3 = 4\left(-\frac{1}{4}\right)^2 = \frac{1}{4}$$

$$a_4 = 4\left(-\frac{1}{4}\right)^3 = -\frac{1}{16}$$

$$a_5 = 4\left(-\frac{1}{4}\right)^4 = \frac{1}{64}$$

Terms: $4, -1, \frac{1}{4}, -\frac{1}{16}, \frac{1}{64}$

27. Write an expression for the nth term of the geometric sequence with $a_1 = 16$ and $a_2 = -8$. Then find the sum of the first 20 terms of the sequence.

Solution:

$$a_1 = 16,\ a_2 = -8 \Rightarrow r = -\frac{1}{2}$$

$$a_n = 16\left(-\frac{1}{2}\right)^{n-1}$$

$$S_{20} = 16\left[\frac{1-(-1/2)^{20}}{1-(-1/2)}\right] \approx 10.667$$

31. A deposit of \$200 is made at the beginning of each month for two years into an account that pays 6%, compounded monthly. What is the balance in the account at the end of two years?

Solution:

$$A = \sum_{i=1}^{24} 200\left(1+\frac{0.06}{12}\right)^i$$

$$= \sum_{i=1}^{24} 200(1.005)^i$$

$$= 200(1.005)\left[\frac{1-(1.005)^{24}}{1-1.005}\right]$$

$$\approx \$5111.82$$

33. Use mathematical induction to prove the formula $1+4+\cdots+(3n-2)=(n/2)(3n-1)$ for every positive integer n.

Solution:

1.) When $n=1$, $1=\frac{1}{2}[3(1)-1]=1$.

2.) Assume that

$$1+4+7+\cdots+(3k-2)=\frac{k}{2}(3k-1).$$

Then,

$$\begin{aligned}1+4+7+\cdots+(3k-2)+[3(k+1)-2] &= [1+4+7+\cdots+(3k-2)]+(3k+1)\\ &= \frac{k}{2}(3k-1)+(3k+1)\\ &= \frac{k(3k-1)}{2}+\frac{2(3k+1)}{2}\\ &= \frac{3k^2+5k+2}{2}\\ &= \frac{(k+1)(3k+2)}{2}\\ &= \frac{(k+1)}{2}[3(k+1)-1].\end{aligned}$$

Thus, the formula holds for all positive integers n.

37. Evaluate ${}_6C_4$.

Solution:

$${}_6C_4=\frac{6!}{(6-4)!4!}=\frac{6\bullet 5}{2}=15$$

43. Use the Binomial Theorem to expand the following binomial. Simplify your answer.

$$\left(\frac{2}{x}-3x\right)^6$$

Solution:

$$\begin{aligned}\left(\frac{2}{x}-3x\right)^6 &= \left(\frac{2}{x}\right)^6+6\left(\frac{2}{x}\right)^5(-3x)+15\left(\frac{2}{x}\right)^4(-3x)^2+20\left(\frac{2}{x}\right)^3(-3x)^3\\ &\quad+15\left(\frac{2}{x}\right)^2(-3x)^4+6\left(\frac{2}{x}\right)(-3x)^5+(-3x)^6\\ &= \frac{64}{x^6}-\frac{576}{x^4}+\frac{2160}{x^2}-4320+4860x^2-2916x^4+729x^6\end{aligned}$$

47. The complexity of interpersonal relationships increases dramatically as the size of a group increases. Determine the number of different two-person relationships in a family of (a) 2, (b) 4, and (c) 6.

Solution:

(a) ${}_2C_2 = 1$ (b) ${}_4C_2 = 6$ (c) ${}_6C_2 = 15$

51. A man has five pairs of socks (no two pairs are the same color). If he randomly selects two socks from the drawer, what is the probability that he gets a matched pair?

Solution:

$$P(\text{pair}) = \tfrac{10}{10} \bullet \tfrac{1}{9} = \tfrac{1}{9}$$

55. Find the probability of obtaining at least one tail when a coin is tossed five times.

Solution:

$$1 - P(HHHHH) = 1 - \left(\frac{1}{2}\right)^5 = \frac{31}{32}$$

57. Five cards are drawn from an ordinary deck of 52 playing cards. Find the probability of getting two pairs. (For example, the hand could be A-A-5-5-Q or 4-4-7-7-K.)

Solution:

$$P(2 \text{ pairs}) = \frac{({}_{13}C_2)({}_4C_2)({}_4C_2)({}_{44}C_1)}{({}_{52}C_5)} = 0.0475$$

Practice Test for Chapter 10

1. Write out the first five terms of the sequence $a_n = \dfrac{2n}{(n+2)!}$.

2. Write an expression for the nth term of the sequence $\{\frac{4}{3}, \frac{5}{9}, \frac{6}{27}, \frac{7}{81}, \frac{8}{243}, \ldots\}$.

3. Find the sum $\displaystyle\sum_{i=1}^{6}(2i-1)$.

4. Write out the first five terms of the arithmetic sequence where $a_1 = 23$ and $d = -2$.

5. Find a_n for the arithmetic sequence with $a_1 = 12$, $d = 3$, and $n = 50$.

6. Find the sum of the first 200 positive integers.

7. Write out the first five terms of the geometric sequence with $a_1 = 7$ and $r = 2$.

8. Evaluate $\displaystyle\sum_{n=0}^{9} 6\left(\frac{2}{3}\right)^n$.

9. Evaluate $\displaystyle\sum_{n=0}^{\infty}(0.03)^n$.

10. Use mathematical induction to prove that $1 + 2 + 3 + 4 + \cdots + n = \dfrac{n(n+1)}{2}$.

11. Use mathematical induction to prove that $n! > 2^n$, $n \geq 4$.

12. Evaluate ${}_{13}C_4$.

13. Expand $(x+3)^5$.

14. Find the term involving x^7 in $(x-2)^{12}$.

15. Evaluate ${}_{30}P_4$.

16. How many ways can six people sit at a table with six chairs?

17. Twelve cars run in a race. How many different ways can they come in first, second, and third place? (Assume that there are no ties.)

18. Two six-sided dice are tossed. Find the probability that the total of the two dice is less than 5.

19. Two cards are selected at random from a deck of 52 playing cards without replacement. Find the probability that the first card is a King and the second card is a black ten.

20. A manufacturer has determined that for every 1000 units it produces, 3 will be faulty. What is the probability that an order of 50 units will have one or more faulty units?

CHAPTER 11

Some Topics in Analytic Geometry

Section 11.1 Lines **447**

Section 11.2 Introduction to Conics: Parabolas **452**

Section 11.3 Ellipses **458**

Section 11.4 Hyperbolas **463**

Section 11.5 Rotation and the General Second-Degree Equation **467**

Section 11.6 Polar Coordinates **472**

Section 11.7 Graphs of Polar Equations **476**

Section 11.8 Polar Equations of Conics **481**

Section 11.9 Plane Curves and Parametric Equations **485**

Review Exercises **492**

Practice Test **498**

SECTION 11.1

Lines

- The **inclination** of a non-horizontal line is the positive angle θ ($\theta < 180°$) measured counterclockwise from the x-axis to the line. A horizontal line has an inclination of zero.
- If a nonvertical line has inclination θ and slope m, then $m = \tan\theta$.
- If two non-perpendicular lines have slopes m_1 and m_2, then the angle between the lines is given by

$$\tan\theta = \left|\frac{m_2 - m_1}{1 + m_1 m_2}\right|.$$

- The distance between a point (x_1, y_1) and a line $Ax + By + C = 0$ is given by

$$d = \frac{|Ax_1 + By_1 + C|}{\sqrt{A^2 + B^2}}.$$

Solutions to Selected Exercises

5. Find the slope of the line with inclination $\theta = 38.2°$.

Solution:

$$m = \tan 38.2° \approx 0.7869$$

11. Find the inclination, θ, of the line with slope $m = \frac{3}{4}$.

Solution:

$$\tfrac{3}{4} = \tan\theta$$

$$\theta = \tan^{-1}\tfrac{3}{4} \approx 36.9°$$

15. Find the inclination, θ, of the line passing through the points $(-2, 20)$ and $(10, 0)$.

Solution:

$$m = \frac{0 - 20}{10 - (-2)} = \frac{-20}{12} = -\frac{5}{3}$$

$$-\frac{5}{3} = \tan\theta$$

$$\theta = 180° + \tan^{-1}\left(-\frac{5}{3}\right) \approx 121.0°$$

19. Find the inclination, θ, of the line $5x + 3y = 0$.

Solution:

$$5x + 3y = 0$$
$$3y = -5x$$
$$y = -\tfrac{5}{3}x \Rightarrow m = -\tfrac{5}{3}$$
$$-\tfrac{5}{3} = \tan\theta$$
$$\theta = 180° + \tan^{-1}\left(-\tfrac{5}{3}\right) \approx 121.0°$$

23. Find the angle, θ, between the lines $x - y = 0$ and $3x - 2y + 1 = 0$.

Solution:

$$x - y = 0 \Rightarrow y = x \Rightarrow m_1 = 1$$
$$3x - 2y + 1 = 0 \Rightarrow y = \frac{3}{2}x + \frac{1}{2} \Rightarrow m_2 = \frac{3}{2}$$
$$\tan\theta = \left|\frac{(3/2) - 1}{1 + (1)(3/2)}\right| = \left|\frac{1/2}{5/2}\right| = \frac{1}{5}$$
$$\theta = \tan^{-1}\left(\frac{1}{5}\right) \approx 11.3°$$

29. Find the angle, θ, between the lines $0.05x - 0.03y - 0.21 = 0$ and $0.07x + 0.02y - 0.16 = 0$.

Solution:

$$0.05x - 0.03y - 0.21 = 0 \Rightarrow 5x - 3y - 21 = 0 \Rightarrow y = \frac{5}{3}x - 7 \Rightarrow m_1 = \frac{5}{3}$$
$$0.07x + 0.02y - 0.16 = 0 \Rightarrow 7x + 2y - 16 = 0 \Rightarrow y = -\frac{7}{2}x + 8 \Rightarrow m_2 = -\frac{7}{2}$$
$$\tan\theta = \left|\frac{-(7/2) - (5/3)}{1 + (5/3)(-7/2)}\right| = \left|\frac{-31/6}{-29/6}\right| = \frac{31}{29}$$
$$\theta = \tan^{-1}\left(\frac{31}{29}\right) \approx 46.9°$$

33. Find the distance between the point (2, 3) and the line $4x + 3y - 10 = 0$.

Solution:

$(2,\ 3) \Rightarrow x_1 = 2$ and $y_1 = 3$

$4x + 3y - 10 = 0 \Rightarrow A = 4,\ B = 3,$ and $C = -10$

$$d = \frac{|4(2) + 3(3) + (-10)|}{\sqrt{4^2 + 3^2}} = \frac{7}{5} = 1.4$$

37. Find the distance between the point (0, 8) and the line $6x - y = 0$.

Solution:

$(0,\ 8) \Rightarrow x_1 = 0$ and $y_1 = 8$

$6x - y = 0 \Rightarrow A = 6,\ B = -1,$ and $C = 0$

$$d = \frac{|6(0) + (-1)(8) + 0|}{\sqrt{6^2 + (-1)^2}} = \frac{8}{\sqrt{37}} = \frac{8\sqrt{37}}{37} \approx 1.3152$$

41. A straight road rises with an inclination of 6.5° from the horizontal. Find the slope of the road and the change in elevation after driving two miles.

Solution:

Slope: $m = \tan 6.5° \approx 0.1139$

Change in elevation: $\sin 6.5° = \dfrac{x}{2}$

$$2 \sin 6.5° = x$$

$$x \approx 0.2264 \text{ mile}$$

$$\approx 0.2264(5280)$$

$$\approx 1195 \text{ feet}$$

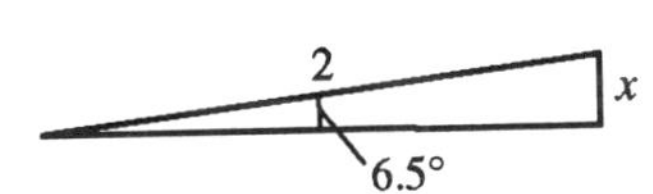

45. There is a rise of two feet for every horizontal change of three feet on a roof. Find the inclination of the roof.

Solution:

$$\tan\theta = \tfrac{2}{3}$$

$$\theta = \tan^{-1}\left(\tfrac{2}{3}\right)$$

$$\approx 33.7°$$

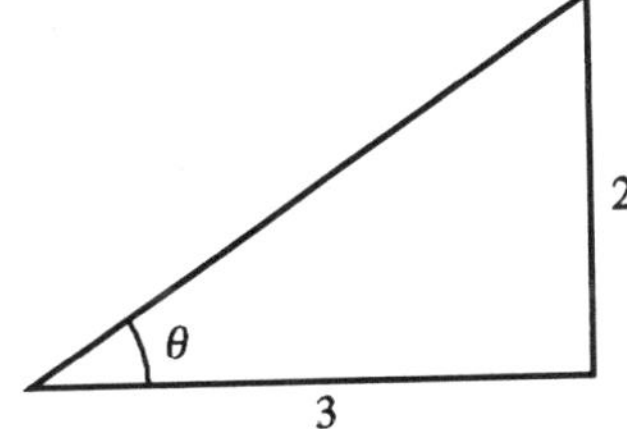

49. (a) Find the altitude from vertex B of a triangle to the side AC, and (b) find the area of the triangle given $A = \left(-\frac{1}{2}, \frac{1}{2}\right)$, $B = (2, 3)$, and $C = \left(\frac{5}{2}, 0\right)$.

Solution:

(a) The slope of the line through AC is

$$m = \frac{0 - (1/2)}{(5/2) - (-1/2)} = -\frac{1}{6}.$$

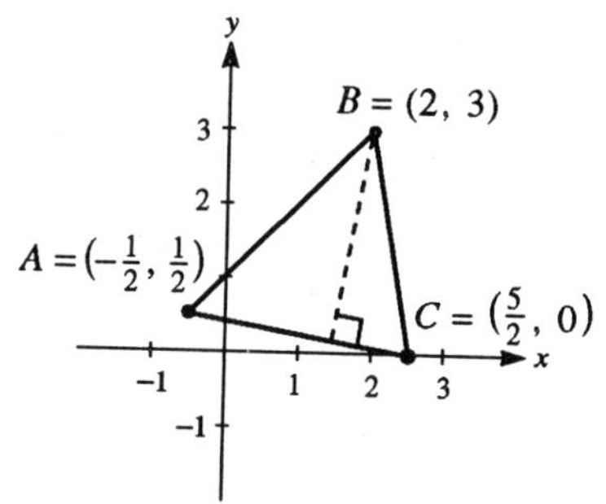

The equation of the line is

$$y - 0 = -\frac{1}{6}\left(x - \frac{5}{2}\right)$$

$$y = -\frac{1}{6}x + \frac{5}{12}$$

$$12y = -2x + 5$$

$$2x + 12y - 5 = 0.$$

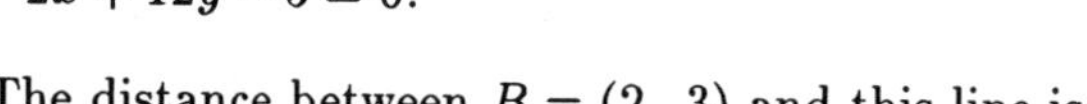
The distance between $B = (2, 3)$ and this line is

$$d = \frac{|2(2) + 12(3) + (-5)|}{\sqrt{2^2 + 12^2}} = \frac{35}{2\sqrt{37}} = \frac{35\sqrt{37}}{74}.$$

(b) The distance between A and C is

$$d = \sqrt{\left(\frac{5}{2} - \left(-\frac{1}{2}\right)\right)^2 + \left(0 - \frac{1}{2}\right)^2} = \sqrt{9 + \frac{1}{4}} = \sqrt{\frac{37}{4}} = \frac{\sqrt{37}}{2}.$$

The area of the triangle is

$$A = \frac{1}{2}bh = \frac{1}{2}\left(\frac{\sqrt{37}}{2}\right)\left(\frac{35\sqrt{37}}{74}\right) = \frac{35}{8} \text{ square units.}$$

53. Find the slope of each side of the triangle and use the slopes to find the magnitude of the interior angles.

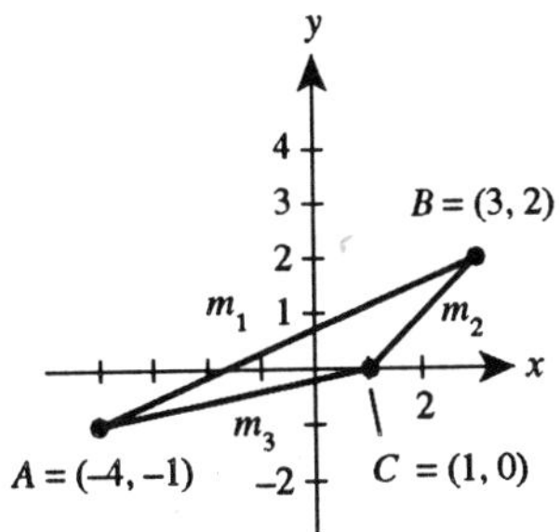

Solution:

Let $A = (-4,\ -1)$, $B = (3,\ 2)$, and $C = (1,\ 0)$. The slopes are as follows.

$$AB:\quad m_1 = \frac{2-(-1)}{3-(-4)} = \frac{3}{7}$$

$$BC:\quad m_2 = \frac{0-2}{1-3} = 1$$

$$AC:\quad m_3 = \frac{0-(-1)}{1-(-4)} = \frac{1}{5}$$

The angle at A is found by

$$\tan A = \left|\frac{m_3 - m_1}{1 + m_1 m_3}\right| = \left|\frac{(1/5)-(3/7)}{1+(3/7)(1/5)}\right| = \frac{4}{19}$$

$$A = \tan^{-1}\frac{4}{19} \approx 11.9°.$$

The angle at B is found by

$$\tan B = \left|\frac{m_2 - m_1}{1 + m_1 m_2}\right| = \left|\frac{1-(3/7)}{1+(3/7)(1)}\right| = \frac{2}{5}$$

$$B = \tan^{-1}\left(\frac{2}{5}\right) \approx 21.8°.$$

The angle at C is found by

$$C = 180° - A - B = 180° - 11.9° - 21.8° = 146.3°.$$

SECTION 11.2

Introduction to Conics: Parabolas

- A *parabola* is the set of all points (x, y) that are equidistant from a fixed line (*directrix*) and a fixed point (*focus*) not on the line.
- The standard equation of a parabola with vertex (h, k) and:

  (a) Vertical axis $x = h$ and directrix $y = k - p$ is:

  $(x - h)^2 = 4p(y - k), \quad p \neq 0$

  (b) Horizontal axis $y = k$ and directrix $x = h - p$ is:

  $(y - k)^2 = 4p(x - h), \quad p \neq 0$

Solutions to Selected Exercises

5. Match $(y - 1)^2 = 4(x - 2)$ with the correct graph.

Solution:

The vertex is at $(2, 1)$ and the axis is horizontal. Therefore, it matches graph (d).

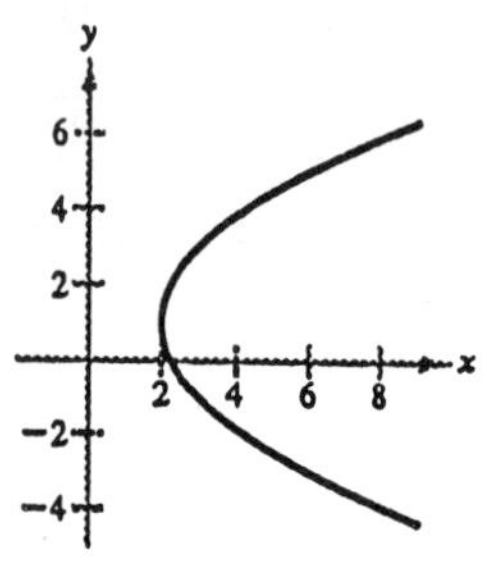

11. Find the vertex, focus, and directrix of the parabola $(x - 1)^2 + 8(y + 2) = 0$ and sketch its graph.

Solution:

$$(x - 1)^2 + 8(y + 2) = 0$$
$$(x - 1)^2 = -8(y + 2)$$
$$(x - 1)^2 = 4(-2)(y + 2)$$

$h = 1,\ k = -2,\ p = -2$

Vertex: $(1, -2)$

Focus: $(1, -2 + (-2))$ or $(1, -4)$

Directrix: $y = -2 - (-2)$ or $y = 0$

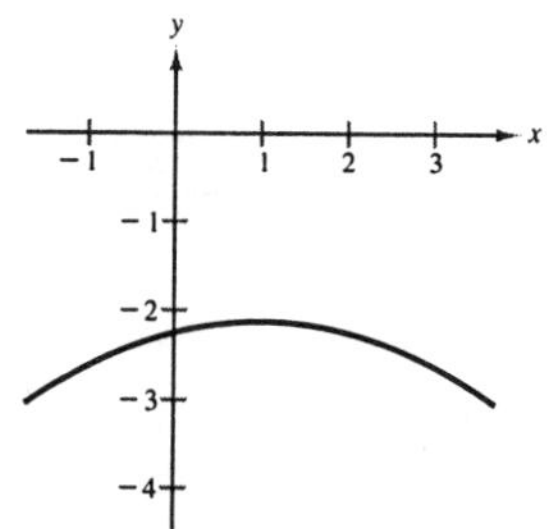

13. Find the vertex, focus, and directrix of the parabola $y = \frac{1}{4}(x^2 - 2x + 5)$ and sketch its graph.

Solution:

$$y = \tfrac{1}{4}(x^2 - 2x + 5)$$
$$4y = x^2 - 2x + 1 - 1 + 5$$
$$4y = (x-1)^2 + 4$$
$$4y - 4 = (x-1)^2$$
$$(x-1)^2 = 4(y-1)$$
$$(x-1)^2 = 4(1)(y-1)$$

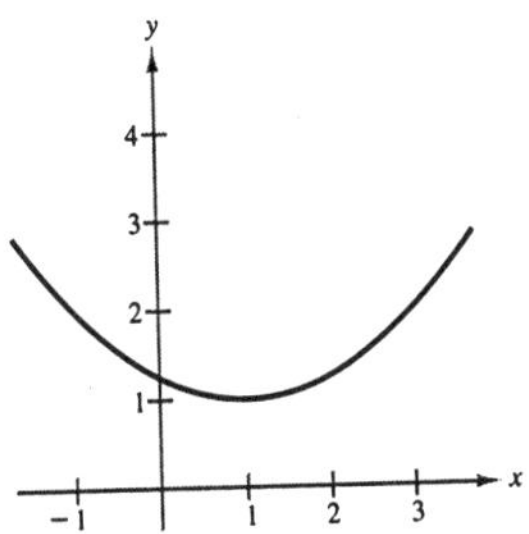

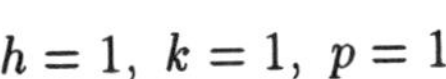
$h = 1,\ k = 1,\ p = 1$

Vertex: $(1,\ 1)$

Focus: $(1,\ 1+1)$ or $(1,\ 2)$

Directrix: $y = 1 - 1$ or $y = 0$

15. Find the vertex, focus, and directrix of the parabola $y^2 + 6y + 8x + 25 = 0$ and sketch its graph.

Solution:

$$y^2 + 6y + 8x + 25 = 0$$
$$y^2 + 6y = -8x - 25$$
$$y^2 + 6y + 9 = -8x - 25 + 9$$
$$(y+3)^2 = -8x - 16$$
$$(y+3)^2 = -8(x+2)$$
$$(y+3)^2 = 4(-2)(x+2)$$

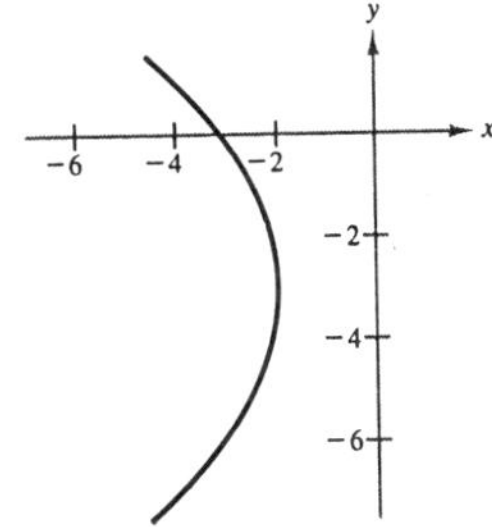

$h = -2,\ k = -3,\ p = -2$

Vertex: $(-2,\ -3)$

Focus: $(-2 + (-2),\ -3)$ or $(-4,\ -3)$

Directrix: $x = -2 - (-2)$ or $x = 0$

17. Find the vertex, focus, and directrix of the parabola $y^2 - 4x - 4 = 0$ and sketch its graph.

Solution:

$$y^2 - 4x - 4 = 0$$
$$y^2 = 4x + 4$$
$$(y - 0)^2 = 4(x + 1)$$
$$(y - 0)^2 = 4(1)(x + 1)$$

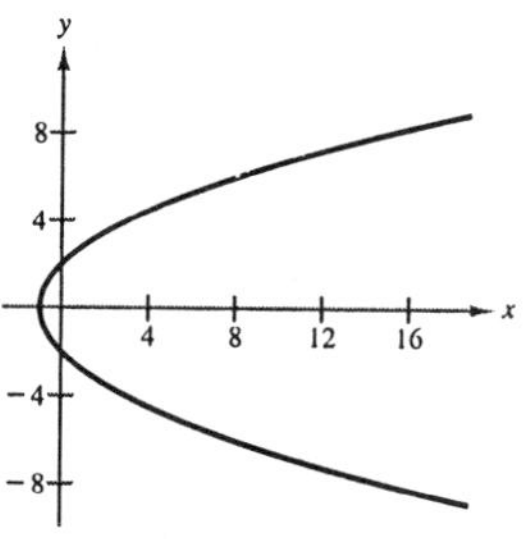

$h = -1, \ k = 0, \ p = 1$
Vertex: $(-1, \ 0)$
Focus: $(-1 + 1, \ 0)$ or $(0, \ 0)$
Directrix: $x = -1 - 1$ or $x = -2$

21. Find an equation of the specified parabola.

Vertex: $(0, \ 0)$
Directrix: $x = 3$

Solution:

The vertex is $(0, \ 0)$ and the axis is horizontal with $p = -3$.

$$(y - 0)^2 = 4(-3)(x - 0)$$
$$y^2 = -12x$$

25. Find an equation of the specified parabola.

Axis: parallel to the y-axis
Passes through the points $(0, \ 3)$, $(3, \ 4)$, and $(4, \ 11)$

Solution:

Since the axis is vertical, we have $(x - h)^2 = 4p(y - k)$. Now, substituting the x- and y-values of the given points into this equation yields:

$(0, \ 3)$: $\quad (0 - h)^2 = 4p(3 - k) \Rightarrow h^2 = 12p - 4pk$
$(3, \ 4)$: $\quad (3 - h)^2 = 4p(4 - k) \Rightarrow (3 - h)^2 = 16p - 4pk$
$(4, \ 11)$: $\quad (4 - h)^2 = 4p(11 - k) \Rightarrow (4 - h)^2 = 44p - 4pk$

By subtraction we have: $h^2 - (3 - h)^2 = (12p - 4pk) - (16p - 4pk)$

$$6h - 9 = -4p$$
$$h^2 - (4 - h)^2 = (12p - 4pk) - (44p - 4pk)$$
$$8h - 16 = -32p$$

–CONTINUED ON NEXT PAGE–

25. –CONTINUED–

Using the method of elimination yields:

$$\begin{aligned} 8h - 16 = -32p \Rightarrow \quad 2h - \quad 4 &= -8p \\ 6h - \ 9 = \ -4p \Rightarrow -12h + \quad 18 &= \ 8p \\ \hline -10h + \quad 14 &= \ 0 \\ h &= \tfrac{-14}{-10} = \tfrac{7}{5} \\ 2\left(\tfrac{7}{5}\right) - 4 &= -8p \\ p &= \tfrac{3}{20} \\ \left(\tfrac{7}{5}\right)^2 &= 12\left(\tfrac{3}{20}\right) - 4\left(\tfrac{3}{20}\right)k \\ k &= -\tfrac{4}{15} \end{aligned}$$

Thus,

$$\left(x - \frac{7}{5}\right)^2 = 4\left(\frac{3}{20}\right)\left(y + \frac{4}{15}\right)$$

$$x^2 - \frac{14}{5}x + \frac{49}{25} = \frac{3}{5}y + \frac{4}{25}$$

$$25x^2 - 70x + 49 = 15y + 4$$

$$25x^2 - 70x - 15y + 45 = 0$$

$$5x^2 - 14x - 3y + 9 = 0$$

27. Find an equation of the specified parabola.

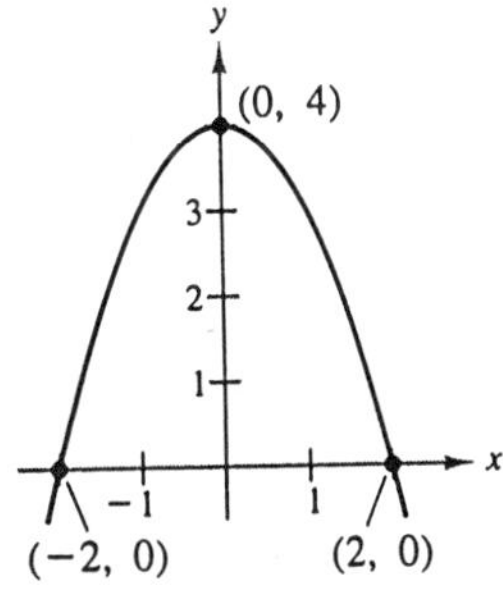

Solution:

The x-intercepts occur at $(\pm 2,\ 0)$ and the parabola opens downward.

$$y = -(x+2)(x-2)$$

$$y = -(x^2 - 4)$$

$$y = 4 - x^2$$

$$x^2 + y - 4 = 0$$

29. The receiver in a parabolic television dish antenna is three feet from the vertex and is located at the focus, as shown in the figure. Find an equation of a cross section of the reflector. (Assume the dish is directed upward and the vertex is at the origin.)

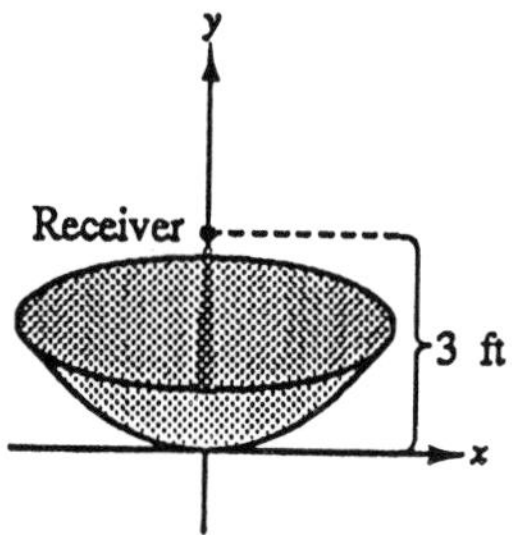

Solution:

The vertex is at (0, 0) and the focus is at (0, 3). Therefore, $p = 3$.

$$(x-0)^2 = 4(3)(y-0)$$
$$x^2 = 12y$$

33. A satellite in a 100-mile-high circular orbit around the earth has a velocity of approximately 17,500 miles per hour. If this velocity is multiplied by $\sqrt{2}$, then the satellite will have the minimum velocity necessary to escape the earth's gravity and it will follow a parabolic path with the center of the earth as the focus (see figure).

(a) Find the escape velocity of the satellite.

(b) Find the equation of its path (assume that the radius of the earth is 4000 miles.)

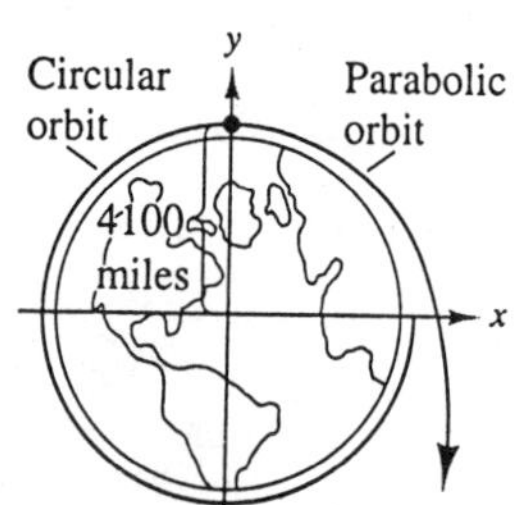

Solution:

(a) $17{,}500\sqrt{2} \approx 24{,}748.737 \approx 24{,}749$ mph

(b) Since the vertex is at (0, 4100) and the focus is at (0, 0), we have

$$(x-0)^2 = 4(-4100)(y-4100)$$
$$x^2 = -16{,}400(y-4100).$$

35. A ball is thrown horizontally from the top of a 75-foot tower with a velocity of 32 feet per second.

(a) Find the equation of the parabolic path.

(b) How far does the ball travel horizontally before striking the ground?

Solution:

(a)

$$\begin{aligned} y &= -\frac{16}{v^2}x^2 + s \\ &= -\frac{16}{(32)^2}x^2 + 75 \\ &= -\frac{x^2}{64} + 75 \text{ or} \end{aligned}$$

$$x^2 + 64y = 4800$$

(b) When $y = 0$, we have

$$\begin{aligned} x^2 &= 4800 \\ x &= \sqrt{4800} \\ &= 40\sqrt{3} \approx 69.3 \text{ feet.} \end{aligned}$$

41. Find the equation of the tangent line to the parabola $y = -2x^2$ at the point $(-1, -2)$ and find the x-intercept of the line.

Solution:

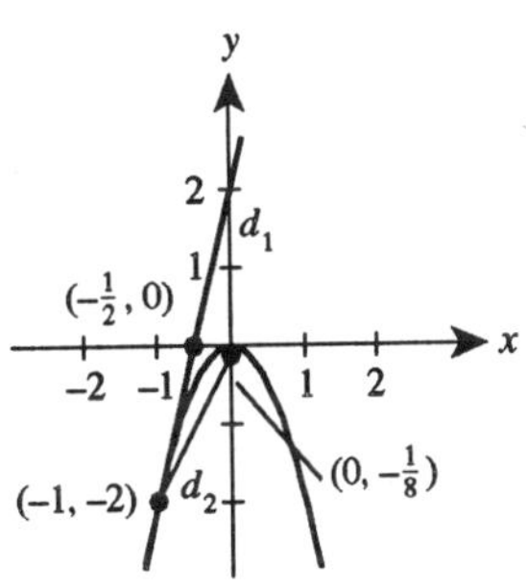

$$y = -2x^2$$

$$x^2 = -\frac{1}{2}y$$

$$x^2 = 4\left(-\frac{1}{8}\right)y \Rightarrow p = -\frac{1}{8}$$

The focus is at $\left(0, -\frac{1}{8}\right)$.

$$d_1 = \frac{1}{8} + b$$

$$d_2 = \sqrt{(-1-0)^2 + \left(-2 - \left(-\frac{1}{8}\right)\right)^2} = \sqrt{1 + \frac{225}{64}} = \frac{17}{8}$$

$$d_1 = d_2$$

$$\frac{1}{8} + b = \frac{17}{8}$$

$$b = 2$$

$$m = \frac{-2-2}{-1-0} = 4$$

$$y = 4x + 2$$

x-intercept: $\left(-\frac{1}{2}, 0\right)$

SECTION 11.3

Ellipses

- An *ellipse* is the set of all points (x, y) the sum of whose distances from two distinct fixed points (*foci*) is constant.
- The standard equation of an ellipse with center (h, k) and major and minor axes of lengths $2a$ and $2b$ is:

  (a) $\dfrac{(x-h)^2}{a^2} + \dfrac{(y-k)^2}{b^2} = 1$ if the major axis is horizontal.

  (b) $\dfrac{(x-h)^2}{b^2} + \dfrac{(y-k)^2}{a^2} = 1$ if the major axis is vertical.
- $c^2 = a^2 - b^2$ where c is the distance from the center to a focus.
- The eccentricity of an ellipse is $e = \dfrac{c}{a}$.

Solutions to Selected Exercises

3. Match the following equation with the correct graph.

$$\frac{x^2}{9} + \frac{y^2}{4} = 1$$

Solution:

$a = 3,\ b = 2$

Center: $(0, 0)$

Major axis is horizontal.

Therefore, it matches graph (c).

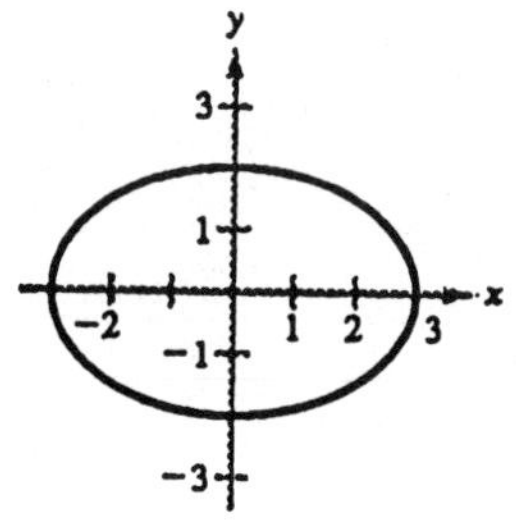

7. Find the center, foci, vertices, and eccentricity of the following ellipse and sketch its graph.

$$\frac{x^2}{25} + \frac{y^2}{16} = 1$$

Solution:

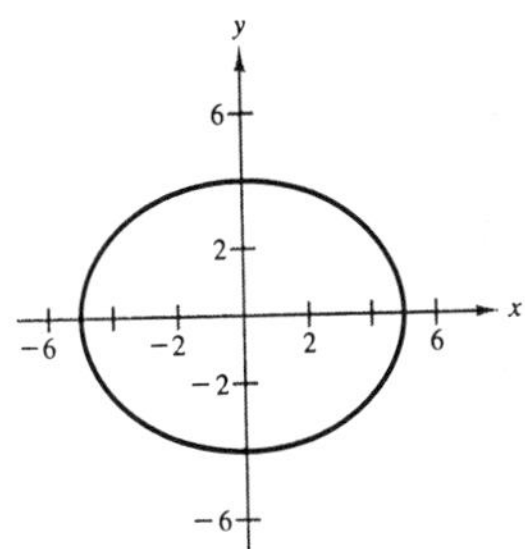

$a^2 = 25,\ b^2 = 16,\ c^2 = 9$
Center: $(0,\ 0)$
Foci: $(\pm 3,\ 0)$
Vertices: $(\pm 5,\ 0)$, $e = \frac{3}{5}$

9. Find the center, foci, vertices, and eccentricity of the following ellipse and sketch its graph.

$$\frac{x^2}{144} + \frac{y^2}{169} = 1$$

Solution:

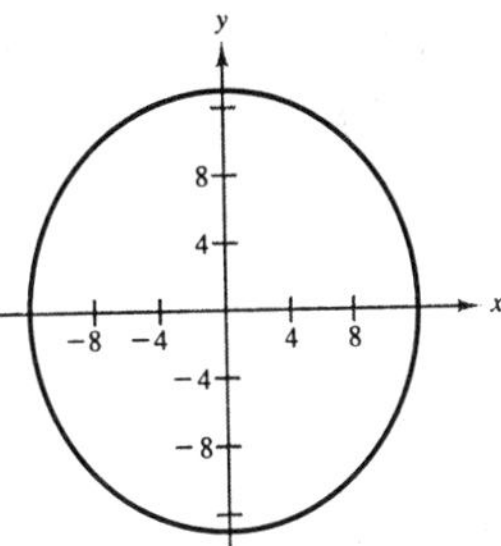

$a^2 = 169,\ b^2 = 144,\ c^2 = 25$
Center: $(0,\ 0)$
Foci: $(0,\ \pm 5)$
Vertices: $(0,\ \pm 13)$, $e = \frac{5}{13}$

11. Find the center, foci, vertices, and eccentricity of the ellipse $3x^2 + 2y^2 = 6$ and sketch its graph.

Solution:

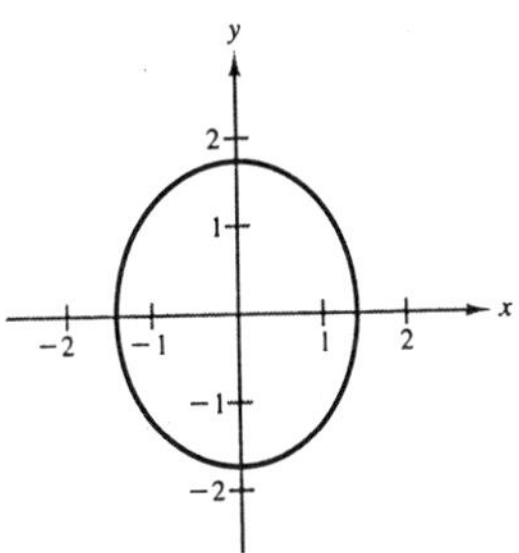

$$3x^2 + 2y^2 = 6$$

$$\frac{x^2}{2} + \frac{y^2}{3} = 1$$

$a^2 = 3,\ b^2 = 2,\ c^2 = 1$
Center: $(0,\ 0)$
Foci: $(0,\ \pm 1)$
Vertices: $(0,\ \pm\sqrt{3})$, $e = \dfrac{1}{\sqrt{3}} = \dfrac{\sqrt{3}}{3}$

13. Find the center, foci, vertices, and eccentricity of the following ellipse and sketch its graph.

$$\frac{(x-1)^2}{9}+\frac{(y-5)^2}{25}=1$$

Solution:

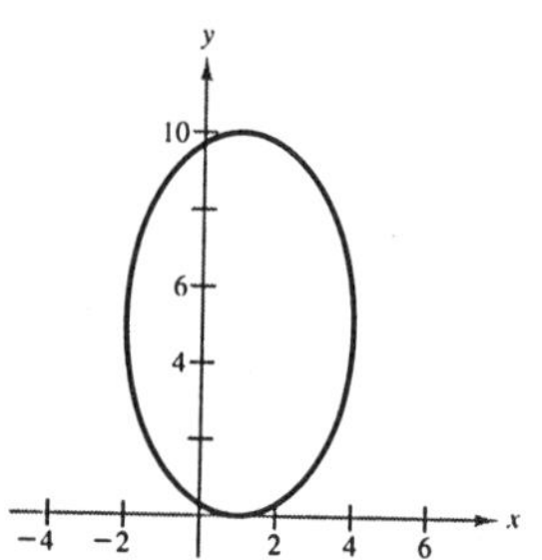

$a^2=25,\ b^2=9,\ c^2=16$
Center: $(1,\ 5)$
Foci: $(1,\ 5\pm 4)$ or $(1,\ 9),\ (1,\ 1)$
Vertices: $(1,\ 5\pm 5)$ or $(1,\ 10),\ (1,\ 0),\quad e=\frac{4}{5}$

17. Find the center, foci, vertices, and eccentricity of the ellipse $16x^2+25y^2-32x+50y+16=0$ and sketch its graph.

Solution:

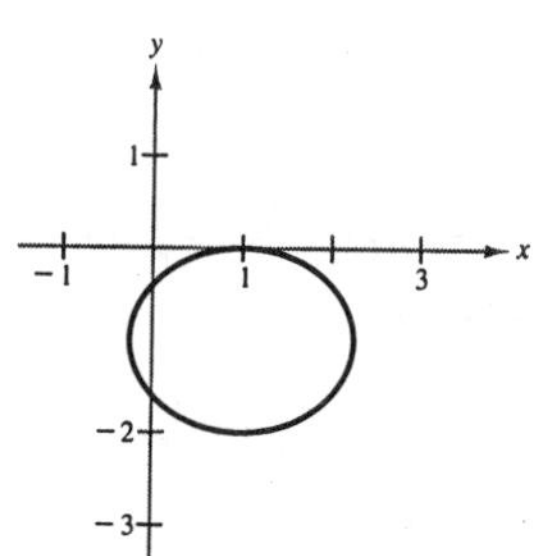

$$16x^2+25y^2-32x+50y+16=0$$
$$16(x^2-2x+1)+25(y^2+2y+1)=-16+16+25$$
$$16(x-1)^2+25(y+1)^2=25$$
$$\frac{(x-1)^2}{25/16}+\frac{(y+1)^2}{1}=1$$

$a^2=\frac{25}{16},\ b^2=1,\ c^2=\frac{9}{16}$
Center: $(1,\ -1)$
Foci: $\left(1\pm\frac{3}{4},\ -1\right)$ or $\left(\frac{1}{4},\ -1\right),\ \left(\frac{7}{4},\ -1\right)$
Vertices: $\left(1\pm\frac{5}{4},\ -1\right)$ or $\left(-\frac{1}{4},\ -1\right),\ \left(\frac{9}{4},\ -1\right),\quad e=\frac{3}{5}$

21. Find an equation of the specified ellipse.

Vertices: $(\pm 6,\ 0)$
Foci: $(\pm 5,\ 0)$

Solution:

The major axis is horizontal with the center at $(0,\ 0)$.

$a=6,\ c=5$ implies $b=\sqrt{11}$.

$$\frac{(x-0)^2}{(6)^2}+\frac{(y-0)^2}{(\sqrt{11})^2}=1$$

$$\frac{x^2}{36}+\frac{y^2}{11}=1$$

25. Find an equation of the specified ellipse.

Foci: $(0,\ 0),\ (0,\ 8)$
Major axis of length 16

Solution:

The major axis is vertical with center at $(0,\ 4)$.

$2a=16\Rightarrow a=8$
$c=4\Rightarrow b=\sqrt{48}=4\sqrt{3}$

$$\frac{(x-0)^2}{(\sqrt{48})^2}+\frac{(y-4)^2}{(8)^2}=1$$

$$\frac{x^2}{48}+\frac{(y-4)^2}{64}=1$$

29. A fireplace arch is to be constructed in the shape of a semi-ellipse. The opening is to have a height of 2 feet at the center and a width of 5 feet along the base, as shown in the figure. The contractor will first draw the form of the ellipse by the method shown in Figure 11.16. Where should the tacks be placed and what should be the length of the piece of string?

Solution:

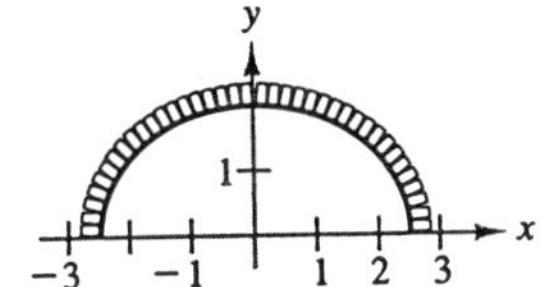

$a = \frac{5}{2},\ b = 2,\ c = \sqrt{\left(\frac{5}{2}\right)^2 - (2)^2} = \frac{3}{2}$

The tacks should be placed 1.5 feet from the center.

$d_1 + d_2 = 2a = 2\left(\frac{5}{2}\right) = 5$

The string should be $2a = 5$ feet.

33. A line segment through a focus with endpoints on the ellipse and perpendicular to the major axis is called a **latus rectum** of the ellipse. Therefore, an ellipse has two latus recta. Knowing the length of the latus recta is helpful in sketching an ellipse because it yields other points on the curve (see figure). The length of each latus rectum is $2b^2/a$. Graph $(x^2/4) + (y^2/1) = 1$ making use of the latus recta.

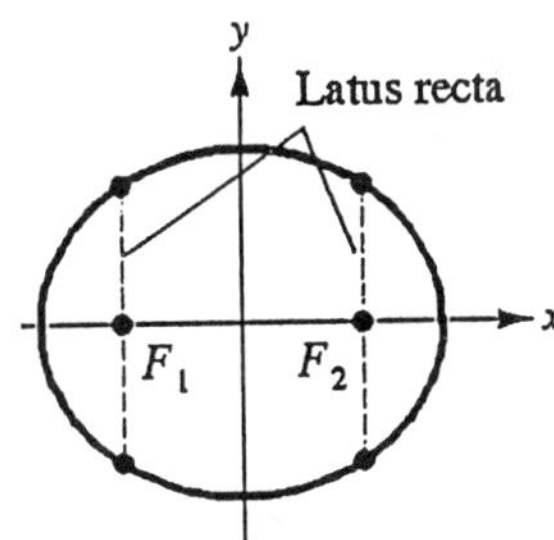

Solution:

Since $a = 2,\ b = 1,\ c = \sqrt{3}$, and $(2b^2)/a = 1$, we have the following graph.

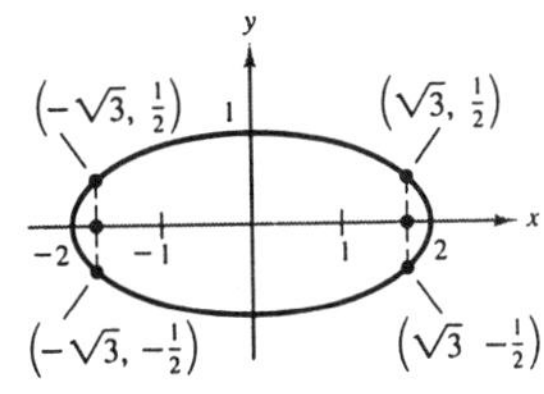

37. The earth moves in an elliptical orbit with the sun at one of the foci. The length of half of the major axis is 92.957×10^6 miles and the eccentricity is 0.017. Find the least and greatest distances of the earth from the sun.

Solution:

$e = \dfrac{c}{a}$, $a = 92.957 \times 10^6$, $e = 0.017$, $0.017 = \dfrac{c}{92.957 \times 10^6}$, $c \approx 1{,}580{,}269$

Least distance: $a - c \approx 91{,}376{,}731$ miles ≈ 91.377 million miles

Greatest distance: $a + c \approx 94{,}537{,}269$ miles ≈ 94.537 million miles

43. Show that the equation of an ellipse can be written as

$$\frac{(x-h)^2}{a^2} + \frac{(y-k)^2}{a^2(1-e^2)} = 1.$$

Note that as e approaches zero, with a remaining fixed, the ellipse approaches a circle of radius a.

Solution:

$$\frac{(x-h)^2}{a^2} + \frac{(y-k)^2}{b^2} = 1$$

$$\frac{(x-h)^2}{a^2} + \frac{(y-k)^2}{a^2(b^2/a^2)} = 1$$

$$\frac{(x-h)^2}{a^2} + \frac{(y-k)^2}{a^2(a^2-c^2)/a^2} = 1$$

$$\frac{(x-h)^2}{a^2} + \frac{(y-k)^2}{a^2(1-e^2)} = 1$$

As $e \Rightarrow 0$, $1-e^2 \Rightarrow 1$ and we have $\dfrac{(x-h)^2}{a^2} + \dfrac{(y-k)^2}{a^2} = 1$ or the circle $(x-h)^2+(y-k)^2 = a^2$.

SECTION 11.4

Hyperbolas

- A *hyperbola* is the set of all points $(x,\ y)$ the difference of whose distances from two distinct fixed points (*foci*) is constant.

- The standard equation of a hyperbola with center $(h,\ k)$ and transverse and conjugate axes of lengths $2a$ and $2b$ is:

  (a) $\dfrac{(x-h)^2}{a^2} - \dfrac{(y-k)^2}{b^2} = 1$ if the transverse axis is horizontal.

  (b) $\dfrac{(y-k)^2}{a^2} - \dfrac{(x-h)^2}{b^2} = 1$ if the transverse axis is vertical.

- $c^2 = a^2 + b^2$ where c is the distance from the center to a focus.

- The asymptotes of a hyperbola are:

  (a) $y = k \pm \dfrac{b}{a}(x-h)$ if the transverse axis is horizontal.

  (b) $y = k \pm \dfrac{a}{b}(x-h)$ if the transverse axis is vertical.

- The eccentricity of a hyperbola is $e = \dfrac{c}{a}$.

- To classify a nondegenerate conic from its general equation $Ax^2+Cy^2+Dx+Ey+F=0$:

  (a) If $A = C$ $(A \neq 0,\ C \neq 0)$, then it is a circle.
  (b) If $AC = 0$ $(A = 0$ or $C = 0$, but not both), then it is a parabola.
  (c) If $AC > 0$, then it is an ellipse.
  (d) If $AC < 0$, then it is a hyperbola.

Solutions to Selected Exercises

5. Match the following equation with its graph.

$$\frac{(x-2)^2}{9} - \frac{y^2}{4} = 1$$

Solution:

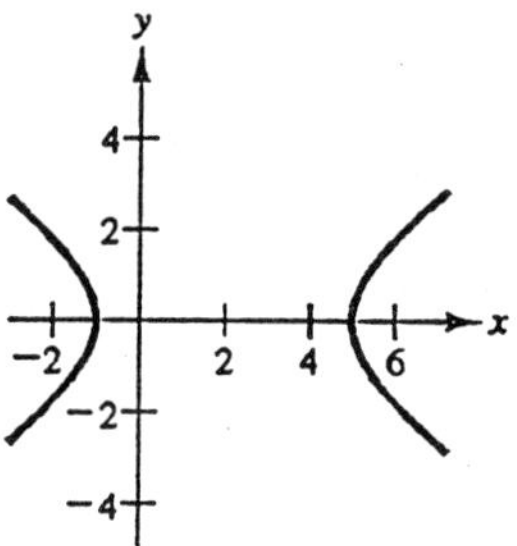

$a = 3,\ b = 2$
Center: $(2,\ 0)$
Horizontal transverse axis
Matches graph (d)

9. Find the center, foci, and vertices of the following hyperbola and sketch its graph, using asymptotes as an aid.

$$\frac{y^2}{25} - \frac{x^2}{144} = 1$$

Solution:

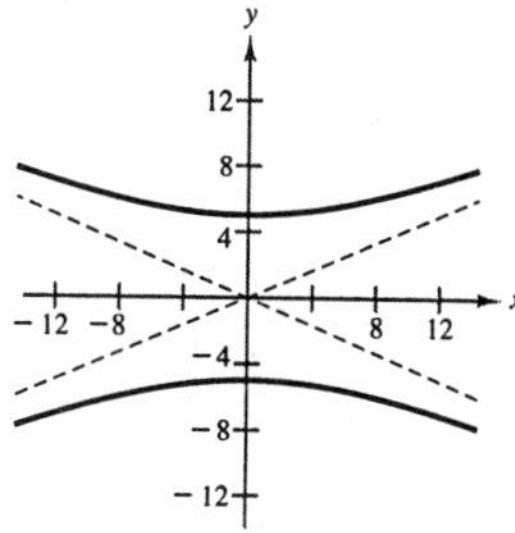

$a = 5,\ b = 12,\ c = \sqrt{25+144} = 13$
Vertical transverse axis
Center: $(0,\ 0)$
Vertices: $(0,\ \pm 5)$
Foci: $(0,\ \pm 13)$
Asymptotes: $y = \pm \frac{5}{12}x$

15. Find the center, foci, and vertices of the hyperbola $(y+6)^2 - (x-2)^2 = 1$ and sketch its graph, using asymptotes as an aid.

Solution:

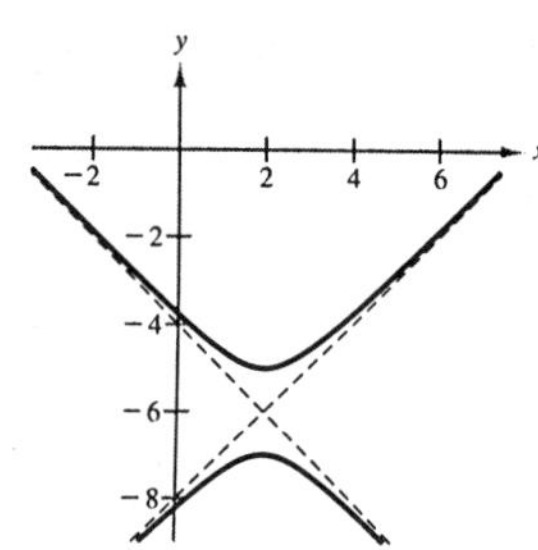

$$\frac{(y+6)^2}{1} - \frac{(x-2)^2}{1} = 1$$

$a = 1,\ b = 1,\ c = \sqrt{2}$
Center: $(2,\ -6)$
Foci: $(2,\ -6 \pm \sqrt{2})$
Vertices: $(2,\ -6 \pm 1)$ or $(2,\ -5),\ (2,\ -7)$
Asymptotes: $y = -6 \pm (x-2)$

17. Find the center, foci, and vertices of the hyperbola $9y^2 - x^2 + 2x + 54y + 62 = 0$ and sketch its graph, using asymptotes as an aid.

Solution:

$$9y^2 - x^2 + 2x + 54y + 62 = 0$$
$$9(y^2 + 6y + 9) - (x^2 - 2x + 1) = -62 - 1 + 81$$
$$\frac{(y+3)^2}{2} - \frac{(x-1)^2}{18} = 1$$

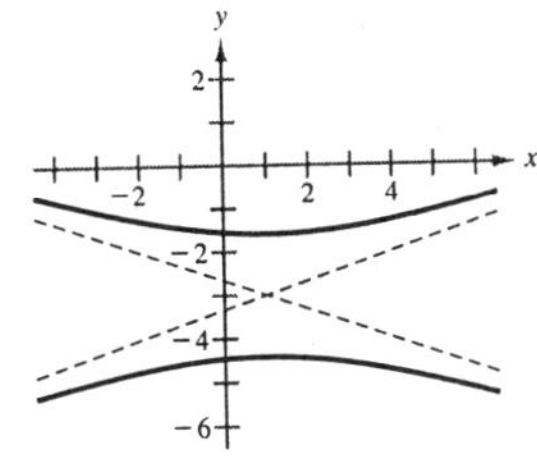

$a = \sqrt{2},\ b = 3\sqrt{2},\ c = 2\sqrt{5}$

Center: $(1,\ -3)$

Foci: $(1,\ -3 \pm 2\sqrt{5})$

Vertices: $(1,\ -3 \pm \sqrt{2})$

Asymptotes: $y = -3 \pm \frac{1}{3}(x - 1)$

21. Find an equation of the specified hyperbola.

Vertices: $(0,\ \pm 2)$

Foci: $(0,\ \pm 4)$

Solution:

The transverse axis is vertical.

Center: $(0,\ 0)$

$a = 2,\ c = 4 \Rightarrow b = \sqrt{16 - 4} = 2\sqrt{3}$

$$\frac{(y-0)^2}{2^2} - \frac{(x-0)^2}{(2\sqrt{3})^2} = 1$$
$$\frac{y^2}{4} - \frac{x^2}{12} = 1$$

25. Find an equation of the specified hyperbola.

Vertices: $(2,\ 0),\ (6,\ 0)$

Foci: $(0,\ 0),\ (8,\ 0)$

Solution:

The transverse axis is horizontal.

Center: $(4,\ 0)$

$a = 2,\ c = 4,\ b = \sqrt{16 - 4} = 2\sqrt{3}$

$$\frac{(x-4)^2}{2^2} - \frac{(y-0)^2}{(2\sqrt{3})^2} = 1$$
$$\frac{(x-4)^2}{4} - \frac{y^2}{12} = 1$$

29. Find an equation of the specified hyperbola.

Vertices: $(0,\ 2),\ (6,\ 2)$

Asymptotes: $y = \frac{2}{3}x$ and $y = 4 - \frac{2}{3}x$

Solution:

The transverse axis is horizontal.

Center: $(3,\ 2)$

$a = 3,\ \pm\dfrac{b}{a} = \pm\dfrac{2}{3} \Rightarrow b = 2$

$$\frac{(x-3)^2}{9} - \frac{(y-2)^2}{4} = 1$$

33. A hyperbolic mirror (used in some telescopes) has the property that a light ray directed at the focus will be reflected to the other focus (see figure). The focus of a hyperbolic mirror has coordinates (12, 0). Find the vertex of the mirror if its mount has coordinates (12, 12).

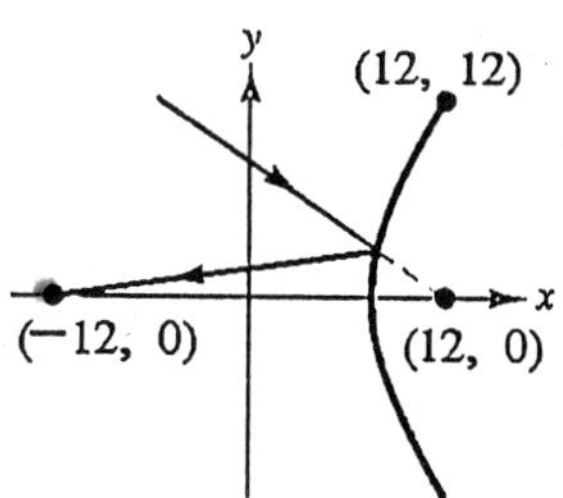

Solution:

The transverse axis is horizontal.

Center: (0, 0)

$c = 12 \Rightarrow b = \sqrt{144 - a^2}$

$$\frac{x^2}{a^2} - \frac{y^2}{144 - a^2} = 1$$

Since the hyperbola passes through (12, 12), we have:

$$\frac{144}{a^2} - \frac{144}{144 - a^2} = 1$$

$$144(144 - a^2) - 144a^2 = a^2(144 - a^2)$$

$$20{,}736 - 288a^2 = 144a^2 - a^4$$

$$a^4 - 432a^2 + 20{,}736 = 0$$

$$a^2 = \frac{432 \pm \sqrt{432^2 - 4(20{,}736)}}{2} = \frac{432 \pm 144\sqrt{5}}{2} = 216 \pm 72\sqrt{5}$$

$$a = \sqrt{216 \pm 72\sqrt{5}}$$

$$a \approx 19.416 \text{ or } a \approx 7.416$$

Since $a < c$ for a hyperbola, we use the smaller value.

Vertex: $\left(\sqrt{216 - 72\sqrt{5}},\ 0\right) \approx (7.42,\ 0)$

37. Classify the graph of the equation $4x^2 - y^2 - 4x - 3 = 0$ as a circle, a parabola, an ellipse, or a hyperbola.

Solution:

$4x^2 - y^2 - 4x - 3 = 0$

$A = 4,\ C = -1$

$AC = (4)(-1) = -4 < 0$

Therefore, the graph is a hyperbola.

41. Classify the graph of the equation $25x^2 - 10x - 200y - 119 = 0$ as a circle, a parabola, an ellipse, or a hyperbola.

Solution:

$25x^2 - 10x - 200y - 119 = 0$

$A = 25,\ C = 0$

$AC = (25)(0) = 0$

Therefore, the graph is a parabola.

SECTION 11.5

Rotation and the General Second-Degree Equation

- The general second-degree equation $Ax^2+Bxy+Cy^2+Dx+Ey+F=0$ can be rewritten as $A'(x')^2+C'(y')^2+D'x'+E'y'+F'=0$ by rotating the coordinate axes through the angle θ, where $\cot 2\theta = (A-C)/B$.
- $x = x'\cos\theta - y'\sin\theta$

  $y = x'\sin\theta + y'\cos\theta$
- The graph of the nondegenerate equation $Ax^2+Bxy+Cy^2+Dx+Ey+F=0$ is:

  (a) An ellipse or circle if $B^2-4AC<0$.
  (b) A parabola if $B^2-4AC=0$.
  (c) A hyperbola if $B^2-4AC>0$.

Solutions to Selected Exercises

5. Rotate the axes to eliminate the xy-term in the equation $xy-2y-4x=0$. Sketch the graph of the resulting equation, showing both sets of axes.

Solution:

$xy-2y-4x=0$

$A=0,\ B=1,\ C=0$

$$\cos 2\theta = \frac{A-C}{B} = 0 \Rightarrow 2\theta = \frac{\pi}{2} \Rightarrow \theta = \frac{\pi}{4}$$

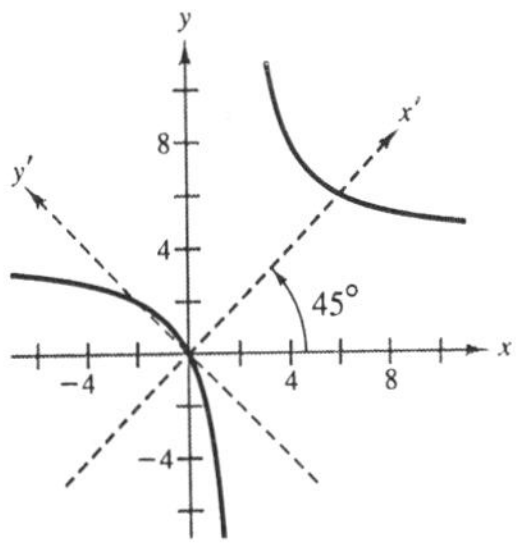

$$x = x'\cos\frac{\pi}{4} - y'\sin\frac{\pi}{4} = x'\left(\frac{\sqrt{2}}{2}\right) - y'\left(\frac{\sqrt{2}}{2}\right) = \frac{x'-y'}{\sqrt{2}}$$

$$y = x'\sin\frac{\pi}{4} + y'\cos\frac{\pi}{4} = x'\left(\frac{\sqrt{2}}{2}\right) + y'\left(\frac{\sqrt{2}}{2}\right) = \frac{x'+y'}{\sqrt{2}}$$

–CONTINUED ON NEXT PAGE–

5. –CONTINUED–

$$xy - 2y - 4x = 0$$

$$\left(\frac{x' - y'}{\sqrt{2}}\right)\left(\frac{x' + y'}{\sqrt{2}}\right) - 2\left(\frac{x' + y'}{\sqrt{2}}\right) - 4\left(\frac{x' - y'}{\sqrt{2}}\right) = 0$$

$$\frac{(x')^2}{2} - \frac{(y')^2}{2} - \sqrt{2}x' - \sqrt{2}y' - 2\sqrt{2}x' + 2\sqrt{2}y' = 0$$

$$\left[(x')^2 - 6\sqrt{2}x' + (3\sqrt{2})^2\right] - \left[(y')^2 - 2\sqrt{2}y' + (\sqrt{2})^2\right] = 0 + (3\sqrt{2})^2 - (\sqrt{2})^2$$

$$(x' - 3\sqrt{2})^2 - (y' - \sqrt{2})^2 = 16$$

$$\frac{(x' - 3\sqrt{2})^2}{16} - \frac{(y' - \sqrt{2})^2}{16} = 1$$

The graph is a hyperbola.

9. Rotate the axes to eliminate the xy-term in the equation $3x^2 - 2\sqrt{3}xy + y^2 + 2x + 2\sqrt{3}y = 0$. Sketch the graph of the resulting equation, showing both sets of axes.

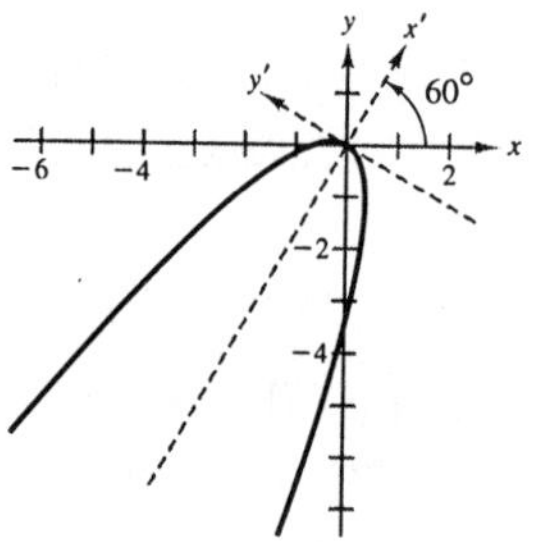

Solution:

$3x^2 - 2\sqrt{3}xy + y^2 + 2x + 2\sqrt{3}y = 0$

$A = 3,\ B = -2\sqrt{3},\ C = 1,\ \cot 2\theta = \dfrac{A - C}{B} = -\dfrac{1}{\sqrt{3}} \Rightarrow \theta = 60°$

$x = x'\cos 60° - y'\sin 60°$

$= x'\left(\frac{1}{2}\right) - y'\left(\frac{\sqrt{3}}{2}\right) = \dfrac{x' - \sqrt{3}y'}{2}$

$y = x'\sin 60° + y'\cos 60°$

$= x'\left(\frac{\sqrt{3}}{2}\right) + y'\left(\frac{1}{2}\right) = \dfrac{\sqrt{3}x' + y'}{2}$

$$3x^2 - 2\sqrt{3}xy + y^2 + 2x + 2\sqrt{3}y = 0$$

$$3\left(\frac{x' - \sqrt{3}y'}{2}\right)^2 - 2\sqrt{3}\left(\frac{x' - \sqrt{3}y'}{2}\right)\left(\frac{\sqrt{3}x' + y'}{2}\right) + \left(\frac{\sqrt{3}x' + y'}{2}\right)^2$$

$$+ 2\left(\frac{x' - \sqrt{3}y'}{2}\right) + 2\sqrt{3}\left(\frac{\sqrt{3}x' + y'}{2}\right) = 0$$

$$\frac{3(x')^2}{4} - \frac{6\sqrt{3}x'y'}{4} + \frac{9(y')^2}{4} - \frac{6(x')^2}{4} + \frac{4\sqrt{3}x'y'}{4} + \frac{6(y')^2}{4} + \frac{3(x')^2}{4} + \frac{2\sqrt{3}x'y'}{4} + \frac{(y')^2}{4}$$

$$+ x' - \sqrt{3}y' + 3x' + \sqrt{3}y' = 0$$

$$4(y')^2 + 4x' = 0$$

$$x' = -(y')^2$$

The graph is a parabola.

11. Rotate the axes to eliminate the xy-term in the equation $9x^2 + 24xy + 16y^2 + 90x - 130y = 0$. Sketch the graph of the resulting equation, showing both sets of axes.

Solution:

$$9x^2 + 24xy + 16y^2 + 90x - 130y = 0$$

$$A = 9, \ B = 24, \ C = 16$$

$$\cot 2\theta = \frac{A - C}{B} = -\frac{7}{24} \Rightarrow \theta \approx 53.13^\circ$$

$$\cos 2\theta = -\frac{7}{25}$$

$$\sin\theta = \sqrt{\frac{1 - \cos 2\theta}{2}} = \sqrt{\frac{1 - (-7/25)}{2}} = \frac{4}{5}$$

$$\cos\theta = \sqrt{\frac{1 + \cos 2\theta}{2}} = \sqrt{\frac{1 + (-7/25)}{2}} = \frac{3}{5}$$

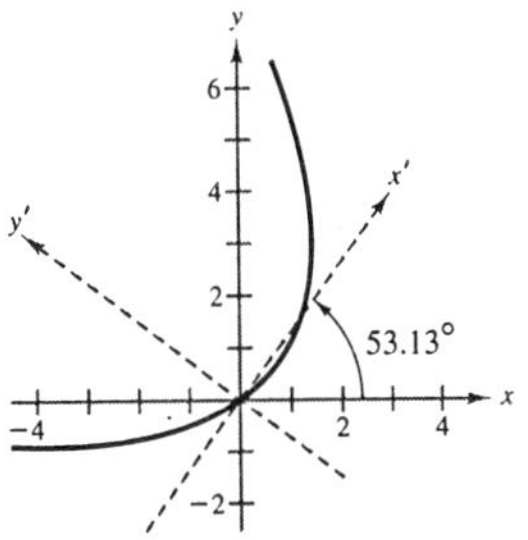

$$x = x'\cos\theta - y'\sin\theta = x'\left(\frac{3}{5}\right) - y'\left(\frac{4}{5}\right) = \frac{3x' - 4y'}{5}$$

$$y = x'\sin\theta + y'\cos\theta = x'\left(\frac{4}{5}\right) + y'\left(\frac{3}{5}\right) = \frac{4x' + 3y'}{5}$$

$$9x^2 + 24xy + 16y^2 + 90x - 130y = 0$$

$$9\left(\frac{3x' - 4y'}{5}\right)^2 + 24\left(\frac{3x' - 4y'}{5}\right)\left(\frac{4x' + 3y'}{5}\right) + 16\left(\frac{4x' + 3y'}{5}\right)^2 + 90\left(\frac{3x' - 4y'}{5}\right) - 130\left(\frac{4x' + 3y'}{5}\right) = 0$$

$$\frac{81(x')^2}{25} - \frac{216x'y'}{25} + \frac{144(y')^2}{25} + \frac{288(x')^2}{25} - \frac{168x'y'}{25} - \frac{288(y')^2}{25} + \frac{256(x')^2}{25} + \frac{384x'y'}{25} + \frac{144(y')^2}{25} + 54x' - 72y' - 104x' - 78y' = 0$$

$$25(x')^2 - 50x' - 150y' = 0$$

$$(x')^2 - 2x' + 1 = 6y' + 1$$

$$(x' - 1)^2 = 4\left(\frac{3}{2}\right)\left(y' + \frac{1}{6}\right)$$

The graph is a parabola.

15. Rotate the axes to eliminate the xy-term in the equation $32x^2 + 50xy + 7y^2 = 52$. Sketch the graph of the resulting equation, showing both sets of axes.

Solution:

$$32x^2 + 50xy + 7y^2 = 52$$

$$A = 32,\ B = 50,\ C = 7$$

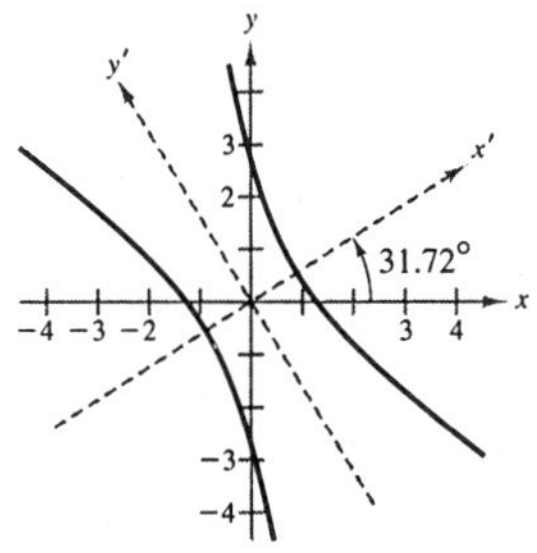

$$\cot 2\theta = \frac{A-C}{B} = \frac{1}{2} \Rightarrow \theta \approx 31.72^\circ$$

$$\cos 2\theta = \frac{1}{\sqrt{5}}$$

$$\sin\theta = \sqrt{\frac{1-\cos 2\theta}{2}} = \sqrt{\frac{1-(1/\sqrt{5})}{2}} = \sqrt{\frac{\sqrt{5}-1}{2\sqrt{5}}}$$

$$\cos\theta = \sqrt{\frac{1+\cos 2\theta}{2}} = \sqrt{\frac{1+(1/\sqrt{5})}{2}} = \sqrt{\frac{\sqrt{5}+1}{2\sqrt{5}}}$$

$$x = x'\cos\theta - y'\sin\theta = x'\sqrt{\frac{\sqrt{5}+1}{2\sqrt{5}}} - y'\sqrt{\frac{\sqrt{5}-1}{2\sqrt{5}}}$$

$$y = x'\sin\theta + y'\cos\theta = x'\sqrt{\frac{\sqrt{5}-1}{2\sqrt{5}}} + y'\sqrt{\frac{\sqrt{5}+1}{2\sqrt{5}}}$$

$$32x^2 + 50xy + 7y^2 = 52$$

$$32\left(x'\sqrt{\frac{\sqrt{5}+1}{2\sqrt{5}}} - y'\sqrt{\frac{\sqrt{5}-1}{2\sqrt{5}}}\right)^2$$

$$+\,50\left(x'\sqrt{\frac{\sqrt{5}+1}{2\sqrt{5}}} - y'\sqrt{\frac{\sqrt{5}-1}{2\sqrt{5}}}\right)\left(x'\sqrt{\frac{\sqrt{5}-1}{2\sqrt{5}}} + y'\sqrt{\frac{\sqrt{5}+1}{2\sqrt{5}}}\right)$$

$$+\,7\left(x'\sqrt{\frac{\sqrt{5}-1}{2\sqrt{5}}} + y'\sqrt{\frac{\sqrt{5}+1}{2\sqrt{5}}}\right)^2 = 52$$

$$47.451(x')^2 - 8.451(y')^2 = 52$$

$$\frac{(x')^2}{1.096} - \frac{(y')^2}{6.153} = 1$$

The graph is a hyperbola.

19. Use the discriminant to determine whether the graph of $13x^2 - 8xy + 7y^2 - 45 = 0$ is a parabola, an ellipse, or a hyperbola.

Solution:

$A = 13,\ B = -8,\ C = 7$
$B^2 - 4AC = (-8)^2 - 4(13)(7) = -300 < 0$
Therefore, the graph is an ellipse or a circle.

23. Use the discriminant to determine whether the graph of $x^2 + 4xy + 4y^2 - 5x - y - 3 = 0$ is a parabola, an ellipse, or a hyperbola.

Solution:

$A = 1,\ B = 4,\ C = 4$
$B^2 - 4AC = (4)^2 - 4(1)(4) = 0$
Therefore, the graph is a parabola.

25. Show that the equation $x^2 + y^2 = r^2$ is invariant under rotation of axes.

Solution:

$$\begin{aligned}(x')^2 + (y')^2 &= [x\cos\theta + y\sin\theta]^2 + [y\cos\theta - x\sin\theta]^2\\ &= x^2\cos^2\theta + 2xy\cos\theta\sin\theta + y^2\sin^2\theta + y^2\cos^2\theta - 2xy\cos\theta\sin\theta + x^2\sin^2\theta\\ &= x^2(\cos^2\theta + \sin^2\theta) + y^2(\sin^2\theta + \cos^2\theta)\\ &= x^2 + y^2 = r^2\end{aligned}$$

SECTION 11.6

Polar Coordinates

- In polar coordinates you do not have unique representation of points. The point $(r,\ \theta)$ can be represented by $(r,\ \theta \pm 2n\pi)$ or by $(-r,\ \theta \pm (2n+1)\pi)$ where n is any integer. The pole is represented by $(0,\ \theta)$ where θ is any angle.
- To convert from polar coordinates to rectangular coordinates, use the following relationships.

  $x = r\cos\theta$
  $y = r\sin\theta$
- To convert from rectangular coordinates to polar coordinates, use the following relationships.

  $r = \pm\sqrt{x^2+y^2}$
  $\tan\theta = y/x$

  If θ is in the same quadrant as the point $(x,\ y)$, then r is positive. If θ is in the opposite quadrant as the point $(x,\ y)$, then r is negative.
- You should be able to convert rectangular equations to polar form and vice versa.

Solutions to Selected Exercises

3. Plot the polar point $(-1,\ 5\pi/4)$ and find the corresponding rectangular coordinates for the point.

Solution:

$r = -1,\ \theta = \dfrac{5\pi}{4}$

$$x = (-1)\cos\frac{5\pi}{4} = (-1)\left(-\frac{\sqrt{2}}{2}\right) = \frac{\sqrt{2}}{2}$$

$$y = (-1)\sin\frac{5\pi}{4} = (-1)\left(-\frac{\sqrt{2}}{2}\right) = \frac{\sqrt{2}}{2}$$

$\left(-1,\ \dfrac{5\pi}{4}\right)$ corresponds to $\left(\dfrac{\sqrt{2}}{2},\ \dfrac{\sqrt{2}}{2}\right)$.

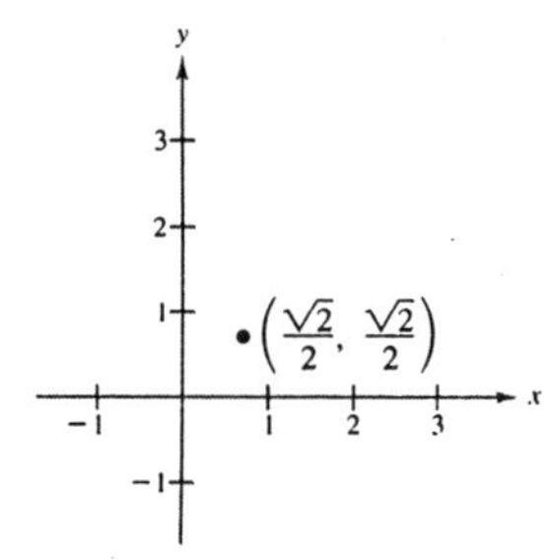

9. Plot the polar point $(\sqrt{2},\ 2.36)$ and find the corresponding rectangular coordinates for the point.

Solution:

$r = \sqrt{2},\ \theta = 2.36$ (in radians)

$$x = \sqrt{2}\cos 2.36 \approx -1.004$$

$$y = \sqrt{2}\sin 2.36 \approx 0.996$$

$(\sqrt{2},\ 2.36)$ corresponds to $(-1.004,\ 0.996)$.

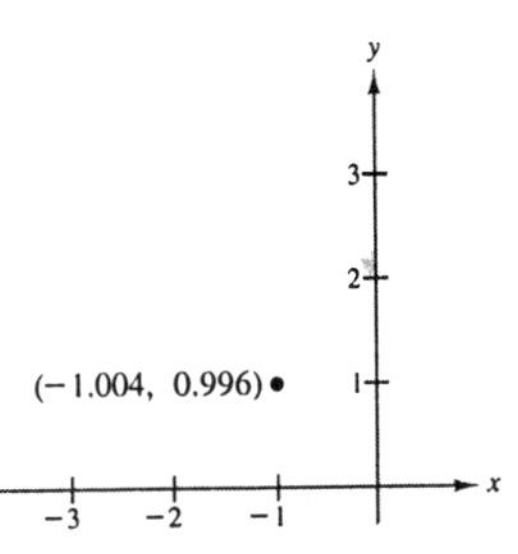

15. Plot the rectangular point $(-3,\ 4)$ and find two sets of polar coordinates for $0 \le \theta < 2\pi$.

Solution:

$x = -3,\ y = 4$

$r = \pm\sqrt{(-3)^2 + (4)^2} = \pm 5$

$\tan\theta = -\frac{4}{3},\ \theta \approx 2.214$

The point $(-3,\ 4)$ is in Quadrant II as is the angle $\theta = 2.214$. Thus, one polar representation is $(5,\ 2.214)$. Another representation is $(-5,\ 2.214 + \pi) \approx (-5,\ 5.356)$.

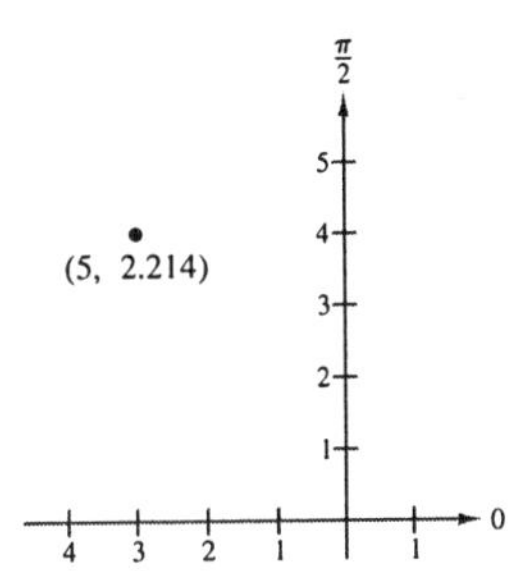

19. Plot the the rectangular point $(4,\ 6)$ and find two sets of polar coordinates for $0 \le \theta < 2\pi$.

Solution:

$x = 4,\ y = 6$

$r = \pm\sqrt{(4)^2 + (6)^2} = \pm 2\sqrt{13}$

$\tan\theta = \frac{6}{4},\ \theta \approx 0.983$

Since $(4,\ 6)$ is in Quadrant I and $\theta = 0.983$ is in Quadrant I, one representation in polar coordinates is $(2\sqrt{13},\ 0.983)$. Another representation is $(-2\sqrt{13},\ 0.983 + \pi) \approx (-2\sqrt{13},\ 4.124)$.

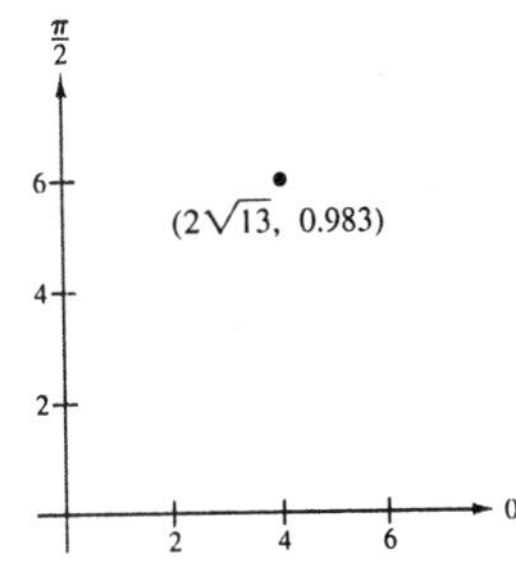

23. Convert the rectangular equation $x^2 + y^2 - 2ax = 0$ to polar form.

Solution:

$$x^2 + y^2 - 2ax = 0$$
$$r^2 - 2ar\cos\theta = 0$$
$$r(r - 2a\cos\theta) = 0$$
$$r = 0 \quad \text{or} \quad r = 2a\cos\theta$$

Since $r = 0$ is the pole and is also on the graph of $r = 2a\cos\theta$, we only have $r = 2a\cos\theta$.

27. Convert the rectangular equation $x = 10$ to polar form.

Solution:

$$x = 10$$
$$r\cos\theta = 10$$
$$r = \frac{10}{\cos\theta}$$
$$r = 10\sec\theta$$

31. Convert the rectangular equation $xy = 4$ to polar form.

Solution:

$$xy = 4$$
$$(r\cos\theta)(r\sin\theta) = 4$$
$$r^2 = \frac{4}{\cos\theta\sin\theta}$$
$$r^2 = 4\sec\theta\csc\theta$$
$$r^2 = 8\csc 2\theta$$

35. Convert the polar equation $r = 4\sin\theta$ to rectangular form.

Solution:

$$r = 4\sin\theta$$
$$r^2 = 4r\sin\theta$$
$$x^2 + y^2 = 4y$$
$$x^2 + y^2 - 4y = 0$$

39. Convert the polar equation $r = 2\csc\theta$ to rectangular form.

Solution:

$$r = 2\csc\theta$$
$$r = \frac{2}{\sin\theta}$$
$$r\sin\theta = 2$$
$$y = 2$$

43. Convert the polar equation $r = \dfrac{6}{2 - 3\sin\theta}$ to rectangular form.

Solution:

$$r = \frac{6}{2 - 3\sin\theta}$$
$$r(2 - 3\sin\theta) = 6$$
$$2r - 3r\sin\theta = 6$$
$$2r = 6 + 3r\sin\theta$$
$$2(\pm\sqrt{x^2 + y^2}) = 6 + 3y$$
$$4(x^2 + y^2) = (6 + 3y)^2$$
$$4x^2 + 4y^2 = 36 + 36y + 9y^2$$
$$4x^2 - 5y^2 - 36y - 36 = 0$$

47. Convert the polar equation $\theta = \pi/4$ to rectangular form and sketch its graph.

Solution:

$$\theta = \frac{\pi}{4}$$

$$\tan\theta = \tan\frac{\pi}{4}$$

$$\frac{y}{x} = 1$$

$$y = x$$

$$x - y = 0$$

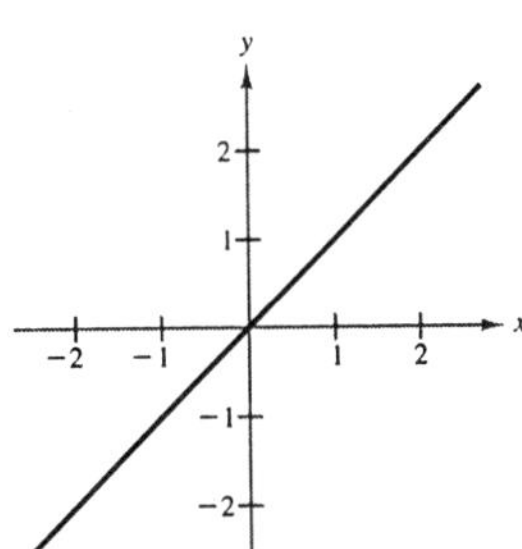

51. Show that the distance between $(r_1,\ \theta_1)$ and $(r_2,\ \theta_2)$ is $\sqrt{{r_1}^2 + {r_2}^2 - 2r_1r_2\cos(\theta_1 - \theta_2)}$.

Solution:

$(r_1,\ \theta_1)$ corresponds to the point $(r_1\cos\theta_1,\ r_1\sin\theta_1)$ in rectangular coordinates. Likewise, $(r_2,\ \theta_2)$ corresponds to the point $(r_2\cos\theta_2,\ r_2\sin\theta_2)$. In rectangular coordinates we use the distance formula, $d = \sqrt{(x_2 - x_1)^2 + (y_2 - y_1)^2}$, to find the distance between two points.

$$d = \sqrt{(r_2\cos\theta_2 - r_1\cos\theta_1)^2 + (r_2\sin\theta_2 - r_1\sin\theta_1)^2}$$

$$= \sqrt{{r_2}^2\cos^2\theta_2 - 2r_1r_2\cos\theta_1\cos\theta_2 + {r_1}^2\cos^2\theta_1 + {r_2}^2\sin^2\theta_2 - 2r_1r_2\sin\theta_1\sin\theta_2 + {r_1}^2\sin^2\theta_1}$$

$$= \sqrt{{r_1}^2(\cos^2\theta_1 + \sin^2\theta_1) + {r_2}^2(\cos^2\theta_2 + \sin^2\theta_2) - 2r_1r_2(\cos\theta_1\cos\theta_2 + \sin\theta_1\sin\theta_2)}$$

$$= \sqrt{{r_1}^2(1) + {r_2}^2(1) - 2r_1r_2\cos(\theta_1 - \theta_2)}$$

$$= \sqrt{{r_1}^2 + {r_2}^2 - 2r_1r_2\cos(\theta_1 - \theta_2)}$$

SECTION 11.7

Graphs of Polar Equations

- When graphing polar equations:
  1. Test for symmetry
     (a) $\theta = \pi/2$: Replace $(r,\ \theta)$ by $(r,\ \pi - \theta)$ or $(-r,\ -\theta)$.
     (b) Polar axis: Replace $(r,\ \theta)$ by $(r,\ -\theta)$ or $(-r,\ \pi - \theta)$.
     (c) Pole: Replace $(r,\ \theta)$ by $(r,\ \pi + \theta)$ or $(-r,\ \theta)$.
     (d) $r = f(\sin\theta)$ is symmetric with respect to the line $\theta = \pi/2$.
     (e) $r = f(\cos\theta)$ is symmetric with respect to the polar axis.
  2. Find the θ values for which $|r|$ is maximum.
  3. Find the θ values for which $r = 0$.
  4. Know the different types of polar graphs.
     (a) Limaçons
     $r = a \pm b\cos\theta$
     $r = a \pm b\sin\theta$
     (b) Rose Curves, $n \geq 2$
     $r = a\cos n\theta$
     $r = a\sin n\theta$
     (c) Circles
     $r = a\cos\theta$
     $r = a\sin\theta$
     $r = a$
     (d) Lemniscates
     $r^2 = a^2\cos 2\theta$
     $r^2 = a^2\sin 2\theta$
  5. Plot additional points.

Solutions to Selected Exercises

3. Test $r = 2/(1 + \sin\theta)$ for symmetry with respect to $\theta = \pi/2$, the polar axis, and the pole.

Solution:

$$\theta = \frac{\pi}{2}: \quad r = \frac{2}{1 + \sin(\pi - \theta)} = \frac{2}{1 + \sin\pi\cos\theta - \cos\pi\sin\theta} = \frac{2}{1 + \sin\theta}$$

The graph is symmetric with respect to the line $\theta = \pi/2$. None of the other substitutions will yield an equivalent equation. The graph is **not** symmetric with respect to the polar axis or the pole.

7. Find the maximum values of $|r|$ and any zeros of r for $r = 5\cos 3\theta$.

Solution:

$|r| = |5\cos 3\theta| = 5|\cos 3\theta| \le 5$

The maximum value of $|r|$ is 5. This occurs when $\cos 3\theta = \pm 1$ or when $\theta = 0,\ \pi/3,\ 2\pi/3$.
$r = 0$ when $\cos 3\theta = 0$. This occurs when $\theta = \pi/6,\ \pi/2,\ 5\pi/6$ (for $0 \le \theta < 2\pi$).

13. Sketch the graph of $\theta = \pi/6$.

Solution:

$$\theta = \frac{\pi}{6}$$

$$\tan\theta = \tan\frac{\pi}{6}$$

$$\frac{y}{x} = \frac{1}{\sqrt{3}}$$

$$\sqrt{3}\,y = x$$

$$y = \frac{x}{\sqrt{3}} \quad \text{Straight line}$$

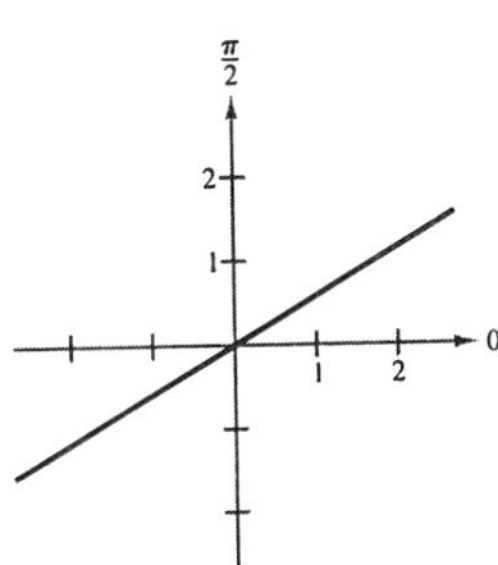

17. Sketch the graph of $r = 4(1 + \sin\theta)$.

Solution:

$r = 4 + 4\sin\theta$

$a/b = 4/4 = 1$, so the graph is a cardioid. Since r is a function of $\sin\theta$, the graph is symmetric with respect to $\theta = \pi/2$. The maximum value of $|r|$ is 8 and occurs when $\theta = \pi/2$. The zero of r occurs when $\theta = 3\pi/2$.

θ	0	$\frac{\pi}{6}$	$\frac{\pi}{2}$	$\frac{5\pi}{6}$	π	$\frac{7\pi}{6}$	$\frac{3\pi}{2}$	$\frac{11\pi}{6}$	2π
r	4	6	8	6	4	2	0	2	4

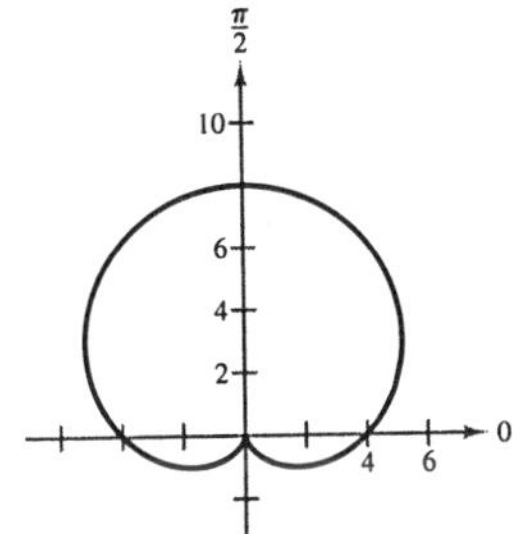

21. Sketch the graph of $r = 3 - 4\cos\theta$.

Solution:

$$r = 3 - 4\cos\theta$$

Since $a/b = 3/4 < 1$, the graph is a limaçon with an inner loop. Also, since r is a function of $\cos\theta$, the graph is symmetric with respect to the polar axis. The maximum value of $|r|$ is 7 and occurs when $\theta = \pi$. The zeros of r occur when $\cos\theta = 3/4$ or when $\theta \approx 0.723,\ 5.560$.

θ	0	$\frac{\pi}{3}$	$\frac{\pi}{2}$	$\frac{2\pi}{3}$	π	$\frac{4\pi}{3}$	$\frac{3\pi}{2}$	$\frac{5\pi}{3}$	2π
r	-1	1	3	5	7	5	3	1	-1

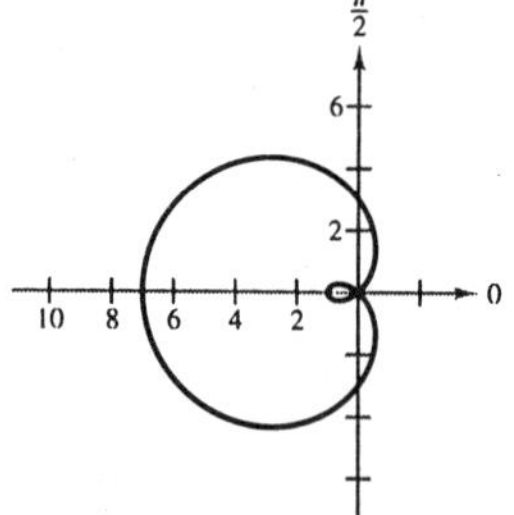

25. Sketch the graph of $r = \dfrac{3}{\sin\theta - 2\cos\theta}$.

Solution:

$$r = \frac{3}{\sin\theta - 2\cos\theta}$$

$$r(\sin\theta - 2\cos\theta) = 3$$

$$r\sin\theta - 2r\cos\theta = 3$$

$$y - 2x = 3$$

$$y = 2x + 3$$

The graph is a line.

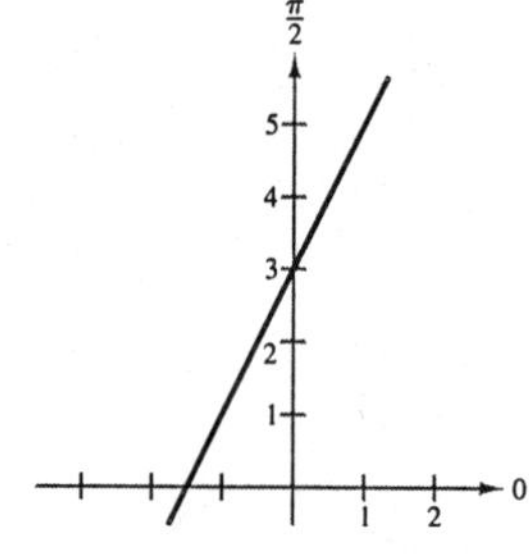

27. Sketch the graph of $r^2 = 4\cos 2\theta$.

Solution:

The graph is a lemniscate. Symmetric to the polar axis, the line $\theta = \pi/2$, and the pole. Maximum value of $|r|$ is 2 and occurs when $\theta = 0$ and $\theta = \pi$. The zeros of r occur when $\theta = \pi/4$, $3\pi/4$, $5\pi/4$, and $7\pi/4$.

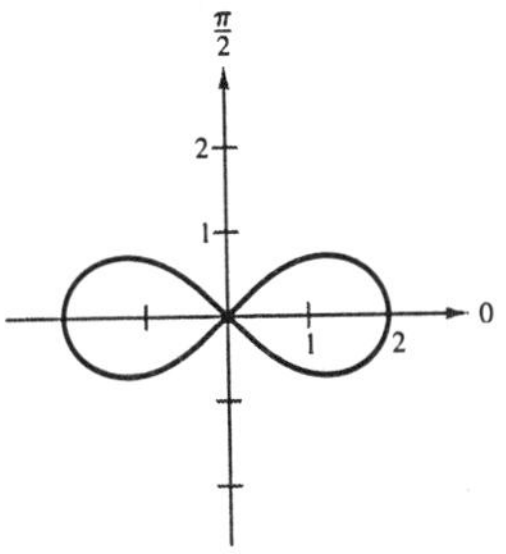

θ	0	$\frac{\pi}{6}$	$\frac{\pi}{4}$
r	± 2	$\pm\sqrt{2}$	0

31. Convert $r = 2 - \sec\theta$ to rectangular form and show that $x = -1$ is an asymptote to the graph.

Solution:

$$r = 2 - \sec\theta = 2 - \frac{1}{\cos\theta}$$

$$r\cos\theta = 2\cos\theta - 1$$

$$r(r\cos\theta) = 2r\cos\theta - r$$

$$(\pm\sqrt{x^2+y^2})x = 2x - (\pm\sqrt{x^2+y^2})$$

$$(\pm\sqrt{x^2+y^2})(x+1) = 2x$$

$$(\pm\sqrt{x^2+y^2}) = \frac{2x}{x+1}$$

$$x^2 + y^2 = \frac{4x^2}{(x+1)^2}$$

$$y^2 = \frac{4x^2}{(x+1)^2} - x^2$$

$$= \frac{4x^2 - x^2(x+1)^2}{(x+1)^2} = \frac{4x^2 - x^2(x^2+2x+1)}{(x+1)^2}$$

$$= \frac{-x^4 - 2x^3 + 3x^2}{(x+1)^2} = \frac{-x^2(x^2+2x-3)}{(x+1)^2}$$

$$y = \pm\sqrt{\frac{x^2(3-2x-x^2)}{(x+1)^2}} = \pm\left|\frac{x}{x+1}\right|\sqrt{3-2x-x^2}$$

The graph has an asymptote at $x = -1$.

35. Write the equation for the limaçon $r = 2 - \sin\theta$ after it has been rotated by the given amount.

(a) $\dfrac{\pi}{4}$ (b) $\dfrac{\pi}{2}$ (c) π (d) $\dfrac{3\pi}{2}$

Solution:

Refer to Exercises 33 and 34.

(a) $r = 2 - \sin\left(\theta - \dfrac{\pi}{4}\right)$

$= 2 - \dfrac{\sqrt{2}}{2}(\sin\theta - \cos\theta)$

(b) $r = 2 - (-\cos\theta) = 2 + \cos\theta$

(c) $r = 2 - (-\sin\theta) = 2 + \sin\theta$

(d) $r = 2 - \cos\theta$

37. Sketch the graphs of the equations.

(a) $r = 1 - \sin\theta$ (b) $r = 1 - \sin\left(\theta - \dfrac{\pi}{4}\right)$

Solution:

(a) Cardioid

θ	0	$\dfrac{\pi}{2}$	π	$\dfrac{3\pi}{2}$
r	1	0	1	2

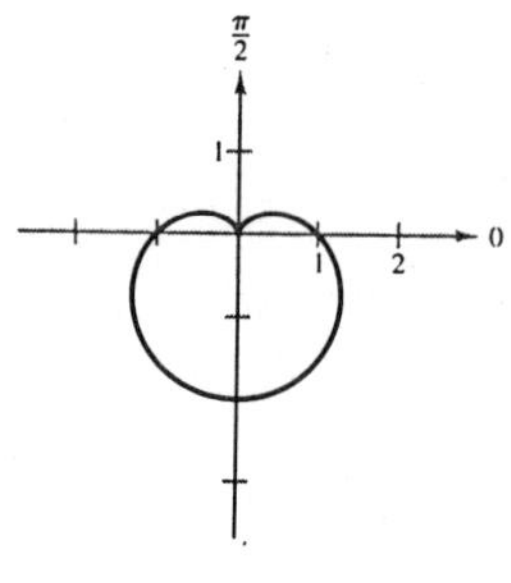

(b) Rotate the graph of $r = 1 - \sin\theta$ through the angle $\pi/4$.

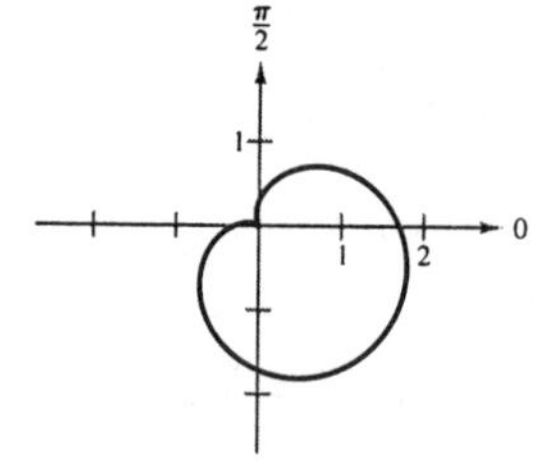

SECTION 11.8

Polar Equations of Conics

- The graph of a polar equation of the form

$$r = \frac{ep}{1 \pm e\cos\theta} \quad \text{or} \quad r = \frac{ep}{1 \pm e\sin\theta}$$

is a conic, where $e > 0$ is the eccentricity and $|p|$ is the distance between the focus (pole) and the directrix.

(a) If $e < 1$, the graph is an ellipse.
(b) If $e = 1$, the graph is a parabola.
(c) If $e > 1$, the graph is a hyperbola.

- Guidelines for finding polar equations of conics:

(a) Horizontal directrix above the pole: $r = \dfrac{ep}{1 + e\sin\theta}$

(b) Horizontal directrix below the pole: $r = \dfrac{ep}{1 - e\sin\theta}$

(c) Vertical directrix to the right of the pole: $r = \dfrac{ep}{1 + e\cos\theta}$

(d) Vertical directrix to the left of the pole: $r = \dfrac{ep}{1 - e\cos\theta}$

Solutions to Selected Exercises

5. Match the polar equation $r = 6/(2 - \sin\theta)$ with the correct graph.

Solution:

$$r = \frac{6}{2 - \sin\theta} = \frac{3}{1 - \frac{1}{2}\sin\theta}$$

$e = \frac{1}{2} < 1 \Rightarrow$ the graph is an ellipse with a horizontal directrix below the pole.

θ	0	$\frac{\pi}{2}$	π	$\frac{3\pi}{2}$
r	3	6	3	2

Matches graph (b).

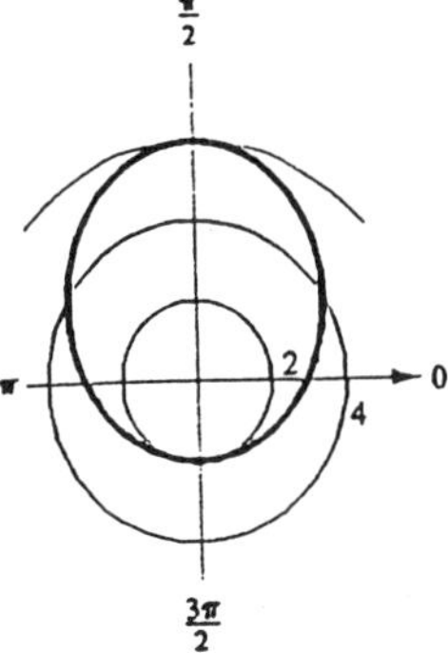

7. Identify and sketch the graph of $r = 2/(1 - \cos\theta)$.

Solution:

$e = 1 \Rightarrow$ the graph is a parabola.
Vertex: $(1,\ \pi)$

θ	$\frac{\pi}{2}$	π	$\frac{3\pi}{2}$
r	2	1	2

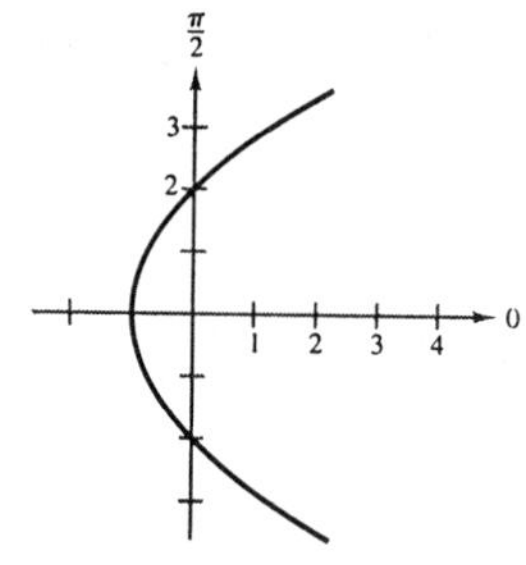

11. Identify and sketch the graph of $r = 2/(2 - \cos\theta)$.

Solution:

$$r = \frac{2}{2 - \cos\theta} = \frac{1}{1 - \frac{1}{2}\cos\theta}$$

$e = \frac{1}{2} < 1 \Rightarrow$ the graph is an ellipse.

θ	0	$\frac{\pi}{2}$	π	$\frac{3\pi}{2}$
r	2	1	$\frac{2}{3}$	1

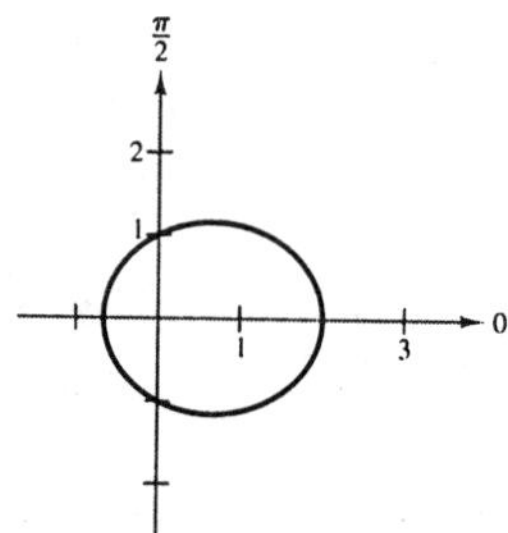

17. Identify and sketch the graph of $r = 3/(2 - 6\cos\theta)$.

Solution:

$$r = \frac{3}{2 - 6\cos\theta}$$

$$r = \frac{3/2}{1 - 3\cos\theta}$$

$e = 3 > 1 \Rightarrow$ the graph is a hyperbola.

θ	0	$\frac{\pi}{2}$	π	$\frac{3\pi}{2}$
r	$-\frac{3}{4}$	$\frac{3}{2}$	$\frac{3}{8}$	$\frac{3}{2}$

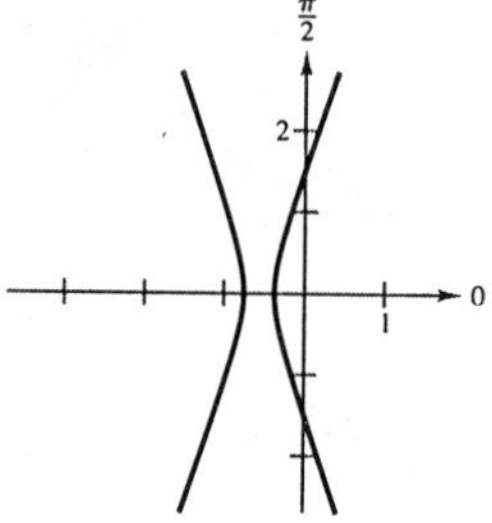

21. Find a polar equation of the ellipse with focus at (0, 0), $e = \frac{1}{2}$, and directrix $y = 1$.

Solution:

$e = \dfrac{1}{2},\ y = 1,\ p = 1$

Horizontal directrix above the pole

$$r = \frac{ep}{1 + e\sin\theta}$$

$$r = \frac{\frac{1}{2}}{1 + \frac{1}{2}\sin\theta}$$

$$r = \frac{1}{2 + \sin\theta}$$

25. Find a polar equation of the parabola with focus at (0, 0) and vertex at $(1, -\pi/2)$.

Solution:

$e = 1,\ p = 2$

Horizontal directrix below the pole

$$r = \frac{ep}{1 - e\sin\theta}$$

$$r = \frac{2}{1 - \sin\theta}$$

27. Find a polar equation of the parabola with focus at (0, 0) and vertex at $(5, \pi)$.

Solution:

Directrix: $x = -10,\ e = 1,\ p = 10$

Vertical directrix to the left of the pole

$$r = \frac{ep}{1 - e\cos\theta}$$

$$r = \frac{10}{1 - \cos\theta}$$

33. Find a polar equation of the hyperbola with focus at (0, 0) and vertices at $(1, 3\pi/2)$, $(9, 3\pi/2)$.

Solution:

Center: $(5, 3\pi/2)$

$c = 5,\ a = 4,\ e = c/a = 5/4$

Horizontal directrix below the pole

$$r = \frac{ep}{1 - e\sin\theta} = \frac{\frac{5}{4}p}{1 - \frac{5}{4}\sin\theta} = \frac{5p}{4 - 5\sin\theta}$$

$$1 = \frac{5p}{4 - 5\sin\frac{3\pi}{2}} = \frac{5p}{9}$$

$$p = \frac{9}{5}$$

$$r = \frac{5\left(\frac{9}{5}\right)}{4 - 5\sin\theta} = \frac{9}{4 - 5\sin\theta}$$

37. Use the results of Exercises 35 and 36 to write the polar form of

$$\frac{x^2}{169} + \frac{y^2}{144} = 1.$$

Solution:

$a = 13,\ b = 12,\ c = 5,\ e = \frac{5}{13}$

$$r^2 = \frac{b^2}{1 - e^2\cos^2\theta} = \frac{144}{1 - \left(\frac{25}{169}\right)\cos^2\theta} = \frac{24{,}336}{169 - 25\cos^2\theta}$$

41. Use the results of Exercises 35 and 36 to write the polar form of the hyperbola with one focus at (5, 0) and vertices at (4, 0), (4, π).

Solution:

Center: (0, 0), $a = 4,\ c = 5,\ b = 3,\ e = \frac{5}{4}$

$$\frac{x^2}{16} - \frac{y^2}{9} = 1$$

$$r^2 = \frac{-b^2}{1 - e^2\cos^2\theta} = \frac{-9}{1 - \left(\frac{25}{16}\right)\cos^2\theta} = \frac{-144}{16 - 25\cos^2\theta} = \frac{144}{25\cos^2\theta - 16}$$

45. The planets travel in elliptical orbits with the sun as a focus. Assume that the focus is at the pole, the major axis lies on the polar axis, and the length of the major axis is $2a$ (see figure). Show that the polar equation of the orbit is given by

$$r = \frac{(1 - e^2)a}{1 - e\cos\theta}$$

where e is the eccentricity.

Solution:

When $\theta = 0$, $r = c + a = ea + a = a(1 + e)$. Therefore,

$$a(1 + e) = \frac{ep}{1 - e}$$

$$a(1 + e)(1 - e) = ep$$

$$a(1 - e^2) = ep.$$

Thus, $r = \dfrac{ep}{1 - e\cos\theta} = \dfrac{(1 - e^2)a}{1 - e\cos\theta}$.

49. A satellite in a 100-mile-high circular orbit around the earth has a velocity of approximately 17,500 miles per hour. If this velocity is multiplied by $\sqrt{2}$, then the satellite will have the minimum velocity necessary to escape the earth's gravity and it will follow a parabolic path with the center of the earth as the focus. Find a polar equation of the parabolic path of the satellite (assume the radius of the earth is 4000 miles).

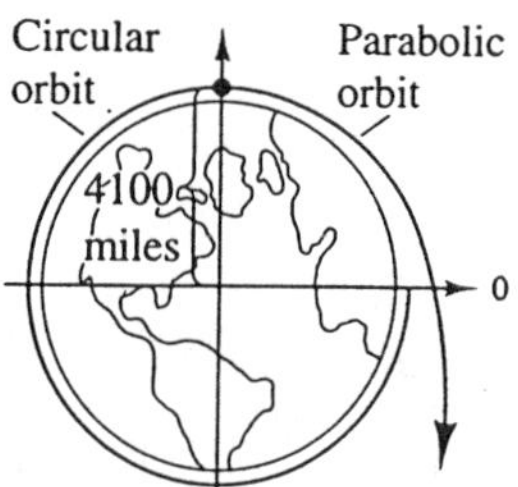

Solution:

Directrix: $y = 8200,\ e = 1,\ p = 8200$

$$r = \frac{ep}{1 + e\sin\theta} = \frac{8200}{1 + \sin\theta}$$

SECTION 11.9

Plane Curves and Parametric Equations

- If f and g are continuous functions of t on an interval I, then the set of ordered pairs $(f(t), g(t))$ is a *plane curve* C. The equations $x = f(t)$ and $y = g(t)$ are *parametric equations* for C and t is the *parameter.*
- To eliminate the parameter:
  (a) Solve for t in one equation and substitute into the second equation.
  (b) Use trigonometric identities.
- You should be able to find the parametric equations for a graph.

Solutions to Selected Exercises

5. Sketch the curve represented by the parametric equations (indicate the direction of the curve), and write the corresponding rectangular equation by eliminating the parameter.

$x = \frac{1}{4}t$

$y = t^2$

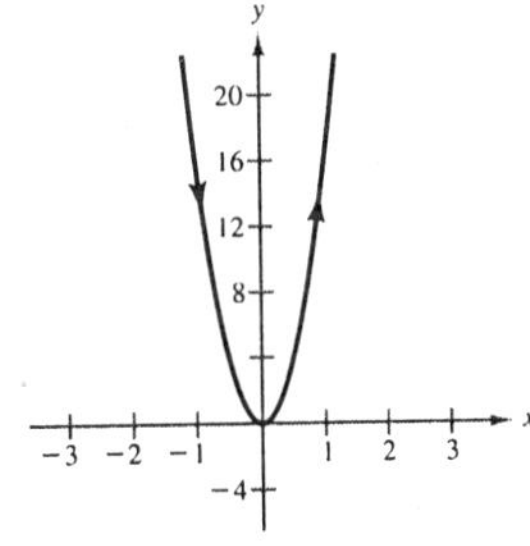

Solution:

$x = \frac{1}{4}t \Rightarrow 4x = t$

$y = t^2 \Rightarrow y = (4x)^2 = 16x^2$

t	-2	-1	0	1	2
x	$-\frac{1}{2}$	$-\frac{1}{4}$	0	$\frac{1}{4}$	$\frac{1}{2}$
y	4	1	0	1	4

9. Sketch the curve represented by the parametric equations (indicate the direction of the curve), and write the corresponding rectangular equation by eliminating the parameter.

$x = t^3, \ y = \dfrac{t}{2}$

Solution:

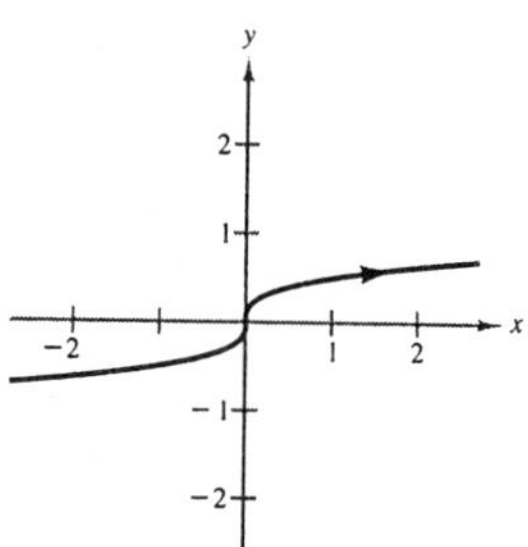

$$y = \frac{t}{2} \Rightarrow 2y = t$$

$$x = t^3 \Rightarrow x = (2y)^3 \Rightarrow x = 8y^3 \Rightarrow y = \frac{\sqrt[3]{x}}{2}$$

t	-2	-1	0	1	2
x	-8	-1	0	1	8
y	-1	$-\frac{1}{2}$	0	$\frac{1}{2}$	1

13. Sketch the curve represented by the parametric equations (indicate the direction of the curve), and write the corresponding rectangular equation by eliminating the parameter.

$x = \cos\theta, \ y = 2\sin^2\theta$

Solution:

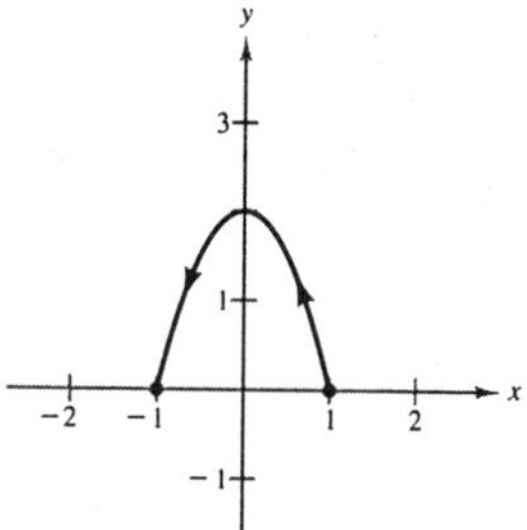

$$x = \cos\theta \Rightarrow x^2 = \cos^2\theta$$

$$y = 2\sin^2\theta \Rightarrow \frac{y}{2} = \sin^2\theta$$

Using the identity $\sin^2\theta + \cos^2\theta = 1$ yields

$$\frac{y}{2} + x^2 = 1$$

$$\frac{y}{2} = 1 - x^2$$

$$y = 2(1 - x^2) = 2 - 2x^2.$$

Also, since $x = \cos\theta$ and $y = 2\sin^2\theta$, we know that $-1 \le x \le 1$ and $0 \le y \le 2$.

θ	0	$\frac{\pi}{2}$	π	$\frac{3\pi}{2}$	2π
x	1	0	-1	0	1
y	0	2	0	2	0

The graph oscillates.

15. Sketch the curve represented by the parametric equations (indicate the direction of the curve), and write the corresponding rectangular equation by eliminating the parameter.

$x = 4 + 2\cos\theta$

$y = -1 + 4\sin\theta$

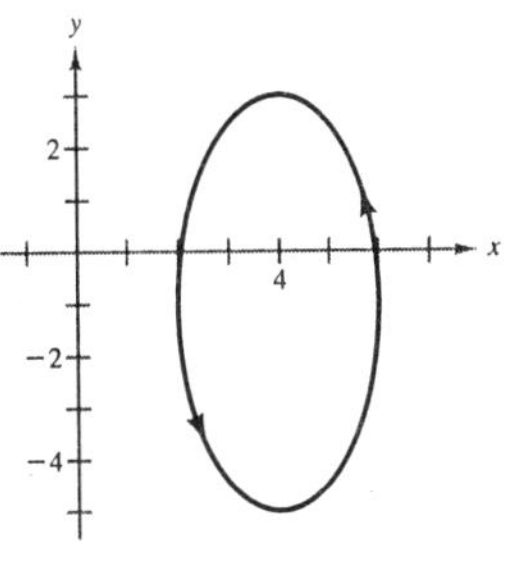

Solution:

$$x = 4 + 2\cos\theta \Rightarrow \frac{x-4}{2} = \cos\theta$$

$$y = -1 + 4\sin\theta \Rightarrow \frac{y+1}{4} = \sin\theta$$

$$\left(\frac{x-4}{2}\right)^2 + \left(\frac{y+1}{4}\right)^2 = 1$$

$$\frac{(x-4)^2}{4} + \frac{(y+1)^2}{16} = 1$$

θ	0	$\frac{\pi}{2}$	π	$\frac{3\pi}{2}$
x	6	4	2	4
y	-1	3	-1	-5

17. Sketch the curve represented by the parametric equations (indicate the direction of the curve), and write the corresponding rectangular equation by eliminating the parameter.

$x = e^{-t}$

$y = e^{3t}$

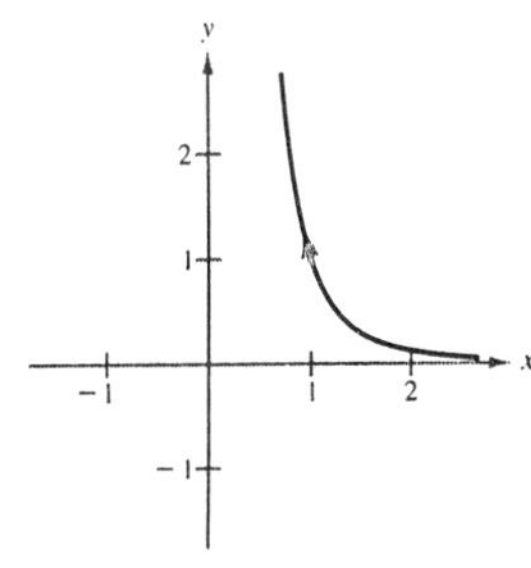

Solution:

Since $x = e^{-t}$ and $y = e^{3t}$, we have $x > 0$ and $y > 0$.

$$x = e^{-t} \Rightarrow \frac{1}{x} = e^t$$

$$y = e^{3t} = (e^t)^3 = \left(\frac{1}{x}\right)^3 = \frac{1}{x^3}$$

where $x > 0$ and $y > 0$.

t	-1	0	1
x	2.718	1	0.368
y	0.050	1	20.086

21. Determine how the plane curves differ from each other.

(a) $x = t$ (b) $x = \cos\theta$ (c) $x = e^{-t}$ (d) $x = e^t$

$y = 2t + 1$ $y = 2\cos\theta + 1$ $y = 2e^{-t} + 1$ $y = 2e^t + 1$

Solution:

By eliminating the parameter, each curve becomes $y = 2x + 1$.

(a) $x = t$

$y = 2t + 1$

There are no restrictions on x and y.

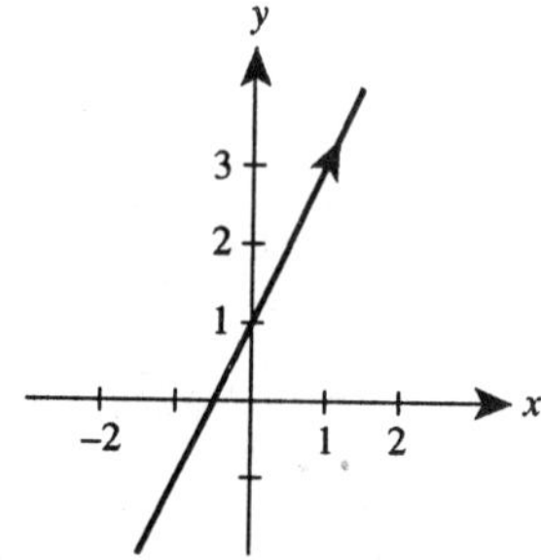

(b) 

$x = \cos\theta \Rightarrow -1 \le x \le 1$

$y = 2\cos\theta + 1 \Rightarrow -1 \le y \le 3$

The graph oscillates.

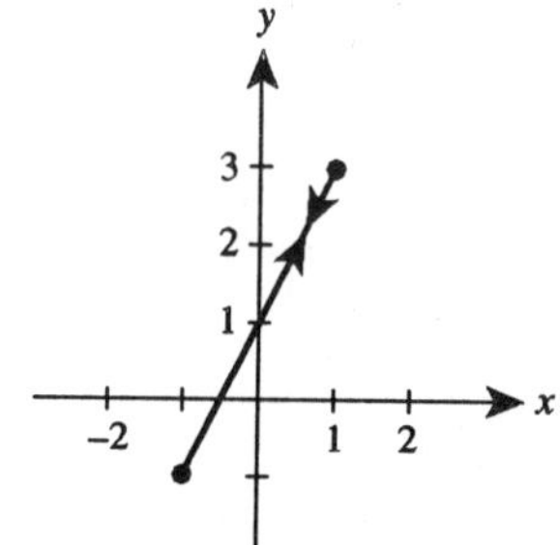

(c) $x = e^{-t} \Rightarrow x > 0$

$y = 2e^{-t} + 1 \Rightarrow y > 1$

This graph is oriented downward.

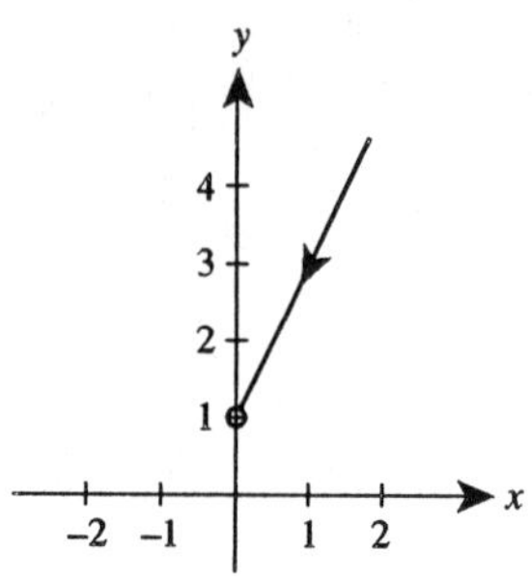

(d) $x = e^t \Rightarrow x > 0$

$y = 2e^t + 1 \Rightarrow y > 1$

This graph is oriented upward.

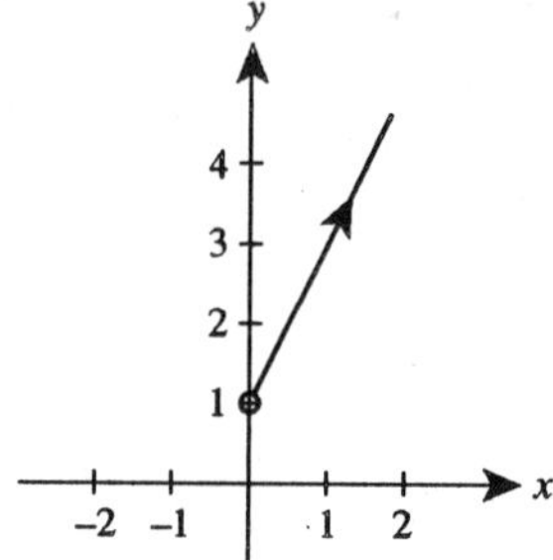

25. Eliminate the parameter and obtain the standard form of the rectangular equation.

Ellipse: $x = h + a\cos\theta,\ y = k + b\sin\theta$

Solution:

$$x = h + a\cos\theta \Rightarrow \frac{x-h}{a} = \cos\theta$$

$$y = k + b\sin\theta \Rightarrow \frac{y-k}{b} = \sin\theta$$

$$\left(\frac{x-h}{a}\right)^2 + \left(\frac{y-k}{b}\right)^2 = 1$$

$$\frac{(x-h)^2}{a^2} + \frac{(y-k)^2}{b^2} = 1$$

29. Find a set of parametric equations for the circle with center at (2, 1) and radius 4.

Solution:

From Exercise 24 we have $x = h + r\cos\theta,\ y = k + r\sin\theta$. Using $h = 2,\ k = 1$ and $r = 4$, we have $x = 2 + 4\cos\theta,\ y = 1 + 4\sin\theta$. This solution is not unique.

33. Find a set of parametric equations for the hyperbola with vertices at $(\pm 4,\ 0)$ and foci at $(\pm 5,\ 0)$.

Solution:

From Exercise 26 we have $x = h + a\sec\theta,\ y = k + b\tan\theta$. Using $(h,\ k) = (0,\ 0),\ a = 4,\ c = 5$, and $b = 3$, we have $x = 4\sec\theta,\ y = 3\tan\theta$. This solution is not unique.

35. Find two different sets of parametric equations for the rectangular equation $y = x^3$.

Solution:

Examples

$x = t,$	$y = t^3$
$x = \sqrt[3]{t},$	$y = t$
$x = \tan t,$	$y = \tan^3 t$
$x = t - 4,$	$y = (t-4)^3$

and so on.

41. Sketch the curve represented by the parametric equations.

Witch of Agnesi: $x = 2\cot\theta,\ y = 2\sin^2\theta$

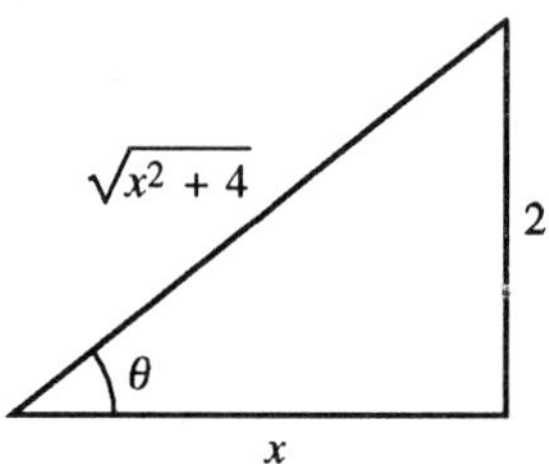

Solution:

$$x = 2\cot\theta \Rightarrow \theta = \operatorname{arccot}\frac{x}{2}$$

$$y = 2\sin^2\theta \Rightarrow y = 2\sin^2\left(\operatorname{arccot}\frac{x}{2}\right)$$

$$y = 2\left(\frac{2}{\sqrt{x^2+4}}\right)^2$$

$$y = \frac{8}{x^2+4}$$

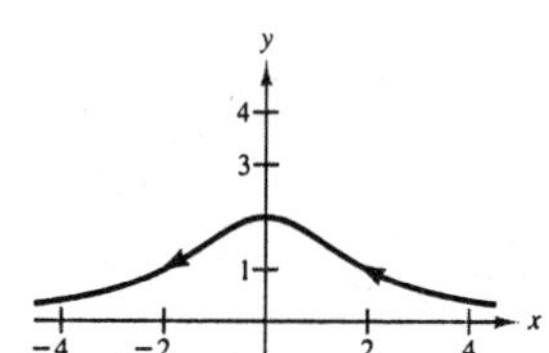

45. Match the equations with the correct graph.

Involute of a Circle: $x = \cos\theta + \theta\sin\theta$

$y = \sin\theta - \theta\cos\theta$

Solution:

$$x = \cos\theta + \theta\sin\theta$$

$$y = \sin\theta - \theta\cos\theta$$

θ	0	$\frac{\pi}{2}$	π	$\frac{3\pi}{2}$	2π
x	1	$\frac{\pi}{2}$	-1	$-\frac{3\pi}{2}$	1
y	0	1	π	-1	-2π

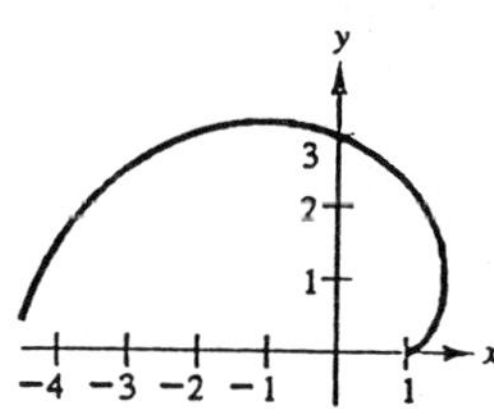

Matches graph (d).

47. A wheel of radius a rolls along a straight line without slipping (see figure). Find the parametric equations for the curve generated by a point P that is b units from the center of the wheel. This curve is called a *curtate cycloid* when $b < a$.

Solution:

When the circle has rolled θ radians, we know that the center is at $(a\theta, a)$.

$$\sin\theta = \sin(180° - \theta) = \frac{|AC|}{b} = \frac{|BD|}{b} \quad \text{or}$$

$$|BD| = b\sin\theta$$

$$\cos\theta = -\cos(180° - \theta) = \frac{|AP|}{-b} \quad \text{or}$$

$$|AP| = -b\cos\theta$$

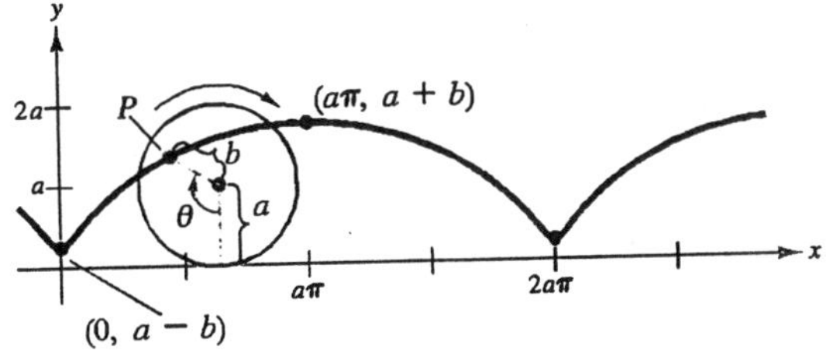

Therefore, $x = a\theta - b\sin\theta$ and $y = a - b\cos\theta$.

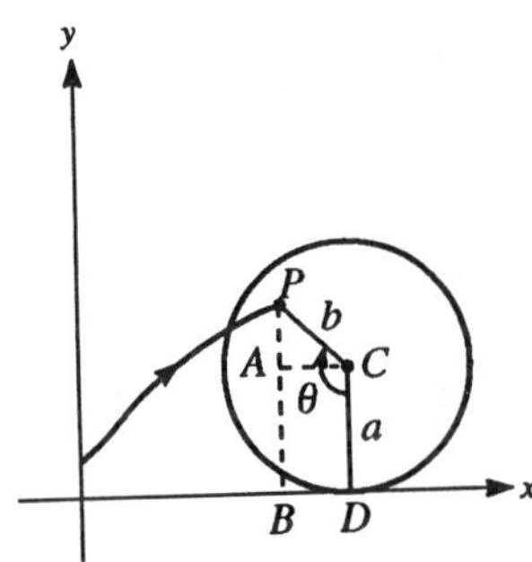

REVIEW EXERCISES FOR CHAPTER 11

Solutions to Selected Exercises

5. Find the distance between the point (1, 2) and the line $x - y - 3 = 0$.

Solution:

$(1,\ 2) \Rightarrow x_1 = 1$ and $y_1 = 2$

$x - y - 3 = 0 \Rightarrow A = 1,\ B = -1,$ and $C = -3$

$$d = \frac{|1(1) + (-1)(2) + (-3)|}{\sqrt{(1)^2 + (-1)^2}} = \frac{4}{\sqrt{2}} = 2\sqrt{2}$$

11. Identify and sketch the graph of the rectangular equation $3x^2 + 2y^2 - 12x + 12y + 29 = 0$.

Solution:

Since $AC = 3(2) = 6 > 0$, the graph is an ellipse.

$$3x^2 + 2y^2 - 12x + 12y + 29 = 0$$

$$3(x^2 - 4x + 4) + 2(y^2 + 6y + 9) = -29 + 12 + 18$$

$$3(x-2)^2 + 2(y+3)^2 = 1$$

$$\frac{(x-2)^2}{1/3} + \frac{(y+3)^2}{1/2} = 1$$

Center: $(2,\ -3)$; Vertices: $\left(2,\ -3 \pm \frac{\sqrt{2}}{2}\right)$

15. Identify and sketch the graph of the rectangular equation
$x^2 + y^2 + 2xy + 2\sqrt{2}\,x - 2\sqrt{2}\,y + 2 = 0$.

Solution:

Since $B^2 - 4AC = 2^2 - 4(1)(1) = 0$, the graph is a parabola.

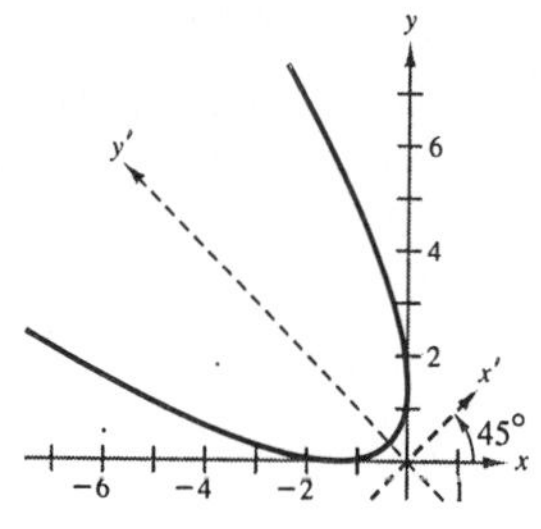

$$\cot 2\theta = \frac{A - C}{B} = 0 \Rightarrow 2\theta = \frac{\pi}{2} \Rightarrow \theta = \frac{\pi}{4}$$

$$x = x'\cos\frac{\pi}{4} - y'\sin\frac{\pi}{4} = \frac{x' - y'}{\sqrt{2}}$$

$$y = x'\sin\frac{\pi}{4} + y'\cos\frac{\pi}{4} = \frac{x' + y'}{\sqrt{2}}$$

–CONTINUED ON NEXT PAGE–

15. –CONTINUED–

$$\left(\frac{x'-y'}{\sqrt{2}}\right)^2+\left(\frac{x'+y'}{\sqrt{2}}\right)^2+2\left(\frac{x'-y'}{\sqrt{2}}\right)\left(\frac{x'+y'}{\sqrt{2}}\right)+2\sqrt{2}\left(\frac{x'-y'}{\sqrt{2}}\right)$$

$$-2\sqrt{2}\left(\frac{x'+y'}{\sqrt{2}}\right)+2=0$$

$$2(x')^2-4y'+2=0$$

$$(x')^2=2y'-1$$

Vertex: $(x',\ y')=\left(0,\ \frac{1}{2}\right),\ \theta=45°$

19. Find a rectangular equation for the parabola with vertex at (0, 2) and directrix $x=-3$.

Solution:

$p=3,\ (h,\ k)=(0,\ 2)$
Horizontal axis

$$(y-k)^2=4p(x-h)$$

$$(y-2)^2=4(3)(x-0)$$

$$(y-2)^2=12x$$

23. Find a rectangular equation for the ellipse with vertices at $(0,\ \pm 6)$ and passes through the point (2, 2).

Solution:

$(h,\ k)=(0,\ 0)$
Vertical major axis with $a=6$

$$\frac{x^2}{b^2}+\frac{y^2}{36}=1$$

Since the graph passes through the point (2, 2), we have:

$$\frac{4}{b^2}+\frac{4}{36}=1$$

$$36+b^2=9b^2$$

$$36=8b^2$$

$$\frac{36}{8}=b^2$$

$$\frac{9}{2}=b^2$$

$$\frac{x^2}{9/2}+\frac{y^2}{36}=1$$

27. Find a rectangular equation of the hyperbola with foci at (0, 0), (8, 0) and asymptotes $y=\pm 2(x-4)$.

Solution:

$(h,\ k)=(4,\ 0)$
The transverse axis is horizontal with $c=4$. Also from the slopes of the asymptotes $\pm b/a=\pm 2$ or $\pm b=\pm 2a$. Now $c^2=a^2+b^2=a^2+(2a)^2=16$. Therefore,

$$a^2=\frac{16}{5},\ b^2=\frac{64}{5}\text{ and }\frac{(x-4)^2}{16/5}-\frac{y^2}{64/5}=1.$$

31. Find an equation of the tangent line to $\frac{x^2}{100} + \frac{y^2}{25} = 1$ at the point $(-8, 3)$. The tangent line to the conic $\frac{x^2}{a^2} \pm \frac{y^2}{b^2} = 1$ at the point (x_0, y_0) is given by $\frac{x_0 x}{a^2} \pm \frac{y_0 y}{b^2} = 1$.

Solution:

$(x_0, y_0) = (-8, 3)$
$a^2 = 100, \ b^2 = 25$

$$\frac{-8x}{100} + \frac{3y}{25} = 1$$

$$-8x + 12y = 100$$

$$-2x + 3y = 25$$

35. Identify and sketch the graph of the polar equation $r = 4$.

Solution:

$r = 4$ is a circle of radius 4 centered at the pole.

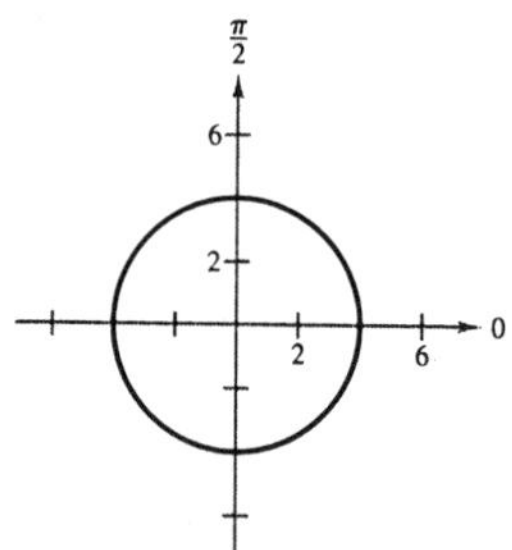

39. Identify and sketch the graph of the polar equation $r = -2(1 + \cos\theta)$.

Solution:

$r = -2(1 + \cos\theta)$ is a cardioid; symmetric to the polar axis.

θ	0	$\frac{\pi}{3}$	$\frac{\pi}{2}$	$\frac{2\pi}{3}$	π
r	-4	-3	-2	-1	0

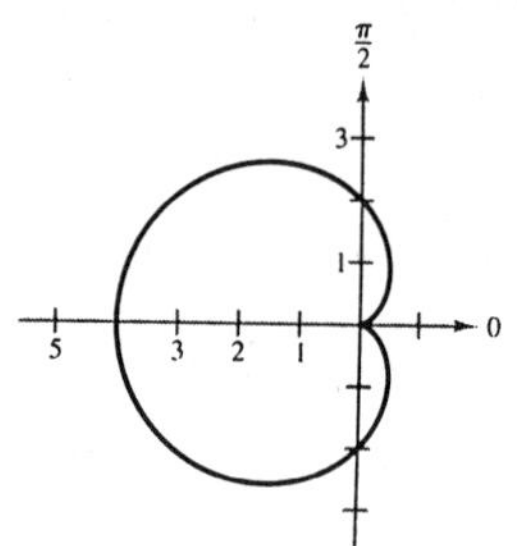

43. Identify and sketch the graph of the polar equation $r = -3\cos 3\theta$.

Solution:

$r = -3\cos 3\theta$ is a rose curve with three petals; symmetric to the polar axis. Maximum value of $|r|$ is 3.

$(-3, 0), \ \left(3, \frac{\pi}{3}\right), \ \left(-3, \frac{2\pi}{3}\right)$

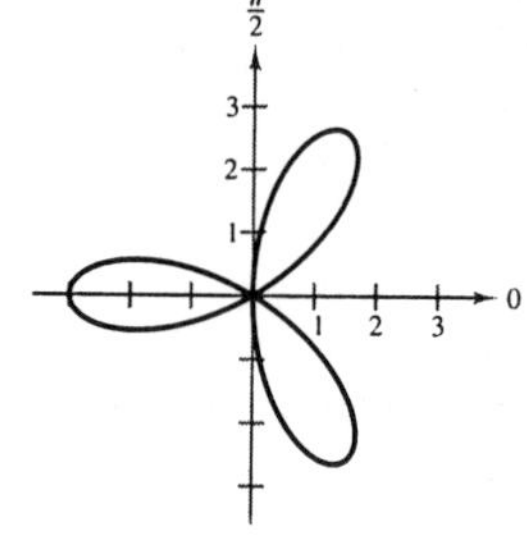

47. Identify and sketch the graph of the polar equation

$$r = \frac{3}{\cos(\theta - (\pi/4))}.$$

Solution:

$$r = \frac{3}{\cos(\theta - (\pi/4))}$$

$$r\cos\left(\theta - \frac{\pi}{4}\right) = 3$$

$$r\left[\cos\theta\cos\frac{\pi}{4} + \sin\theta\sin\frac{\pi}{4}\right] = 3$$

$$r\left[\frac{\sqrt{2}}{2}\cos\theta + \frac{\sqrt{2}}{2}\sin\theta\right] = 3$$

$$\frac{\sqrt{2}}{2}r\cos\theta + \frac{\sqrt{2}}{2}r\sin\theta = 3$$

$$r\cos\theta + r\sin\theta = 3\sqrt{2}$$

$$x + y = 3\sqrt{2}$$

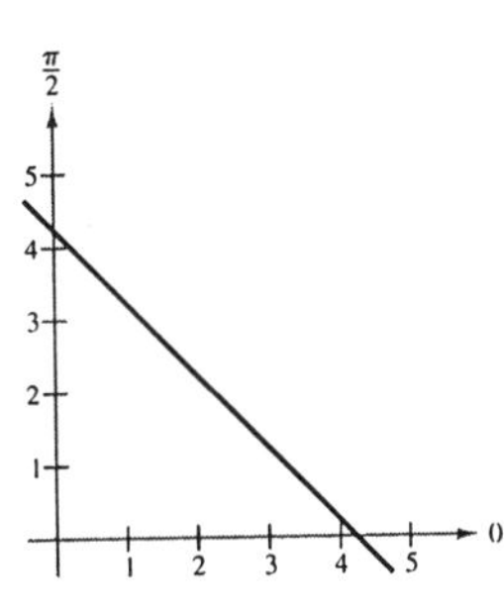

The graph is a line.

51. Convert $r = 3\cos\theta$ to rectangular form.

Solution:

$$r = 3\cos\theta$$

$$r^2 = 3r\cos\theta$$

$$x^2 + y^2 = 3x$$

$$x^2 + y^2 - 3x = 0$$

55. Convert $r^2 = \cos 2\theta$ to rectangular form.

Solution:

$$r^2 = \cos 2\theta$$

$$r^2 = 2\cos^2\theta - 1$$

$$x^2 + y^2 = 2\left(\frac{x^2}{x^2 + y^2}\right) - 1$$

$$(x^2 + y^2)^2 = 2x^2 - (x^2 + y^2)$$

$$(x^2 + y^2)^2 - x^2 + y^2 = 0$$

59. Find a polar equation for a circle with center at $(5, \pi/2)$ and passes through $(0, 0)$.

Solution:

The radius is 5.

$$x^2 + (y-5)^2 = 25$$

$$x^2 + y^2 - 10y = 0$$

$$r^2 - 10r\sin\theta = 0$$

$$r(r - 10\sin\theta) = 0$$

$$r = 10\sin\theta$$

61. Find a polar equation for a parabola with vertex at $(2, \pi)$ and focus at $(0, 0)$.

Solution:

$e = 1,\ p = 4$

Vertical directrix to the left of the pole

$$r = \frac{ep}{1 - e\cos\theta}$$

$$r = \frac{4}{1 - \cos\theta}$$

67. Sketch the curve represented by the parametric equations, $x = 1+4t$, $y = 2-3t$, and where possible, write the corresponding rectangular equation by eliminating the parameter.

Solution:

$$x = 1 + 4t \Rightarrow t = \frac{x-1}{4}$$

$$y = 2 - 3t \Rightarrow y = 2 - 3\left(\frac{x-1}{4}\right)$$

$$y = -\frac{3}{4}x + \frac{11}{4}$$

$$3x + 4y = 11$$

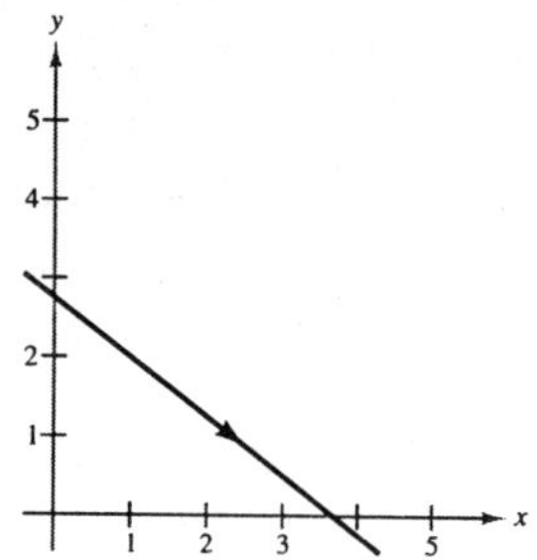

71. Sketch the curve represented by the parametric equations, $x = 6\cos\theta$, $y = 6\sin\theta$, and where possible, write the corresponding rectangular equation by eliminating the parameter.

Solution:

$$x = 6\cos\theta \Rightarrow \cos\theta = \frac{x}{6}$$

$$y = 6\sin\theta \Rightarrow \sin\theta = \frac{y}{6}$$

$$\left(\frac{x}{6}\right)^2 + \left(\frac{y}{6}\right)^2 = 1$$

$$x^2 + y^2 = 36$$

$$\frac{x^2}{36} + \frac{y^2}{36} = 1$$

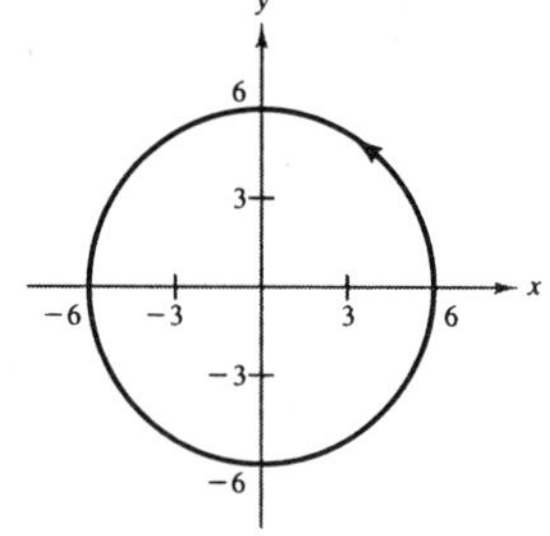

73. Sketch the curve represented by the parametric equations, $x = \cos^3\theta,\ y = 4\sin^3\theta$, and where possible, write the corresponding rectangular equation by eliminating the parameter.

Solution:

$$x = \cos^3\theta \Rightarrow \cos\theta = x^{1/3}$$

$$y = 4\sin^3\theta \Rightarrow \sin\theta = \left(\frac{y}{4}\right)^{1/3}$$

$$(x^{1/3})^2 + \left[\left(\frac{y}{4}\right)^{1/3}\right]^2 = 1$$

$$x^{2/3} + \left(\frac{y}{4}\right)^{2/3} = 1$$

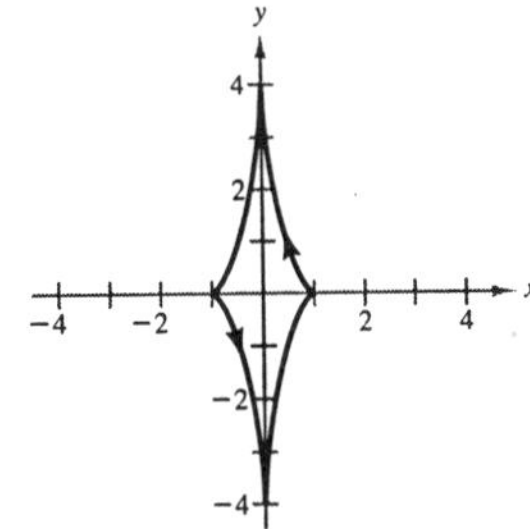

79. Show that the Cartesian equation of a cycloid is

$$x = a\arccos\left(\frac{a-y}{a}\right) \pm \sqrt{2ay - y^2}.$$

Solution:

The parametric equations of a cycloid are $x = a(\theta - \sin\theta)$ and $y = a(1-\cos\theta)$. See Example 5 in Section 11.9.

$$\cos\theta = \frac{a-y}{a}$$

$$\theta = \arccos\left(\frac{a-y}{a}\right)$$

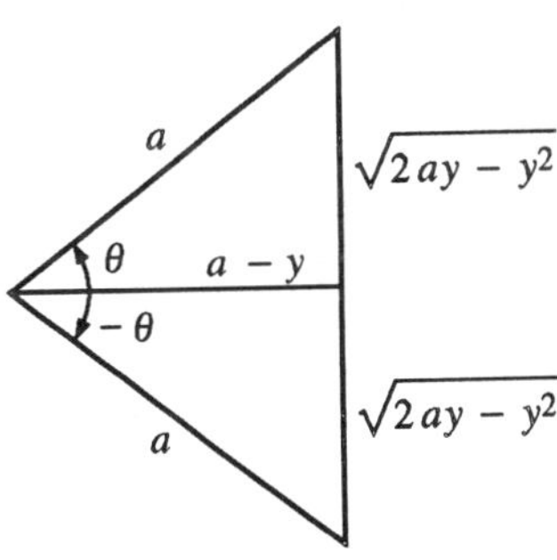

Thus,

$$\begin{aligned} x &= a(\theta - \sin\theta) \\ &= a\left\{\arccos\left(\frac{a-y}{a}\right) - \sin\left[\arccos\left(\frac{a-y}{a}\right)\right]\right\} \\ &= a\arccos\left(\frac{a-y}{a}\right) \pm a\left[\frac{\sqrt{2ay-y^2}}{a}\right] \\ &= a\arccos\left(\frac{a-y}{a}\right) \pm \sqrt{2ay-y^2} \end{aligned}$$

Practice Test for Chapter 11

1. Find the angle, θ, between the lines $3x + 4y = 12$ and $4x - 3y = 12$.

2. Find the distance between the point $(5, -9)$ and the line $3x - 7y = 21$.

3. Find the vertex, focus, and directrix of the parabola $x^2 - 6x - 4y + 1 = 0$.

4. Find an equation of the parabola with its vertex at $(2, -5)$ and focus at $(2, -6)$.

5. Find the center, foci, vertices, and eccentricity of the ellipse $x^2 + 4y^2 - 2x + 32y + 61 = 0$.

6. Find an equation of the ellipse with vertices $(0, \pm 6)$ and eccentricity $e = \frac{1}{2}$.

7. Find the center, vertices, foci, and asymptotes of the hyperbola $16y^2 - x^2 - 6x - 128y + 231 = 0$.

8. Find an equation of the hyperbola with vertices at $(\pm 3, 2)$ and foci at $(\pm 5, 2)$.

9. Rotate the axes to eliminate the xy-term. Sketch the graph of the resulting equation, showing both sets of axes.

$5x^2 + 2xy + 5y^2 - 10 = 0$

10. Use the discriminant to determine whether the graph of the equation is a parabola, ellipse, or hyperbola.

(a) $6x^2 - 2xy + y^2 = 0$

(b) $x^2 + 4xy + 4y^2 - x - y + 17 = 0$

11. Convert the polar point $(\sqrt{2}, (3\pi)/4)$ to rectangular coordinates.

12. Convert the rectangular point $(\sqrt{3}, -1)$ to polar coordinates.

13. Convert the rectangular equation $4x - 3y = 12$ to polar form.

14. Convert the polar equation $r = 5\cos\theta$ to rectangular form.

15. Sketch the graph of $r = 1 - \cos\theta$.

16. Sketch the graph of $r = 5\sin 2\theta$.

17. Sketch the graph of $r = \dfrac{3}{6 - \cos\theta}$.

18. Find a polar equation of the parabola with its vertex at $(6, \pi/2)$ and focus at $(0, 0)$.

For Exercises 19 and 20, eliminate the parameter and write the corresponding rectangular equation.

19. $x = 3 - 2\sin\theta, \quad y = 1 + 5\cos\theta$

20. $x = e^{2t}, \quad y = e^{4t}$

CHAPTER 1

Practice Test Solutions

1. $4 + 3(18 - 11) = 4 + 3(7)$
$= 4 + 21$
$= 25$

2. $\left(\frac{4}{15} \div 2\right) - \left(5 \times \frac{8}{15}\right) = \frac{4}{15} \cdot \frac{1}{2} - \frac{5}{1} \cdot \frac{8}{15}$
$= \frac{2}{15} - \frac{8}{3}$
$= \frac{2}{15} - \frac{40}{15}$
$= -\frac{38}{15}$

3. $|x - (-6)| < 4$
$|x + 6| < 4$

4. $0.0000439 = 4.39 \times 10^{-5}$

5. $(3x^2y^{-1})^2(4x^{-2}y)^{-1} = 9x^4y^{-2}4^{-1}x^2y^{-1}$
$= \dfrac{9x^6}{4y^3}$

6. $\sqrt[3]{81x^5y^6} = \sqrt[3]{27x^3y^6 3x^2}$
$= 3xy^2\sqrt[3]{3x^2}$

7. $\dfrac{4}{\sqrt[3]{2}} = \dfrac{4}{\sqrt[3]{2}} \cdot \dfrac{\sqrt[3]{2}}{\sqrt[3]{2}} \cdot \dfrac{\sqrt[3]{2}}{\sqrt[3]{2}} = \dfrac{4\sqrt[3]{4}}{\sqrt[3]{8}} = \dfrac{4\sqrt[3]{4}}{2} = 2\sqrt[3]{4}$

8. $(x + 3)(x^2 - 4x - 7) = (x + 3)(x^2) + (x + 3)(-4x) + (x + 3)(-7)$
$= x^3 + 3x^2 - 4x^2 - 12x - 7x - 21$
$= x^3 - x^2 - 19x - 21$

9. $x^4 - 81 = (x^2 + 9)(x^2 - 9) = (x^2 + 9)(x + 3)(x - 3)$

10. $x^5 - 4x^3 - x^2 + 4 = x^3(x^2 - 4) - (x^2 - 4)$
$= (x^3 - 1)(x^2 - 4)$
$= (x - 1)(x^2 + x + 1)(x - 2)(x + 2)$

11. $8x^2 + 6x - 9 = (2x + 3)(4x - 3)$

12. $\dfrac{8x^3 + 8x^2y}{x^2y + xy^2} = \dfrac{8x^2(x + y)}{xy(x + y)} = \dfrac{8x}{y}$

13. $\dfrac{3x}{x^2 - x - 6} - \dfrac{2}{x - 3} = \dfrac{3x}{(x + 2)(x - 3)} - \dfrac{2}{x - 3} \cdot \dfrac{x + 2}{x + 2}$
$= \dfrac{3x - 2(x + 2)}{(x + 2)(x - 3)} = \dfrac{3x - 2x - 4}{(x + 2)(x - 3)} = \dfrac{x - 4}{(x + 2)(x - 3)}$

14. $\dfrac{\dfrac{1}{x+1}-\dfrac{1}{x}}{\dfrac{1}{x^2+x}} = \dfrac{\dfrac{x-(x+1)}{x(x+1)}}{\dfrac{1}{x(x+1)}} = \dfrac{-1}{x(x+1)} \cdot \dfrac{x(x+1)}{1} = -1$

15. $\dfrac{x^2-49}{x^2+6x-7} \cdot \dfrac{x^2-1}{x} = \dfrac{(x+7)(x-7)}{(x+7)(x-1)} \cdot \dfrac{(x+1)(x-1)}{x} = \dfrac{(x-7)(x+1)}{x}$

16.
$$\frac{1}{x+2} - \frac{3}{x-4} = \frac{5}{x^2-2x-8}$$
$$(x+2)(x-4)\left[\frac{1}{x+2} - \frac{3}{x-4}\right] = (x+2)(x-4)\left[\frac{5}{(x+2)(x-4)}\right]$$
$$(x-4) - 3(x+2) = 5$$
$$x - 4 - 3x - 6 = 5$$
$$-2x - 10 = 5$$
$$-2x = 15$$
$$x = -\frac{15}{2}$$

17. $(x+12)^2 = 20$
$$x + 12 = \pm\sqrt{20}$$
$$x + 12 = \pm 2\sqrt{5}$$
$$x = -12 \pm 2\sqrt{5}$$

18. $3x^2 + 6x + 2 = 0$
$$x^2 + 2x + \frac{2}{3} = 0$$
$$x^2 + 2x + 1 = -\frac{2}{3} + 1$$
$$(x+1)^2 = \frac{1}{3}$$
$$x + 1 = \pm\sqrt{\frac{1}{3}} = \pm\frac{\sqrt{3}}{3}$$
$$x = -1 \pm \frac{\sqrt{3}}{3} = \frac{-3 \pm \sqrt{3}}{3}$$

19. $2x^2 - 3x - 5 = 0$

$a = 2,\ b = -3,\ c = -5$
$$x = \frac{-(-3) \pm \sqrt{(-3)^2 - 4(2)(-5)}}{2(2)} = \frac{3 \pm \sqrt{9+40}}{4} = \frac{3 \pm \sqrt{49}}{4} = \frac{3 \pm 7}{4}$$
$$x = \frac{3+7}{4} = \frac{10}{4} = \frac{5}{2}$$
$$x = \frac{3-7}{4} = \frac{-4}{4} = -1$$

20.
$$-3 \le \frac{4-x}{2} \le 5$$
$$-6 \le 4-x \le 10$$
$$-10 \le -x \le 6$$
$$10 \ge x \ge -6 \quad \text{or} \quad -6 \le x \le 10$$

21. $|x-15| \ge 10$
$$x - 15 \le -10 \quad \text{or} \quad x - 15 \ge 10$$
$$x \le 5 \quad \text{or} \quad x \ge 25$$

22.
$$x^3 - 9x \le 0$$
$$x(x^2-9) \le 0$$
$$x(x+3)(x-3) \le 0$$

Critical numbers: $x = 0,\ x = -3,\ x = 3$

Test intervals: $(-\infty,\ -3),\ (-3,\ 0),\ (0,\ 3),\ (3,\ \infty)$

$(-)(-)(-) < 0$	$(-)(+)(-) > 0$	$(+)(+)(-) < 0$	$(+)(+)(+) > 0$
YES	NO	YES	NO

Number line points: -3, 0, 3

Solution: $(-\infty,\ -3] \cup [0,\ 3]$

23.
$$\frac{4}{x+3} > \frac{-1}{x-2}$$
$$\frac{4}{x+3} + \frac{1}{x-2} > 0$$
$$\frac{4(x-2)+(x+3)}{(x+3)(x-2)} > 0$$
$$\frac{4x-8+x+3}{(x+3)(x-2)} > 0$$
$$\frac{5x-5}{(x+3)(x-2)} > 0$$
$$\frac{5(x-1)}{(x+3)(x-2)} > 0$$

Critical numbers: $x = 1,\ x = -3,\ x = 2$

Test intervals: $(-\infty,\ -3),\ (-3,\ 1),\ (1,\ 2),\ (2,\ \infty)$

$\frac{(-)}{(-)(-)} < 0$	$\frac{(-)}{(+)(-)} > 0$	$\frac{(+)}{(+)(-)} < 0$	$\frac{(+)}{(+)(+)} > 0$
NO	YES	NO	YES

Number line points: -3, -2, -1, 0, 1, 2

Solution: $(-3,\ 1) \cup (2,\ \infty)$

24. $x^2(1-2x)^{4/3} - 3x(1-2x)^{1/3} = x(1-2x)^{1/3}[x(1-2x) - 3]$

$$= x(1-2x)^{1/3}(x - 2x^2 - 3) = x(1-2x)^{1/3}(-2x^2 + x - 3)$$

25. $\sqrt{36 + x^2} = 6 + x$ is false.

$\sqrt{36 + x^2}$ cannot be reduced any further.

$\sqrt{36 + 12x + x^2} = \sqrt{(6 + x)^2} = |6 + x|$

CHAPTER 2

Practice Test Solutions

1. $d = \sqrt{(4-0)^2 + (-1-3)^2}$

$= \sqrt{16+16}$

$= \sqrt{32}$

$= 4\sqrt{2}$

2. Midpoint: $\left(\frac{4+0}{2}, \frac{-1+3}{2}\right) = (2, 1)$

3. $6 = \sqrt{(x-0)^2 + (-2-0)^2}$

$6 = \sqrt{x^2+4}$

$36 = x^2 + 4$

$x^2 = 32$

$x = \pm\sqrt{32}$

$x = \pm 4\sqrt{2}$

4. x-intercept: Let $y = 0$; $0 = \frac{x-2}{x+3}$

$0 = x - 2$

$x = 2 \quad (2, 0)$

y-intercept: Let $x = 0$; $y = \frac{0-2}{0+3}$

$y = -\frac{2}{3} \quad \left(0, -\frac{2}{3}\right)$

5. $xy^2 = 6$

$x(-y)^2 = 6 \Rightarrow xy^2 = 6$ x-axis symmetry

$(-x)y^2 = 6 \Rightarrow xy^2 = -6$ No y-axis symmetry

$(-x)(-y)^2 = 6 \Rightarrow xy^2 = -6$ No origin symmetry

6. x-intercepts: $(0, 0)$, $(2, 0)$, $(-2, 0)$
Origin symmetry:

x	0	1	-1	2	-2	3
y	0	-3	3	0	0	15

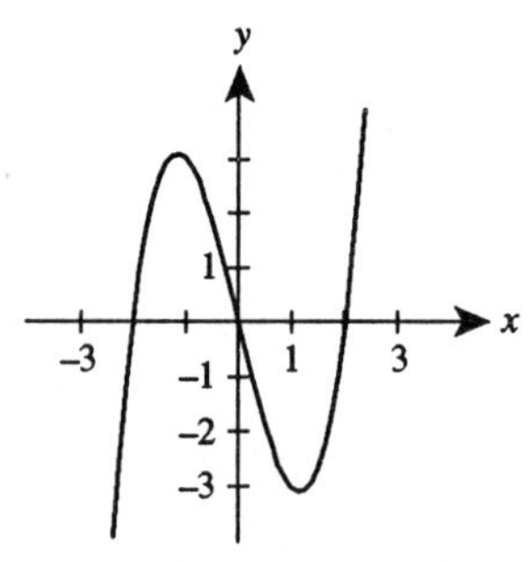

7. $$x^2 + y^2 - 6x + 2y + 6 = 0$$
$$x^2 - 6x + \underline{9} + y^2 + 2y + \underline{1} = -6 + 9 + 1$$
$$(x-3)^2 + (y+1)^2 = 4$$

Center: $(3, \ -1)$
Radius: 2

8. $f(x-3) = (x-3)^2 - 2(x-3) + 1$
$$= x^2 - 6x + 9 - 2x + 6 + 1$$
$$= x^2 - 8x + 16$$

9. $$f(3) = 12 - 11 = 1$$
$$\frac{f(x) - f(3)}{x - 3} = \frac{(4x - 11) - 1}{x - 3}$$
$$= \frac{4x - 12}{x - 3} = \frac{4(x-3)}{x-3} = 4$$

10. $f(x) = \sqrt{36 - x^2} = \sqrt{(6+x)(6-x)}$

Domain: $[-6, \ 6]$

Range: $[0, \ 6]$

11. (a) $6x - 5y + 4 = 0$
$$y = \frac{6x + 4}{5} \quad \text{function}$$

(b) $x^2 + y^2 = 9$
$$y = \pm\sqrt{9 - x^2} \quad \text{not a function}$$

(c) $y^3 = x^2 + 6$
$$y = \sqrt[3]{x^2 + 6} \quad \text{function}$$

12. Parabola: Vertex $(0, \ -5)$
Intercepts: $(0, \ -5)$, $(\pm\sqrt{5}, \ 0)$
y-axis symmetry

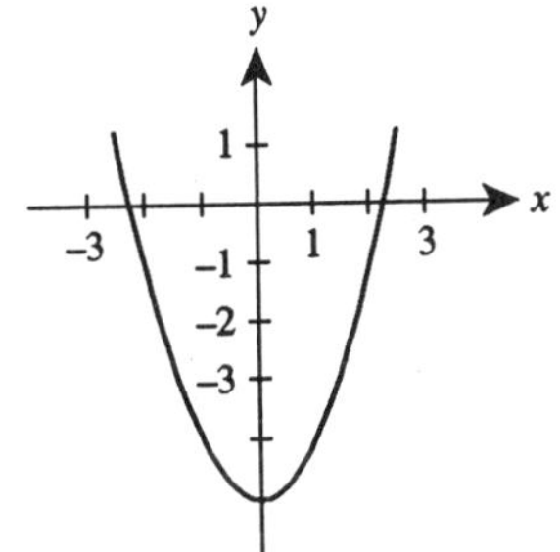

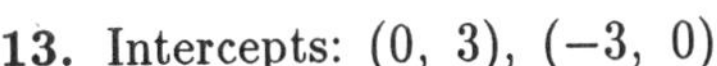

13. Intercepts: $(0, \ 3)$, $(-3, \ 0)$

x	0	1	-1	2	-2	-3	-4
y	3	4	2	5	1	0	1

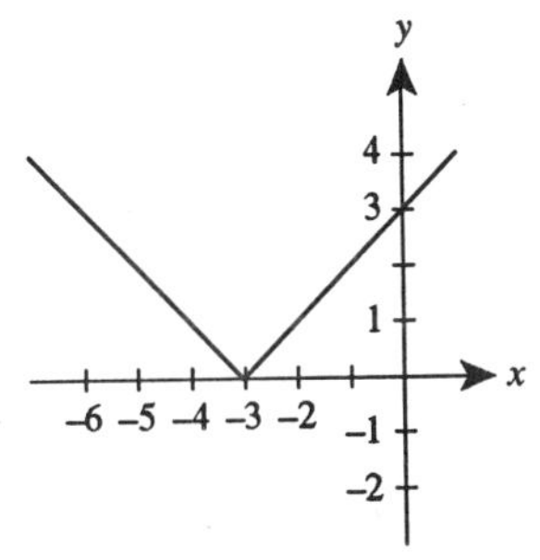

14.

x	0	1	2	3	-1	-2	-3
y	1	3	5	7	2	6	12

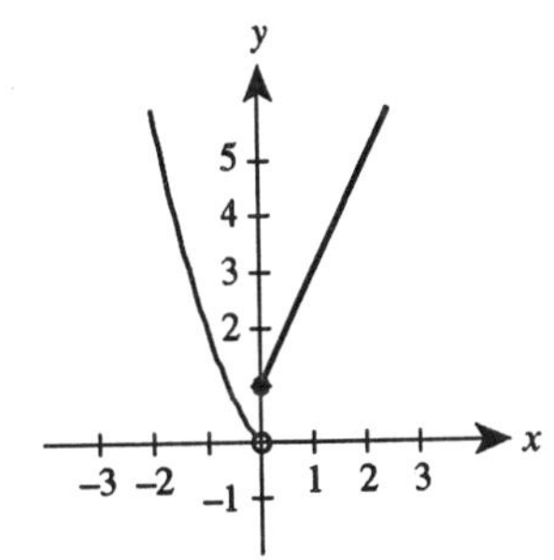

15. $m = \dfrac{-1-4}{3-2} = -5$

$y - 4 = -5(x-2)$

$y - 4 = -5x + 10$

$y = -5x + 14$

16. $y = \frac{4}{3}x - 3$

17. $2x + 3y = 0$

$y = -\frac{2}{3}x$

$m_1 = -\frac{2}{3}$

$\perp m_2 = \frac{3}{2}$ through (4, 1)

$y - 1 = \frac{3}{2}(x-4)$

$y - 1 = \frac{3}{2}x - 6$

$y = \frac{3}{2}x - 5$

18. (5, 32) and (9, 44)

$m = \dfrac{44-32}{9-5} = \dfrac{12}{4} = 3$

$y - 32 = 3(x-5)$

$y - 32 = 3x - 15$

$y = 3x + 17$

When $x = 20$, $y = 3(20) + 17$

$y = \$77.$

19. $f(g(x)) = f(2x+3)$

$= (2x+3)^2 - 2(2x+3) + 16$

$= 4x^2 + 12x + 9 - 4x - 6 + 16$

$= 4x^2 + 8x + 19$

20. $f(x) = x^3 + 7$

$y = x^3 + 7$

$x = y^3 + 7$

$x - 7 = y^3$

$\sqrt[3]{x-7} = y$

$f^{-1}(x) = \sqrt[3]{x-7}$

21. (a) $f(x) = |x-6|$ is not one-to-one.
For example, $f(0) = 6$ and $f(12) = 6$.
(b) $f(x) = ax + b$, $a \neq 0$ is one-to-one.
(c) $f(x) = x^3 - 19$ is one-to-one.

22. $f(x) = \sqrt{\dfrac{3-x}{x}}, \quad 0 < x \le 3$

$$y = \sqrt{\frac{3-x}{x}}$$

$$x = \sqrt{\frac{3-y}{y}}$$

$$x^2 = \frac{3-y}{y}$$

$$x^2 y = 3 - y$$

$$x^2 y + y = 3$$

$$y(x^2+1) = 3$$

$$y = \frac{3}{x^2+1}$$

$$f^{-1}(x) = \frac{3}{x^2+1}$$

23. $y = kx$ and $y = 30$ when $x = 5$.

$$30 = k(5)$$

$$6 = k$$

$$y = 6x$$

24. $y = \dfrac{k}{x}$ and $y = 0.5$ when $x = 14$.

$$0.5 = \frac{k}{14}$$

$$7 = k$$

$$y = \frac{7}{x}$$

25. $z = \dfrac{kx^2}{y}$ and $z = 3$ when $x = 3,\ y = -6$.

$$3 = \frac{k(3)^2}{-6}$$

$$-18 = 9k$$

$$-2 = k$$

$$z = \frac{-2x^2}{y}$$

CHAPTER 3

Practice Test Solutions

1. $f(x) = x^2 - 9$

Vertex: $(0, -9)$

x-intercepts: $(3, 0)$, $(-3, 0)$

y-intercept: $(0, -9)$

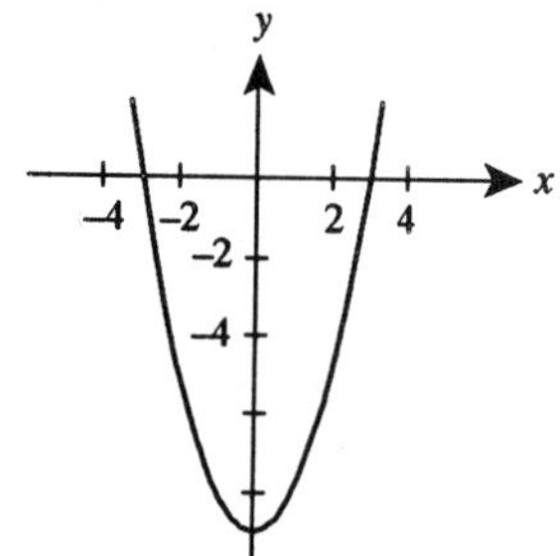

2. $g(x) = 3x^2 - 5x - 28$

Vertex: $-\dfrac{b}{2a} = \dfrac{5}{6}$, $g\left(\dfrac{5}{6}\right) = -\dfrac{361}{12}$

$\left(\dfrac{5}{6}, -\dfrac{361}{12}\right)$

x-intercepts: $0 = 3x^2 - 5x - 28$

$0 = (3x + 7)(x - 4)$

$x = -\dfrac{7}{3}$, $x = 4$

$\left(-\dfrac{7}{3}, 0\right)$, $(4, 0)$

y-intercept: $(0, -28)$

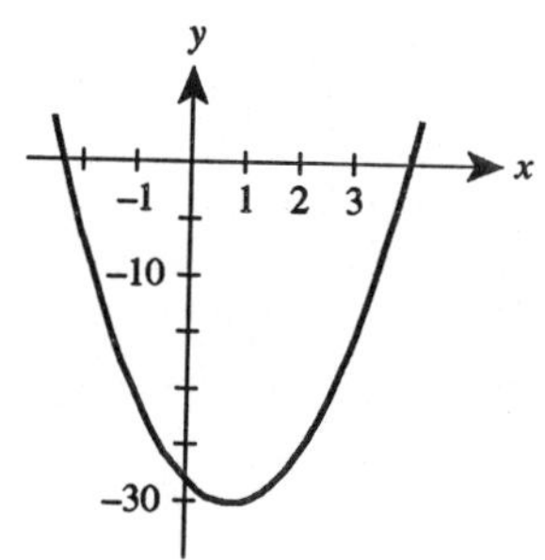

3. Vertex: $(-2, 1)$

Point: $(3, 4)$

$f(x) = a(x + 2)^2 + 1$

$4 = a(3 + 2)^2 + 1$

$3 = 25a \Rightarrow a = \frac{3}{25}$

$f(x) = \frac{3}{25}(x + 2)^2 + 1$

4.

$$f(x) = x^4 - 5x^2 + 4$$

$$x^4 - 5x^2 + 4 = 0$$

$$(x^2 - 1)(x^2 - 4) = 0$$

$$(x + 1)(x - 1)(x + 2)(x - 2) = 0$$

Zeros: ± 1, ± 2

5. Zeros: -1, $\sqrt{3}$ and $-\sqrt{3}$

$f(x) = (x - (-1))(x - \sqrt{3})(x - (-\sqrt{3}))$

$= (x + 1)(x - \sqrt{3})(x + \sqrt{3}) = (x + 1)(x^2 - 3) = x^3 + x^2 - 3x - 3$

6. $f(x) = x^3 - 9x = x(x+3)(x-3)$
Intercepts: $(0,\ 0)$, $(3,\ 0)$, $(-3,\ 0)$
Origin symmetry

x	-4	-2	-1	1	2	4
y	-28	10	8	-8	-10	28

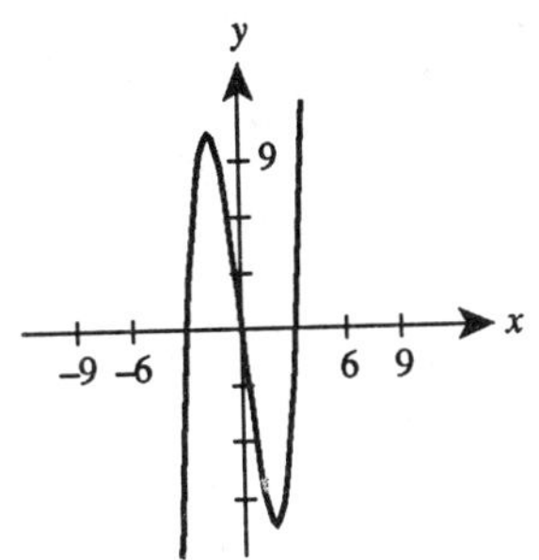

7. 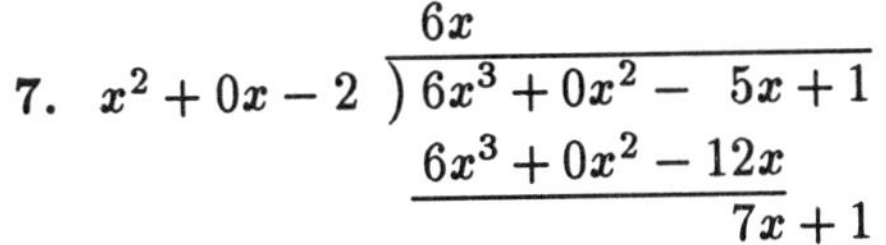

$$\frac{6x^3 - 5x + 1}{x^2 - 2} = 6x + \frac{7x + 1}{x^2 - 2}$$

8.
$$\begin{array}{r|rrrrr} -3 & 1 & 3 & -9 & 0 & 1 \\ & & -3 & 0 & 27 & -81 \\ \hline & 1 & 0 & -9 & 27 & -80 \end{array}$$

$$\frac{x^4 + 3x^3 - 9x^2 + 1}{x + 3} = x^3 - 9x + 27 - \frac{80}{x + 3}$$

9.
$$\begin{array}{r|rrrrrrr} -2 & 1 & 0 & 0 & -3 & 0 & 1 & -5 \\ & & -2 & 4 & -8 & 22 & -44 & 86 \\ \hline & 1 & -2 & 4 & -11 & 22 & -43 & 81 \end{array}$$

$f(-2) = 81$

10. $0 = x^3 - 2x^2 - 5x + 6$
Possible rational zeros: ± 1, ± 2, ± 3, ± 6

$$\begin{array}{r|rrrr} 1 & 1 & -2 & -5 & 6 \\ & & 1 & -1 & -6 \\ \hline & 1 & -1 & -6 & 0 \end{array}$$

$$x^3 - 2x^2 - 5x + 6 = (x - 1)(x^2 - x - 6)$$
$$0 = (x - 1)(x - 3)(x + 2)$$

Zeros: 1, 3, -2

11. $0 = x^3 + x^2 - 5x - 6$
Possible rational zeros: ± 1, ± 2, ± 3, ± 6

$$\begin{array}{r|rrrr} -2 & 1 & 1 & -5 & -6 \\ & & -2 & 2 & 6 \\ \hline & 1 & -1 & -3 & 0 \end{array}$$

$x^3 + x^2 - 5x - 6 = (x + 2)(x^2 - x - 3)$
Zeros: $x = -2$

$$x = \frac{-(-1) \pm \sqrt{(-1)^2 - 4(1)(-3)}}{2(1)} = \frac{1 \pm \sqrt{13}}{2}$$

12. $p(x) = 4x^3 - 7x^2 + x - 20$
Possible rational zeros: ± 1, ± 2, ± 4, ± 5, ± 10, ± 20, $\pm\frac{1}{2}$, $\pm\frac{1}{4}$, $\pm\frac{5}{2}$, $\pm\frac{5}{4}$

13. $5i^3 - (\sqrt{-9})^2 = -5i - (-9) = 9 - 5i$

14. $$\begin{aligned}(6+5i)(-2+9i) &= -12 + 54i - 10i + 45i^2 \\ &= -12 - 45 + (54-10)i \\ &= -57 + 44i\end{aligned}$$

15. $$\begin{aligned}\frac{3+2i}{4-7i} &= \frac{3+2i}{4-7i} \bullet \frac{4+7i}{4+7i} \\ &= \frac{12 + 21i + 8i + 14i^2}{16+49} \\ &= \frac{-2+29i}{65} \\ &= -\frac{2}{65} + \frac{29}{65}i\end{aligned}$$

16. $f(x) = x^2 - 3x + 5$
$$\begin{aligned}x &= \frac{-(-3) \pm \sqrt{(-3)^2 - 4(1)(5)}}{2(1)} \\ &= \frac{3 \pm \sqrt{9-20}}{2} \\ &= \frac{3 \pm \sqrt{11}\,i}{2} \\ &= \frac{3}{2} \pm \frac{\sqrt{11}}{2}i\end{aligned}$$

17. $p(x) = x^5 + x^3 - 8x^2 - 8$
$$\begin{aligned}0 &= x^3(x^2+1) - 8(x^2+1) \\ 0 &= (x^3 - 8)(x^2+1) \\ 0 &= (x-2)(x^2+2x+4)(x^2+1)\end{aligned}$$
$x = 2$
$$\begin{aligned}x &= \frac{-2 \pm \sqrt{(2)^2 - 4(1)(4)}}{2(1)} \\ &= \frac{-2 \pm \sqrt{-12}}{2} \\ &= \frac{-2 \pm 2\sqrt{3}\,i}{2} = -1 \pm \sqrt{3}\,i\end{aligned}$$
$x = \pm i$

Zeros: $x = 2$, $x = -1 \pm \sqrt{3}\,i$, $x = \pm i$

18. Zeros: 0, 4, $2 + i$
Since $2 + i$ is a zero, so is $2 - i$.
$$\begin{aligned}p(x) &= (x-0)(x-4)[x-(2+i)][x-(2-i)] \\ &= x(x-4)(x-2-i)(x-2+i) \\ &= (x^2-4x)(x^2-4x+5) \\ &= x^4 - 8x^3 + 21x^2 - 20x\end{aligned}$$

19. $h(x) = \dfrac{x^2+1}{x^2-16}$
Vertical asymptotes:
$x = 4$, $x = -4$
Horizontal asymptote:
$y = 1$

20. $f(x) = \dfrac{x+3}{x-2}$

x-intercept: $(-3,\ 0)$

y-intercept: $(0,\ -\frac{3}{2})$

Vertical asymptote: $x = 2$

Horizontal asymptote: $y = 1$

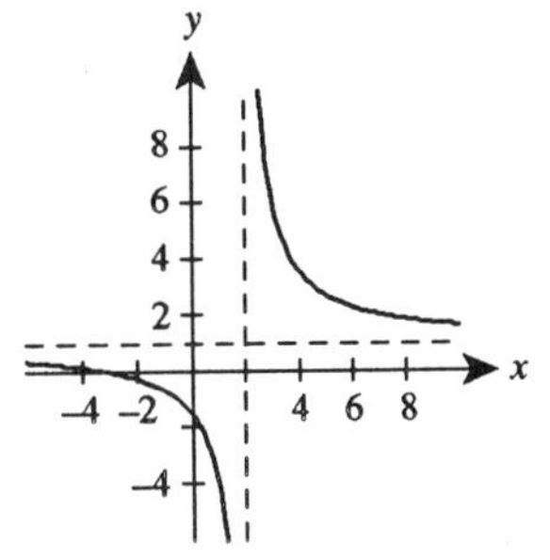

21. $g(x) = \dfrac{x}{x^2-1}$

x-intercept: $(0,\ 0)$

Vertical asymptotes: $x = 1,\ x = -1$

Horizontal asymptote: $y = 0$

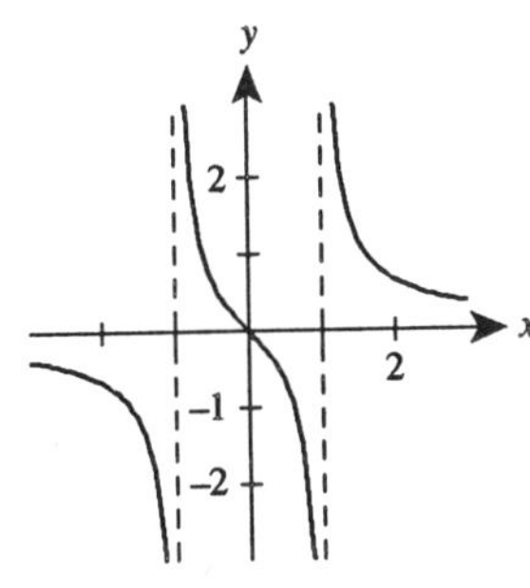

22. $h(x) = \dfrac{x^2+6x-7}{x+3}$

$$\begin{array}{r}
x + 3 \\
x+3 \;\overline{)\, x^2+6x-7} \\
\underline{x^2+3x} \\
3x-7 \\
\underline{3x+9} \\
-16
\end{array}$$

$$\frac{x^2+6x-7}{x+3} = x+3-\frac{16}{x+3}$$

Slant asymptote: $y = x + 3$

23. $\dfrac{8x-2}{x^2-2x-8} = \dfrac{8x-2}{(x+2)(x-4)} = \dfrac{A}{x+2} + \dfrac{B}{x-4}$

$8x - 2 = A(x-4) + B(x+2)$

Let $x = -2$, $-18 = -6A$, $A = 3$.

Let $x = 4$, $30 = 6B$, $B = 5$.

$\dfrac{8x-2}{x^2-2x-8} = \dfrac{3}{x+2} + \dfrac{5}{x-4}$

24. $\dfrac{3x^2+15}{x^3+3x} = \dfrac{3x^2+15}{x(x^2+3)} = \dfrac{A}{x} + \dfrac{Bx+C}{x^2+3}$

$$3x^2+15 = A(x^2+3) + (Bx+C)x$$

$$3x^2+15 = Ax^2+3A+Bx^2+Cx$$

$$3x^2+0x+15 = (A+B)x^2+Cx+3A$$

$$3 = A+B$$

$$0 = C$$

$$15 = 3A \Rightarrow A = 5,\quad B = -2$$

$$\frac{3x^2+15}{x^3+3x} = \frac{5}{x} + \frac{-2x+0}{x^2+3} = \frac{5}{x} - \frac{2x}{x^2+3}$$

25. $$\begin{array}{r} 2x \\ x^2+4x+4 \overline{\smash{)}\,2x^3+8x^2+9x-1} \\ \underline{2x^3+8x^2+8x} \\ x-1 \end{array}$$

$$\frac{2x^3+8x^2+9x-1}{x^2+4x+4} = 2x + \frac{x-1}{x^2+4x+4}$$

$$\frac{x-1}{(x+2)^2} = \frac{A}{x+2} + \frac{B}{(x+2)^2}$$

$$x-1 = A(x+2)+B$$

$$x-1 = Ax+2A+B$$

$$1 = A$$

$$-1 = 2A+B, \quad B=-3$$

$$\frac{2x^3+8x^2+9x-1}{x^2+4x+4} = 2x + \frac{1}{x+2} - \frac{3}{(x+2)^2}$$

CHAPTER 4

Practice Test Solutions

1. $x^{3/5} = 8$

$$x = 8^{5/3} = (\sqrt[3]{8})^5 = 2^5 = 32$$

2. $3^{x-1} = \frac{1}{81}$

$$3^{x-1} = 3^{-4}$$

$$x - 1 = -4$$

$$x = -3$$

3. $f(x) = 2^{-x} = \left(\frac{1}{2}\right)^x$

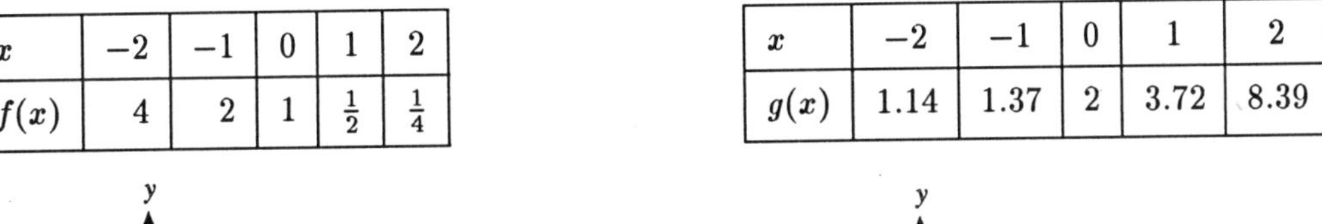

x	-2	-1	0	1	2
$f(x)$	4	2	1	$\frac{1}{2}$	$\frac{1}{4}$

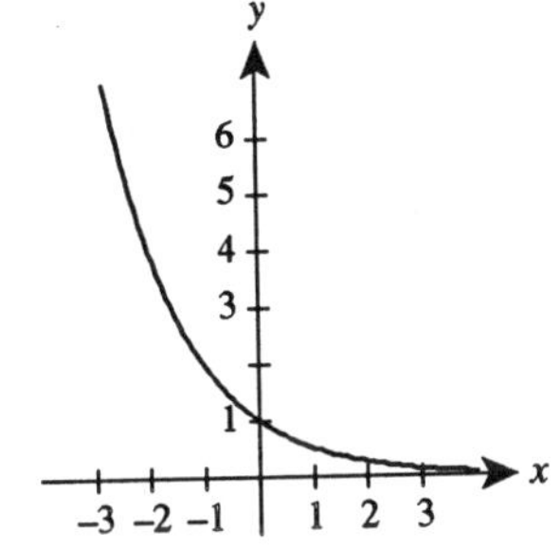

4. $g(x) = e^x + 1$

x	-2	-1	0	1	2
$g(x)$	1.14	1.37	2	3.72	8.39

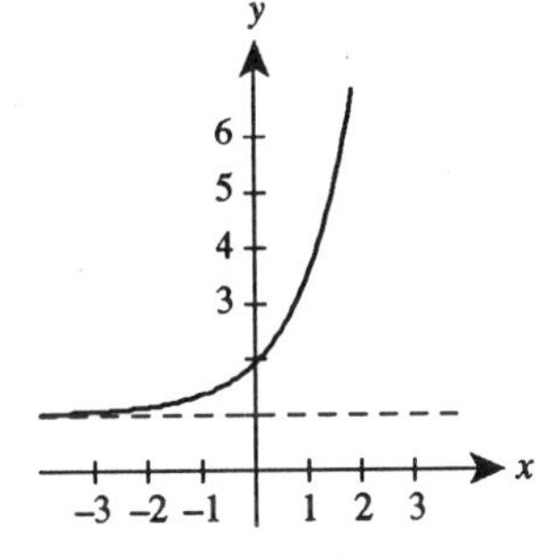

5. $A = P\left(1 + \frac{r}{n}\right)^{nt}$

(a) $A = 5000\left(1 + \frac{0.09}{12}\right)^{12(3)} \approx \$6543.23$

(b) $A = 5000\left(1 + \frac{0.09}{4}\right)^{4(3)} \approx \$6530.25$

(c) $A = 5000e^{(0.09)(3)} \approx \$6549.82$

6. $7^{-2} = \frac{1}{49}$

$$\log_7 \tfrac{1}{49} = -2$$

7. $x - 4 = \log_2 \frac{1}{64}$

$$2^{x-4} = \frac{1}{64}$$
$$2^{x-4} = 2^{-6}$$
$$x - 4 = -6$$
$$x = -2$$

8. $\log_b \sqrt[4]{\frac{8}{25}} = \frac{1}{4} \log_b \frac{8}{25}$

$$= \frac{1}{4}[\log_b 8 - \log_b 25]$$
$$= \frac{1}{4}[\log_b 2^3 - \log_b 5^2]$$
$$= \frac{1}{4}[3 \log_b 2 - 2 \log_b 5]$$
$$= \frac{1}{4}[3(0.3562) - 2(0.8271)]$$
$$= -0.1464$$

9. $5 \ln x - \dfrac{1}{2} \ln y + 6 \ln z = \ln x^5 - \ln \sqrt{y} + \ln z^6 = \ln \left(\dfrac{x^5 z^6}{\sqrt{y}} \right)$

10. $\log_9 28 = \dfrac{\log 28}{\log 9} \approx 1.5166$

11. $\log N = 0.6646$

$$N = 10^{0.6646} \approx 4.62$$

12.

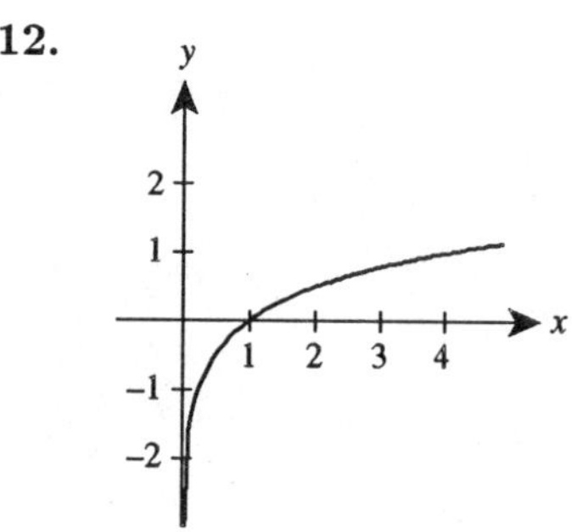

13. Domain:

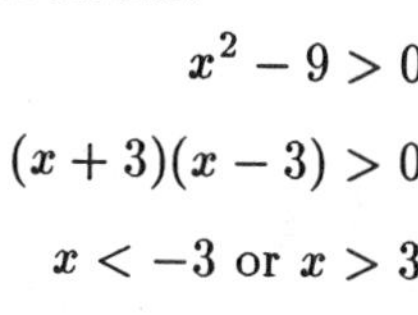

$$x^2 - 9 > 0$$
$$(x + 3)(x - 3) > 0$$
$$x < -3 \text{ or } x > 3$$

14.

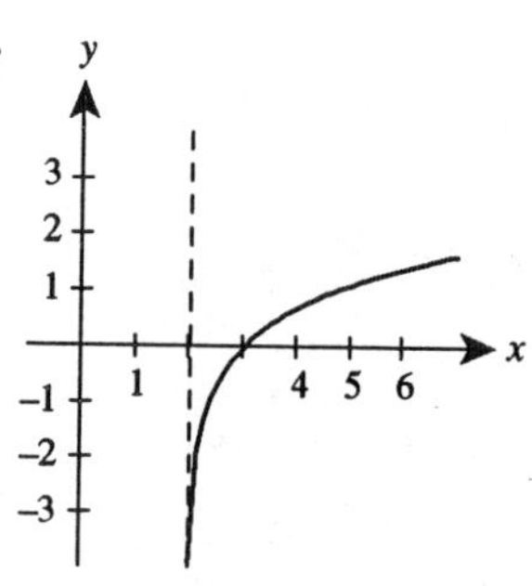

15. $\dfrac{\ln x}{\ln y} \neq \ln(x - y)$ since $\dfrac{\ln x}{\ln y} = \log_y x$

16. $5^x = 41$

$$x = \log_5 41 = \frac{\ln 41}{\ln 5} \approx 2.3074$$

17. $x - x^2 = \log_5 \frac{1}{25}$

$$5^{x-x^2} = \frac{1}{25}$$
$$5^{x-x^2} = 5^{-2}$$
$$x - x^2 = -2$$
$$0 = x^2 - x - 2$$
$$0 = (x + 1)(x - 2)$$
$$x = -1 \text{ or } x = 2$$

18. $\log_2 x + \log_2(x - 3) = 2$

$$\log_2[x(x - 3)] = 2$$
$$x(x - 3) = 2^2$$
$$x^2 - 3x = 4$$
$$x^2 - 3x - 4 = 0$$
$$(x + 1)(x - 4) = 0$$
$$x = 4$$
$$x = -1$$

(extraneous solution)

19. $$\frac{e^x + e^{-x}}{3} = 4$$

$$e^x(e^x + e^{-x}) = 12e^x$$

$$e^{2x} + 1 = 12e^x$$

$$e^{2x} - 12e^x + 1 = 0$$

$$e^x = \frac{12 \pm \sqrt{144 - 4}}{2}$$

$e^x = 11.9161$ or $e^x = 0.0839$

$x = \ln 11.9161$ $\qquad$ $x = \ln 0.0839$

$x \approx 2.4779$ $\qquad$ $x \approx -2.4779$

20. $$A = Pe^{rt}$$

$$12{,}000 = 6000e^{0.13t}$$

$$2 = e^{0.13t}$$

$$0.13t = \ln 2$$

$$t = \frac{\ln 2}{0.13}$$

$t \approx 5.3319$ years or

5 years 4 months

CHAPTER 5

Practice Test Solutions

1. (a) $350° = 350\left(\frac{\pi}{180}\right) = \frac{35\pi}{18}$

(b) $\frac{5\pi}{9} = \frac{5\pi}{9} \cdot \frac{180}{\pi} = 100°$

2. (a) $135°14'12'' = \left(135 + \frac{14}{60} + \frac{12}{3600}\right)°$

$\approx 135.2367°$

(b) $-22.569° = -(22° + 0.569(60)')$

$= -22°34.14'$

$= -(22°34' + 0.14(60)'')$

$\approx -22°34'8''$

3. (a) $\frac{5\pi}{6}$ corresponds to the point $\left(-\frac{\sqrt{3}}{2}, \frac{1}{2}\right)$.

$\sin\frac{5\pi}{6} = y = \frac{1}{2}$

(b) $\frac{5\pi}{4}$ corresponds to the point $\left(-\frac{\sqrt{2}}{2}, -\frac{\sqrt{2}}{2}\right)$.

$\tan\frac{5\pi}{4} = \frac{y}{x} = 1$

4. (a) $\sin 7\pi = \sin(6\pi + \pi) = \sin\pi = 0$

(b) $\cos\left(-\frac{13\pi}{3}\right) = \cos\left(-4\pi - \frac{\pi}{3}\right)$

$= \cos\left(-\frac{\pi}{3}\right)$

$= \cos\frac{\pi}{3} = \frac{1}{2}$

5. $\cos\theta = \frac{2}{3}$

$x = 2,\ r = 3,\ y = \sqrt{9-4} = \sqrt{5}$

$\tan\theta = \frac{y}{x} = \frac{\sqrt{5}}{2}$

6. $\sin\theta = 0.9063$

$\theta = \arcsin(0.9063)$

$\theta \approx 65°$ or $\frac{13\pi}{36}$

7. $\tan 20° = \frac{35}{x}$

$x = \frac{35}{\tan 20°} \approx 96.1617$

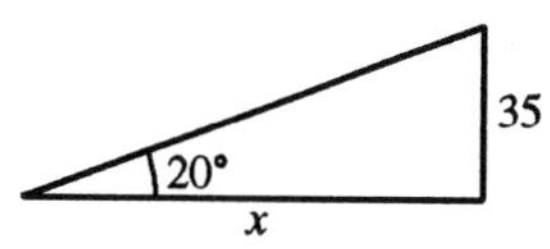

8. $\theta = \frac{6\pi}{5}$, θ is in Quadrant III.

Reference angle: $\frac{6\pi}{5} - \pi = \frac{\pi}{5}$ or $36°$

9. $\csc 3.92 = \dfrac{1}{\sin 3.92} \approx -1.4242$

10. $\tan\theta = 6 = \dfrac{6}{1}$, θ lies in Quadrant III.

$y = -6,\ x = -1,\ r = \sqrt{36+1} = \sqrt{37}$,

so $\sec\theta = \dfrac{\sqrt{37}}{-1} \approx -6.0828$.

11. Period: 4π
Amplitude: 3

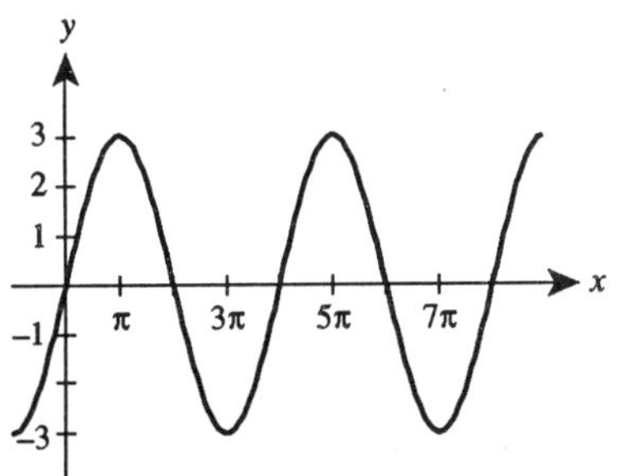

12. Period: 2π
Amplitude: 2

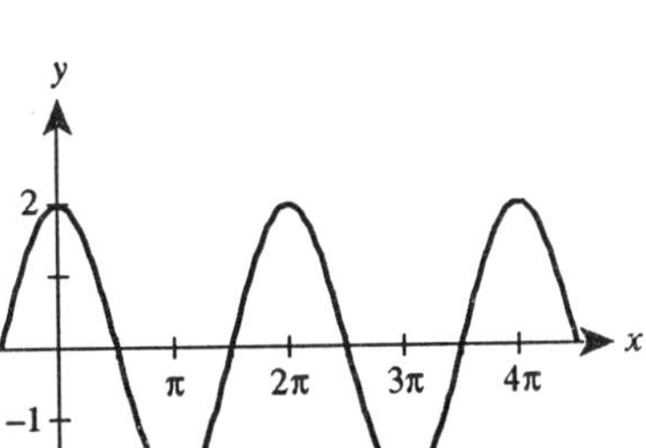

13. Period: $\dfrac{\pi}{2}$

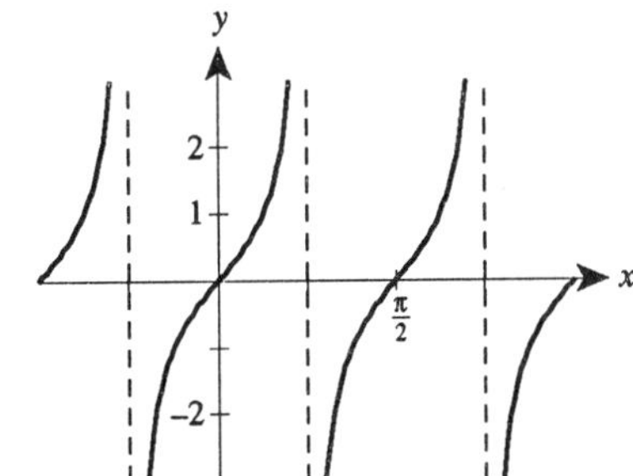

14. Period: 2π

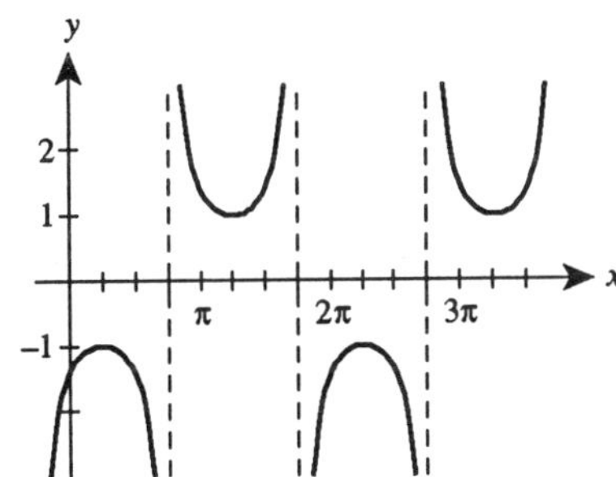

15.

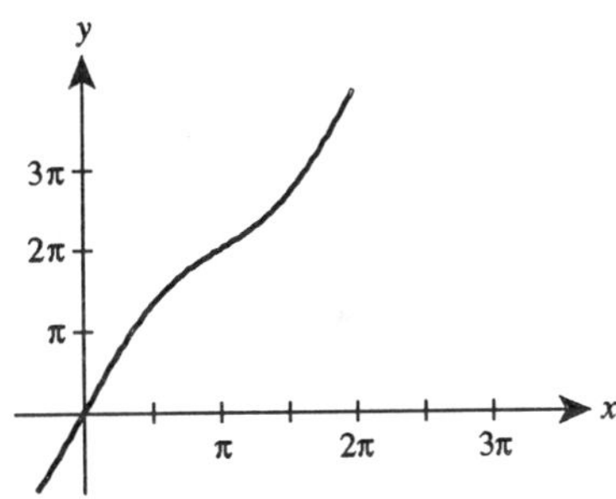

16.

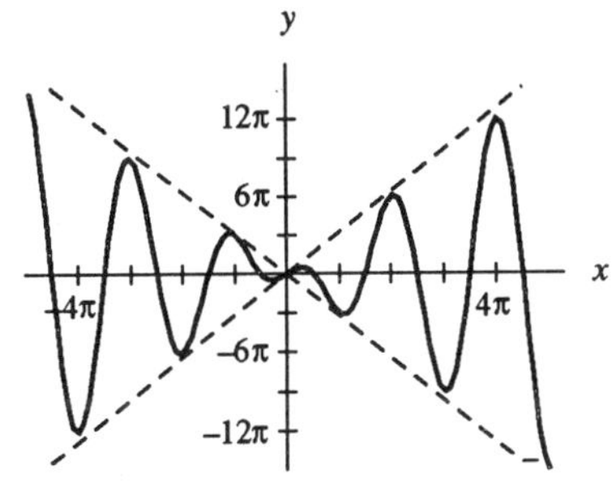

17. Let $\theta = \arcsin 1$.

$\sin\theta = 1$

$\theta = \dfrac{\pi}{2}$

18. $\arctan(-3) = -\arctan 3$

$\tan\theta = -3$

$\theta \approx -1.249 \approx -71.565°$

19. $\sin\left(\arccos\dfrac{4}{\sqrt{35}}\right)$

$\sin\theta = \dfrac{\sqrt{19}}{\sqrt{35}} \approx 0.7368$

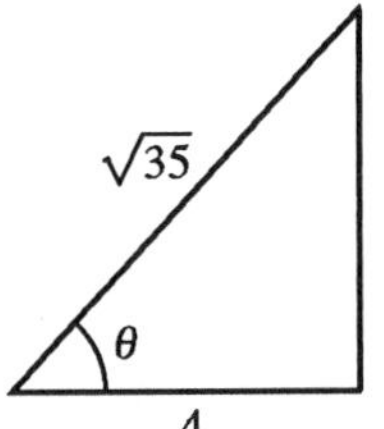

20. $\cos\left(\arcsin\dfrac{x}{4}\right)$

$\cos\theta = \dfrac{\sqrt{16-x^2}}{4}$

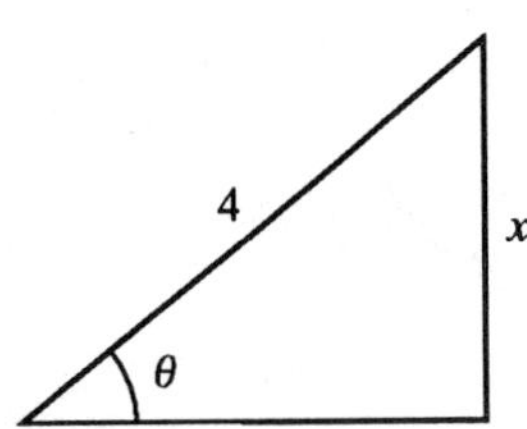

21. Given $A = 40°,\ c = 12$

$B = 90° - 40° = 50°$

$\sin 40° = \dfrac{a}{12}$

$a = 12\sin 40° \approx 7.713$

$\cos 40° = \dfrac{b}{12}$

$b = 12\cos 40° \approx 9.192$

22. Given $B = 6.84°,\ a = 21.3$

$A = 90° - 6.84° = 83.16°$

$\sin 83.16° = \dfrac{21.3}{c}$

$c = \dfrac{21.3}{\sin 83.16°} \approx 21.453$

$\tan 83.16° = \dfrac{21.3}{b}$

$b = \dfrac{21.3}{\tan 83.16°} \approx 2.555$

23. Given $a = 5,\ b = 9$

$c = \sqrt{25+81} = \sqrt{106} \approx 10.296$

$\tan A = \frac{5}{9}$

$A = \arctan\frac{5}{9} \approx 29.055°$

$B = 90° - 29.055° = 60.945°$

24. $\sin 67° = \dfrac{x}{20}$

$x = 20 \sin 67° \approx 18.41$ feet

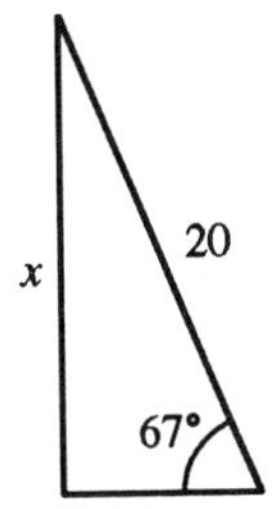

25. $\tan 5° = \dfrac{250}{x}$

$$x = \frac{250}{\tan 5°}$$

$$\approx 2857.513 \text{ ft}$$

$$\approx 0.541 \text{ mi}$$

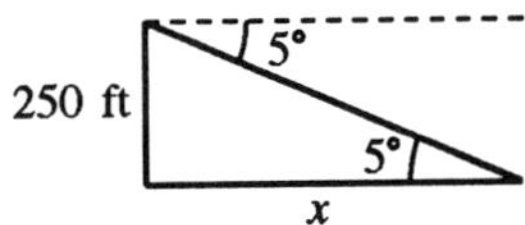

CHAPTER 6

Practice Test Solutions

1. $\tan x = \frac{4}{11}$, $\sec x < 0 \Rightarrow x$ is in Quadrant III.

   $y = -4,\ x = -11,\ r = \sqrt{16+121} = \sqrt{137}$

   $$\sin x = -\frac{4}{\sqrt{137}} = -\frac{4\sqrt{137}}{137} \qquad \csc x = -\frac{\sqrt{137}}{4}$$
   $$\cos x = -\frac{11}{\sqrt{137}} = -\frac{11\sqrt{137}}{137} \qquad \sec x = -\frac{\sqrt{137}}{11}$$
   $$\tan x = \frac{4}{11} \qquad \cot x = \frac{11}{4}$$

2. $$\frac{\sec^2 x + \csc^2 x}{\csc^2 x(1+\tan^2 x)} = \frac{\sec^2 x + \csc^2 x}{\csc^2 x + (\csc^2 x)\tan^2 x} = \frac{\sec^2 x + \csc^2 x}{\csc^2 x + \frac{1}{\sin^2 x} \cdot \frac{\sin^2 x}{\cos^2 x}}$$
   $$= \frac{\sec^2 x + \csc^2 x}{\csc^2 x + \frac{1}{\cos^2 x}} = \frac{\sec^2 x + \csc^2 x}{\csc^2 x + \sec^2 x} = 1$$

3. $$\ln|\tan\theta| - \ln|\cot\theta| = \ln\frac{|\tan\theta|}{|\cot\theta|} = \ln\left|\frac{\sin\theta/\cos\theta}{\cos\theta/\sin\theta}\right| = \ln\left|\frac{\sin^2\theta}{\cos^2\theta}\right| = \ln|\tan^2\theta| = 2\ln|\tan\theta|$$

4. $\cos\left(\frac{\pi}{2} - x\right) = \frac{1}{\csc x}$ is true since $\cos\left(\frac{\pi}{2} - x\right) = \sin x = \frac{1}{\csc x}$.

5. $\sin^4 x + (\sin^2 x)\cos^2 x = \sin^2 x(\sin^2 x + \cos^2 x) = \sin^2 x(1) = \sin^2 x$

6. $(\csc x + 1)(\csc x - 1) = \csc^2 x - 1 = \cot^2 x$

7. $$\frac{\cos^2 x}{1-\sin x} \cdot \frac{1+\sin x}{1+\sin x} = \frac{\cos^2 x(1+\sin x)}{1-\sin^2 x} = \frac{\cos^2 x(1+\sin x)}{\cos^2 x} = 1 + \sin x$$

8. $$\frac{1+\cos\theta}{\sin\theta} + \frac{\sin\theta}{1+\cos\theta} = \frac{(1+\cos\theta)^2 + \sin^2\theta}{\sin\theta(1+\cos\theta)}$$
   $$= \frac{1 + 2\cos\theta + \cos^2\theta + \sin^2\theta}{\sin\theta(1+\cos\theta)} = \frac{2+2\cos\theta}{\sin\theta(1+\cos\theta)} = \frac{2}{\sin\theta} = 2\csc\theta$$

9. $\tan^4 x + 2\tan^2 x + 1 = (\tan^2 x + 1)^2 = (\sec^2 x)^2 = \sec^4 x$

10. (a) $\sin 105° = \sin(60° + 45°) = \sin 60° \cos 45° + \cos 60° \sin 45°$

$$= \frac{\sqrt{3}}{2} \cdot \frac{\sqrt{2}}{2} + \frac{1}{2} \cdot \frac{\sqrt{2}}{2} = \frac{\sqrt{2}}{4}(\sqrt{3} + 1)$$

(b) $\tan 15° = \tan(60° - 45°) = \dfrac{\tan 60° - \tan 45°}{1 + \tan 60° \tan 45°}$

$$= \frac{\sqrt{3} - 1}{1 + \sqrt{3}} \cdot \frac{1 - \sqrt{3}}{1 - \sqrt{3}} = \frac{2\sqrt{3} - 1 - 3}{1 - 3} = \frac{2\sqrt{3} - 4}{-2} = 2 - \sqrt{3}$$

11. $(\sin 42°) \cos 38° - (\cos 42°) \sin 38° = \sin(42° - 38°) = \sin 4°$

12. $\tan\left(\theta + \dfrac{\pi}{4}\right) = \dfrac{\tan\theta + \tan(\pi/4)}{1 - (\tan\theta)\tan(\pi/4)} = \dfrac{\tan\theta + 1}{1 - \tan\theta(1)} = \dfrac{1 + \tan\theta}{1 - \tan\theta}$

13. $\sin(\arcsin x - \arccos x) = \sin(\arcsin x)\cos(\arccos x) - \cos(\arcsin x)\sin(\arccos x)$

$$= (x)(x) - (\sqrt{1 - x^2})(\sqrt{1 - x^2}) = x^2 - (1 - x^2) = 2x^2 - 1$$

14. (a) $\cos(120°) = \cos[2(60°)] = 2\cos^2 60° - 1 = 2\left(\dfrac{1}{2}\right)^2 - 1 = -\dfrac{1}{2}$

(b) $\tan(300°) = \tan[2(150°)] = \dfrac{2\tan 150°}{1 - \tan^2 150°} = \dfrac{-2\sqrt{3}/3}{1 - (1/3)} = -\sqrt{3}$

15. (a) $\sin 22.5° = \sin\dfrac{45°}{2} = \sqrt{\dfrac{1 - \cos 45°}{2}} = \sqrt{\dfrac{1 - \sqrt{2}/2}{2}} = \dfrac{\sqrt{2 - \sqrt{2}}}{2}$

(b) $\tan\dfrac{\pi}{12} = \tan\dfrac{\pi/6}{2} = \dfrac{\sin(\pi/6)}{1 + \cos(\pi/6)} = \dfrac{1/2}{1 + \sqrt{3}/2} = \dfrac{1}{2 + \sqrt{3}} = 2 - \sqrt{3}$

16. $\sin\theta = \dfrac{4}{5}$, θ lies in Quadrant II $\Rightarrow \cos\theta = -\dfrac{3}{5}$.

$$\cos\frac{\theta}{2} = \sqrt{\frac{1 + \cos\theta}{2}} = \sqrt{\frac{1 - 3/5}{2}} = \sqrt{\frac{2}{10}} = \frac{1}{\sqrt{5}} = \frac{\sqrt{5}}{5}$$

17. $(\sin^2 x)\cos^2 x = \dfrac{1 - \cos 2x}{2} \cdot \dfrac{1 + \cos 2x}{2} = \dfrac{1}{4}[1 - \cos^2 2x] = \dfrac{1}{4}\left[1 - \dfrac{1 + \cos 4x}{2}\right]$

$$= \frac{1}{8}[2 - (1 + \cos 4x)] = \frac{1}{8}[1 - \cos 4x]$$

18. $6(\sin 5\theta)\cos 2\theta = 6\{\tfrac{1}{2}[\sin(5\theta + 2\theta) + \sin(5\theta - 2\theta)]\} = 3[\sin 7\theta + \sin 3\theta]$

19. $\sin(x + \pi) + \sin(x - \pi) = 2\left(\sin\dfrac{[(x + \pi) + (x - \pi)]}{2}\right)\cos\dfrac{[(x + \pi) - (x - \pi)]}{2}$

$$= 2(\sin x)\cos\pi = -2\sin x$$

20. $\dfrac{\sin 9x + \sin 5x}{\cos 9x - \cos 5x} = \dfrac{2 \sin 7x \cos 2x}{-2 \sin 7x \sin 2x} = -\dfrac{\cos 2x}{\sin 2x} = -\cot 2x$

21. $\frac{1}{2}[\sin(u+v) - \sin(u-v)] = \frac{1}{2}\{(\sin u)\cos v + (\cos u)\sin v - [(\sin u)\cos v - (\cos u)\sin v]\}$

$$= \tfrac{1}{2}[2(\cos u)\sin v] = (\cos u)\sin v$$

22. $4\sin^2 x = 1$

$$\sin^2 x = \frac{1}{4}$$

$$\sin x = \pm\frac{1}{2}$$

$\sin x = \dfrac{1}{2}$ or $\sin x = -\dfrac{1}{2}$

$x = \dfrac{\pi}{6}$ or $\dfrac{5\pi}{6}$ $\qquad x = \dfrac{7\pi}{6}$ or $\dfrac{11\pi}{6}$

23. $\tan^2\theta + (\sqrt{3} - 1)\tan\theta - \sqrt{3} = 0$

$$(\tan\theta - 1)(\tan\theta + \sqrt{3}) = 0$$

$\tan\theta = 1$ or $\tan\theta = -\sqrt{3}$

$\theta = \dfrac{\pi}{4}$ or $\dfrac{5\pi}{4}$ $\qquad \theta = \dfrac{2\pi}{3}$ or $\dfrac{5\pi}{3}$

24. $\sin 2x = \cos x$

$$2(\sin x)\cos x - \cos x = 0$$

$$\cos x(2\sin x - 1) = 0$$

$\cos x = 0$ or $\sin x = \dfrac{1}{2}$

$x = \dfrac{\pi}{2}$ or $\dfrac{3\pi}{2}$ $\qquad x = \dfrac{\pi}{6}$ or $\dfrac{5\pi}{6}$

25. $\tan^2 x - 6\tan x + 4 = 0$

$$\tan x = \frac{-(-6) \pm \sqrt{(-6)^2 - 4(1)(4)}}{2(1)}$$

$$\tan x = \frac{6 \pm \sqrt{20}}{2} = 3 \pm \sqrt{5}$$

$\tan x = 3 + \sqrt{5}$ or $\tan x = 3 - \sqrt{5}$

$x \approx 1.3821$ or 4.5237 $\qquad x \approx 0.6524$ or 3.7940

CHAPTER 7

Practice Test Solutions

1. $C = 180° - (40° + 12°) = 128°$

$$a = \sin 40° \left(\frac{100}{\sin 12°}\right) \approx 309.164$$

$$c = \sin 128° \left(\frac{100}{\sin 12°}\right) \approx 379.012$$

2. $\sin A = 5\left(\frac{\sin 150°}{20}\right) = 0.125$

$A \approx 7.181°$

$B \approx 180° - (150° + 7.181°) = 22.819°$

$$b = \sin 22.819° \left(\frac{20}{\sin 150°}\right) \approx 15.513$$

3. $\text{Area} = \frac{1}{2}ab\sin C$

$= \frac{1}{2}(3)(5)\sin 130°$

≈ 5.745 square units

4. $h = b\sin A$

$= 35\sin 22.5°$

≈ 13.394

$a = 10$

Since $a < h$ and A is acute, the triangle has no solution.

5. $$\cos A = \frac{(53)^2 + (38)^2 - (49)^2}{2(53)(38)} \approx 0.4598$$

$A \approx 62.627°$

$$\cos B = \frac{(49)^2 + (38)^2 - (53)^2}{2(49)(38)} \approx 0.2782$$

$B \approx 73.847°$

$C \approx 180° - (62.627° + 73.847°)$

$= 43.526°$

6. $c^2 = (100)^2 + (300)^2 - 2(100)(300)\cos 29°$

≈ 47522.8176

$c \approx 218$

$$\cos A = \frac{(300)^2 + (218)^2 - (100)^2}{2(300)(218)} \approx 0.97495$$

$A \approx 12.85°$

$B \approx 180° - (12.85° + 29°) = 138.15°$

7. $$s = \frac{a+b+c}{2} = \frac{4.1 + 6.8 + 5.5}{2} = 8.2$$

$\text{Area} = \sqrt{s(s-a)(s-b)(s-c)}$

$= \sqrt{8.2(8.2-4.1)(8.2-6.8)(8.2-5.5)}$

≈ 11.273 square units

8. $x^2 = (40)^2 + (70)^2 - 2(40)(70)\cos 168°$

≈ 11977.6266

$x \approx 109.442$ miles

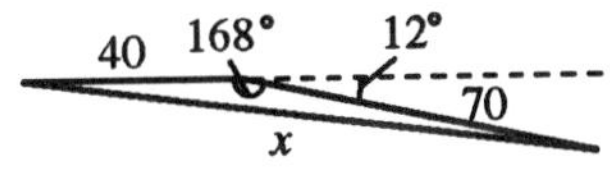

9. $\mathbf{w} = 4(3\mathbf{i} + \mathbf{j}) - 7(-\mathbf{i} + 2\mathbf{j})$

$= 19\mathbf{i} - 10\mathbf{j}$

10. $\dfrac{\mathbf{v}}{\|\mathbf{v}\|} = \dfrac{5\mathbf{i} - 3\mathbf{j}}{\sqrt{25+9}} = \dfrac{5}{\sqrt{34}}\mathbf{i} - \dfrac{3}{\sqrt{34}}\mathbf{j}$

$= \dfrac{5\sqrt{34}}{34}\mathbf{i} - \dfrac{3\sqrt{34}}{34}\mathbf{j}$

11. $\mathbf{u} = 6\mathbf{i} + 5\mathbf{j} \qquad \mathbf{v} = 2\mathbf{i} - 3\mathbf{j} \qquad \mathbf{w} = -4\mathbf{i} - 8\mathbf{j}$

$\|\mathbf{u}\| = \sqrt{61} \qquad \|\mathbf{v}\| = \sqrt{13} \qquad \|\mathbf{w}\| = \sqrt{80}$

$\cos\theta = \dfrac{61 + 13 - 80}{2\sqrt{61}\sqrt{13}}$

$\theta \approx 96.116°$

12. $\tan 30° = \dfrac{y}{x} = \dfrac{1}{\sqrt{3}}$

$\mathbf{u} = \sqrt{3}\mathbf{i} + \mathbf{j}$ but $\|\mathbf{u}\| = 2$

$\mathbf{v} = 2\mathbf{u} = 2\sqrt{3}\mathbf{i} + 2\mathbf{j}$

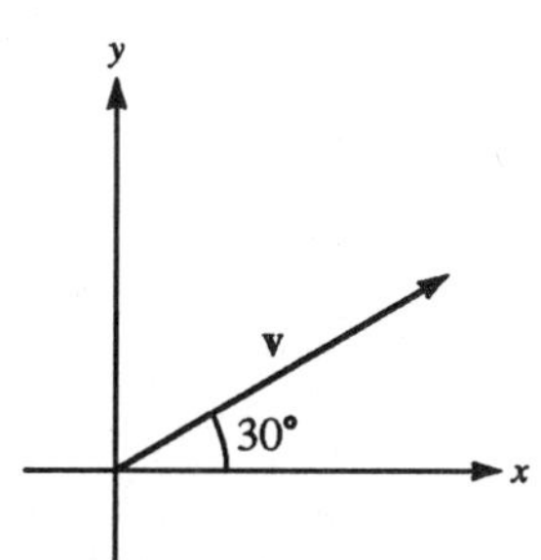

13. $r = \sqrt{25 + 25} = \sqrt{50} = 5\sqrt{2}$

$\tan\theta = \dfrac{-5}{5} = -1$

Since z is in Quadrant IV,

$\theta = 315°$

$z = 5\sqrt{2}(\cos 315° + i\sin 315°)$

14. $\cos 225° = -\dfrac{\sqrt{2}}{2} \qquad \sin 225° = -\dfrac{\sqrt{2}}{2}$

$z = 6\left(-\dfrac{\sqrt{2}}{2} - i\dfrac{\sqrt{2}}{2}\right)$

$= -3\sqrt{2} - 3\sqrt{2}\,i$

15. $[7(\cos 23° + i\sin 23°)][4(\cos 7° + i\sin 7°)] = 7(4)[\cos(23° + 7°) + i\sin(23° + 7°)]$

$= 28(\cos 30° + i\sin 30°)$

16. $\dfrac{9\left(\cos\dfrac{5\pi}{4} + i\sin\dfrac{5\pi}{4}\right)}{3(\cos\pi + i\sin\pi)} = \dfrac{9}{3}\left[\cos\left(\dfrac{5\pi}{4} - \pi\right) + i\sin\left(\dfrac{5\pi}{4} - \pi\right)\right] = 3\left(\cos\dfrac{\pi}{4} + i\sin\dfrac{\pi}{4}\right)$

17. $(2 + 2i)^8 = [2\sqrt{2}(\cos 45° + i\sin 45°)]^8 = (2\sqrt{2})^8[\cos(8)(45°) + i\sin(8)(45°)]$

$= 4096[\cos 360° + i\sin 360°] = 4096$

18. $z = 8\left(\cos\frac{\pi}{3} + i\sin\frac{\pi}{3}\right), \quad n = 3$

The cube roots of z are:

For $k = 0$, $\sqrt[3]{8}\left[\cos\frac{\pi/3}{3} + i\sin\frac{\pi/3}{3}\right] = 2\left(\cos\frac{\pi}{9} + i\sin\frac{\pi}{9}\right)$.

For $k = 1$, $\sqrt[3]{8}\left[\cos\frac{\pi/3 + 2\pi}{3} + i\sin\frac{\pi/3 + 2\pi}{3}\right] = 2\left(\cos\frac{7\pi}{9} + i\sin\frac{7\pi}{9}\right)$.

For $k = 2$, $\sqrt[3]{8}\left[\cos\frac{\pi/3 + 4\pi}{3} + i\sin\frac{\pi/3 + 4\pi}{3}\right] = 2\left(\cos\frac{13\pi}{9} + i\sin\frac{13\pi}{9}\right)$.

19. $x^3 = -125 = 125(\cos\pi + i\sin\pi)$

For $k = 0$, $\sqrt[3]{125}\left(\cos\frac{\pi}{3} + i\sin\frac{\pi}{3}\right) = 5\left(\cos\frac{\pi}{3} + i\sin\frac{\pi}{3}\right)$.

For $k = 1$, $\sqrt[3]{125}\left(\cos\frac{\pi + 2\pi}{3} + i\sin\frac{\pi + 2\pi}{3}\right) = -5$.

For $k = 2$, $\sqrt[3]{125}\left(\cos\frac{\pi + 4\pi}{3} + i\sin\frac{\pi + 4\pi}{3}\right) = 5\left(\cos\frac{5\pi}{3} + i\sin\frac{5\pi}{3}\right)$.

20. $x^4 = -i = 1\left(\cos\frac{3\pi}{2} + i\sin\frac{3\pi}{2}\right)$

For $k = 0$, $\cos\frac{3\pi/2}{4} + i\sin\frac{3\pi/2}{4} = \cos\frac{3\pi}{8} + i\sin\frac{3\pi}{8}$.

For $k = 1$, $\cos\frac{3\pi/2 + 2\pi}{4} + i\sin\frac{3\pi/2 + 2\pi}{4} = \cos\frac{7\pi}{8} + i\sin\frac{7\pi}{8}$.

For $k = 2$, $\cos\frac{3\pi/2 + 4\pi}{4} + i\sin\frac{3\pi/2 + 4\pi}{4} = \cos\frac{11\pi}{8} + i\sin\frac{11\pi}{8}$.

For $k = 3$, $\cos\frac{3\pi/2 + 6\pi}{4} + i\sin\frac{3\pi/2 + 6\pi}{4} = \cos\frac{15\pi}{8} + i\sin\frac{15\pi}{8}$.

CHAPTER 8

Practice Test Solutions

1. $x + y = 1$

$3x - y = 15 \Rightarrow y = 3x - 15$

$x + (3x - 15) = 1$

$4x = 16$

$x = 4$

$y = -3$

2. $x - 3y = -3 \Rightarrow x = 3y - 3$

$x^2 + 6y = 5$

$(3y - 3)^2 + 6y = 5$

$9y^2 - 18y + 9 + 6y = 5$

$9y^2 - 12y + 4 = 0$

$(3y - 2)^2 = 0$

$y = \frac{2}{3}$

$x = -1$

3. $x + y + z = 6 \Rightarrow z = 6 - x - y$

$2x - y + 3z = 0 \qquad 2x - y + 3(6 - x - y) = 0 \Rightarrow -x - 4y = -18$

$5x + 2y - z = -3 \qquad 5x + 2y - (6 - x - y) = -3 \Rightarrow 6x + 3y = 3$

$x = 18 - 4y$

$6(18 - 4y) + 3y = 3$

$-21y = -105$

$y = 5$

$x = 18 - 4y = -2$

$z = 6 - x - y = 3$

4. $x + y = 110 \Rightarrow y = 110 - x$

$xy = 2800$

$x(110 - x) = 2800$

$0 = x^2 - 110x + 2800$

$0 = (x - 40)(x - 70)$

$x = 40$ or $x = 70$

$y = 70 \qquad y = 40$

5. $2x + 2y = 170 \Rightarrow y = \dfrac{170 - 2x}{2} = 85 - x$

$xy = 2800$

$x(85 - x) = 2800$

$0 = x^2 - 85x + 2800$

$0 = (x - 25)(x - 60)$

$x = 25$ or $x = 60$

$y = 60 \qquad y = 25$

Dimensions: $60' \times 25'$

6. $$\begin{aligned} 2x + 15y &= 4 \Rightarrow 2x + 15y = 4 \\ x - 3y &= 23 \Rightarrow \underline{5x - 15y = 115} \\ &\quad 7x = 119 \\ x &= 17 \\ y &= \frac{x-23}{3} \\ &= -2 \end{aligned}$$

7. $$\begin{aligned} x + y &= 2 \Rightarrow 19x + 19y = 38 \\ 38x - 19y &= 7 \Rightarrow \underline{38x - 19y = 7} \\ &\quad 57x = 45 \end{aligned}$$

$$x = \tfrac{45}{57} = \tfrac{15}{19}$$

$$y = 2 - x = \tfrac{38}{19} - \tfrac{15}{19} = \tfrac{23}{19}$$

8. $$\begin{aligned} 0.4x + 0.5y &= 0.112 \Rightarrow 0.28x + 0.35y = 0.0784 \\ 0.3x - 0.7y &= -0.131 \Rightarrow \underline{0.15x - 0.35y = -0.0655} \\ &\quad 0.43x = 0.0129 \end{aligned}$$

$$x = \frac{0.0129}{0.43} = 0.03$$

$$y = \frac{0.112 - 0.4x}{0.5} = 0.20$$

9. Let x = amount in 11% fund and y = amount in 13% fund.

$$x + y = 17000 \Rightarrow y = 17000 - x$$

$$0.11x + 0.13y = 2080$$

$$\begin{aligned} 0.11x + 0.13(17000 - x) &= 2080 \\ -0.02x &= -130 \\ x &= \$6500 \\ y &= \$10{,}500 \end{aligned}$$

10. $(4, 3)$, $(1, 1)$, $(-1, -2)$, $(-2, -1)$

$$n = 4,\ \sum_{i=1}^{4} x_i = 2,\ \sum_{i=1}^{4} y_i = 1,\ \sum_{i=1}^{4} x_i^2 = 22,\ \sum_{i=1}^{4} x_i y_i = 17$$

$$\begin{aligned} 4b + 2a &= 1 \Rightarrow 4b + 2a = 1 \\ 2b + 22a &= 17 \Rightarrow \underline{-4b - 44a = -34} \\ &\quad -42a = -33 \end{aligned}$$

$$a = \tfrac{33}{42} = \tfrac{11}{14}$$

$$b = \tfrac{1}{4}\left(1 - 2\left(\tfrac{33}{42}\right)\right) = -\tfrac{1}{7}$$

$$y = ax + b = \tfrac{11}{14}x - \tfrac{1}{7}$$

11. $\begin{aligned} x + y \quad &= -2 \\ 2x - y + z &= 11 \\ 4y - 3z &= -20 \end{aligned}$ $\Rightarrow$ $\begin{aligned} -2x - 2y \quad &= 4 \\ 2x - y + z &= 11 \\ \hline -3y + z &= 15 \end{aligned}$ $\quad$ $\begin{aligned} -9y + 3z &= 45 \\ 4y - 3z &= -20 \\ \hline -5y \quad &= 25 \\ y &= -5 \\ x &= 3 \\ z &= 0 \end{aligned}$

12. $\begin{aligned} 4x - y + 5z &= 4 \Rightarrow & 4x - y + 5z &= 4 \\ 2x + y - z &= 0 \Rightarrow & -4x - 2y + 2z &= 0 \\ \hline 2x + 4y + 8z &= 0 & -3y + 7z &= 4 \end{aligned}$

$$\begin{aligned} 2x + 4y + 8z &= 0 \\ -2x - y + z &= 0 \\ \hline 3y + 9z &= 0 \\ -3y + 7z &= 4 \\ \hline 16z &= 4 \\ z &= \tfrac{1}{4} \\ y &= -\tfrac{3}{4} \\ x &= \tfrac{1}{2} \end{aligned}$$

13. $\begin{aligned} 3x + 2y - z &= 5 \Rightarrow & 6x + 4y - 2z &= 10 \\ 6x - y + 5z &= 2 \Rightarrow & -6x + y - 5z &= -2 \\ \hline & & 5y - 7z &= 8 \\ & & y &= \frac{8 + 7z}{5} \end{aligned}$

$$\begin{aligned} 3x + 2y - z &= 5 \\ 12x - 2y + 10z &= 4 \\ \hline 15x \quad + 9z &= 9 \\ x = \frac{9 - 9z}{15} &= \frac{3 - 3z}{5} \end{aligned}$$

Let $z = a$, then $x = \dfrac{3 - 3a}{5}$ and $y = \dfrac{8 + 7a}{5}$.

14. $y = ax^2 + bx + c$ passes through (0, −1), (1, 4), and (2, 13).

$$\begin{array}{lllll}
\text{At } (0,\ -1): & -1 = a(0)^2 + b(0) + c & \Rightarrow & c = -1 & \\
\text{At } (1,\ 4): & 4 = a(1)^2 + b(1) - 1 & \Rightarrow & 5 = a + b & \Rightarrow \quad 5 = a + b \\
\text{At } (2,\ 13): & 13 = a(2)^2 + b(2) - 1 & \Rightarrow & 14 = 4a + 2b & \Rightarrow \quad \underline{-7 = -2a - b}
\end{array}$$

$$\begin{aligned}
-2 &= -a \\
a &= 2 \\
b &= 3
\end{aligned}$$

Thus, $y = 2x^2 + 3x - 1$.

15. $s = \frac{1}{2}at^2 + v_0t + s_0$ passes through (1, 12), (2, 5), and (3, 4).

$$\begin{array}{llll}
\text{At } (1,\ 12): & 12 = \frac{1}{2}a + v_0 + s_0 & \Rightarrow & 24 = a + 2v_0 + 2s_0 \\
\text{At } (2,\ 5): & 5 = 2a + 2v_0 + s_0 & \Rightarrow & \underline{-5 = -2a - 2v_0 - s_0} \\
\text{At } (3,\ 4): & 4 = \frac{9}{2}a + 3v_0 + s_0 & & 19 = -a + s_0
\end{array}$$

$$\begin{aligned}
15 &= 6a + 6v_0 + 3s_0 \\
\underline{-8} &\underline{= -9a - 6v_0 - 2s_0} \\
7 &= -3a + s_0 \\
\underline{-19} &\underline{= a - s_0} \\
-12 &= -2a \\
a &= 6 \\
s_0 &= 25 \\
v_0 &= -16
\end{aligned}$$

Thus, $s = \frac{1}{2}(6)t^2 - 16t + 25 = 3t^2 - 16t + 25$.

16. $x^2 + y^2 \geq 9$

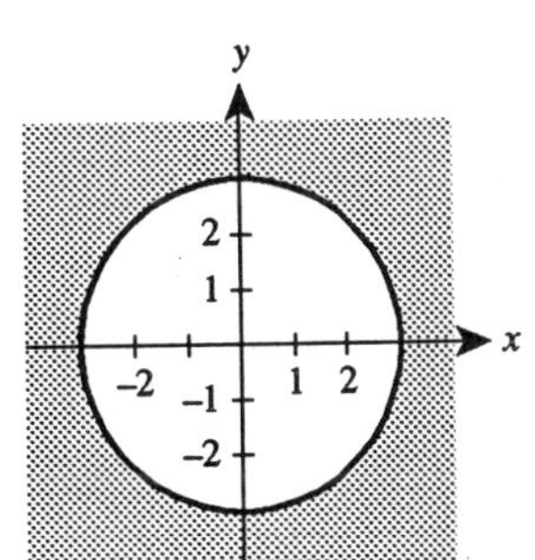

17. $x + y \leq 6$

$x \geq 2$

$y \geq 0$

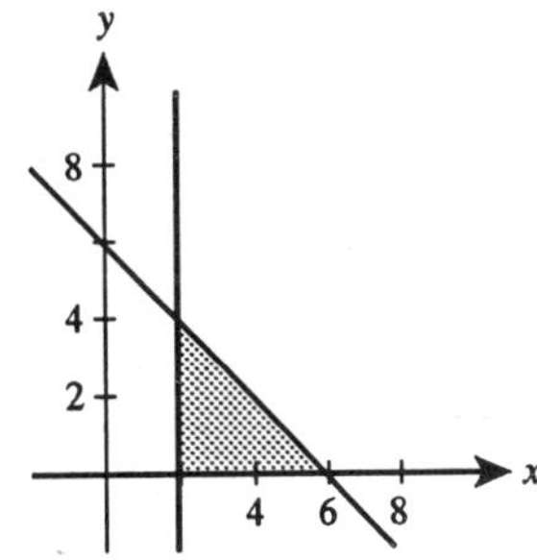

18. Line through (0, 0) and (0, 7):

$x = 0$

Line through (0, 0) and (2, 3):

$y = \frac{3}{2}x$ or $3x - 2y = 0$

Line through (0, 7) and (2, 3):

$y = -2x + 7$ or $2x + y = 7$

Inequalities: $x \geq 0$

$3x - 2y \leq 0$

$2x + y \leq 7$

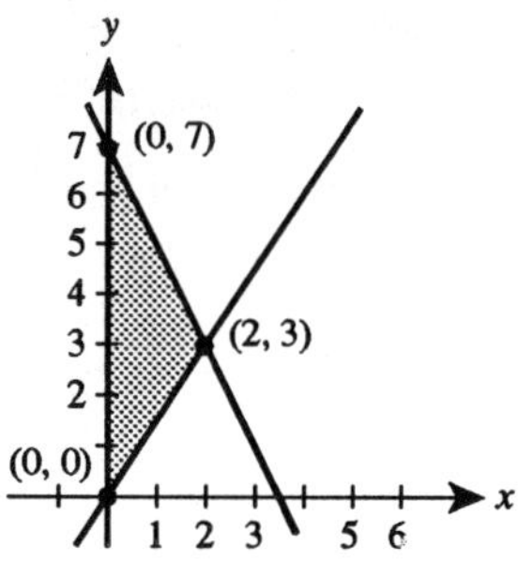

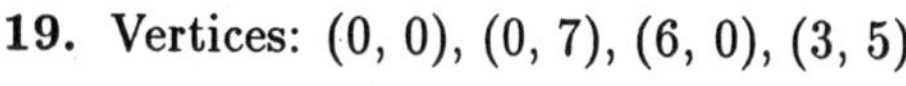

19. Vertices: (0, 0), (0, 7), (6, 0), (3, 5)

$z = 30x + 26y$

At (0, 0): $z = 0$

At (0, 7): $z = 182$

At (6, 0): $z = 180$

At (3, 5): $z = 220$

The maximum value of z is 220.

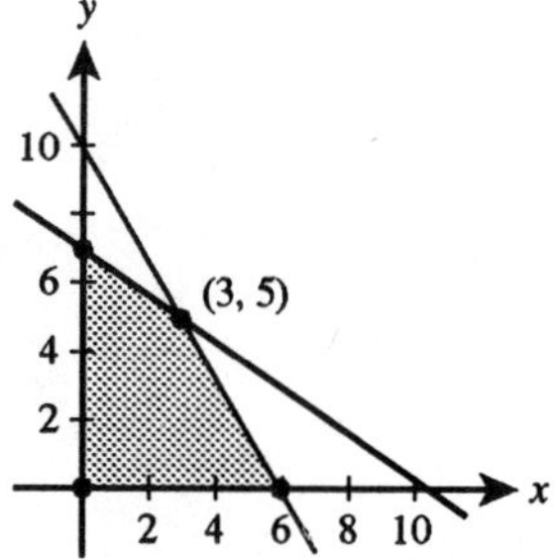

20. $x^2 + y^2 \leq 4$

$(x - 2)^2 + y^2 \geq 4$

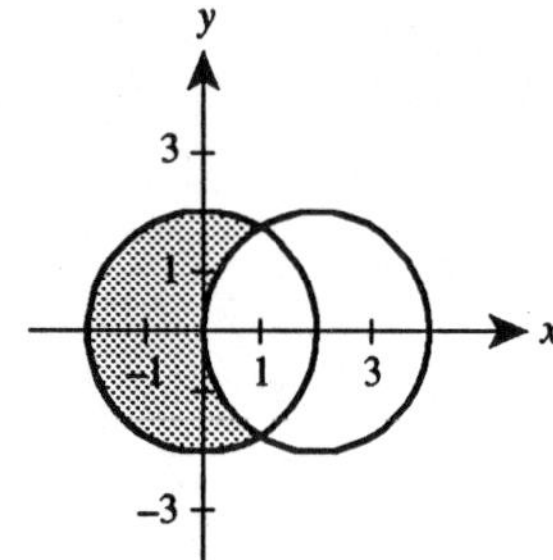

CHAPTER 9

Practice Test Solutions

1. $\begin{bmatrix} 1 & -2 & 4 \\ 3 & -5 & 9 \end{bmatrix} -3R_1 + R_2 \rightarrow \begin{bmatrix} 1 & -2 & 4 \\ 0 & 1 & -3 \end{bmatrix} 2R_2 + R_1 \rightarrow \begin{bmatrix} 1 & 0 & -2 \\ 0 & 1 & -3 \end{bmatrix}$

2. $3x + 5y = 3$

$2x - y = -11$

$$\left[\begin{array}{cc:c} 3 & 5 & 3 \\ 2 & -1 & -11 \end{array}\right] \quad -R_2 + R_1 \rightarrow \left[\begin{array}{cc:c} 1 & 6 & 14 \\ 2 & -1 & -11 \end{array}\right]$$

$$-2R_1 + R_2 \rightarrow \left[\begin{array}{cc:c} 1 & 6 & 14 \\ 0 & -13 & -39 \end{array}\right]$$

$$-\tfrac{1}{13}R_2 \rightarrow \left[\begin{array}{cc:c} 1 & 6 & 14 \\ 0 & 1 & 3 \end{array}\right]$$

$$-6R_2 + R_1 \rightarrow \left[\begin{array}{cc:c} 1 & 0 & -4 \\ 0 & 1 & 3 \end{array}\right]$$

Answer: $x = -4, \ y = 3$

3. $2x + 3y = -3$

$3x + 2y = 8$

$x + y = 1$

$$\left[\begin{array}{cc:c} 2 & 3 & -3 \\ 3 & 2 & 8 \\ 1 & 1 & 1 \end{array}\right] \quad R_3 \leftrightarrow R_1 \quad \left[\begin{array}{cc:c} 1 & 1 & 1 \\ 3 & 2 & 8 \\ 2 & 3 & -3 \end{array}\right]$$

$$\begin{array}{r} \\ -3R_1 + R_2 \rightarrow \\ -2R_1 + R_3 \rightarrow \end{array} \left[\begin{array}{cc:c} 1 & 1 & 1 \\ 0 & -1 & 5 \\ 0 & 1 & -5 \end{array}\right]$$

$$\begin{array}{r} R_2 + R_1 \rightarrow \\ -R_2 \rightarrow \\ -R_2 + R_3 \rightarrow \end{array} \left[\begin{array}{cc:c} 1 & 0 & 6 \\ 0 & 1 & -5 \\ 0 & 0 & 0 \end{array}\right]$$

Answer: $x = 6, \ y = -5$

4. $$\begin{aligned} x \qquad + 3z &= -5 \\ 2x + y \qquad &= 0 \\ 3x + y - z &= 3 \end{aligned}$$

$$\left[\begin{array}{ccc:c} 1 & 0 & 3 & -5 \\ 2 & 1 & 0 & 0 \\ 3 & 1 & -1 & 3 \end{array}\right] \qquad \begin{array}{r} \\ -2R_1 + R_2 \rightarrow \\ -3R_1 + R_3 \rightarrow \end{array} \left[\begin{array}{ccc:c} 1 & 0 & 3 & -5 \\ 0 & 1 & -6 & 10 \\ 0 & 1 & -10 & 18 \end{array}\right]$$

$$\begin{array}{r} \\ \\ -R_2 + R_3 \rightarrow \end{array} \left[\begin{array}{ccc:c} 1 & 0 & 3 & -5 \\ 0 & 1 & -6 & 10 \\ 0 & 0 & -4 & 8 \end{array}\right]$$

$$\begin{array}{r} -3R_3 + R_1 \rightarrow \\ 6R_3 + R_2 \rightarrow \\ -\frac{1}{4}R_4 \rightarrow \end{array} \left[\begin{array}{ccc:c} 1 & 0 & 0 & 1 \\ 0 & 1 & 0 & -2 \\ 0 & 0 & 1 & -2 \end{array}\right]$$

Answer: $x = 1,\ y = -2,\ z = -2$

5. $$\begin{bmatrix} 1 & 4 & 5 \\ 2 & 0 & -3 \end{bmatrix} \begin{bmatrix} 1 & 6 \\ 0 & -7 \\ -1 & 2 \end{bmatrix} = \begin{bmatrix} -4 & -12 \\ 5 & 6 \end{bmatrix}$$

6. $$\begin{aligned} 3A - 5B &= 3\begin{bmatrix} 9 & 1 \\ -4 & 8 \end{bmatrix} - 5\begin{bmatrix} 6 & -2 \\ 3 & 5 \end{bmatrix} \\ &= \begin{bmatrix} 27 & 3 \\ -12 & 24 \end{bmatrix} - \begin{bmatrix} 30 & -10 \\ 15 & 25 \end{bmatrix} \\ &= \begin{bmatrix} -3 & 13 \\ -27 & -1 \end{bmatrix} \end{aligned}$$

7. $$\begin{aligned} f(A) &= \begin{bmatrix} 3 & 0 \\ 7 & 1 \end{bmatrix}^2 - 7\begin{bmatrix} 3 & 0 \\ 7 & 1 \end{bmatrix} + 8\begin{bmatrix} 1 & 0 \\ 0 & 1 \end{bmatrix} \\ &= \begin{bmatrix} 3 & 0 \\ 7 & 1 \end{bmatrix}\begin{bmatrix} 3 & 0 \\ 7 & 1 \end{bmatrix} - \begin{bmatrix} 21 & 0 \\ 49 & 7 \end{bmatrix} + \begin{bmatrix} 8 & 0 \\ 0 & 8 \end{bmatrix} \\ &= \begin{bmatrix} 9 & 0 \\ 28 & 1 \end{bmatrix} - \begin{bmatrix} 21 & 0 \\ 49 & 7 \end{bmatrix} + \begin{bmatrix} 8 & 0 \\ 0 & 8 \end{bmatrix} \\ &= \begin{bmatrix} -4 & 0 \\ -21 & 2 \end{bmatrix} \end{aligned}$$

8. False since

$$\begin{aligned} (A + B)(A + 3B) &= A(A + 3B) + B(A + 3B) \\ &= A^2 + 3AB + BA + 3B^2. \end{aligned}$$

9. $\begin{bmatrix} 1 & 2 & \vdots & 1 & 0 \\ 3 & 5 & \vdots & 0 & 1 \end{bmatrix}$

$$\begin{array}{r} -3R_1 + R_2 \rightarrow \end{array} \begin{bmatrix} 1 & 2 & \vdots & 1 & 0 \\ 0 & -1 & \vdots & -3 & 1 \end{bmatrix}$$

$$\begin{array}{r} 2R_2 + R_1 \rightarrow \\ -R_2 \rightarrow \end{array} \begin{bmatrix} 1 & 0 & \vdots & -5 & 2 \\ 0 & 1 & \vdots & 3 & -1 \end{bmatrix}$$

$$A^{-1} = \begin{bmatrix} -5 & 2 \\ 3 & -1 \end{bmatrix}$$

10. $\begin{bmatrix} 1 & 1 & 1 & \vdots & 1 & 0 & 0 \\ 3 & 6 & 5 & \vdots & 0 & 1 & 0 \\ 6 & 10 & 8 & \vdots & 0 & 0 & 1 \end{bmatrix}$

$$\begin{array}{r} \\ -3R_1 + R_2 \rightarrow \\ -6R_1 + R_3 \rightarrow \end{array} \begin{bmatrix} 1 & 1 & 1 & \vdots & 1 & 0 & 0 \\ 0 & 3 & 2 & \vdots & -3 & 1 & 0 \\ 0 & 4 & 2 & \vdots & -6 & 0 & 1 \end{bmatrix}$$

$$\begin{array}{r} -R_2 + R_1 \rightarrow \\ \frac{1}{3}R_2 \rightarrow \\ -4R_2 + R_3 \rightarrow \end{array} \begin{bmatrix} 1 & 0 & \frac{1}{3} & \vdots & 2 & -\frac{1}{3} & 0 \\ 0 & 1 & \frac{2}{3} & \vdots & -1 & \frac{1}{3} & 0 \\ 0 & 0 & -\frac{2}{3} & \vdots & -2 & -\frac{4}{3} & 1 \end{bmatrix}$$

$$\begin{array}{r} \frac{1}{2}R_3 + R_1 \rightarrow \\ R_3 + R_2 \rightarrow \\ -\frac{3}{2}R_3 \rightarrow \end{array} \begin{bmatrix} 1 & 0 & 0 & \vdots & 1 & -1 & \frac{1}{2} \\ 0 & 1 & 0 & \vdots & -3 & -1 & 1 \\ 0 & 0 & 1 & \vdots & 3 & 2 & -\frac{3}{2} \end{bmatrix}$$

$$A^{-1} = \begin{bmatrix} 1 & -1 & \frac{1}{2} \\ -3 & -1 & 1 \\ 3 & 2 & -\frac{3}{2} \end{bmatrix}$$

11. (a) $x + 2y = 4$

$3x + 5y = 1$

$\begin{bmatrix} 1 & 2 & \vdots & 1 & 0 \\ 3 & 5 & \vdots & 0 & 1 \end{bmatrix}$

$$\begin{array}{r} -3R_1 + R_2 \rightarrow \end{array} \begin{bmatrix} 1 & 2 & \vdots & 1 & 0 \\ 0 & -1 & \vdots & -3 & 1 \end{bmatrix}$$

$$\begin{array}{r} -2R_2 + R_1 \rightarrow \\ -R_2 \rightarrow \end{array} \begin{bmatrix} 1 & 0 & \vdots & -5 & 2 \\ 0 & 1 & \vdots & 3 & -1 \end{bmatrix}$$

$$X = A^{-1}B = \begin{bmatrix} -5 & 2 \\ 3 & -1 \end{bmatrix} \begin{bmatrix} 4 \\ 1 \end{bmatrix} = \begin{bmatrix} -18 \\ 11 \end{bmatrix}$$

$x = -18, \; y = 11$

(b) $x + 2y = 3$

$3x + 5y = -2$

$$X = A^{-1}B = \begin{bmatrix} -5 & 2 \\ 3 & -1 \end{bmatrix} \begin{bmatrix} 3 \\ -2 \end{bmatrix} = \begin{bmatrix} -19 \\ 11 \end{bmatrix}$$

$x = -19, \; y = 11$

12. $\begin{vmatrix} 6 & -1 \\ 3 & 4 \end{vmatrix} = 24 - (-3) = 27$

13. $\begin{vmatrix} 1 & 3 & -1 \\ 5 & 9 & 0 \\ 6 & 2 & -5 \end{vmatrix}\begin{matrix} 1 & 3 \\ 5 & 9 \\ 6 & 2 \end{matrix} = (-45 + 0 - 10) - (-54 + 0 - 75) = 74$

14. $\begin{vmatrix} 1 & 4 & 2 & 3 \\ 0 & 1 & -2 & 0 \\ 3 & 5 & -1 & 1 \\ 2 & 0 & 6 & 1 \end{vmatrix} = \begin{vmatrix} 1 & 2 & 3 \\ 3 & -1 & 1 \\ 2 & 6 & 1 \end{vmatrix} + 2\begin{vmatrix} 1 & 4 & 3 \\ 3 & 5 & 1 \\ 2 & 0 & 1 \end{vmatrix}$

$= 51 + 2(-29) = -7$ Expansion along Row 2.

15. $\begin{vmatrix} 3 & 0 & 0 \\ 0 & 3 & 0 \\ 0 & 0 & 3 \end{vmatrix} = 3(3)(3)\begin{vmatrix} 1 & 0 & 0 \\ 0 & 1 & 0 \\ 0 & 0 & 1 \end{vmatrix} = -3^3\begin{vmatrix} 1 & 0 & 0 \\ 0 & 0 & 1 \\ 0 & 1 & 0 \end{vmatrix}$

True

16. $\begin{vmatrix} 6 & 4 & 3 & 0 & 6 \\ 0 & 5 & 1 & 4 & 8 \\ 0 & 0 & 2 & 7 & 3 \\ 0 & 0 & 0 & 9 & 2 \\ 0 & 0 & 0 & 0 & 1 \end{vmatrix} = 6(5)(2)(9)(1) = 540$

17. Area $= \dfrac{1}{2}\begin{vmatrix} 0 & 7 & 1 \\ 5 & 0 & 1 \\ 3 & 9 & 1 \end{vmatrix}$

$= \dfrac{1}{2}(31)$

$= 15.5$ square units

18. $x = \dfrac{\begin{vmatrix} 4 & -7 \\ 11 & 5 \end{vmatrix}}{\begin{vmatrix} 6 & -7 \\ 2 & 5 \end{vmatrix}} = \dfrac{97}{44}$

19. $z = \dfrac{\begin{vmatrix} 3 & 0 & 1 \\ 0 & 1 & 3 \\ 1 & -1 & 2 \end{vmatrix}}{\begin{vmatrix} 3 & 0 & 1 \\ 0 & 1 & 4 \\ 1 & -1 & 0 \end{vmatrix}} = \dfrac{14}{11}$

20. $y = \dfrac{\begin{vmatrix} 721.4 & 33.77 \\ 45.9 & 19.85 \end{vmatrix}}{\begin{vmatrix} 721.4 & -29.1 \\ 45.9 & 105.6 \end{vmatrix}}$

$= \dfrac{12{,}769.747}{77{,}515.530} \approx 0.1647$

CHAPTER 10

Practice Test Solutions

1. $a_n = \dfrac{2n}{(n+2)!}$

$a_1 = \dfrac{2(1)}{3!} = \dfrac{2}{6} = \dfrac{1}{3}$

$a_2 = \dfrac{2(2)}{4!} = \dfrac{4}{24} = \dfrac{1}{6}$

$a_3 = \dfrac{2(3)}{5!} = \dfrac{6}{120} = \dfrac{1}{20}$

$a_4 = \dfrac{2(4)}{6!} = \dfrac{8}{720} = \dfrac{1}{90}$

$a_5 = \dfrac{2(5)}{7!} = \dfrac{10}{5040} = \dfrac{1}{504}$

Terms: $\frac{1}{3}, \frac{1}{6}, \frac{1}{20}, \frac{1}{90}, \frac{1}{504}$

2. $a_n = \dfrac{n+3}{3^n}$

3. $\displaystyle\sum_{i=1}^{6}(2i-1) = 1+3+5+7+9+11$

$= 36$

4. $a_1 = 23,\ d = -2$

$a_2 = a_1 + d = 21$

$a_3 = a_2 + d = 19$

$a_4 = a_3 + d = 17$

$a_5 = a_4 + d = 15$

Terms: 23, 21, 19, 17, 15

5. $a_1 = 12,\ d = 3,\ n = 50$

$a_n = a_1 + (n-1)d$

$a_{50} = 12 + (50-1)3 = 159$

6. $a_1 = 1$

$a_{200} = 200$

$S_n = \dfrac{n}{2}(a_1 + a_n)$

$S_{200} = \dfrac{200}{2}(1 + 200) = 20{,}100$

7. $a_1 = 7,\ r = 2$

$a_2 = a_1 r = 14$

$a_3 = a_1 r^2 = 28$

$a_4 = a_1 r^3 = 56$

$a_5 = a_1 r^4 = 112$

Terms: 7, 14, 28, 56, 112

8. $\sum_{n=0}^{9} 6\left(\frac{2}{3}\right)^n$, $a_1 = 6,\ r = \frac{2}{3},\ n = 10$

$$S_n = \frac{a_1(1 - r^n)}{1 - r}$$

$$= \frac{6(1 - (2/3)^{10})}{1 - (2/3)} \approx 17.6879$$

9. $\sum_{n=0}^{\infty}(0.03)^n$, $a_1 = 1,\ r = 0.03$

$$S = \frac{a_1}{1 - r}$$

$$= \frac{1}{1 - 0.03} = \frac{1}{0.97} = \frac{100}{97} \approx 1.0309$$

10. For $n = 1,\ 1 = \dfrac{1(1+1)}{2}$.

Assume that $1 + 2 + 3 + 4 + \cdots + k = \dfrac{k(k+1)}{2}$.

Now for $n = k + 1$,

$$1 + 2 + 3 + 4 + \cdots + k + (k+1) = \frac{k(k+1)}{2} + k + 1 = \frac{k(k+1)}{2} + \frac{2(k+1)}{2} = \frac{(k+1)(k+2)}{2}.$$

Thus, $1 + 2 + 3 + 4 + \cdots + n = \dfrac{n(n+1)}{2}$ for all integers $n \geq 1$.

11. For $n = 4,\ 4! > 2^4$.

Assume that $k! > 2^k$. Then

$$(k+1)! = (k+1)(k!) > (k+1)2^k > 2 \bullet 2^k$$
$$= 2^{k+1}.$$

Thus, $n! > 2^n$ for all integers $n \geq 4$.

12. ${}_{13}C_4 = \dfrac{13!}{(13-4)!4!} = 715$

13. $(x+3)^5 = x^5 + 5x^4(3) + 10x^3(3)^2 + 10x^2(3)^3 + 5x(3)^4 + (3)^5$

$$= x^5 + 15x^4 + 90x^3 + 270x^2 + 405x + 243$$

14. ${}_{12}C_5 x^7(-2)^5 = -25{,}344x^7$

15. ${}_{30}P_4 = \dfrac{30!}{(30-4)!}$

$$= 657{,}720$$

16. $6! = 720$ ways

17. ${}_{12}P_3 = 1320$

18. $P(2) + P(3) + P(4) = \frac{1}{36} + \frac{2}{36} + \frac{3}{36}$

$$= \tfrac{6}{36} = \tfrac{1}{6}$$

19. $P(K,\ B10) = \frac{4}{52} \bullet \frac{2}{51} = \frac{2}{663}$

20. Let A = probability of no faulty units.

$$P(A) = \left(\frac{997}{1000}\right)^{50} \approx 0.8605$$

$$P(A') = 1 - P(A) \approx 0.1395$$

CHAPTER 11

Practice Test Solutions

1. $3x + 4y = 12 \Rightarrow y = -\frac{3}{4}x + 3 \Rightarrow m_1 = -\frac{3}{4}$

$4x - 3y = 12 \Rightarrow y = \frac{4}{3}x - 4 \Rightarrow m_2 = \frac{4}{3}$

$$\tan\theta = \left|\frac{\frac{4}{3} - \left(-\frac{3}{4}\right)}{1 + \left(\frac{4}{3}\right)\left(-\frac{3}{4}\right)}\right| = \left|\frac{\frac{25}{12}}{0}\right|$$

Since $\tan\theta$ is undefined, the lines are perpendicular (note that $m_2 = -1/m_1$) and $\theta = 90°$.

2. $x_1 = 5,\ x_2 = -9,\ A = 3,\ B = -7,\ C = -21$

$$d = \frac{|3(5) + (-7)(-9) + (-21)|}{\sqrt{3^2 + (-7)^2}} = \frac{57}{\sqrt{58}} \approx 7.484$$

3. $x^2 - 6x - 4y + 1 = 0$

$$x^2 - 6x + 9 = 4y - 1 + 9$$
$$(x-3)^2 = 4y + 8$$
$$(x-3)^2 = 4(1)(y+2) \Rightarrow p = 1$$

Vertex: $(3,\ -2)$
Focus: $(3,\ -1)$
Directrix: $y = -3$

4. Vertex: $(2,\ -5)$
Focus: $(2,\ -6)$
Vertical axis;
opens downward with $p = -1$

$$(x-h)^2 = 4p(y-k)$$
$$(x-2)^2 = 4(-1)(y+5)$$
$$x^2 - 4x + 4 = -4y - 20$$
$$x^2 - 4x + 4y + 24 = 0$$

5. $x^2 + 4y^2 - 2x + 32y + 61 = 0$

$$(x^2 - 2x + 1) + 4(y^2 + 8y + 16) = -61 + 1 + 64$$
$$(x-1)^2 + 4(y+4)^2 = 4$$
$$\frac{(x-1)^2}{4} + \frac{(y+4)^2}{1} = 1$$

$a = 2,\ b = 1,\ c = \sqrt{3}$
Horizontal major axis
Center: $(1,\ -4)$
Foci: $(1 \pm \sqrt{3},\ -4)$
Vertices: $(3,\ -4),\ (-1,\ -4)$
Eccentricity: $e = \sqrt{3}/2$

6. Vertices: $(0,\ \pm 6)$
Eccentricity: $e = 1/2$
Center: $(0,\ 0)$
Vertical major axis

$a = 6,\ e = \frac{c}{a} = \frac{c}{6} = \frac{1}{2} \Rightarrow c = 3$

$b^2 = (6)^2 - (3)^2 = 27$

$$\frac{x^2}{27} + \frac{y^2}{36} = 1$$

7. $16y^2 - x^2 - 6x - 128y + 231 = 0$

$$16(y^2 - 8y + 16) - (x^2 + 6x + 9) = -231 + 256 - 9$$
$$16(y-4)^2 - (x+3)^2 = 16$$
$$\frac{(y-4)^2}{1} - \frac{(x+3)^2}{16} = 1$$

$a = 1,\ b = 4,\ c = \sqrt{17}$

Center: $(-3,\ 4)$

Vertical transverse axis

Vertices: $(-3,\ 5),\ (-3,\ 3)$

Foci: $(-3,\ 4 \pm \sqrt{17})$

Asymptotes: $y = 4 \pm \frac{1}{4}(x+3)$

8. Vertices: $(\pm 3,\ 2)$

Foci: $(\pm 5,\ 2)$

Center: $(0,\ 2)$

Horizontal transverse axis

$a = 3,\ c = 5,\ b = 4$

$$\frac{(x-0)^2}{9} - \frac{(y-2)^2}{16} = 1$$
$$\frac{x^2}{9} - \frac{(y-2)^2}{16} = 1$$

9. $5x^2 + 2xy + 5y^2 - 10 = 0$

$A = 5,\ B = 2,\ C = 5$

$$\cot 2\theta = \frac{5-5}{2} = 0$$
$$2\theta = \frac{\pi}{2} \Rightarrow \theta = \frac{\pi}{4}$$

$$x = x'\cos\frac{\pi}{4} - y'\sin\frac{\pi}{4} = \frac{x'-y'}{\sqrt{2}} \qquad y = x'\sin\frac{\pi}{4} + y'\cos\frac{\pi}{4} = \frac{x'+y'}{\sqrt{2}}$$

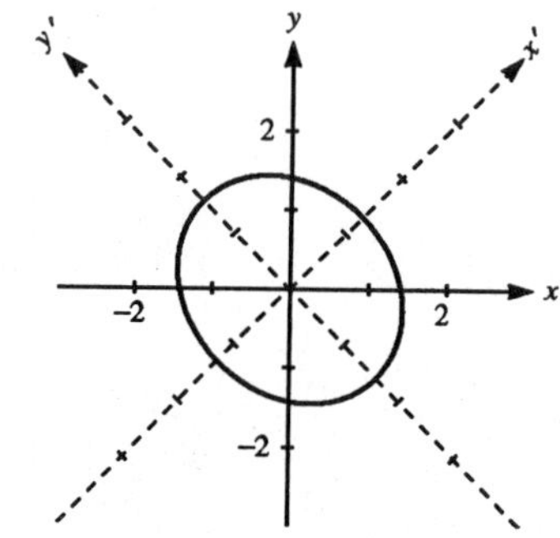

$$5\left(\frac{x'-y'}{\sqrt{2}}\right)^2 + 2\left(\frac{x'-y'}{\sqrt{2}}\right)\left(\frac{x'+y'}{\sqrt{2}}\right) + 5\left(\frac{x'+y'}{\sqrt{2}}\right)^2 - 10 = 0$$
$$\frac{5(x')^2}{2} - \frac{10x'y'}{2} + \frac{5(y')^2}{2} + (x')^2 - (y')^2 + \frac{5(x')^2}{2} + \frac{10x'y'}{2} + \frac{5(y')^2}{2} - 10 = 0$$
$$6(x')^2 + 4(y')^2 - 10 = 0$$
$$\frac{3(x')^2}{5} + \frac{2(y')^2}{5} = 1$$
$$\frac{(x')^2}{5/3} + \frac{(y')^2}{5/2} = 1$$

Ellipse centered at the origin

10. (a) $6x^2 - 2xy + y^2 = 0$

$A = 6,\ B = -2,\ C = 1$

$B^2 - 4AC = (-2)^2 - 4(6)(1) = -20 < 0$

Ellipse

(b) $x^2 + 4xy + 4y^2 - x - y + 17 = 0$

$A = 1,\ B = 4,\ C = 4$

$B^2 - 4AC = (4)^2 - 4(1)(4) = 0$

Parabola

11. Polar: $\left(\sqrt{2},\ \frac{3\pi}{4}\right)$

$$x = \sqrt{2}\cos\frac{3\pi}{4} = \sqrt{2}\left(-\frac{1}{\sqrt{2}}\right) = -1$$

$$y = \sqrt{2}\sin\frac{3\pi}{4} = \sqrt{2}\left(\frac{1}{\sqrt{2}}\right) = 1$$

Rectangular: $(-1,\ 1)$

12. Rectangular: $(\sqrt{3},\ -1)$

$$r = \pm\sqrt{(\sqrt{3})^2 + (-1)^2} = \pm 2$$

$$\tan\theta = \frac{\sqrt{3}}{-1} = -\sqrt{3}$$

$$\theta = \frac{2\pi}{3} \quad \text{or} \quad \theta = \frac{5\pi}{3}$$

Polar: $\left(-2,\ \frac{2\pi}{3}\right)$ or $\left(2,\ \frac{5\pi}{3}\right)$

13. Rectangular: $4x - 3y = 12$

Polar:

$$4r\cos\theta - 3r\sin\theta = 12$$

$$r(4\cos\theta - 3\sin\theta) = 12$$

$$r = \frac{12}{4\cos\theta - 3\sin\theta}$$

14. Polar: $r = 5\cos\theta$

$$r^2 = 5r\cos\theta$$

Rectangular:

$$x^2 + y^2 = 5x$$

$$x^2 + y^2 - 5x = 0$$

15. $r = 1 - \cos\theta$

Cardioid

Symmetry: Polar axis

Maximum value of $|r|$:

$r = 2$ when $\theta = \pi$

Zero of r: $r = 0$ when $\theta = 0$

θ	0	$\frac{\pi}{2}$	π	$\frac{3\pi}{2}$
r	0	1	2	1

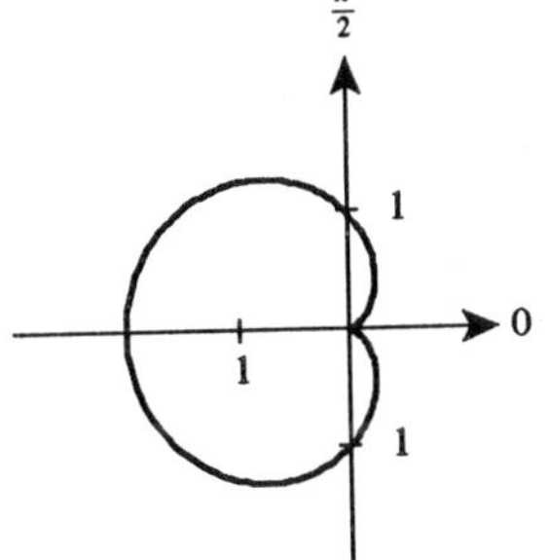

16. $r = 5\sin 2\theta$

Rose curve with four petals

Symmetry: Polar axis, $\theta = \dfrac{\pi}{2}$, and pole

Maximum value of $|r|$: $|r| = 5$ when $\theta = \dfrac{\pi}{4}, \dfrac{3\pi}{4}, \dfrac{5\pi}{4}, \dfrac{7\pi}{4}$

Zeros of r: $r = 0$ when $\theta = 0, \dfrac{\pi}{2}, \pi, \dfrac{3\pi}{2}$

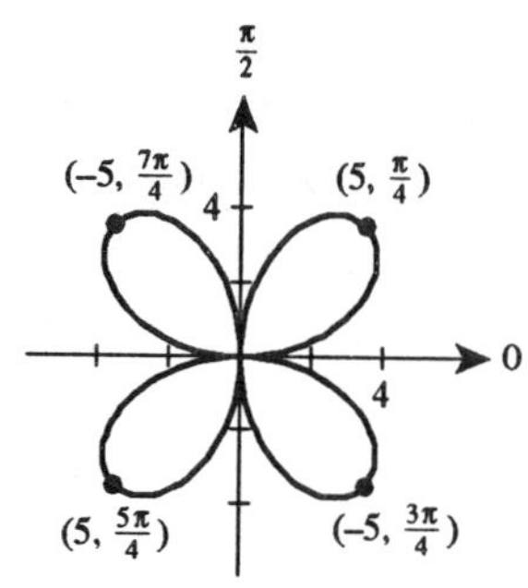

17. $r = \dfrac{3}{6 - \cos\theta}$

$r = \dfrac{\frac{1}{2}}{1 - \frac{1}{6}\cos\theta}$

$e = \frac{1}{6} < 1$, so the graph is an ellipse.

θ	0	$\frac{\pi}{2}$	π	$\frac{3\pi}{2}$
r	$\frac{3}{5}$	$\frac{1}{2}$	$\frac{3}{7}$	$\frac{1}{2}$

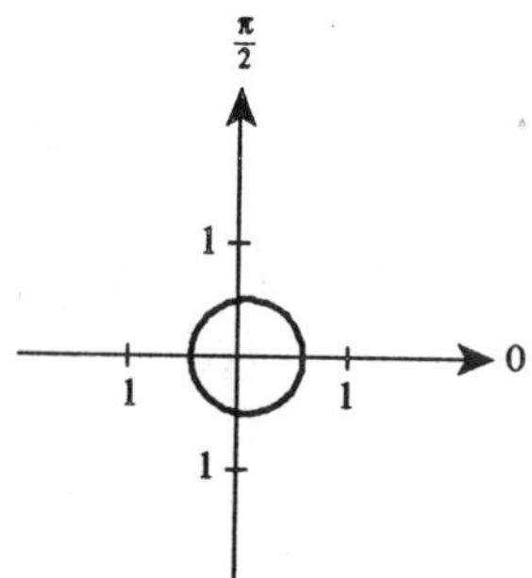

18. Parabola

Vertex: $\left(6, \dfrac{\pi}{2}\right)$

Focus: $(0, 0)$

$e = 1$

$r = \dfrac{ep}{1 + e\sin\theta}$

$r = \dfrac{p}{1 + \sin\theta}$

$6 = \dfrac{p}{1 + \sin(\pi/2)}$

$6 = \dfrac{p}{2}$

$12 = p$

$r = \dfrac{12}{1 + \sin\theta}$

19. $x = 3 - 2\sin\theta, \quad y = 1 + 5\cos\theta$

$\dfrac{x-3}{-2} = \sin\theta, \quad \dfrac{y-1}{5} = \cos\theta$

$\left(\dfrac{x-3}{-2}\right)^2 + \left(\dfrac{y-1}{5}\right)^2 = 1$

$\dfrac{(x-3)^2}{4} + \dfrac{(y-1)^2}{25} = 1$

20. $x = e^{2t}, \quad y = e^{4t}$

$x > 0, \quad y > 0$

$x = e^{2t} \Rightarrow \ln x = 2t \Rightarrow t = \frac{1}{2}\ln x$

$y = e^{4t} = e^{4(1/2\ln x)} = e^{2\ln x} = e^{\ln x^2} = x^2$

$y = x^2, \quad x > 0, \quad y > 0$

Alternate solution:

$y = e^{4t} = (e^{2t})^2 = x^2, \quad x > 0, \quad y > 0$